Intermediate Mathematics:

Algebra II

Intermediate Mathematics:

ALGEBRA II

UNDER THE EDITORIAL DIRECTION OF
ROY DUBISCH AND ISABELLE P. RUCKER

RONALD J. CLARK

JOHN E. YARNELLE

BENZIGER, INC. New York, Beverly Hills
in Association with
JOHN WILEY & SONS, INC. New York, London,
Sidney, Toronto

EDITORS

ROY DUBISCH. Professor of Mathematics at the University of Washington, has been writing for both teachers and students for many years. He has been Associate Editor of the *American Mathematical Monthly* and Editor of the *Mathematics Magazine*. He has several books to his credit including *The Teaching of Mathematics* (Wiley, 1963) and is the author of numerous journal articles. Since 1961 Professor Dubisch has served on the Mathematics Steering Committee of the African Education Project and has participated in several workshops and institutes in various African countries. He has also directed and lectured at several NSF summer institutes. He is a past Vice President of the National Council of Teachers of Mathematics and has served on the Teacher Training Panel of the Committee on the Undergraduate Program in Mathematics, the Advisory Board and Panel on Supplementary Publications of the School Mathematics Study Group, and the Board of Governors of the Mathematical Association of America. He has also been active in many other professional organizations.

ISABELLE P. RUCKER. Supervisor of Mathematics of the State Board of Education of Virginia, is a former teacher of both elementary and secondary school mathematics. She has participated in summer mathematics institutes and an NSF Academic Year Institute at the University of Virginia and has published numerous articles and book reviews in professional journals. In addition to being active in other professional organizations, Mrs. Rucker is a member of both the Advisory Board and the Executive Committee of the School Mathematics Study Group, is Chairman of the Committee on Plans and Proposals of the National Council of Teachers of Mathematics, and has held offices in the Association of State Supervisors of Mathematics.

Copyright © 1971 by John Wiley and Sons, Inc.

All rights reserved. Published simultaneously in Canada.

No part of this book may be reproduced by any means, nor transmitted, nor translated in to a machine language without the written permission of the publisher.

Library of Congress Catalogue Card Number: 76-139276

ISBN 0-471-94553-6

Printed in the United States of America

10 9 8 7 6 5 4 3 2 1

AUTHORS

JOHN E. YARNELLE. Chairman of the Department of Mathematics at Hanover College, Hanover, Indiana, is a former writer for the School Mathematics Study Group and the Committee on Educational Media. He previously co-authored two junior high school texts, has directed four NSF institutes throughout Indiana, and was a Visiting Professor at Purdue University. Professor Yarnelle is a past member of the Board of Governors of the Mathematical Association of America and in 1965 participated in the African Mathematics Workshop for Educational Services, Inc.

RONALD J. CLARK. Charles Levy Master in Mathematics at St. Paul's School, Concord, New Hampshire, has been teaching mathematics since 1939 where he has served as Head of the Mathematics Department. He is a former writer for the School Mathematics Study Group and, for five years, was Chairman of the SMSG Panel on Supplementary Publications. In 1960, he taped 120 half-hour television lessons in calculus, which were presented on local educational stations, in the New England area and in other areas of the United States. He is a past Director of the New England Association of Teachers of Mathematics. He has served as a consultant on mathematics to the New Hampshire Board of Education. He has been a member of the Publications Committee of the National Council of Teachers of Mathematics.

PREFACE

INTERMEDIATE MATHEMATICS is one of a series of mathematics textbooks written for junior and senior high school students. It is designed for a one year course for students with a background of elementary algebra and geometry such as that included in ALGEBRA and GEOMETRY of this series.

The title "Intermediate Mathematics" is intended to convey the idea that the content is not strictly that of a second course in algebra, per se. Interwoven into the content are not only algebraic functions but also trigonometric functions; not only synthetic plane and solid geometric interpretations of algebra but also analytic geometry interpretations and their extensions to vectors and vector spaces. The content is considered to be appropriate either for a terminal course in academic high school mathematics or as a prerequisite to further study of mathematics either in high school or in college.

A glance at the table of contents reveals that many traditional topics have been retained. Now, as always, topics such as quadratic equations involving real and complex numbers, for example, are important. The presentation of these and other topics, however, is not always a traditional one. Whether the presentation is a traditional one or an innovative one, the authors have tried to present the mathematics in an honest and straightforward manner that reflects the opinions of practicing mathematicians.

It can be noted throughout the book that review of geometric and algebraic concepts and skills is handled as needed in the development of a given topic new to the student. The authors believe that such an approach to review helps students to think about mathematics as a logical body of knowledge, and not as a series of disjointed topics and/or courses.

CONTENTS

Chapter 1
Sequences and Series   1

1.1 The Function Concept 1   /   **1.2** Sequence Functions 8   /
1.3 Arithmetic Progression 13   /   **1.4** Geometric Progressions
17   /   **1.5** The Sum of A Sequence 22   /   **1.6** Series and
Sigma Notation 26   /   Chapter Summary 30   /   Review
Exercises 31   /   Going Further: Reading and Research 32

Chapter 2
Natural Numbers and Mathematical Induction   34

2.1 A Computing Formula 35   /   **2.2** The Natural Numbers 37
/   **2.3** Mathematical Induction 40   /   **2.4** Further Applications of Mathematical Induction 47   /   **2.5** Induction and
Well-Ordering 51   /   Chapter Summary 55   /   Review
Exercises 55   /   Going Further: Reading and Research 56

Chapter 3
Real Numbers and Related Systems   60

3.1 A Preview 61   /   **3.2** Fields 61   /   **3.3** Ordered Fields
67   /   **3.4** A Complete Ordered Field 77   /   **3.5** Rational
Numbers and the Completeness Property 80   /   **3.6** Subfields
83   /   **3.7** The Smallest Subfield, the Rational Numbers 87   /
3.8 Integral Domains and the Integers 92   /   Chapter
Summary 96   /   Review Exercises 97   /   Going Further:
Reading and Research 98

Chapter 4
Complex Numbers   100

4.1 Quadratic Equations 101   /   **4.2** A New Set of Numbers
111   /   **4.3** Operations on Complex Numbers 118   /   **4.4**
The Square Root of Complex Numbers 123   /   **4.5** The Field
of Complex Numbers 127   /   **4.6** Quadratic Equations Over
C 133   /   **4.7** Argand Diagrams 141   /   **4.8** Representation
of Complex Numbers as Ordered Pairs 144   /   **4.9** Matrix

Representation of Complex Numbers 150 / Chapter Summary 158 / Review Exercises 159 / Going Further: Reading and Research 160

Chapter 5
Polynomial Functions 162

5.1 Polynomials 163 / **5.2** Zeros of Polynomials 171 / **5.3** Synthetic Division 181 / **5.4** Solving Equations by Synthetic Division 186 / **5.5** A Fundamental Theorem 193 / Chapter Summary 198 / Review Exercises 198 / Going Further: Reading and Research 199

Chapter 6
Coordinate Systems 202

6.1 Linear Graphs 203 / **6.2** Slope 206 / **6.3** Families of Lines 211 / **6.4** Distance 216 / **6.5** Graphs of Polynomial Equations 220 / **6.6** Absolute Values and Inequalities 225 / **6.7** Approximate Solutions 231 / **6.8** Algebraic Proofs of Geometric Theorems 240 / **6.9** Coordinates in Three Dimensions 247 / Chapter Summary 254 / Review Exercises 255 / Going Further: Reading and Research 256

Chapter 7
Conic Sections and Their Graphs 258

7.1 Introduction 259 / **7.2** The Circle 264 / **7.3** The Ellipse 268 / **7.4** The Parabola 284 / **7.5** The Hyperbola 292 / **7.6** Translation in the Plane 301 / **7.7** Quadratic Functions and Inequalities 308 / **7.8** Special Cases of the Conic Sections 314 / Chapter Summary 318 / Review Exercises 321 / Going Further: Reading and Research 322

Chapter 8
Systems of Equations 324

8.1 Linear Equations 325 / **8.2** One Linear and One Second-Degree Equation 333 / **8.3** Two Second-Degree Equations 340 / **8.4** Systems of Inequalities 349 / **8.5** Systems of Linear Equations 354 / Chapter Summary 369 / Review Exercises 370 / Going Further: Reading and Research 372

Chapter 9

Arrangements, Subsets, and The Binomial Theorem 376

9.1 A Problem of Communication 377 / 9.2 Arrangements
and Subsets 383 / 9.3 Some Probability Applications 394 /
9.4 The Binomial Theorem 397 / 9.5 More on the Binomial
Theorem 407 / 9.6 The Binomial Theorem and Induction
410 / Chapter Summary 415 / Review Exercises 417 /
Going Further: Reading and Research 418

Chapter 10

Circular Functions   420

10.1 The Wrapping Function 421 / 10.2 The Circular
Functions 433 / 10.3 Graphs of the Circular Functions 447
/ 10.4 The Graphs of Composite Functions 452 / 10.5
Addition Formulas 459 / 10.6 Equations and Identities 471
/ Chapter Summary 477 / Review Exercises 479 /
Going Further: Reading and Research 481

Chapter 11

Exponential and Logarithmic Functions   484

11.1 Integral Exponents 485 / 11.2 Rational Exponents 492
/ 11.3 Real Number Exponents and Exponential Functions
499 / 11.4 Linear Interpolation 507 / 11.5 Inverse
Functions 515 / 11.6 Graphs of Inverse Functions 527 /
11.7 Logarithmic Functions 529 / 11.8 Properties and
Applications of Logarithms 535 / Chapter Summary 547 /
Review Exercises 549 / Going Further: Reading and Research 551

Chapter 12

Trigonometric Functions   554

12.1 Angles and Angle Measure 555 / 12.2 Trigonometric
Functions 563 / 12.3 Solution of Right Triangles 569 /
12.4 The Angle Between Two Lines 580 / 12.5 Law of
Cosines 586 / 12.6 Law of Sines 591 / 12.7 The Inverse
Trigonometric Functions 596 / Chapter Summary 607 /
Review Exercises 609 / Going Further: Reading and Re-
search 610

Chapter 13

Vectors   614

13.1 Directed Line Segments 615   /   **13.2** Vectors 620   /
13.3 Addition of Vectors 625   /   **13.4** Scalar Multiplication of
Vectors 633   /   **13.5** Parallel and Perpendicular Vectors 638
/   **13.6** Basis Vectors 645   /   **13.7** Vector Spaces 649   /
Chapter Summary 652   /   Review Exercises 654   /   Going
Further: Reading and Research 655

Chapter 14

Powers and Roots of Complex Numbers   658

14.1 Polar Form of Complex Numbers 659   /   **14.2** Products
and Powers of Complex Numbers 672   /   **14.3** Roots of Order
n 681   /   **14.4** The nth Roots of 1 691   /   Chapter Summary
697   /   Review Exercises 698   /   Going Further: Reading
and Research 699

Tables   703

Index   713

CHAPTER ONE
SEQUENCES AND SERIES

1.1 THE FUNCTION CONCEPT

The function concept is one of the most important in mathematics, not only because it appears so often but also because it unites so many topics which, at first glance, may seem unrelated. Hence we shall briefly review it before moving to some new ideas that use this concept.

Definition 1.1 If with each element of set A there is associated in some way exactly one element of a set B, then this association is called a **function** from A to B. The set A is called the **domain** of the function. The member of the set B that is associated with a member x of the domain is called the **image** of x under the function. The set of all images is called the **range** of the function.

Example 1 The table below lists the squares of all natural (counting) numbers from 1 to 10.

Number	Square	Number	Square
1	1	6	36
2	4	7	49
3	9	8	64
4	16	9	81
5	25	10	100

The table defines a function. The domain is the set of the first ten natural numbers. The range is the set of the first ten perfect squares.

Example 2 The equation $y = 2x + 3$ can be used to define a function by choosing a domain. Thus, for example, if the domain D is the set of nonnegative rational numbers, then any $x \in D$ determines a rational number y that is greater than or equal to 3. If we solve for x in terms of y, we have $x = \dfrac{y-3}{2}$. This yields a nonnegative rational number whenever $y \geq 3$. Hence the range of the function is the set of all rational numbers equal to or greater than 3.

1

Example 3 Let s be the measure of the side of a square. Let A be the measure of the area of the square. With each positive real number s we have associated a positive real number A such that $A = s^2$. Thus we have a function whose domain and range are both the set of all positive real numbers.

_____

Frequently, we use letters as names of functions. We speak of function f, or of function g, or of function F. Suppose we have a function f. If x is an element in the domain of f, then the image of x is called $f(x)$ (read "the value of f at x" or simply "f of x"). Hence $f(x)$ is the unique element in the range of f associated by the function f with x. Since the function f associates with every x in the domain D of f a unique element $f(x)$ in the range R of f, the function f defines a set of ordered pairs $\{(x, f(x)) : x \in D\}$. Thus in Example 1 the function defines the set

$$\{(1, 1), (2, 4), (3, 9), (4, 16), (5, 25), (6, 36), (7, 49),$$
$$(8, 64), (9, 81), (10, 100)\}$$

of ordered pairs. Conversely, certain sets of ordered pairs define functions. For example, suppose we have the set

$$\{(1, 3), (2, 5), (3, 7), (4, 9), (5, 11)\}.$$

The set defines a function whose domain is the set

$$\{x : x \text{ is a natural number and } 1 \leq x \leq 5\}$$

and for which

$$f(x) = 2x + 1.$$

Not all sets of ordered pairs, however, define functions. For example, the set

$$\{(1, 1), (1, -1), (2, \sqrt{2}), (2, -\sqrt{2}), (3, \sqrt{3})\}$$

does not define a function since it is not true that the first element of each ordered pair is associated with exactly one number as our definition requires. The number 3 is associated with just one number, $\sqrt{3}$, but the number 1 is associated with both 1 and -1 and the number 2 with both $\sqrt{2}$ and $-\sqrt{2}$.

The $f(x)$ notation helps us to describe a function. If we wish to speak about a function f which associates with each number x the number $2x + 1$, we can write

$$f(x) = 2x + 1.$$

Accordingly, each number x is paired with another number, $f(x) = 2x + 1$. Thus we pair 3 with $2(3) + 1 = 7$, 4 with $2(4) + 1 = 9$, and $-\frac{5}{2}$ with $2(-\frac{5}{2}) + 1 = -4$. We write

$$f(3) = 2(3) + 1 = 7, \qquad f(4) = 2(4) + 1 = 9,$$

and

$$f\left(-\frac{5}{2}\right) = 2\left(-\frac{5}{2}\right) + 1 = -4.$$

If a and b are in the domain of f, we may write

$$f(a) = 2a + 1 \qquad \text{and} \qquad f(b) = 2b + 1.$$

Thus five members of the set of ordered pairs defined by f are $(3, 7)$, $(4, 9)$, $(-\frac{5}{2}, -4)$, $(a, 2a + 1)$, and $(b, 2b + 1)$.

As an abbreviation for "f, the function defined by $f(x) = 2x + 1$," we sometimes write

$$f : f(x) = 2x + 1.$$

Example 4 Let $D = \{0, 1, 2, 3\}$. If $f(x) = x^2 - 1$ for each $x \in D$, find $f(0)$, $f(1)$, $f(2)$, and $f(3)$.

Solution:

$$f(0) = (0)^2 - 1 = -1 \qquad f(2) = (2)^2 - 1 = 3$$
$$f(1) = (1)^2 - 1 = 0 \qquad f(3) = (3)^2 - 1 = 8.$$

If D is the domain of the function f, then the range of the function is $\{-1, 0, 3, 8\}$.

Example 5 Let F be a function whose domain is the set of real numbers and such that $F(x) = 3$. Find $F(2)$, $F(3)$, and $F(\sqrt{2})$. Give the range of the function F.

Solution:

$$F(2) = 3, \qquad F(3) = 3, \qquad \text{and} \qquad F(\sqrt{2}) = 3.$$

The range is the set $\{3\}$. The function F is an example of a **constant** function.

Example 6 Let R be the set of real numbers. If $g(x) = x^2 + 1$ for each $x \in R$, find $g(0)$, $g(-1)$, $g(\pi)$, $g(a)$, and $g(a + 1)$ for $a \in R$. What is the range of the function g?

Solution:

$$g(0) = (0)^2 + 1 = 1$$
$$g(-1) = (-1)^2 + 1 = 2$$
$$g(\pi) = \pi^2 + 1$$
$$g(a) = a^2 + 1$$
$$g(a + 1) = (a + 1)^2 + 1 = a^2 + 2a + 2.$$

For all values of x, $x^2 \geq 0$. Hence $x^2 + 1 \geq 1$ and thus $g(x) \geq 1$. The range is the set of all real numbers greater than or equal to 1.

It is often helpful to picture a function as a **mapping** showing the elements of the domain and the range as points, and the function as a set of arrows from points representing elements of the domain to points representing elements of the range as in Figure 1-1.

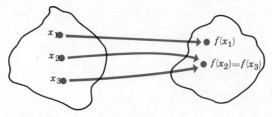

Figure 1-1

Figure 1-2 does not illustrate a function. Why not?

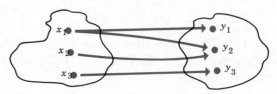

Figure 1-2

A graph of a function can also be considered as representing a mapping. For example, suppose that we have the graph of $f : f(x) = 2x - 4$ as in Figure 1-3. Any point on the graph, such as (4, 4), is an element of the set

$$\{(x, f(x)) : f(x) = 2x - 4\}.$$

Not all graphs are graphs of functions. Consider the graph in

Figure 1-4. Is it possible to have two points such as (0, 3) and (0, −3) on the graph of a function?

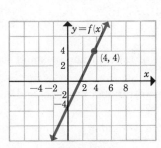

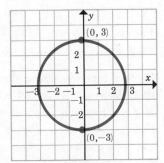

Figure 1-3   Figure 1-4

EXERCISES 1.1

◀ In Exercises 1–10, the domain of each function is {1, 2, 3, 4}. Find, in each exercise, the range of the function.

1. $f : f(x) = 3x$

2. $g : g(x) = 2x - 1$

3. $h : h(x) = 3x^2$

4. $F : F(x) = x$

5. $G : G(x) = 3$

6. $f : f(y) = y^2 - 2$

7. $g : g(w) = 2w - 3$

8. $F : F(t) = \dfrac{t^2 - 2}{2}$

9. $G : G(y) = y^2 + 2y + 1$

10. $F : F(s) = \dfrac{1}{2} s^2$

11. If $f(x) = 3x + 1$, find $f(0)$, $f(3)$, and $f(-2)$.

12. If $g(y) = y^2 - 3$, find $g(0)$, $g(-2)$, and $g(\sqrt{3})$.

13. If $h(t) = t^3 - 1$, find $h(-2)$, $h(3)$, and $h(\sqrt[3]{2})$.

14. If $F(x) = \sqrt{x^2 - 4}$, find $F(2)$, $F(4)$, and $F(\sqrt{5})$.

15. If $G(x) = |x|$, find $G(-2)$, $G\left(\dfrac{1}{2}\right)$, and $G(\pi)$.

16. Let g be a function defined by $g(x) = 2x + 4$ for all integers x and let h be a function defined by $h(t) = 2t + 4$ for all integers t. Are g and h the same function? Explain.

17. **CHALLENGE PROBLEM.** Let F be a function defined by $F(x) = x + 2$ for all integers x and let G be a function defined by $G(t) = \dfrac{t^2 - 4}{t - 2}$ for all integers $t \neq 2$ and $G(2) = 4$. Are F and G the same function? Explain.

◀ Exercises 18–22 refer to the function $g : g(x) = x^2 + 1$, with the domain of g the set of all real numbers.

18. Find $g(3)$, $g(4)$, and $g(7)$. Does $g(3) + g(4) = g(7)$?
19. Find $g(-2)$, $g(3)$, and $g(1)$. Does $g(3) + g(-2) = g(1)$?
20. Fing $g(-2)$ and $g(2)$. Does $g(2) + g(-2) = g(0)$?
21. Find $g(3)$, $g(4)$, and $g(12)$. Does $g(3) \cdot g(4) = g(12)$?
22. Find $g(-2)$, $g(-3)$, and $g(6)$. Does $g(-2) \cdot g(-3) = g(6)$?

23. If f is a function whose domain is the set of rational numbers and such that $f(x) = 2x + 1$, what is the range of f?
24. If g is a function whose domain is the set of rational numbers and such that $g(x) = 3x - 1$, what is the range of g?
25. Let R be the set of real numbers. If R is the domain of $g : g(x) = x^2 - 1$, what is the range of g?
26. Let R be the set of real numbers. If R is the domain of $F : F(x) = \sqrt{x^2}$, what is the range of F? Can you give another rule that defines the same function?

◀Exercises 27–30 refer to a function F such that $F(x) = 2x^2 - x + 1$ for all $x \in R$, the set of real numbers.

27. Find $F(a)$, $(a \in R)$.
28. Find $F(a + 1)$, $(a \in R)$.
29. Find $F(a^2 + 1)$, $(a \in R)$.
30. Find $F[F(2)]$. (*Hint:* Let $F(2) = y$. Find $F(2)$; then find $F(y)$.)

◀In Exercises 31–35, a statement is given that implies a function. Define the function by naming the domain and by giving a rule such as an equation or table that defines the association. Then determine the range.

31. The circumference, C, of a circle is π times the measure, d, of its diameter.
32. Every real number has an absolute value.
33. The highest temperature on each of the first ten days of July was $63°$, $82°$, $59°$, $86°$, $84°$, $81°$, $73°$, $78°$, $69°$, and $72°$.
34. Each positive even number is twice a natural number.
35. The set $\{1, \sqrt{2}, \sqrt{3}, 2, \sqrt{5}, \sqrt{6}, \sqrt{7}, \sqrt{8}, 3, \sqrt{10}\}$ has as its members the positive square root of the natural numbers from 1 to 10.

◀In Exercises 36–40, a set of ordered pairs is listed. If the set defines a function, give an equation that also defines the function. If the set does not define a function, explain why not.

36. $S = \{(1, 1), (2, 2), (3, 3), (4, 4), (5, 5)\}$
37. $T = \{(1, 9), (2, 9), (3, 9), (4, 9), (5, 9)\}$
38. $R = \{(1, 2), (2, 3), (3, 4), (4, 5), (5, 6)\}$
39. $M = \{(1, 2), (2, 1), (2, 3), (3, 2), (3, 4)\}$
40. $N = \{(1, 1), (2, 4), (3, 9), (4, 16), (5, 25)\}$

41. The first ten prime numbers are 2, 3, 5, 7, 11, 13, 17, 19, 23, and 29. These prime numbers can be arranged in a table.

1	2	3	4	5	6	7	8	9	10
2	3	5	7	11	13	17	19	23	29

Does the table define a function? Is the set of prime numbers finite or infinite?

42. Let N be the set of natural numbers and let $G(x) = 2x - 1$ for all $x \in N$. List ten ordered pairs defined by the function G. What is the range of the function G?

43. Let N be the set of natural numbers and let

$$F(x) = \frac{3x}{2} + \frac{7}{2}$$

for all $x \in N$. Find $F(1)$, $F(5)$, $F(20)$, and $F(30)$. If $F(x) = 14$, what is x?

44. A function F whose domain is the set of real numbers is defined by

$$F(x) = \begin{cases} x & \text{if } x \geq 0, \\ -x & \text{if } x < 0. \end{cases}$$

Find $F(0)$, $F(2)$, $F(-2)$, $F(2 - \sqrt{3})$, $F(\sqrt{3} - 2)$, and $F(\pi)$. Can you give a name for this function?

45. Let $H(x) = 2^x$ with the domain of H the set

$$\{x : x \text{ is a natural number and } 0 < x \leq 10\}.$$

Write the set of ordered pairs defined by H. What is the range of the function H?

46. Two of the following four diagrams are illustrations of functions; two are not. Determine which are not illustrations of functions and explain why not.

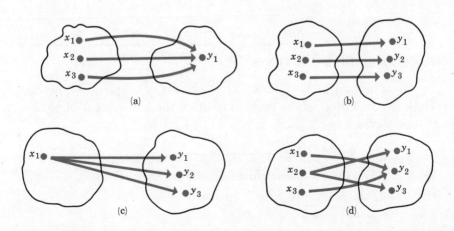

(a)   (b)   (c)   (d)

47. Two of the following four diagrams are graphs of functions; two are not. Determine which are not graphs of functions and explain why not.

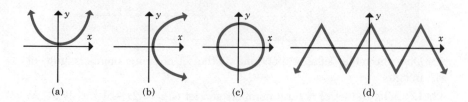

(a)   (b)   (c)   (d)

48. A function f is called a **linear function** if $f(x) = mx + b$ and $m, b \in R$. The **slope** of the linear function is m. Each of the following equations defines a linear function f where $y = f(x)$. Find the slope of each.

(a) $y = 2x - 3$ (c) $6x = 4y + 2$

(b) $2x - 3y = 4$ (d) $8x = 3y$

CHALLENGE PROBLEMS. A function g, defined on the range of a function f, is called the **inverse** of f if every element of $f(x)$ in the range of f is (a) the image of only one x in the domain of f and (b) is mapped into x under g, that is,

$$g[f(x)] = x.$$

Determine if f and g are inverses in Exercises 49 and 50.

49. $f : f(x) = 2x - 1$; $g : g(x) = \dfrac{1}{2}x + 1$.   (Domain of f is the set of real numbers; domain of g is range of f.)

50. $f : f(x) = x^2 + 1$; $g : g(x) = \sqrt{x} - 1$.   (Domain of f is the set of real numbers; domain of g is range of f.)

1.2   SEQUENCE FUNCTIONS

An important class of functions is the set of **sequence functions**. These are functions whose domain is either the entire set N of natural numbers or a subset, $\{1, 2, 3, 4, \ldots, n\}$, of N.

Example 1 Let $D = \{1, 2, 3, 4, 5\}$ be the domain of $f : f(x) = 2x + 3$. Then f is a sequence function and we have $f(1) = 5, f(2) = 7, f(3) = 9, f(4) = 11$, and $f(5) = 13$.

Example 2 Let g be a function such that $g(n) = n^2$ for $n \in N$. Then g is a sequence function and we have

$$g(1) = 1, \ g(2) = 4, \ g(3) = 9, \ldots .$$

Any sequence function f yields an ordered arrangement of numbers corresponding to $f(1), f(2), f(3), f(4), f(5)$, and so on. Thus in Example 1 we have 5, 7, 9, 11, 13. In Example 2, we have 1, 4, 9, 16, 25, ... Such an ordered arrangement, $f(1), f(2), f(3), \ldots$, arising from a sequence function is called a **sequence.**

A sequence that has a finite number of terms, such as the one defined by f in Example 1, is called a **finite** sequence. A sequence, such as the one defined by g in Example 2, that has an infinite number of terms is called an **infinite** sequence.

We can, of course, represent a sequence as $f(1), f(2), f(3), \ldots$ using functional notation. It is customary, however, in writing functional values for sequence functions to omit the parentheses and to write the numbers from the domain N as subscripts. Thus, for example, we write

$$f_1, f_2, f_3, f_4, \ldots$$

in place of $f(1), f(2), f(3), f(4), \ldots .$ Of course, any letter can be used in place of f for a sequence function. We shall use the letter t to remind us of the word "term" and represent a sequence of n terms by

$$t_1, t_2, t_3, \ldots, t_n.$$

In this notation, then, the sequence of Example 1 can be written as

$$t_1 = 5, \qquad t_2 = 7, \qquad t_3 = 9, \qquad t_4 = 11, \qquad t_5 = 13,$$

and the sequence of Example 2 as

$$t_1 = 1, \qquad t_2 = 4, \qquad t_3 = 9, \qquad t_4 = 16, \ldots .$$

A sequence can be defined in various ways. For example, given time enough we could list all the terms of any finite sequence as we did for the sequence 5, 7, 9, 11, 13. On the other hand, sometimes a sequence is indicated by listing only the first few terms. Three such listings are

$$2, 4, 6, 8, 10, \ldots ,$$

$$1, 1, 2, 3, 5, 8, 13, \ldots , \qquad \text{and}$$

$$1, -\frac{1}{2}, \frac{1}{4}, -\frac{1}{8}, \frac{1}{16}, \ldots .$$

Can you see a pattern in each example? If the sequences were continued, what do you think would t_{10} be in each of the sequences?

If only a few terms of a sequence are known, there is always more than one pattern that fits the sequence. For example, suppose that we have the first three terms, 3, 5, 7, of a sequence. Is the next term 9? The next term is certainly 9 if the sequence is one of consecutive odd numbers. On the other hand, if the sequence is one of consecutive prime numbers, then the next number in the sequence 3, 5, 7 is 11. These three numbers could also be the first three terms of a sequence defined by the expression $n^3 - 6n^2 + 13n - 5$, as you can verify by replacing n by 1, 2, and 3 successively. The fact that 3, 5, 7 are the first three terms of a sequence is not enough information to define the sequence. If a sequence is infinite, then no matter how many terms we list, we have not uniquely defined the sequence.

A second method of defining a sequence is to give a formula that specifies the nth term, t_n, in terms of n.

Example 3   Give the first five terms of a sequence defined by $t_n = 2n + 3$.

Solution: We have $n = 1, 2, 3, 4,$ and 5. Substituting 1 for n in the formula $t_n = 2n + 3$, we have

$$t_1 = 2(1) + 3 = 5.$$

Likewise, substituting 2 for n, we have

$$t_2 = 2(2) + 3 = 7.$$

When we have substituted the five numbers 1, 2, 3, 4, 5 for n, we have the first five terms of the sequence, namely 5, 7, 9, 11, 13.

Example 4   Find the first five terms of the sequence defined by the formula

$$t_n = \frac{(n + 1)^2}{n}.$$

Solution: Substituting 1 for n, we have $t_1 = \dfrac{(1 + 1)^2}{1} = 4.$ Next,

$$t_2 = \frac{(2 + 1)^2}{2} = \frac{9}{2}.$$

The other three of the first five terms of the sequence, found in a similar manner, are $\frac{16}{3}$, $\frac{25}{4}$, and $\frac{36}{5}$.

A third method of defining a sequence is by giving one or more of the first terms and an **iteration formula** (also called a **recursion formula**) that gives the relationship between two or more consecutive terms. For example, if $t_1 = 10$ and $t_{n+1} = t_n + 4$ for $n \geq 1$, the first four terms of the sequence are

$$10, \ 14, \ 18, \text{ and } 22$$

since

$$t_2 = t_1 + 4 = 10 + 4 = 14,$$
$$t_3 = t_2 + 4 = 14 + 4 = 18,$$

and

$$t_4 = t_3 + 4 = 18 + 4 = 22.$$

If $t_1 = 5$, $t_2 = 1$, and $t_{n+2} = t_{n+1} - t_n$ for $n \geq 1$, we have

$$t_3 = t_2 - t_1 = 1 - 5 = -4,$$
$$t_4 = t_3 - t_2 = -4 - 1 = -5,$$
$$t_5 = t_4 - t_3 = -5 - (-4) = -1,$$

and so on.

In the Exercises that follow and in the rest of the chapter you should assume, unless otherwise specified, that the domain of each sequence function is the set N of natural numbers.

EXERCISES 1.2

◀In Exercises 1–10, continue each sequence through seven terms. Give the rule you used to add new terms.

1. 2, 7, 12, 17

2. -1, -4, -7, -10

3. 2, -4, 8, -16

4. 2, 6, 18, 54

5. $\dfrac{1}{2}$, $\dfrac{2}{3}$, $\dfrac{3}{4}$, $\dfrac{4}{5}$

6. $\dfrac{3}{4}$, $\dfrac{6}{7}$, $\dfrac{9}{10}$, $\dfrac{12}{13}$

7. $\sqrt{2}$, 2, $2\sqrt{2}$, 4

8. 2, 3, 6, 18

9. $\sqrt[3]{2}$, $\sqrt[3]{4}$, 2, $2\sqrt[3]{2}$

10. b, $\sqrt[3]{b^2}$, $\sqrt[5]{b^3}$, $\sqrt[7]{b^4}$

◀In Exercises 11–20, write the first six terms of the sequence defined by the given function.

11. $f : f(x) = 2x - 1$

16. $f : f(y) = \dfrac{16}{y}$

12. $g : g(x) = x^2$

17. $g : g(y) = \dfrac{1}{y} + y$

13. $F : F(x) = \dfrac{x + 1}{x}$

18. $F : F(t) = \dfrac{2t}{3t - 1}$

14. $G : G(x) = 2^x$

19. $h : h(x) = x^2 + \dfrac{2}{x}$

15. $f : f(x) = x^2 + x + 1$

20. $G : G(y) = (y - 1)^3$

◀ In Exercises 21–26, find the first five terms of each sequence defined by the given rule.

21. $t_n = n + 4$

24. $t_n = \dfrac{2n}{3n - 1}$

22. $t_n = 3 - 2n$

25. $t_n = \dfrac{1}{2} n(n + 1)$

23. $t_n = \dfrac{n + 1}{n}$

26. $t_n = 2^n$

◀ In Exercises 27–30, find the first five terms of each of the given sequences.

27. $t_1 = -1$ and $t_{n+1} = t_n + 2$ for $n \geq 1$
28. $t_1 = 1$ and $t_{n+1} = t_n - 5$ for $n \geq 1$
29. $t_1 = 2$, $t_2 = 3$, and $t_{n+2} = 2t_{n+1} + t_n$
30. $t_1 = -1$, $t_2 = -2$, and $t_{n+2} = t_n - 2t_{n+1}$

◀ In Exercises 31–40, find a sequence function that will produce a sequence whose first four terms are as listed.

31. $1, 2, 3, 4, \ldots$

36. $-1, -\dfrac{1}{2}, -\dfrac{1}{4}, -\dfrac{1}{8}, \ldots$

32. $10, 20, 30, 40, \ldots$

37. $1, 1\dfrac{3}{4}, 2\dfrac{1}{2}, 3\dfrac{1}{4}, \ldots$

33. $0, \dfrac{1}{2}, 1, \dfrac{3}{2}, \ldots$

38. $3, 3, 3, 3, \ldots$

34. $-1, 1, -1, 1, \ldots$

39. $1, 4, 9, 16, \ldots$

35. $-10, -7, -4, -1, \ldots$

40. $8, 27, 64, 125, \ldots$

41. **CHALLENGE PROBLEM.** Find an iteration formula that will give the Fibonacci sequence $1, 1, 2, 3, 5, 8, 13, \ldots$. (The Fibonacci sequence is named after an Italian mathematician who lived in the early part of the thirteenth century.)

1.3 ARITHMETIC PROGRESSION

Some sequences have special names. One such important sequence is an arithmetic sequence.

> *Definition* 1.2 A sequence, t_1, t_2, t_3, . . . , is called an **arithmetic sequence** if the difference, d, between any two successive terms is constant. We call this difference the **common** difference and have $t_{n+1} - t_n = d$ for $n \in N$.

Example 1   In the arithmetic sequence 2, 4, 6, 8, 10, . . . , we have $d = 2$.

Example 2   In the arithmetic sequence 1, $\frac{5}{2}$, 4, $\frac{11}{2}$, 7, . . . , we have $d = \frac{3}{2}$.

Example 3   In the arithmetic sequence 1, $-\frac{1}{2}$, -2, . . . , we have $d = -\frac{3}{2}$.

Example 4   Suppose that in drilling for water the cost is \$150 for the first unit of 10 ft. or less and increases \$40 for each additional unit of 10 ft. or less. Then the costs, in dollars, for successive units of 10 ft. form an arithmetic progression: 150, 190, 230, 270, ...

An arithmetic sequence is also called an **arithmetic progression** (A.P.) and we say that the terms of the sequence are in arithmetic progression.

From Definition 1.2, it follows that if t_1 is the first term of an arithmetic progression, we have

$$t_2 = t_1 + d,$$
$$t_3 = (t_1 + d) + d = t_1 + 2d,$$
$$t_4 = (t_1 + 2d) + d = t_1 + 3d,$$
$$\cdots$$
$$t_n = t_1 + (n - 1)d.$$

The last line of this list of equalities provides a useful formula for finding particular terms of an arithmetic progression.

Example 5 Find (1) the tenth and (2) the fifteenth term of an A.P. for which $t_1 = 3$ and $d = 4$.

Solution:

$$
\begin{aligned}
1. \quad t_{10} &= t_1 + (10 - 1)d \\
t_{10} &= 3 + (9)4 = 39. \\
2. \quad t_{15} &= t_1 + (15 - 1)d \\
t_{15} &= 3 + 14(4) = 59.
\end{aligned}
$$

Example 6 If the first two terms of an A.P. are 3 and 7, find the 100th term.

Solution:

$$
\begin{aligned}
d &= t_2 - t_1 = 7 - 3 = 4 \\
t_{100} &= t_1 + (100 - 1)d \\
t_{100} &= 3 + 99(4) = 399.
\end{aligned}
$$

Example 7 If the tenth term of an A.P. is 32 and the sixth term is 12, find the first term and the common difference.

Solution: We know that $t_{10} = t_1 + 9d$. We have, then,

$$32 = t_1 + 9d.$$

Likewise, we have

$$12 = t_1 + 5d.$$

To solve this system of two linear equations, we can write

$$
\begin{aligned}
32 &= t_1 + 9d \\
12 &= t_1 + 5d \\
\hline
20 &= \qquad 4d \\
d &= 5.
\end{aligned}
$$

By substitution in the second equation, we have $12 = t_1 + 25$, and so $t_1 = -13$. Would substitution in the first equation also give us $t_1 = -13$?

If three numbers a, b, c form an arithmetic progression, then b is called **the arithmetic mean** between a and c. If more than three numbers form an arithmetic progression, then the terms between any two given terms of this arithmetic progression are called **arithmetic means** between the given terms.

For example, in the arithmetic sequence 1, 4, 7, 10, 13, 16, 19, . . . , we say that 4 is the arithmetic mean between 1 and 7; 7 and 10 are called arithmetic means between 4 and 13; and 10, 13, and 16 are called arithmetic means between 7 and 19.

Example 8 Insert a single arithmetic mean between 3 and -7.

Solution: If we insert exactly one arithmetic mean between 3 and -7, we have three numbers in arithmetic sequence. These we represent as 3, $3 + d$, -7 or, alternatively, as 3, $3 + d$, $3 + 2d$. We note that $-7 = 3 + 2d$ and obtain $d = -5$. Hence the three numbers in arithmetic sequence are 3, -2, -7 and -2 is the arithmetic mean between 3 and -7.

Example 9 Insert two arithmetic means between 16 and 21.

Solution: If we insert two means between 16 and 21, we shall have four numbers in arithmetic sequence. These we can represent as 16, $16 + d$, $16 + 2d$, 21. We see that

$$21 = 16 + 3d.$$

Hence we have $d = \frac{5}{3}$ and the four numbers are 16, $17\frac{2}{3}$, $19\frac{1}{3}$, and 21. The two means are $17\frac{2}{3}$ and $19\frac{1}{3}$.

EXERCISES 1.3

◀ In Exercises 1–10, determine which sequences are arithmetic. For those that are arithmetic, find the common difference and the next two terms.

1. 4, 10, 16, 22, . . .

2. 2, $\dfrac{5}{2}$, 3, $\dfrac{7}{2}$, . . .

3. 2, 3.2, 5.2, 7.2, . . .

4. -10, -6, -2, 2, . . .

5. 3, $3 + \sqrt{2}$, $3 + 2\sqrt{2}$, . . .

6. 1, -1, -3, -5, . . .

7. 3, 3, 3, 3, . . .

8. $\dfrac{1}{2}$, $\dfrac{1}{2}$, $\dfrac{1}{2}$, $\dfrac{1}{2}$, . . .

9. 2, 4, 8, 16, . . .

10. 1, 3, 9, 27, . . .

◀ In Exercises 11–19, find the indicated term of the given arithmetic progression.

11. Find the tenth term given 5, 7, 9, ...

12. Find the twentieth term given -1, $-\dfrac{1}{2}$, 0, ...

13. Find the fifteenth term given $\dfrac{5}{2}, \dfrac{13}{2}, \dfrac{21}{2}, \ldots$

14. Find the twelfth term given $-3, -\dfrac{1}{2}, 2, \ldots$

15. Find the sixteenth term given $10, 7, 4, \ldots$

16. Find the twelfth term given that the first term is 7 and the common difference is 2.

17. Find the tenth term given that the first term is -3 and the common difference is 4.

18. Find the fifteenth term given that the third term is 5 and the common difference is -3.

19. Find the fourteenth term given that the third term is -2 and the common difference is $\dfrac{1}{2}$.

20. Write the first five terms of an arithmetic sequence in which the second term is 4 and the fourth term is 10.

21. Write the first five terms of an arithmetic sequence in which the second term is 2 and the fourth term is 11.

22. Write the first six terms of an arithmetic sequence in which the third term is 5 and the sixth term is 13.

23. Write the first six terms of an arithmetic sequence in which the third term is 2 and the fourth term is 17.

24. A person earns $3.00 on the first day of the month and then increases his earnings by $4.00 on each working day. How much money will he earn on the 30th working day?

25. If the cost of drilling for water is $150 for the first unit of 10 ft. and increases $40 for each additional unit of 10 ft. or less, what is the cost of drilling from 60 ft. to 70 ft.?

26. A man saved $5.00 the first week of the year and then increased his weekly savings by $1.75 each week. In what week did he save $20.75?

27. A ball rolling down an inclined plane travels 8 ft. during the first second and each second thereafter increases the distance it rolls by 6 ft. In what second will it roll 56 ft.?

28. Find x so that $3 - x$, x, and $7 - 2x$, taken in the given order, are in arithmetic progression.

29. Find y so that $y - 2$, $1 - 2y$, and $3y$, taken in the given order, are in arithmetic progression.

30. Insert a single arithmetic mean between 5 and 14.

31. Insert two arithmetic means between 3 and $4\frac{1}{2}$.

32. Insert four arithmetic means between -1 and 15.

33. Insert four arithmetic means between $3\frac{1}{2}$ and $8\frac{1}{2}$.

34. Prove that the arithmetic mean between a and b is $\dfrac{a+b}{2}$.

35. CHALLENGE PROBLEM. Find a formula for the sum of the first n terms of an arithmetic progression.

36. CHALLENGE PROBLEM. If $\dfrac{1}{a+b}$, $\dfrac{1}{b+c}$, and $\dfrac{1}{c+a}$ form an arithmetic sequence, show that a^2, b^2, and c^2 also form an arithmetic sequence.

1.4 GEOMETRIC PROGRESSIONS

A second type of sequence that has been given a special name is the geometric sequence.

> **Definition 1.3** A sequence t_1, t_2, t_3, . . . is called a **geometric sequence** if the ratio between two successive terms is constant. We call this ratio, r, the **common ratio**
>
> and have $r = \dfrac{t_{n+1}}{t_n}$ for $n \in N$.

Example 1 In the geometric sequence 2, 4, 8, 16, 32, . . . we have $r = 2$.

Example 2 In the geometric sequence 2, -2, 2, -2, . . . we have $r = -1$.

Example 3 In the geometric sequence -3, $-\frac{3}{2}$, $-\frac{3}{4}$, $-\frac{3}{8}$, . . . we have $r = \frac{1}{2}$.

Example 4 A ball is dropped from a height of 21 ft. On each rebound it reaches a height $\frac{2}{3}$ of that from which it fell. Then the successive heights form a geometric progression: 21, 14, $\frac{28}{3}$, $\frac{56}{9}$, . . . with $r = \frac{2}{3}$.

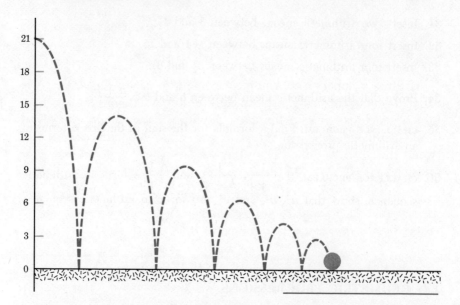

Geometric sequences are also referred to as **geometric progressions** (G.P.) and the terms of the sequence are said to be in geometric progression. Note that no term of a geometric progression can be zero (Why?) and that $r \neq 0$.

If the first term of a geometric progression is t_1 and the ratio is r, it follows from Definition 1.3 that

$$t_2 = t_1 r$$
$$t_3 = (t_1 r)r = t_1 r^2$$
$$t_4 = (t_1 r^2)r = t_1 r^3$$
$$\cdots$$
$$t_n = t_1 r^{n-1}.$$

The last line of this list of equalities provides a simple formula for finding particular terms of a geometric sequence.

Example 5 If a G.P. has a constant ratio 3 and the fourth term is 54, find the first term.

Solution: We know $t_4 = t_1 r^3$. Hence $54 = t_1 (3)^3$ and $t_1 = 2$.

Example 6 The fourth term of a G.P. is 1 and the seventh term is $-\frac{1}{8}$. Find the first term and the common ratio.

Solution: We have $t_4 = t_1 r^3$ and $t_7 = t_1 r^6$. Notice that

$$t_1 r^6 = (t_1 r^3)r^3. \qquad \text{(Why?)}$$

Since $t_4 = t_1 r^3$, we can write $t_1 r^6 = t_4 r^3$. Substituting given values into the equation $t_7 = t_4 r^3$, we have $-\frac{1}{8} = 1 \cdot r^3$. This gives us $r^3 = -\frac{1}{8}$, so that $r = -\frac{1}{2}$. Using $t_4 = t_1 r^3$, we find $t_1 = -8$.

If three numbers a, b, c form a geometric progression, then b is called a **geometric mean** between a and c. If more than three numbers form a geometric progression, then the terms between any two given terms of this geometric progression are called **geometric means** between the given terms. For example, if we have a geometric progression

$$t_1,\ t_2,\ t_3,\ t_4,\ t_5,\ t_6,$$

we say that t_2, t_3, and t_4 are geometric means between t_1 and t_5. Likewise, we say that t_4 is a geometric mean between t_3 and t_5.

Example 7 Insert a geometric mean between 3 and $\frac{4}{3}$.

Solution: Let $3 = t_1$. Then $\frac{4}{3} = t_3$. We have $t_3 = t_1 r^2$. Hence

$$r^2 = \frac{\frac{4}{3}}{3} = \frac{4}{9}.$$

Thus $r = \frac{2}{3}$ or $r = -\frac{2}{3}$. These two values of r give two numbers each of which is a geometric mean. These are 2 and -2, since

$$3 \cdot \frac{2}{3} = 2 \quad \text{and} \quad 3 \cdot \left(-\frac{2}{3}\right) = -2.$$

Is it always possible to find two numbers each of which is a geometric mean between two given numbers? Is there any situation in which a geometric mean does not exist?

When two such numbers, one the negative of the other, can each serve as a geometric mean, we call the positive one *the* geometric mean. Thus we can say, in Example 7, that *the* geometric mean of 3 and $\frac{4}{3}$ is 2.

Example 8 Insert three geometric means between 6 and $\frac{3}{8}$.

Solution: Let $t_1 = 6$. Then $t_5 = \frac{3}{8}$. We have $t_5 = t_1 r^4$. Hence

$$r^4 = \frac{t_5}{t_1} = \frac{\frac{3}{8}}{6} = \frac{3}{48} = \frac{1}{16}.$$

Thus $r = \frac{1}{2}$ or $r = -\frac{1}{2}$. Using these values for r, we find two possible sequences:

$$6,\ 3,\ \frac{3}{2},\ \frac{3}{4},\ \frac{3}{8} \quad \text{and} \quad 6,\ -3,\ \frac{3}{2},\ -\frac{3}{4},\ \frac{3}{8}.$$

EXERCISES 1.4

◀ In Exercises 1–10, the first few terms of a geometric sequence are given. Find the common ratio and give the next three terms of each sequence.

1. 1, 2, 4, 8, . . .

2. 1, -1, 1, -1, . . .

3. 2, 1, $\dfrac{1}{2}$, $\dfrac{1}{4}$, . . .

4. -1, $\dfrac{1}{3}$, $-\dfrac{1}{9}$, $\dfrac{1}{27}$, . . .

5. 2, 2, 2, 2, . . .

6. $\sqrt{3}$, 3, $3\sqrt{3}$, 9, . . .

7. $\sqrt{2}$, $\sqrt{6}$, $3\sqrt{2}$, . . .

8. $\dfrac{10}{3}$, 1, $\dfrac{3}{10}$, $\dfrac{9}{100}$, . . .

9. π, π^2, π^3, . . .

10. 3, $3\sqrt[3]{2}$, $3\sqrt[3]{4}$, 6, . . .

11. Find x so that $-\frac{3}{2}$, x, and $-\frac{8}{27}$, taken in the given order, form a geometric progression.

12. Find x so that $\frac{2}{3}$, x, and $\frac{27}{8}$, taken in the given order, form a geometric progression.

13. For what values of y will $3y + 1$, $y + 2$, and $4y + 8$, taken in the given order, form a geometric progression?

14. For what values of y will $2y + 3$, $y - 2$, and $3y - 6$, taken in the given order, form a geometric progression?

15. Insert two geometric means between 2 and 54.

16. Insert two geometric means between 8 and 1.

17. Insert three positive geometric means between 2 and 512.

18. Insert three positive geometric means between 27 and $\frac{24}{3}$.

◀ In Exercises 19–22, find the geometric mean of the pair of numbers.

19. 3 and 12   21. 5 and 7

20. 4 and 36   22. 8 and 13

23. Prove that the geometric mean of a and b ($a > 0$ and $b > 0$) is $\sqrt{ab}$.

24. A golf ball rebounds $\frac{2}{3}$ of the distance through which it falls when it lands on a hard surface. If the golf ball is dropped from 21 ft., how far will it rebound after it hits for the sixth time?

25. If a man earns \$1.00 the first day and on each succeeding day doubles his earnings, how much will he earn on the tenth day?

26. The population of a city is increasing at the rate of 8% a year. If the city now has a population of 100,000, what will be its population 3 years later?

27. The first three terms of a sequence of four terms form an arithmetic progression and the last three terms form a geometric progression. If the first term is 2 and the fourth term is 9, find the middle two terms.

28. The first three terms of a sequence of four terms form an arithmetic progression and the last three form a geometric progression. The sum of the first and third terms is 28. What are the four terms of the sequence?

29. The starting salary for a job is $5000 a year. What will be the salary after 4 years if there is a 5% increase in salary each year?

30. The sum of the second and third terms of a geometric sequence is 10 and their product is 16. What are the first four terms of the sequence?

31. Is the sequence

$$1, 1, 1, 1, 1, 1, \ldots$$

both an arithmetic sequence and a geometric sequence? Explain.

◀ In Exercises 32–37, we have given that $t_1, t_2, t_3, \ldots, t_n, \ldots$ is a geometric progression and that a is a constant. Determine if the given sequence in each exercise is a geometric progression.

32. $t_n, t_{n-1}, \ldots, t_2, t_1$

33. $t_1 + a, t_2 + a, \ldots, t_n + a$

34. $at_1, at_2, \ldots, at_n$

35. $t_1^2, t_2^2, t_3^2, \ldots, t_n^2$

36. $t_1, t_3, t_5, \ldots, t_{2n+1}$

37. $t_2, t_4, t_6, \ldots, t_{2n}$

38. Find the number of terms in the geometric sequence

$$32, 16, 8, \ldots, \frac{1}{256}.$$

39. Find the number of terms in the geometric sequence

$$\frac{1}{27}, -\frac{1}{9}, \frac{1}{3}, \ldots, 729.$$

40. **CHALLENGE PROBLEM.** Find a number x that will make the following three numbers a geometric sequence in the given order:

$$\sqrt{2 - x}, \qquad \sqrt{20 - x}, \qquad \sqrt{18 - 9x}.$$

41. **CHALLENGE PROBLEM.** Find all sets of three integers in geometric progression whose product is -216 and the sum of whose squares is 189.

1.5   THE SUM OF A SEQUENCE

The symbol S_n commonly represents the sum of the first n terms of a sequence. For example, suppose we have the arithmetic sequence

$$2, 5, 8, 11, 14, 17, \ldots.$$

Then

$$S_1 = 2,$$
$$S_2 = 2 + 5 = 7,$$
$$S_3 = 2 + 5 + 8 = 15,$$
$$S_4 = 2 + 5 + 8 + 11 = 26.$$

If we have the geometric sequence

$$1, 2, 4, 8, 16, 32, \ldots,$$

then

$$S_1 = 1,$$
$$S_2 = 1 + 2 = 3,$$
$$S_3 = 1 + 2 + 4 = 7,$$
$$S_4 = 1 + 2 + 4 + 8 = 15.$$

In general, $S_n = t_1 + t_2 + t_3 + \cdots + t_{n-1} + t_n$.

It is useful to have a formula for the sum of the first n terms of a sequence. We shall now develop such a formula for an arithmetic sequence. We have

$$(1) \qquad S_n = t_1 + (t_1 + d) + (t_1 + 2d) + \cdots + t_n.$$

By reversing the order of the terms, we have

$$(2) \qquad S_n = t_n + (t_n - d) + (t_n - 2d) + \cdots + t_1.$$

By addition of Equation 1 to Equation 2 we have

$$2S_n = (t_1 + t_n) + (t_1 + t_n) + \cdots + (t_1 + t_n).$$

The right-hand member of this last equality consists of n terms, each of which is $(t_1 + t_n)$. Thus

$$2S_n = n(t_1 + t_n)$$

or

$$S_n = \frac{n}{2}(t_1 + t_n).$$

Since $t_n = t_1 + (n - 1)d$, we also have

$$S_n = \frac{n}{2}[2t_1 + (n - 1)d].$$

We state the result as a theorem.

THEOREM 1.1

The sum, S_n, of the first n terms of an arithmetic progression in which the first term is t_1 and the common difference is d is

$$S_n = \frac{n}{2}[2t_1 + (n - 1)d].$$

Example 1   Find the sum of the first ten terms of an A. P. if $t_1 = 3$ and $d = 2$.

Solution:

$$S_{10} = \frac{10}{2}[2(3) + (10 - 1)2]$$

$$= 5(6 + 18) = 120.$$

A formula for the sum of the first n terms of a geometric sequence can be developed in a similar fashion. We have

$$(3) \qquad S_n = t_1 + t_1 r + t_1 r^2 + \cdots + t_1 r^{n-1}.$$

If we multiply all members of this equation by r, we have

$$(4) \qquad rS_n = t_1 r + t_1 r^2 + \cdots + t_1 r^{n-1} + t_1 r^n.$$

By subtraction of Equation 4 from Equation 3 we have

$$S_n - rS_n = t_1 - t_1 r^n$$

so that

$$S_n(1 - r) = t_1 - t_1 r^n.$$

If $r \neq 1$, we can divide by $1 - r$ and obtain

$$S_n = \frac{t_1 - t_1 r^n}{1 - r} = \frac{t_1 r^n - t_1}{r - 1} \qquad \text{(if } r \neq 1\text{)}.$$

Again we state the result as a theorem.

THEOREM 1.2

The sum, S_n, of the first n terms of a geometric progression whose first term is t_1 and common ratio is r ($r \neq 1$) is

$$S_n = \frac{t_1 r^n - t_1}{r - 1} = t_1\left(\frac{r^n - 1}{r - 1}\right).$$

Example 2 Find the sum of the first eight terms of the geometric sequence 1, 2, 4, 8, ...

Solution:

$$t_1 = 1, \qquad r = \frac{2}{1} = 2, \qquad n = 8.$$

$$S_8 = \frac{1(2^8) - 1}{2 - 1} = \frac{256 - 1}{1} = 255.$$

Example 3 Find the sum of the first five terms of the geometric sequence -81, 54, -36, ...

Solution:

$$r = \frac{54}{-81} = -\frac{2}{3}, \qquad t_1 = -81.$$

$$S_5 = \frac{-81(-\frac{2}{3})^5 + 81}{-\frac{2}{3} - 1} = \frac{\frac{32}{3} + 81}{-\frac{5}{3}} = -55.$$

What is the situation when $r = 1$ in a geometric sequence?

EXERCISES 1.5

◀In Exercises 1–8, find the sum of the terms of the given arithmetic progression.

1. 2, 4, 6, 8, . . . , 30

2. -4, -1, 2, 5, . . . , 29

3. 1, $\frac{3}{2}$, 2, $\frac{5}{2}$, . . . , 29

4. -1, $-\frac{1}{4}$, $\frac{1}{2}$, $\frac{5}{4}$, . . . , 11

5. The arithmetic progression with $t_1 = -10$, $d = 4$, and $n = 10$.

6. The arithmetic progression with $t_1 = 5$, $d = -3$, and $n = 12$.

7. The arithmetic progression with $t_1 = 13$, $t_n = 89$, and $n = 19$.

8. The arithmetic progression with $t_1 = -4$, $t_n = -55$, and $n = 17$.

9. Find the sum of the first 20 even natural numbers.

◀In Exercises 10–17, find the sum of the terms of the given geometric progression.

10. -5, 15, -45, . . . , 1215

11. $4, 2, 1, \ldots, \dfrac{1}{64}$

12. $9, -3, 1, \ldots, -\dfrac{1}{243}$

13. The geometric progression with $t_1 = 64$, $r = -\dfrac{1}{2}$, and $n = 8$.

14. The geometric progression with $t_1 = \dfrac{1}{2}$, $r = \dfrac{1}{3}$, and $n = 6$.

15. The geometric progression with $t_1 = 1$, $r = -1$, and $n = 30$.

16. The geometric progression with $t_1 = -27$, $t_n = 27$, and $n = 8$.

17. The geometric progression with $t_1 = \dfrac{1}{4}$, $t_n = 256$, and $n = 11$.

18. The front row of a school assembly hall has 20 seats, the second row has 22 seats, the third row 24 seats, and so on. How many seats are there in the first 15 rows?

19. A freely falling body falls 16 ft. in the first second, 48 ft. in the second second, and 32 ft. more in each succeeding second than it did in the one preceding. How far does a stone, dropped from a very high bridge, fall in 5 seconds?

20. A student saves 10 cents the first week, 15 cents the second week, 20 cents the third week, and so on. How long will it take the student to save more than $10.00?

21. Faye and Isabelle are both reading a long historical novel. They both start the same day. Faye reads 50 pages every day. Isabelle reads 10 pages the first day, 20 the second, 30 the third, and so on. The novel has 1200 pages. Who completes the novel first?

22. How many ancestors does a person have in the ten generations preceding him if we assume no intermarriages (that is, no marriages between cousins)?

23. A golf ball rebounds $\frac{2}{3}$ of the distance through which it falls when it lands on a hard surface. If the golf ball is dropped from 21 ft., how far will it have traveled when it hits the ground for the sixth time? (Compare your answer with that of Exercise 24 of Exercises 1.4.)

24. If a man earns $1.00 the first day and on each succeeding day doubles his earnings, how much will he have earned at the end of the tenth day? (Compare your answer with that of Exercise 25 of Exercises 1.4.)

25. The sum of the first and second terms of a geometric progression is -3 and the sum of the fifth and sixth terms is $-\frac{3}{16}$. Find the sum of the first ten terms.

26. The sum of the second and third terms of a geometric progression is $-\frac{2}{3}$; the sum of the fifth and sixth terms is 18. Find the sum of the first eight terms.

◀ In Exercises 27–31, the given data refer to geometric progressions.

27. If $r = \dfrac{2}{3}$ and $t_5 = \dfrac{4}{9}$, find t_7.

28. If $S_1 = \dfrac{1}{3}$ and $S_2 = \dfrac{1}{2}$, find r and t_1.

29. CHALLENGE PROBLEM. If $r = \dfrac{3}{2}$, $t_1 = 32$, and $S_n = 665$, find t_n and n.

30. CHALLENGE PROBLEM. If $r = -3$, $t_n = -189$, and $S_n = -140$, find n and t_1.

31. CHALLENGE PROBLEM. Two lines intersect at a 45° angle. From a point 4 in. from the intersection and on one of the lines, a perpendicular is dropped to a point, P, on the other line. From P a perpendicular is drawn back to a point, Q, on the first line. From Q a perpendicular is drawn back to a second line, etc. Find the sum of the lengths of the first ten perpendicular segments.

1.6 SERIES AND SIGMA NOTATION

The indicated sum of a sequence is often called a **series.** If we have an arithmetic sequence 2, 4, 6, 8, 10, then the indicated sum of the sequence

$$2 + 4 + 6 + 8 + 10$$

is called an **arithmetic series.** The word "series" refers to the form. We do not say, for example, that 30 is an arithmetic series even though $2 + 4 + 6 + 8 + 10 = 30$.

The indicated sum of a geometric sequence such as $1, \frac{1}{2}, \frac{1}{4}, \frac{1}{8}, \frac{1}{16}$ is called a **geometric series.** Hence

$$1 + \frac{1}{2} + \frac{1}{4} + \frac{1}{8} + \frac{1}{16}$$

is a geometric series.

In the series, $t_1 + t_2 + \cdots + t_n$, the numbers $t_1, t_2, \ldots, t_n$ are called the **terms** of the series.

It is interesting to note that the idea of a series leads to that of another type of sequence called a **sequence of partial sums.** For example, from the arithmetic sequence 2, 4, 6, 8, . . . , we have

$$S_1 = 2, \qquad S_2 = 2 + 4 = 6, \qquad S_3 = 2 + 4 + 6 = 12,$$
$$S_4 = 2 + 4 + 6 + 8 = 20, \ldots$$

which gives the sequence 2, 6, 12, 20, ... Is S_1, S_2, S_3, S_4, . . . an arithmetic sequence? A geometric sequence?

The Greek letter Σ (sigma) is the symbol generally used to denote the sum of a sequence. For example, the sum

$$1 + 2 + 3 + 4 + 5 + 6 + 7 + 8$$

can be expressed as

$$\sum_{n=1}^{8} n.$$

(Read "the summation from 1 to 8 of n.")

Other examples of the use of sigma notation are

$$\sum_{k=2}^{6} 3k = 3 \cdot 2 + 3 \cdot 3 + 3 \cdot 4 + 3 \cdot 5 + 3 \cdot 6 = 60,$$

$$\sum_{j=1}^{5} 2^j = 2^1 + 2^2 + 2^3 + 2^4 + 2^5 = 62,$$

and

$$\sum_{i=3}^{5} (2i - 1)^2 = [(2 \cdot 3) - 1]^2 + [(2 \cdot 4) - 1]^2 + [(2 \cdot 5) - 1]^2$$

$$= 25 + 49 + 81 = 155.$$

When we write

$$\sum_{i=1}^{m} t_i = t_1 + t_2 + \cdots + t_m,$$

we call the letter Σ the **summation sign**; the expression t_i the **summand**; and the letter i the **index**.

The sigma notation provides an alternate method of expressing certain sums.

Example 1 Find $\displaystyle\sum_{k=1}^{20} (2k - 10)$.

Solution:

$$\sum_{k=1}^{20} (2k - 10) = (2 \cdot 1 - 10) + (2 \cdot 2 - 10) + (2 \cdot 3 - 10) + \cdots$$

$$+ (2 \cdot 19 - 10) + (2 \cdot 20 - 10)$$

$$= (-8) + (-6) + (-4) + \cdots + 28 + 30.$$

The numbers $-8, -6, -4, \ldots$ form an arithmetic sequence in which $t_1 = -8$ and $d = 2$. We use the formula

$$S_n = \frac{n}{2}(t_1 + t_n)$$

and obtain

$$S_{20} = \frac{20}{2}(-8 + 30) = 220.$$

Hence

$$\sum_{k=1}^{20} (2k - 10) = 220.$$

Example 2   Find $\displaystyle\sum_{j=3}^{8} 3 \cdot 2^{j-2}$.

Solution:

$$\sum_{j=3}^{8} 3 \cdot 2^{j-2} = 3 \cdot 2^{3-2} + 3 \cdot 2^{4-2} + 3 \cdot 2^{5-2} + 3 \cdot 2^{6-2} + 3 \cdot 2^{7-2}$$

$$+ 3 \cdot 2^{8-2}$$

$$= 3 \cdot 2 + 3 \cdot 2^2 + 3 \cdot 2^3 + 3 \cdot 2^4 + 3 \cdot 2^5 + 3 \cdot 2^6.$$

The numbers $3 \cdot 2, 3 \cdot 2^2, \ldots, 3 \cdot 2^6$ form a geometric sequence in which $t_1 = 3 \cdot 2 = 6$ and $r = 2$. We use the formula

$$S_n = t_1 \frac{r^n - 1}{r - 1}$$

and note that $n = (8 - 3) + 1 = 6$. We have

$$S_n = 6 \cdot \frac{2^6 - 1}{2 - 1} = 6 \cdot 63 = 378.$$

Hence

$$\sum_{j=3}^{8} 3 \cdot 2^{j-2} = 378.$$

EXERCISES 1.6

◀ In Exercises 1–10, write out the series designated by the sigma notation.

1. $\displaystyle\sum_{k=1}^{7} k$ **6.** $\displaystyle\sum_{k=1}^{2} (-k)^3$

2. $\displaystyle\sum_{n=1}^{7} n$ **7.** $\displaystyle\sum_{k=1}^{4} (2^k - k)$

3. $\displaystyle\sum_{i=1}^{6} t_i$ **8.** $\displaystyle\sum_{k=1}^{10} (t_1 + kd)$

4. $\displaystyle\sum_{n=3}^{3} n$ **9.** $\displaystyle\sum_{i=1}^{8} 2^i$

5. $\displaystyle\sum_{j=1}^{5} j^2$ **10.** $\displaystyle\sum_{i=1}^{4} (2 - i)^i$

◀ In Exercises 11–17, find the sum.

11. $\displaystyle\sum_{n=1}^{8} n$ **15.** $\displaystyle\sum_{m=1}^{8} (4 - 2m)$

12. $\displaystyle\sum_{k=1}^{2n} (-1)^k$ **16.** $\displaystyle\sum_{n=1}^{6} 2^{n-1}$

13. $\displaystyle\sum_{k=1}^{20} (k - 8)$ **17.** $\displaystyle\sum_{n=1}^{6} 32\left(\frac{1}{2}\right)^{n-1}$

14. $\displaystyle\sum_{k=1}^{10} t_1 + (k - 1)d$

18. Show, by writing out the series, that

$$\sum_{k=1}^{6} ca_k = c \sum_{k=1}^{6} a_k.$$

19. Show, by writing out the series, that

$$\sum_{k=1}^{6} (a_k + b_k) = \sum_{k=1}^{6} a_k + \sum_{k=1}^{6} b_k.$$

20. Show that

$$\sum_{n=1}^{4} \frac{n+6}{n+3} = \sum_{n=4}^{7} \frac{n+3}{n}.$$

CHAPTER SUMMARY

A FUNCTION is a correspondence between two sets that associates with each element of the first set exactly one element of the second set. The first set is called the DOMAIN of the function. For each element, x, of the domain, the corresponding element of the second set is called the IMAGE of x under the function. The set of all images is called the RANGE of the function.

Single letters are frequently used as names of functions. If a function is designated by the letter f and x is an element in the domain of the function f, then we write $f(x)$ for the image of x. Given the function $f: f(x) = 2x + 1$, we have $f(2) = 5$.

A SEQUENCE FUNCTION is a function whose domain is the set N of natural numbers or a subset $\{1, 2, 3, 4, \ldots, n\}$ of the natural numbers. The ordered arrangement of functional values is a SEQUENCE. The TERMS of a FINITE sequence of n terms are written as

$$t_1, t_2, t_3, \ldots, t_n.$$

An ARITHMETIC SEQUENCE, also called an ARITHMETIC PROGRESSION, is a sequence in which the difference d between two successive terms is constant. In an arithmetic sequence

$$t_n = t_1 + (n-1)d.$$

A GEOMETRIC SEQUENCE, also called a GEOMETRIC PROGRESSION, is a sequence in which the ratio r between two successive terms is constant. In a geometric sequence

$$t_n = t_1 r^{n-1}.$$

A SERIES is the indicated sum of two or more terms of a sequence. SIGMA NOTATION is sometimes used to denote a series. Thus

$$\sum_{i=1}^{n} t_i = t_1 + t_2 + t_3 + \cdots + t_n.$$

The formula for the sum, S_n, of the first n terms of an arithmetic sequence is

$$S_n = \frac{n}{2}(t_1 + t_n) = \frac{n}{2}[2t_1 + (n-1)d].$$

The formula for the sum, S_n, of the first n terms of a geometric sequence is

$$S_n = \frac{t_1 r^n - t_1}{r-1} = t_1\left(\frac{r^n - 1}{r-1}\right) \qquad (r \neq 1).$$

REVIEW EXERCISES

1. Let f be a function with domain $\{-1, 0, 1, 2, 3, 4\}$ such that

$$f(x) = \frac{x}{x+2}.$$

Find the range of f.

2. Let $f : f(x) = x^2 + 2$ be a function with domain the set of real numbers. Find $f(0)$, $f(1)$, $f(-1)$, $f\left(\frac{1}{2}\right)$, $f(\sqrt{2})$, and give the range of f.

3. Write the first six terms of the sequence defined by the function $F : F(x) = 2x + 3$.

4. Write the first six terms of the sequence defined by $t_n = 3n - 1$.

5. Write the first six terms of the sequence defined by $t_1 = 3$ and $t_{n+1} = t_n - 2$ for $n \geq 1$.

6. Find the fiftieth term of the A. P.: $-2, 7, 16, 25, \ldots$.

7. Write the first six terms of an arithmetic sequence in which the second term is 4 and the fourth term is 14.

8. Insert three arithmetic means between -3 and 13.

9. Find the sum of an arithmetic progression in which $t_1 = -3$, $d = \frac{3}{2}$, and $n = 25$.

10. Find the tenth term of the G. P.: $-2, 3, -\frac{9}{2}, \frac{27}{4}, \ldots$.

11. Insert three positive geometric means between 5 and 405.

12. Find the sum of the first eight terms of the geometric sequence for which $t_1 = -3$ and $r = \frac{2}{3}$.

13. Find x so that $x + 2$, x, and $3x + 5$, in that order, form an arithmetic progression.

14. Find y so that 16, $\dfrac{2}{y+2}$, and $2y + 4$, in that order, form a geometric progression.

15. In a geometric progression it is given that $r = \frac{3}{2}$ and $t_4 = \frac{9}{4}$. Find t_7 and S_7.

16. Find $\displaystyle\sum_{i=3}^{10} (3i + 2)$.

17. Find $\displaystyle\sum_{i=2}^{6} 2 \cdot 3^{i-1}$.

18. Find $\displaystyle\sum_{j=1}^{5} (2^j + j)$.

19. Several men played 18 holes of golf and made scores that varied from 80 to 96. When they arranged their scores in order from low to high, they noted that their scores were in arithmetic progression and totaled 440. How many players were there and what was the score of each?

20. A boy started a chain letter by writing to two friends and asking that each send a copy to two other friends. If the chain remained unbroken when the tenth set was mailed, how much was spent for postage at 6 cents per letter?

GOING FURTHER: READING AND RESEARCH

Perhaps you have heard the statement "No matter how thin you slice it, it is still baloney." This comment has interesting mathematical overtones involving series!

Let us look at the infinite series

(1) $$1 + \frac{1}{2} + \frac{1}{4} + \frac{1}{8} + \frac{1}{16} + \cdots$$

whose nth term is $\dfrac{1}{2^{n-1}}$. Is there a smallest number which is a term of this series? Let us change the question—is there a number which is less than any term of the series? What is the largest number which is less than any term of the series?

In what basic way does the series given above differ from the series

(2) $$1 + 2 + 4 + 8 + 16 + \cdots?$$

What about the series

(3) $$1 + (-1) + 1 + (-1) + 1 + (-1) + \cdots?$$

The first of these series is said to **converge** because the points on the number line that correspond to the partial sums $S_1 = 1$, $S_2 = 1\frac{1}{2}$, $S_3 = 1\frac{3}{4}$, ... get closer and closer to a fixed point, namely the point corresponding to the number 2 as indicated in the figure.

The other two series are called **divergent** because there is no single point such as 2 to which the partial sums get closer and closer.

Thus in (2) the partial sums are $S_1 = 1$, $S_2 = 3$, $S_3 = 7$, $S_4 = 15$, . . . and increase beyond bound. On the other hand, in (3) the partial sums are $S_1 = 1$, $S_2 = 0$, $S_3 = 1$, $S_4 = 0$, . . . and alternate between 1 and 0. Clearly there is no single point to which the partial sums converge.

Try guessing whether the following series converge or diverge.

1. $1 + (-2) + 3 + (-4) + 5 + \cdots$; nth term is $(-1)^{n+1}n$.

2. $\dfrac{4}{3} + \dfrac{2}{3} + \dfrac{1}{3} + \cdots$; nth term is $\left(\dfrac{1}{3}\right)\left(\dfrac{1}{2}\right)^{n-3}$.

3. $\sqrt{2} + 1 + \dfrac{1}{\sqrt{2}} + \dfrac{1}{2} + \cdots$; nth term is $\left(\dfrac{1}{\sqrt{2}}\right)^{n-2}$.

4. $0.3 + 0.03 + 0.003 + \cdots$; nth term is $3\left(\dfrac{1}{10}\right)^{n}$.

5. $(-1) + 2 + (-3) + 4 + (-5) + \cdots$; nth term is $(-1)^{n}n$.

6. $1 + \dfrac{1}{2} + \dfrac{1}{3} + \dfrac{1}{4} + \cdots$; nth term is $\dfrac{1}{n}$.

7. $1 + \left(-\dfrac{1}{2}\right) + \dfrac{1}{3} + \left(-\dfrac{1}{4}\right) + \cdots$; nth term is $(-1)^{n+1}\dfrac{1}{n}$.

8. $1 + \dfrac{1}{3} + \dfrac{1}{5} + \dfrac{1}{7} + \cdots$; nth term is $\dfrac{1}{2n+1}$.

9. $1 + \left(-\dfrac{1}{3}\right) + \dfrac{1}{5} + \left(-\dfrac{1}{7}\right) + \cdots$; nth term is $(-1)^{n+1}\dfrac{1}{2n-1}$.

10. $\dfrac{1}{4} + \dfrac{1}{9} + \dfrac{1}{16} + \dfrac{1}{25} + \cdots$; nth term is $\dfrac{1}{(n+1)^2}$.

If you want to read more about convergence and divergence, here are some suggestions.

DOLCIANI, M. P. AND EDWIN F. BECKENBACH, et al. *Modern Introductory Analysis,* Boston, Mass.: Houghton Mifflin, 1967.

GLICKSMAN, A. M. AND H. D. RUDERMAN. *Fundamentals for Advanced Mathematics,* New York: Holt, Rinehart and Winston, 1964.

JAMES, R. L. *University Mathematics,* Belmont, Calif.: Wadsworth, 1962.

BELL, E. T. *Men of Mathematics* (Chaps. 14 and 22). New York: Simon and Schuster, 1937.

BELL, E. T. *Mathematics, Queen and Servant of Science,* New York: McGraw-Hill, 1951.

NEWMAN, J. R. *The World of Mathematics,* Vol. 1, New York: Simon and Schuster, 1956.

Insights Into Modern Mathematics, Chapter VII, Thirteenth Yearbook of the National Council of Teachers of Mathematics, Washington, D.C.: National Council of Teachers of Mathematics, 1957.

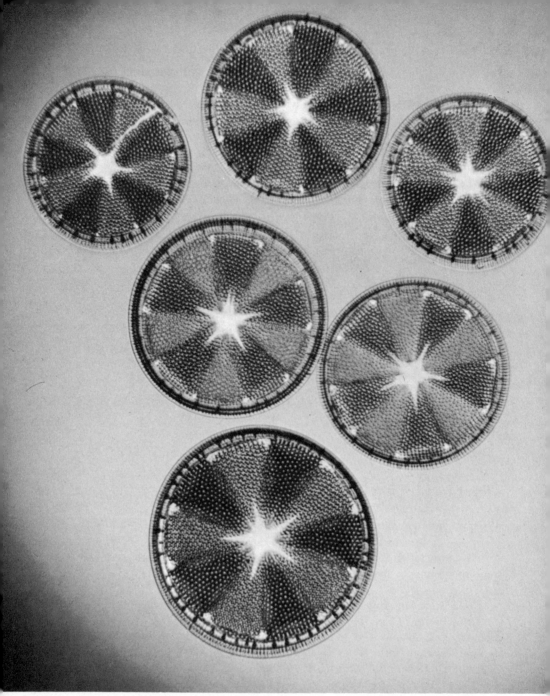

CHAPTER TWO
NATURAL NUMBERS AND
MATHEMATICAL INDUCTION

2.1 A COMPUTING FORMULA

Here is a problem for you. Find the sum of the squares of the natural numbers from 1 to 20. *Answer:* 2870. To compute this sum by squaring each number and adding would probably take several minutes. (Unless, of course, you are some kind of lightning calculator!) Using a shortcut, however, the job can be done in about 10 seconds—by a nonlightning calculator. What is the catch?

Actually this shortcut involves the use of the mathematical formula

$$S_n = \frac{1}{6} n(n + 1)(2n + 1).$$

Here, as in Chapter 1, n is the number of terms in the sequence $1^2, 2^2, 3^2, \ldots, n^2$ and S_n is their sum. In our example, n is 20. Hence

$$S_{20} = \frac{1}{6}(20)(21)(41) = 2870.$$

If you are skeptical, you can check this result by computing the squares and adding.

For the sum of the first 25 squares, we have

$$S_{25} = \frac{1}{6}(25)(26)(51) = 5525.$$

Run a similar check on this.

The two verifications should give you some confidence in the formula. A few more examples may increase this confidence.

Can you be sure, however, that the same formula would work for $n = 1000$ or $n = 5000$? You would need a computer to verify the results in these two cases! But would these and a few other examples really assure you that the formula holds for any natural number n?

Look, for example, at the formula

$$1 + 2 + 3 + \cdots + n = \frac{1}{2}(3n^2 - 13n + 24).$$

Testing this formula for $n = 3$, we have

$$1 + 2 + 3 = 6$$

and

$$\frac{1}{2}[3(3^2) - 13(3) + 24] = \frac{1}{2}(12) = 6.$$

For $n = 4$ we have

$$1 + 2 + 3 + 4 = 10$$

and

$$\frac{1}{2}[3(4^2) - 13(4) + 24] = \frac{1}{2}(20) = 10.$$

So far so good!

But now let $n = 5$. Here we see that

$$1 + 2 + 3 + 4 + 5 = 15.$$

However, $\frac{1}{2}[3(5^2) - 13(5) + 24] = \frac{1}{2}(34) = 17$. The moral is that a formula cannot be validated on the basis of a few examples, nor, for that matter, by a hundred or any finite number of examples. How, then, might we prove the general validity of such formulas?

This job and others like it can be accomplished by means of a remarkable mathematical concept called the Principle of Mathematical Induction. In this chapter we shall carefully examine this principle.

EXERCISES 2.1

◀ In Exercises 1–10, test the given formula for $n = 5$, $n = 8$, and $n = 11$.

1. $1 + 2 + 3 + \cdots + n = \dfrac{1}{2}n(n + 1)$

2. $1^2 + 2^2 + 3^2 + \cdots + n^2 = \dfrac{1}{6}n(n + 1)(2n + 1)$

3. $1 + 3 + 5 + \cdots + (2n - 1) = n^2$

4. $\dfrac{1}{1 \cdot 2} + \dfrac{1}{2 \cdot 3} + \dfrac{1}{3 \cdot 4} + \cdots + \dfrac{1}{n(n + 1)} = \dfrac{n}{n + 1}$

5. $1 + 7 + 13 + \cdots + (6n - 5) = 3n^2 - n - 11$

6. $2 + 2^2 + 2^3 + \cdots + 2^n = 2(2^n - 1)$

7. $2^2 + 4^2 + 6^2 + \cdots + (2n)^2 = \dfrac{1}{2}n(n + 1)(2n + 1)$

8. $1 \cdot 3 + 2 \cdot 4 + 3 \cdot 5 + \cdots + n(n + 2) = \dfrac{n}{6}(n + 1)(2n + 7)$

9. $2 + 4 + 6 + \cdots + 2n = 2n^2 - 12n + 40$

10. $1 + 5 + 9 + \cdots + (4n - 3) = n(2n - 1)$

11. Test the proposition that $3^{2n} - 1$ is divisible by 8 for $n = 3$ and $n = 5$.

12. Test the formula

$$1^4 + 2^4 + \cdots + n^4 = \frac{n}{30}(n + 1)(2n + 1)(3n^2 + 3n - 1)$$

for $n = 3$ and $n = 5$.

2.2 THE NATURAL NUMBERS

The Principle of Mathematical Induction mentioned in Section 2.1 is a property of the system of natural numbers. Our first task, then, is to examine this system in some depth.

The set of natural numbers

$$N = \{1, 2, 3, \ldots\}$$

under the operations of addition and multiplication constitutes probably the most familiar of all mathematical systems. At the same time, since it is with this system that most of you had your earliest mathematical encounters, it would appear to be one of the simplest. Appearances can be deceiving. An examination of the properties of the natural numbers brings to light some far-reaching and highly sophisticated mathematical ideas.

To conduct our examination we shall begin with a listing of properties. Undoubtedly you are already acquainted with many of these properties since most of them are also properties of other familiar number systems such as the system of rational numbers and the system of real numbers. However, the natural number system is essentially the only number system which has *all* of these properties.

Note carefully, then, the following characterization of the natural numbers.

The system of natural numbers consists of a set N of elements equipped with two binary operations, denoted by $+$ and $\cdot$, for which the following properties hold.

1. Closure Property under $+$ and $\cdot$

If a and b are any elements in N, then $a + b$ and $a \cdot b$ are also in N.

2. Commutative Property of $+$ and $\cdot$

If a and b are any elements in N, then

$$a + b = b + a \qquad \text{and} \qquad a \cdot b = b \cdot a.$$

3. **Associative Property of + and ·**

   If a, b, and c are any elements in N, then

   $$(a + b) + c = a + (b + c) \quad \text{and} \quad (a \cdot b) \cdot c = a \cdot (b \cdot c).$$

4. **Identity Property for ·**

   There exists an element 1 in N such that $a \cdot 1 = a$ for all a in N.

5. **Cancellation Property of + and ·**

   If a, b, and c are any elements in N and if $c + a = c + b$, then $a = b$. If $c \cdot a = c \cdot b$, then $a = b$.

6. **Distributive Properties**

   If a, b, and c are any elements in N, then

   $$a \cdot (b + c) = (a \cdot b) + (a \cdot c)$$

   and

   $$(a + b) \cdot c = (a \cdot c) + (b \cdot c).$$

7. **Trichotomy Property**

   For any a and b in N one and only one of the following alternatives holds: (1) $a = b$; (2) there exists an x in N such that $a + x = b$; or (3) there exists a y in N such that $b + y = a$.

8. **Mathematical Induction Property**

   If S is any subset of N such that

   (A) $1 \in S$

   and

   (B)   if $k \in S$, then $k + 1 \in S$,

   it follows that $S = N$.

   Properties 1 through 6 with a small adjustment in Property 5 are also familiar properties of the system of rational numbers and the system of real numbers. These properties together with others are discussed in greater detail in Chapter 3.

   As an illustration of Property 7, part 2, let $a = 5$ and $b = 9$. Then since $5 + 4 = 9$, the x in question is, of course, 4. But there is no natural number y such that $9 + y = 4$. If $a = 7$ and $b = 2$, the situation is that of part 3 and the y in question is, of course, 5. In this case there is no natural number x such that $7 + x = 2$.

If there exists an $x \in N$ such that $a + x = b$, we say, as you know, that $a < b$ or, equivalently, $b > a$.

In this chapter we shall be specially interested in Property 8, the Induction Property. The following Exercises, however, serve as a review as well as a strengthening of your understanding of some of the other properties.

EXERCISES 2.2

◀In Exercises 1 and 2, illustrate the Distributive Properties for natural numbers by completing the computations.

1. $17(21 + 19) = \boxed{?}$, $(17 \cdot 21) + (17 \cdot 19) = \boxed{?}$
2. $(35 + 27)18 = \boxed{?}$, $(35 \cdot 18) + (27 \cdot 18) = \boxed{?}$

3. Show how the Distributive Properties can be used to simplify the computations.
   (a) $(25 \cdot 41) + (25 \cdot 59)$ and
   (b) $(121 \cdot 83) + (79 \cdot 83)$

◀In Exercises 4 and 5, illustrate the Associative Properties by completing the computations.

4. $(12 + 28) + 14 = \boxed{?} + 14 = \boxed{?}$
   $12 + (28 + 14) = 12 + \boxed{?} = \boxed{?}$
5. $(9 \cdot 13) \cdot 15 = \boxed{?} \cdot 15 = \boxed{?}$
   $9 \cdot (13 \cdot 15) = 9 \cdot \boxed{?} = \boxed{?}$

6. Consider the set S of even natural numbers, that is, $S = \{2, 4, 6, 8, \ldots\}$, and the operations of addition and multiplication. Does Property 1 hold for this set? What about Property 4?
7. Does the Closure Property hold for the set of odd natural numbers under addition? Under multiplication?
8. Does Property 4 hold for the set of odd natural numbers?
9. Consider the set of all natural numbers which are multiples of 3, that is, $\{3, 6, 9, \ldots\}$, and the operations of addition and multiplication. Does Property 1 hold for this system?
10. Does Property 1 hold for the set of all natural numbers greater than 10?
11. Does Property 1 hold for the set of all natural numbers less than or equal to 100?

12. For the following pairs of natural numbers a and b, determine which of the alternatives of the Trichotomy Principle hold.
   (a) $a = 9$, $b = 5$
   (b) $a = 6$, $b = 6$
   (c) $a = 2$, $b = 9$

2.3   MATHEMATICAL INDUCTION

Let us imagine an endless line of chalkboard erasers all standing on end as in the drawing below.

Suppose we know that the arrangement is such that if any eraser is tipped over to the right, it will automatically knock over its right-hand neighbor. If we also know that the number one eraser on the left is soon to be tipped to the right, what can we predict about all the erasers?

Consider next an endless flight of stairs. A tennis ball is placed on the top step. Suppose we know that if it falls to any step the momentum will automatically carry it down to the next step. Assume, then, that it is pushed off the number one step. How many additional steps will it traverse?

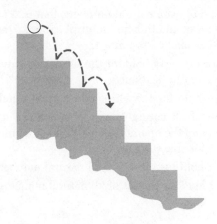

Keep these two pictures in mind as we return now to a consideration of Property 8, the Mathematical Induction Property, which, for handy reference, we repeat here.

If S is any subset of N such that

$$\text{(A)} \qquad 1 \in S$$

and

$$\text{(B)} \qquad \text{if } k \in S, \text{ then } k + 1 \in S,$$

it follows that $S = N$.

Can you see the analogy between the Induction Property and the case of the falling erasers or that of the descending tennis ball?

Suppose that we are considering a subset S of the set N of natural numbers. We may know that S contains *some* of the natural numbers but not necessarily know that it contains all of them. The Induction Property provides a method of determining whether or not S contains *all* the natural numbers, that is, whether or not $S = N$.

As you can see, the method involves two tests. Test A is usually quite easy to administer. We merely have to show that the set under consideration contains the number 1. Test B, however, usually takes a bit more doing. Test B requires us to show that if k is any number in S, then $k + 1$ (often called the *successor* of k) must also be in S.

To demonstrate how the two parts of the test work, let us investigate a particular example. Suppose it is claimed that the sum, $1 + 2 + 3 + \cdots + n$, of the first n natural numbers is equal to $\frac{1}{2}n(n + 1)$, that is, it is claimed that

$$1 + 2 + 3 + \cdots + n = \frac{1}{2} n(n + 1)$$

for all natural numbers n.

As mathematicians we should react to this claim with a certain amount of skepticism. Is it valid? Check it for $n = 5$. We see that

$$1 + 2 + 3 + 4 + 5 = 15$$

and

$$\frac{1}{2}(5)(5 + 1) = \frac{1}{2}(5)(6) = 15.$$

How about $n = 8$? In this case

$$1 + 2 + 3 + 4 + 5 + 6 + 7 + 8 = 36$$

and

$$\frac{1}{2}(8)(8 + 1) = \frac{1}{2}(8)(9) = 36.$$

We now have evidence that the formula holds for $n = 5$ and $n = 8$. Our job, then, is to find out if it holds for *all* natural numbers.

This is where the Induction Property enters the scene. We define S to be the set of all natural numbers for which the equation

(1) $$1 + 2 + 3 + \cdots + n = \frac{1}{2}n(n + 1)$$

is a true sentence. That is, we let

$$S = \{n : n \in N \text{ and } 1 + 2 + 3 + \cdots + n = \frac{1}{2}n(n + 1)\}.$$

Our research has already assured us that $5 \in S$ and $8 \in S$.

Question: Does $S = N$? Recall that Test A is the determination of whether or not S contains 1. In our case this becomes the question: Is Equation (1) satisfied if 1 is substituted for n? Since

$$1 = \frac{1}{2}(1)(1 + 1),$$

Test A is passed.

Test B is not as easy to apply. Read the following discussion very carefully, perhaps several times.

We want to find out if $k \in S$ implies that $k + 1 \in S$. Now what, exactly, does it mean to say that $k \in S$? And what does it mean to say that $k + 1 \in S$? Since

$$S = \{n : n \in N \text{ and } 1 + 2 + 3 + \cdots + n = \frac{1}{2}n(n + 1)\},$$

we conclude that $k \in S$ if and only if

(2) $$1 + 2 + 3 + \cdots + k = \frac{1}{2}k(k + 1)$$

and $k + 1 \in S$ if and only if

(3) $$1 + 2 + 3 + \cdots + k + (k + 1)$$
$$= \frac{1}{2}(k + 1)[(k + 1) + 1] = \frac{1}{2}(k + 1)(k + 2).$$

Thus to say that $k \in S$ implies $k + 1 \in S$ means that Equation (2) implies (3), that is, that we can derive Equation (3) from (2).

Now to the task. Consider the sum

$$1 + 2 + 3 + \cdots + k + (k + 1).$$

If $k \in S$, then we may, by (2), replace $1 + 2 + 3 + \cdots + k$ by $\frac{1}{2}k(k + 1)$, which gives us

$$1 + 2 + 3 + \cdots + k + (k + 1)$$

$$= \frac{1}{2}k(k + 1) + (k + 1)$$

$$= (k + 1)\left(\frac{1}{2}k + 1\right) \qquad \text{[using the Distributive Property to "factor out" } (k + 1)]$$

$$= (k + 1)\frac{1}{2}(k + 2)$$

$$= \frac{1}{2}(k + 1)(k + 2).$$

That is, we have shown that if $k \in S$, then

$$1 + 2 + 3 + \cdots + k + (k + 1) = \frac{1}{2}(k + 1)(k + 2).$$

But this is precisely Equation (3). Thus we have shown that Equation (2) implies (3), Test B is passed, and our job is done.

We conclude, then, that $S = N$ and hence that

$$1 + 2 + 3 + \cdots + n = \frac{1}{2}n(n + 1)$$

for *every* natural number n.

As a second illustration of the use of the Induction Property, let us go back to the example given at the beginning of the chapter, the formula for the sum of squares

(4) $\qquad 1^2 + 2^2 + 3^2 + \cdots + n^2 = \frac{1}{6}n(n + 1)(2n + 1).$

You have already checked this in Section 2.1 for $n = 20$ and for $n = 25$.

Now, as before, we let S be the set of all natural numbers for which the formula is valid, that is, in this case we take

$$S = \{n : n \in N \text{ and } 1^2 + 2^2 + 3^2 + \cdots + n^2$$

$$= \frac{1}{6}n(n + 1)(2n + 1)\}.$$

Test A: Is $1 \in S$? Replacing n by 1 in the formula (4) we have

$$1^2 = \frac{1}{6}(1)(1 + 1)[2(1) + 1].$$

Since $1^2 = 1$ and $\frac{1}{6}(1 + 1)[2(1) + 1] = \frac{1}{6} \cdot 2 \cdot 3 = 1$, Test A is passed.
Now for Test B. Assume that $k \in S$. This means that

(5) $\qquad 1^2 + 2^2 + 3^2 + \cdots + k^2 = \frac{1}{6}k(k + 1)(2k + 1).$

We wish to show that this implies that $k + 1 \in S$. Now $k + 1 \in S$
if and only if

(6) $\qquad 1^2 + 2^2 + 3^2 + \cdots + k^2 + (k + 1)^2$

$$= \frac{1}{6}(k + 1)[(k + 1) + 1][2(k + 1) + 1]$$

$$= \frac{1}{6}(k + 1)(k + 2)(2k + 3).$$

Consider now the sum

$$1^2 + 2^2 + 3^2 + \cdots + k^2 + (k + 1)^2.$$

If $k \in S$, we may use Equation (5) and substitute $\frac{1}{6}k(k + 1)(2k + 1)$
for

$$1^2 + 2^2 + 3^2 + \cdots + k^2.$$

This gives us

$$1^2 + 2^2 + 3^2 + \cdots + k^2 + (k + 1)^2$$

$$= \frac{1}{6}k(k + 1)(2k + 1) + (k + 1)^2.$$

Working with the right member of this equation, we see that

$$\frac{1}{6}k(k + 1)(2k + 1) + (k + 1)^2$$

$$= (k + 1)\left[\frac{1}{6}k(2k + 1) + (k + 1)\right] \qquad \text{(using the Distributive Property again)}$$

$$= (k + 1)\frac{1}{6}[k(2k + 1) + 6(k + 1)]$$

$$= \frac{1}{6}(k + 1)(2k^2 + 7k + 6)$$

$$= \frac{1}{6}(k + 1)(k + 2)(2k + 3).$$

From this we conclude that

$$1^2 + 2^2 + 3^2 + \cdots + k^2 + (k + 1)^2 = \frac{1}{6}(k + 1)(k + 2)(2k + 3).$$

But this is Equation (6), which tells us that $k + 1 \in S$. Hence $k \in S$ implies $k + 1 \in S$ and our test is complete; $S = N$. We can now be assured that

$$1^2 + 2^2 + 3^2 + \cdots + n^2 = \frac{1}{6}n(n + 1)(2n + 1)$$

for every natural number n.

It takes considerable practice to acquire facility and understanding in using the Induction Property. Induction proofs appear in many of the Exercises in the remainder of this chapter.

EXERCISES 2.3

◀ In Exercises 1–5, use the formula $1 + 2 + 3 + \cdots + n = \frac{1}{2}n(n + 1)$ to calculate the given sums.

1. The sum of the first 50 natural numbers.
2. The sum of the first 200 natural numbers.
3. The sum of the first 500 natural numbers.
4. The sum of all natural numbers from 20 to 60 inclusive. (*Hint:* Subtract the sum of the first 19 natural numbers from the sum of the first 60.)
5. The sum of all the natural numbers n such that $100 < n \leq 400$.

◀ In Exercises 6–8, calculate the given sums by use of the formula $1^2 + 2^2 + 3^2 + \cdots + n^2 = \frac{1}{6}n(n + 1)(2n + 1)$.

6. The sum of the squares of the first 15 natural numbers.
7. The sum of the squares of the natural numbers from 10 to 30 inclusive.
8. The sum of the squares of the first 50 natural numbers.
9. The sum of the cubes of the first n natural numbers is given by the formula

$$1^3 + 2^3 + 3^3 + \cdots + n^3 = \frac{1}{4}n^2(n + 1)^2.$$

Check this formula for $n = 3$, 5, and 8.

10. Copy and complete the steps in the following proof by mathematical induction of the validity of the formula in Exercise 9.

(a) Let $S = \{n : n \in N \text{ and } 1^3 + 2^3 + 3^3 + \cdots + n^3 = \frac{1}{4}n^2(n+1)^2\}$.

Test A is passed since $1^3 = 1$ and $\frac{1}{4}(\boxed{?})^2 \, (\boxed{?})^2 = \boxed{?}$.

(b) For Test B assume that $k \in S$. This means that

$$1^3 + 2^3 + 3^3 + \cdots + k^3 = \frac{1}{4}(\boxed{?})^2(\boxed{?})^2.$$

(c) We wish to see if the equation in (b) implies that $k + 1 \in S$. But $k + 1 \in S$ if and only if $(1^3 + 2^3 + 3^3 + \cdots + k^3) + (\boxed{?})^3 = \frac{1}{4}(k+1)^2 \, (\boxed{?})^2$.

(d) Consider the sum

$$1^3 + 2^3 + 3^3 + \cdots + k^3 + (k+1)^3.$$

By the assumption that $k \in S$, we can substitute $\frac{1}{4}k^2(k+1)^2$ for $1^3 + 2^3 + 3^3 + \cdots + k^3$. This gives us

$$1^3 + 2^3 + 3^3 + \cdots + k^3 + (k+1)^3 = \frac{1}{4}k^2(k+1)^2 + (\boxed{?})^3.$$

(e) Looking at the right member of this equation, we see that

$$\frac{1}{4}k^2(k+1)^2 + (k+1)^3 = (k+1)^2\left[\frac{1}{4}k^2 + (k+1)\right] \qquad \text{by the } \boxed{?} \text{ Property}$$

$$= (k+1)^2\,\frac{1}{4}[k^2 + 4(k+1)]$$

$$= \frac{1}{4}(k+1)^2 \, (\boxed{?})$$

$$= \frac{1}{4}(k+1)^2(k+2)^2.$$

Thus we see that

$$1^3 + 2^3 + 3^3 + \cdots + k^3 + (k+1)^3 = \frac{1}{4}(\boxed{?})^2(\boxed{?})^2,$$

which assures us that $k \in S$ implies that $(\boxed{?}) \in S$. Hence $S = \boxed{?}$.

11. Prove by induction that the sum of the first n odd numbers is equal to n^2, that is, that $1 + 3 + 5 + \cdots + (2n - 1) = n^2$.

12. What is the sum of the first 50 odd numbers? (See Exercise 11.)

13. Prove by induction that the sum of the first n even numbers is $n(n + 1)$, that is, that $2 + 4 + 6 + \cdots + 2n = n(n + 1)$.

14. What is the sum of the first 75 even numbers? (See Exercise 13.)

15. **CHALLENGE PROBLEM.** Use induction to prove that

$$\frac{1}{1 \cdot 2} + \frac{1}{2 \cdot 3} + \frac{1}{3 \cdot 4} + \cdots + \frac{1}{n(n+1)} = \frac{n}{n+1}$$

for all natural numbers n.

16. **CHALLENGE PROBLEM.** Prove by induction the formula for the sum of the first n terms of an arithmetic sequence

$$t_1 + (t_1 + d) + (t_1 + 2d) + \cdots + [t_1 + (n-1)d] = \frac{n}{2}[2t_1 + (n-1)d].$$

17. **CHALLENGE PROBLEM.** Prove by induction the formula for the sum of the first n terms of a geometric sequence

$$t_1 + rt_1 + r^2 t_1 + \cdots + r^{n-1} t_1 = \frac{t_1(r^n - 1)}{r - 1} \qquad \text{if } r \neq 1.$$

18. **CHALLENGE PROBLEM.** Construct a formula which is valid for $n = 1$, 2, 3, 4, and 5, but which fails for $n = 6$.

2.4 FURTHER APPLICATIONS OF MATHEMATICAL INDUCTION

Thus far we have used mathematical induction only to prove the validity of formulas relating to sums of the terms of a sequence. As we shall now see, the Induction Property may be used in a variety of different ways.

First, however, let us devise a somewhat more convenient way of stating the property. We can accomplish this by means of the following definition.

> **Definition 2.1.** Let S be any subset of N. We say that S is an **inductive** set if for every $k \in S$ it is true that $k + 1$ is also in S.

As an example, the set of all natural numbers greater than 10 is an inductive set. The set, T, of all natural numbers from 1 through 50 is, however, not an inductive set since, although $50 \in T$, $50 + 1 = 51 \notin T$.

With this definition in mind we can state the Induction Property as follows.

> Let S be any subset of N. If (A) $1 \in S$ and if (B) S is an inductive set, then $S = N$.

Let us now see if we can establish the following proposition: For any natural number n,

$$n^4 + 2n^3 + n^2 \text{ is divisible by 4.}$$

First we note that another way of saying that a natural number y is divisible by 4 is to assert that $y = 4t$ for some natural number t. Let us use induction, then, to show that for any $n \in N$,

$$n^4 + 2n^3 + n^2 = 4t$$

for some $t \in N$.

Test A is, as usual, quite simple. We know that

$$1^4 + 2(1)^3 + 1^2 = 4.$$

In this instance what is t? For Test B let

$$S = \{n : n \in N \text{ and } n^4 + 2n^3 + n^2 = 4t \text{ for some } t \in N\}.$$

Assume that $k \in S$. Under this assumption

$$k^4 + 2k^3 + k^2 = 4t \qquad \text{for some } t \in N.$$

Question: Does this imply that $k + 1 \in S$? To answer this question we must determine whether or not $k \in S$ implies that

$$(k + 1)^4 + 2(k + 1)^3 + (k + 1)^2 = 4r,$$

where r is a natural number. Since

$$(k + 1)^2 = k^2 + 2k + 1,$$

it follows that

$$
\begin{aligned}
(k + 1)^3 &= (k + 1)(k + 1)^2 \\
&= (k + 1)(k^2 + 2k + 1) \\
&= k(k^2 + 2k + 1) + 1 \cdot (k^2 + 2k + 1) \\
&= k^3 + 2k^2 + k + k^2 + 2k + 1 \\
&= k^3 + 3k^2 + 3k + 1.
\end{aligned}
$$

Similarly we can show, as you should check, that

$$(k + 1)^4 = k^4 + 4k^3 + 6k^2 + 4k + 1.$$

Hence

$$
\begin{aligned}
(k + 1)^4 &+ 2(k + 1)^3 + (k + 1)^2 \\
&= (k^4 + 4k^3 + 6k^2 + 4k + 1) + 2(k^3 + 3k^2 + 3k + 1) \\
&\quad + (k^2 + 2k + 1)
\end{aligned}
$$

$$= k^4 + 4k^3 + 6k^2 + 4k + 1 + 2k^3 + 6k^2 + 6k + 2 + k^2$$
$$+ 2k + 1$$
$$= k^4 + 6k^3 + 13k^2 + 12k + 4.$$

Because of our assumption that $k \in S$, we can substitute $4t$ for $k^4 + 2k^3 + k^2$. Thus if we rewrite the polynomial $k^4 + 6k^3 + 13k^2 + 12k + 4$ as

$$(k^4 + 2k^3 + k^2) + 4k^3 + 12k^2 + 12k + 4,$$

we can substitute $4t$ for the terms in parentheses and obtain

$$4t + 4k^3 + 12k^2 + 12k + 4,$$

which is equal to

$$4(t + k^3 + 3k^2 + 3k + 1).$$

We know that the expression in parentheses is a natural number. (Why?) We have therefore shown that

$$(k + 1)^4 + 2(k + 1)^3 + k^2 = 4(t + k^3 + 3k^2 + 3k + 1) = 4r$$

for some natural number r. Thus $k \in S$ implies that $k + 1 \in S$ and so S is an inductive set. We therefore conclude that $S = N$. Hence for all $n \in N$, $n^4 + 2n^3 + n^2$ is divisible by 4.

As a second example, let us prove that for any natural number n, $4^n - 1$ is divisible by 3, that is, for any $n \in N$,

$$4^n - 1 = 3t$$

for some natural number t.

We invite your cooperation in completing this proof! Let

$$S = \{n : n \in N \text{ and } 4^n - 1 = 3t \text{ for some } t \in N\}.$$

We must first show that $1 \in \boxed{?}$. This is true since

$$4^{\boxed{?}} - 1 = 3t \qquad \text{for } t = 1.$$

Now assume that $k \in S$. This means that

$$4^k - 1 = 3t$$

for some natural number t.

We next wish to determine whether $k \in S$ implies that $\boxed{?} \in S$. By our assumption that

$$4^k - 1 = 3t,$$

it follows that

(1) $$4^k = 3t + 1$$

and hence that

$$4 \cdot (4^k) = 4(3t + 1)$$

where we have multiplied both sides of Equation (1) by $\boxed{?}$. This last equation may be written as

$$4 \cdot (4^k) = 12t + 4.$$

If we subtract 1 from both sides, we get

(2) $$4(4^k) - 1 = 12t + 3.$$

But we know that for any number y,

$$y \cdot (y^k) = y^{k+1}.$$

Hence $4(4^k) = \boxed{?}$. Thus from (2) we have

$$4^{k+1} - 1 = 12t + 3 = 3(4t + 1).$$

From this we conclude that

$$4^{k+1} - 1 \text{ is divisible by } \boxed{?}.$$

Hence $k + 1 \in \boxed{?}$ and S is an $\boxed{?}$ set.

Therefore since $1 \in S$ and S is an inductive set, we conclude that $S = \boxed{?}$, and the proof is complete.

EXERCISES 2.4

◀ Prove by induction that each of the following statements is true for all natural numbers n.

1. $n^2 + n$ is divisible by 2.
2. $n^3 + 2n$ is divisible by 3.
3. $2 + 2^2 + 2^3 + \cdots + 2^n = 2(2^n - 1)$
4. $1 + 5 + 9 + \cdots + (4n - 3) = n(2n - 1)$
5. $5^{2n} - 1$ is divisible by 24.
6. $2^2 + 4^2 + 6^2 + \cdots + (2n)^2 = \dfrac{2}{3}n(n + 1)(2n + 1)$
7. $\dfrac{1}{1 \cdot 3} + \dfrac{1}{3 \cdot 5} + \dfrac{1}{5 \cdot 7} + \cdots + \dfrac{1}{(2n - 1)(2n + 1)} = \dfrac{n}{2n + 1}$
8. $6^n - 1$ is divisible by 5.

9. $(1 \cdot 2) + (2 \cdot 3) + (3 \cdot 4) + \cdots + n(n + 1) = \frac{n}{3}(n + 1)(n + 2)$

10. CHALLENGE PROBLEM. $(1 + 2 + 3 + \cdots + n)^2 = 1^3 + 2^3 + \cdots + n^3$

11. CHALLENGE PROBLEM. $\frac{1}{2} + \frac{1}{2^2} + \frac{1}{2^3} + \cdots + \frac{1}{2^n} = 1 - \frac{1}{2^n}$

12. CHALLENGE PROBLEM. $1^4 + 2^4 + 3^4 + \cdots + n^4 =$

$\frac{n}{30}(n + 1)(2n + 1)(3n^2 + 3n - 1)$

13. CHALLENGE PROBLEM. $1^5 + 2^5 + 3^5 + \cdots + n^5 =$

$\frac{n^2}{12}(n + 1)^2(2n^2 + 2n - 1)$

14. CHALLENGE PROBLEM. $9^n - 8n - 1$ is divisible by 64.

15. CHALLENGE PROBLEM. $1^2 + 4^2 + 7^2 + \cdots + (3n - 2)^2 =$

$\frac{n}{2}(6n^2 - 3n - 1)$

2.5 INDUCTION AND WELL-ORDERING (Optional)

As we have seen, the Induction Property forms the basis for a substantial number of proofs. What about a proof of the validity of the property itself?

In our formal characterization of the natural numbers, the Induction Property was presented as a postulate, that is, a property that was accepted without proof and that therefore constituted one of the defining components of the natural numbers. The Induction Property can be proved, however, by means of an alternative axiom, or postulate, for the natural numbers. This axiom is called the **Well-Ordering Property.** It, too, is the basis for the proof of many important mathematical theorems.

First we shall discuss the property. Then we shall show that this property is "equivalent" to the Induction Property, where by "equivalent" we mean that each property can be proved by means of the other property, assuming Postulates 1 to 7.

9. Well-Ordering Property

If S is any nonempty subset of N, then S has a least element.

This means that if S is any nonempty subset of N, then S contains a natural number y such that if $x \in N$ and $x < y$, then $x \notin S$. Note that the property specifies that *all* nonempty subsets of N have a least element.

The set of rational numbers does not have this Well-Ordering Property even though many subsets of the rational numbers do have least elements. For example, let Q be the set of rational numbers. Then the subset S of Q where

$$S = \{r : r \in Q \text{ and } r \geq 5\}$$

does have a least element, namely 5. However, if we consider the subset T of Q where $T = \{r : r \in Q \text{ and } r > 0\}$, then S has no least element since, given any rational number $s > 0$, the rational number $\frac{s}{2}$ has the property that $\frac{s}{2} \in T$ and $\frac{s}{2} < s$.

Now let us turn to the task of showing the equivalence of Properties 8 (Induction) and 9 (Well-Ordering). We shall first prove that (9) implies (8). That is, we will prove the following theorem.

> *Hypothesis:* The set of natural numbers has the Well-Ordering Property.
> *Conclusion:* The set of natural numbers has the Induction Property.

Our argument will be an indirect one. We know that the set of natural numbers either has the Induction Property (our desired conclusion) or else it does not. Thus if we show that the assumption that it does *not* have the Induction Property contradicts the hypothesis of well-ordering, then we must conclude that the Induction Property does hold if the Well-Ordering Property holds.

We begin the proof by noting that if the set of natural numbers is well-ordered and at the same time does *not* possess the Induction Property, then it follows that

1. There is a subset S of N such that
   (A) $1 \in S$,
   (B) if $k \in S$, then $k + 1 \in S$,
   but
2. $S \neq N$.
3. Let T be the set of elements in N which are *not* in S. Then, since $S \neq N$ by assumption 2, we know that T is not the $\boxed{?}$ set.
4. Therefore, by the Well-Ordering Property, T has a $\boxed{?}$ element; call this element x.
5. Since x is the least element in T, then $x - 1$ is $\boxed{?}$ in T.
6. If there is a natural number k such that $k = x - 1$, then $k \in \boxed{?}$.

7. Furthermore, we know that $x - 1$ is a natural number unless $x = 1$.
8. But we have assumed that $1 \in S$. Therefore $1 \notin T$, and since $x \in T$, then $x \boxed{?} 1$.
9. So k is a natural number and $k \in S$.
10. But by assumption B of (1) we know that if k is any element in S, then $k + \boxed{?}$ is also in S.
11. But since $k = x - 1$, then $k + 1 = x$, and x was assumed to be in T. So we have a contradiction: $x \in T$ and $x \in \boxed{?}$. But, by assumption, T is the set of all elements in N which are $\boxed{?}$ in S. We have shown, then, that T must be the empty set. Therefore $S = \boxed{?}$, and the Induction Property holds.

The foregoing proof shows that the Induction Property, so essential to much of our work in this chapter, can be actually deduced if we accept the alternative axiom that the set of natural numbers is well-ordered. We saw that the proof rests squarely on the requirement that every nonempty subset of N must have a least element. This, of course, includes the set N itself. What is the least element of N?

Now let us go in the opposite direction. We shall now assume that the Induction Property holds for N and proceed to show that every subset of N must have a least element.

We shall again use a proof by contradiction.

1. Suppose S is a *nonempty* subset of N that has *no* least element. This implies that S does not contain 1. (Why?)
2. Now let T be the set of all elements in N that are smaller than every element s in S.
3. We know that 1 is smaller than any element in S and hence we know that $1 \in T$.
4. Let k be any element in T and let y be any element in S.
5. By the definition of T, $k < y$.
6. But if $k < y$, it follows that either $k + 1 = y$ or else $k + 1 < y$ since it is not possible to have $k < y < k + 1$ because there is no natural number between k and $k + 1$.
7. Now we know that if $k + 1 = y$, then y must be the least element of S since every number less than $k + 1$ is equal to or less than k, and hence is *not* in S.
8. If y is a least element of S, then this contradicts (1).
9. Hence if $k + 1 \neq y$, then it follows by (6) that $k + 1 \boxed{?} y$.
10. But this means that $k + 1 \in \boxed{?}$.

11. We have shown that $1 \in T$ and that if k is any element in T, then $k + 1 \in T$. Hence by induction, $T = \boxed{?}$.
12. This means that S is the empty set, which contradicts the assumption (1) that S is nonempty.
13. Therefore we conclude that every nonempty subset of N has a $\boxed{?}$ element and that the natural numbers have the $\boxed{?}$ property.

In the proof that induction implied well-ordering we made two very plausible assumptions:

(a) that 1 is the least element of N and
(b) that there is no natural number y such that for any $k \in N$

$$k < y < k + 1.$$

Actually these two assumptions can be formally proved by induction. Try to construct these proofs in the following Exercises.

EXERCISES 2.5

1. Prove that 1 is the least element of N by using induction. (*Hint:* Let $S = \{n : n \in N \text{ and } n \geq 1\}$; then show that $S = N$.)

2. Prove that for any natural number x there is no $k \in N$ such that $x < k < x + 1$. (*Hint:* Assume the existence of such a k. Then there exists a natural number y such that $x + y = k$. Substitute and show a contradiction with the results of Exercise 1.)

3. Copy and complete the statements to prove the following theorem: If S is a subset of N such that (a) $1 \in S$ and (b) $k \in S$ for $k < n$ implies $n \in S$, then $S = N$.

   (1) Let T be the set of natural numbers that are *not* in S. Then if we show that $T = \emptyset$, we can conclude that $S = \boxed{?}$.
   (2) If $T \neq \emptyset$, it will, by the well-ordering principle, have a $\boxed{?}$ element. Call it n.
   (3) $n \neq 1$ for, by (a), $1 \in \boxed{?}$.
   (4) Since n is the least element of T, it follows that if $k < n$, then $k \notin T$ and hence $k \in \boxed{?}$.
   (5) But, by (b), if $1 \in S$ (which it is by ($\boxed{?}$)) and if $k \in S$ for $k < n$ (which is true by ($\boxed{?}$)), then $n \in \boxed{?}$ and so $n \notin T$.
   (6) Thus T cannot have a $\boxed{?}$ element n and so $T = \emptyset$; hence $S = \boxed{?}$.

CHAPTER SUMMARY

In this chapter we have characterized the natural numbers as a set, N, of elements equipped with two binary operations $+$ and $\cdot$ having the following properties.

1. CLOSURE PROPERTY of $+$ and $\cdot$
2. COMMUTATIVE PROPERTY of $+$ and $\cdot$
3. ASSOCIATIVE PROPERTY of $+$ and $\cdot$
4. IDENTITY PROPERTY for $\cdot$
5. CANCELLATION PROPERTY of $+$ and $\cdot$
6. DISTRIBUTIVE PROPERTIES
7. TRICHOTOMY PROPERTY
8. MATHEMATICAL INDUCTION PROPERTY

The focus of this chapter has been on Property 8 which we stated as follows:

If S is any subset of N such that

and

(A) $1 \in S$

(B)   if $k \in S$, then $k + 1 \in S$,

it follows that $S = N$. This property was used to validate a variety of formulas.

A subset of N that has the property that $k \in S$ implies $k + 1 \in S$ was called an INDUCTIVE SET. This definition enabled us to restate the Principle of Mathematical Induction as:

> If S is a subset of N such that (A) $1 \in S$ and (B) S is an inductive set, then $S = N$.

We saw that Property 8 is equivalent to the WELL-ORDERING PROPERTY for natural numbers, which states that if S is any nonempty subset of N, then S has a least element.

REVIEW EXERCISES

1. Test the formula

$$1^2 + 2^2 + 3^2 + \cdots + n^2 = \frac{1}{6}n(n + 1)(2n + 1)$$

for $n = 7$ and $n = 13$.

2. Test the proposition that $3^{2n} - 1$ is divisible by 8 for $n = 4$.

3. Illustrate the Distributive Property by completing the following computations.

$22 \cdot (17 + 33) = 22 \cdot \boxed{?} = \boxed{?}$

$22 \cdot (17 + 33) = (22 \cdot 17) + (22 \cdot 33) = \boxed{?} + \boxed{?} = \boxed{?}$

4. Illustrate the Associative Properties by completing the following computations.

(a) $(28 + 54) + 42 = \boxed{?} + 42 = \boxed{?}$

$28 + (54 + 42) = 28 + \boxed{?} = \boxed{?}$

(b) $(17 \cdot 16) \cdot 23 = \boxed{?} \cdot 23 = \boxed{?}$

$17 \cdot (16 \cdot 23) = 17 \cdot \boxed{?} = \boxed{?}$

5. Use the formula $1 + 2 + 3 + \cdots + n = \frac{1}{2}n(n + 1)$ to calculate

(a) the sum of the first 100 natural numbers;

(b) the sum of the first 300 natural numbers;

(c) the sum of all the natural numbers n such that $50 < n \leq 150$.

6. Use the formula in Exercise 1 to calculate the sum of the squares of the first 25 natural numbers.

7. Use the formula in Exercise 1 to calculate the sum of the squares of all natural numbers n such that $20 < n \leq 40$.

8. Use the formula $1^3 + 2^3 + 3^3 + \cdots + n^3 = \frac{1}{4}n^2(n + 1)^2$ to calculate the sum of the cubes of the first 10 natural numbers.

9. Given the formula

$$1 + 2 + 3 + \cdots + n = \frac{1}{2}(2n^3 - 17n^2 + 53n + 48).$$

Test this formula for $n = 2$, $n = 3$, and $n = 4$.

10. Test the formula in Exercise 9 for $n = 1$ and for $n = 5$.

11. Prove by induction that

$$3 + 6 + 9 + \cdots + 3n = \frac{1}{2}(3n^2 + 3n)$$

for every natural number n.

12. Prove by induction that, for any natural number n, $5^n - 1$ is divisible by 4.

GOING FURTHER: READING AND RESEARCH

The natural numbers have been defined in this chapter as a set with two operations and a fairly sizable number of properties. You may have wondered whether there might not be a simpler, more basic way of defining the natural numbers, and, as a matter of fact, there is. The natural numbers can actually be developed on the basis of four postulates and recursive definitions of addition and multiplication. From these all the properties that we have been examining can be deduced.

This approach to the natural numbers was originated by an Italian mathematician and logician named Guiseppe Peano (1858–1932).

The basic idea behind Peano's approach is the concept of a **successor**. The system of natural numbers, as defined by Peano, consists of a set of elements N subject to the following postulates or axioms.

1. Every element $n \in N$ has a unique successor n' which is another element in N.
2. There is a unique element in N called "1" which is the successor of no element of N.
3. If n', the successor of n, is equal to m', the successor of m, then $n = m$ (that is, different elements have different successors).
4. If a subset S of N contains 1 and also contains the successor of every one of its elements, then $S = N$.

Postulate 4 should sound familiar. What name would you give this postulate?

The operation $+$ is defined as follows.

1. $n + 1 = n'$.
2. If $n + m$ is defined, then

$$n + m' = (n + m)'.$$

The operation $\cdot$ is defined as follows.

1. $n \cdot 1 = n$.
2. If $n \cdot m$ is defined, then

$$n \cdot m' = (n \cdot m) + n.$$

It is interesting to see that if we label the elements of N in the usual way, that is, if we let 2 symbolize the successor of 1, 3 the successor of 2, 4 the successor of 3, etc., we can determine any sum, say $3 + 4$, or product, say $3 \cdot 4$, by merely using the definitions. For example, what is the sum $3 + 4$? Let us assume that we do not know the answer. We do know, however, that $3 + 1 = 4$. What is $3 + 2$? Since $3 + 2 = 3 + 1'$, we have, from (2) for addition, that

$$3 + 1' = (3 + 1)'.$$

Hence

$$3 + 2 = (3 + 1)' = 4' = 5.$$

Now that $3 + 2 = 5$ is defined, we get

$$3 + 3 = 3 + 2' = (3 + 2)' = 5' = 6$$

and finally

$$3 + 4 = 3 + 3' = (3 + 3)' = 6' = 7.$$

Thus we can construct, given enough time, an addition table of any size if we merely know the name of each successor!

See if you can find the product $3 \cdot 4$ assuming only that you know the successor of each number and the definition of $\cdot$.

You might also try to prove that addition is commutative or that multiplication is associative. There is much to be gained from a further study of the Peano approach to natural numbers. The following are recommended books on the subject.

EVES, HOWARD and CARROLL NEWSOM, *An Introduction to the Foundations and Fundamental Concepts of Mathematics*, New York: Holt, Rinehart and Winston, 1958. Rev. Edition 1965.

LANDAU, E., *Foundations of Analysis*. New York: Chelsea, 1951.

LEVI, HOWARD, *Elements of Algebra*, Second Edition. New York: Chelsea Publishing Company, 1956. Pages 150–153. Fourth Edition, 1968.

STABLER, E. R., *An Introduction to Mathematical Thought*. Reading, Mass.: Addison Wesley, 1953.

JOHNSON, R. E., *First Course in Abstract Algebra*. Englewood, New Jersey: Prentice-Hall, 1953.

NASA

CHAPTER THREE
REAL NUMBERS AND
RELATED SYSTEMS

3.1 A PREVIEW

Each vocation has its special set of raw materials. For the artist these are his paints and brushes; for the chemist, these are the basic elements all the way from actinium to zirconium; and for the poet, his words.

What, might we say, are the "raw materials" of the mathematician? Certainly among these one would have to include as fundamental ingredients the real numbers.

Everyone is familiar to some degree with the raw materials used by the artist, the chemist, and the poet. But what, exactly, do we mean when we speak of the set or, more precisely, the system of real numbers? By now you have doubtless had considerable experience working with real numbers. How, then, would you describe these particular "raw materials"?

In Chapter 2 the system of natural numbers was defined as a set of elements equipped with two binary operations, $+$ and $\cdot$, and a specified collection of properties.

The real numbers constitute another such system, a system which again consists of a set of elements, two binary operations, $+$ and $\cdot$, and a collection of properties. Many of the properties of the real numbers are also properties of the natural numbers, as you will see. There are, however, some significant differences.

To begin with, the real numbers are an example of what is known as a *field,* a term that may be familiar to you. Because of the importance of the concept of a field in mathematics, we shall precede our examination of the real numbers by considering the idea of a field in very general, and hence, abstract terms.

3.2 FIELDS

For convenience in framing our definition of a field we shall use the symbols $+$ and $\cdot$ to denote the two operations in the field. Keep in mind, however, that these symbols do not necessarily imply addition and multiplication in the conventional sense.

> ***Definition* 3.1** A field is a mathematical system consisting of a set F of at least two elements on which are defined two binary operations, $+$ and $\cdot$, such that the following properties hold.

61

(Many of these properties have already appeared in characterizing natural numbers in Chapter 2. Watch for similarities and differences.)

1. Closure Property under + and ·

If a and b are any elements in F, then $a + b$ and $a \cdot b$ are also in F.

2. Commutative Property of + and ·

If a and b are any elements in F, then

$$a + b = b + a \qquad \text{and} \qquad a \cdot b = b \cdot a.$$

3. Associative Property of + and ·

If a, b, and c are any elements in F, then

$$(a + b) + c = a + (b + c) \qquad \text{and} \qquad (a \cdot b) \cdot c = a \cdot (b \cdot c).$$

4. Identity Properties

There are in F *unique* identity elements for + and ·. If z is the identity for +, then for any element a in F,

$$a + z = a.$$

If u is the identity for ·, then for any element a in F,

$$a \cdot u = a.$$

Furthermore, $u \neq z$.

(In the field of rational numbers the element corresponding to z is, of course, 0, and the element corresponding to u is 1.)

5. Inverse Properties

If a is any element in F, there exists a *unique* element b in F such that

$$a + b = z,$$

and b is called the inverse of a with respect to the operation +. If c is any element in F, other than the identity for + (which we have called z), then there exists a *unique* element d in F such that

$$c \cdot d = u,$$

where u, as above, denotes the identity for the operation ·. The element d is called the inverse of c with respect to the operation ·.

6. Distributive Property

If a, b, and c are any elements in F, then

$$a \cdot (b + c) = (a \cdot b) + (a \cdot c).$$

To simplify notation, we shall often write $a \cdot b$ as ab. We shall also use the words "addition" and "multiplication" for the operations of $+$ and $\cdot$, respectively, even though these operations should be considered as abstract operations and not necessarily those associated with elementary arithmetic. Thus we can, without loss of generality, refer to additive or multiplicative identities and additive or multiplicative inverses. We shall designate the additive inverse of a by $-a$ and the multiplicative inverse of a by a^{-1}. Note that, because of the Commutative Properties of addition and multiplication, we can also conclude that

$$z + a = a + z = a,$$
$$u \cdot a = a \cdot u = a,$$
$$(-a) + a = a + (-a) = z,$$
$$a \cdot a^{-1} = a^{-1} \cdot a = u,$$

and

$$(a + b) \cdot c = (a \cdot c) + (b \cdot c).$$

The system of rational numbers constitutes perhaps the best known example of a field. To illustrate the more general concept, however, let us examine a special and somewhat less familiar field.

Consider the set of numbers

$$S = \{0, 1, 2, 3, 4, 5, 6, 7, 8, 9, 10\}$$

consisting of eleven elements. Now define a binary operation, *, on this set as follows.

For any $a \in S$ and $b \in S$ let

$$\begin{cases} a * b = a + b & \text{if } a + b < 11, \\ a * b = (a + b) - 11 & \text{if } a + b \geq 11, \end{cases}$$

where the $+$ and $-$ symbols indicate ordinary addition and subtraction of whole numbers.

Thus, for example,

$$3 * 5 = 3 + 5 = 8 \quad \text{since } 3 + 5 < 11.$$
$$7 * 6 = (7 + 6) - 11 = 2 \quad \text{since } 7 + 6 > 11.$$

Define a second operation, ∘, in the following way:

$$a \circ b = ab - 11n,$$

where n is the largest nonnegative integer such that

$$0 \leq ab - 11n < 11.$$

To illustrate this operation let us compute, for example, $3 \circ 9$. We first need to find the largest nonnegative integer n such that

$$0 \leq 27 - 11n < 11.$$

We note that $27 - (11 \cdot 2) = 5$ and $0 < 5 < 11$. On the other hand, $27 - (11 \cdot 3) = -6$ and $-6 < 0$. Thus the proper choice for n is 2. This gives us

$$3 \circ 9 = (3 \cdot 9) - (11 \cdot 2) = 5.$$

As another example, we have

$$4 \circ 2 = 4 \cdot 2 - 11n.$$

If we let $n = 1$, we get $(4 \cdot 2) - (11 \cdot 1) = -3$. But

$$-3 < 0.$$

In this case, then, the proper choice for n is 0 and we have

$$4 \circ 2 = (4 \cdot 2) - (11 \cdot 0) = 8.$$

You should experiment with further illustrations of the two operations * and ∘ to facilitate complete understanding. Is it clear, for example, that $7 * 7 = 3$ and $7 \circ 7 = 5$?

We now assert that the set of elements $\{0, 1, 2, \ldots, 10\}$ together with the two operations * and ∘, as defined, do constitute a field with * as $+$ and ∘ as $\cdot$. For convenience we shall designate this field as

$$B = \{0, 1, 2, 3, \ldots, 10; *, \circ\}.$$

In the Exercises which follow you are asked to test this assertion. That is, you are asked to make plausible by various means (illustrative examples and a few informal proofs) the statement that the field properties do hold for the system

$$B = \{0, 1, 2, 3, \ldots, 10; *, \circ\}$$

with operations * and ∘ defined as above and * considered as $+$ and ∘ as $\cdot$.

EXERCISES 3.2

◀ In Exercises 1–20, which element in the set $\{0, 1, 2, 3, \ldots, 10\}$ does the exercise name?

1. $3 * 9$	**11.** $10 * 10$
2. $4 * 6$	**12.** $8 * 9$
3. $5 * 10$	**13.** $9 * 8$
4. $3 \circ 9$	**14.** $8 \circ 9$
5. $4 \circ 6$	**15.** $9 \circ 8$
6. $5 \circ 10$	**16.** $9 \circ 10$
7. $8 * 8$	**17.** $7 \circ 6$
8. $8 \circ 8$	**18.** $6 \circ 7$
9. $5 * 6$	**19.** $9 \circ 9$
10. $5 \circ 6$	**20.** $8 \circ 7$

21. Give the identity element for the operation $*$.

22. Give the identity element for the operation $\circ$.

23. Do the results of Exercises 1–20 strengthen the claim that the Closure Properties and the Commutative Properties hold for $B = \{0, 1, 2, 3, \ldots, 10; *, \circ\}$? Explain your answer.

24. What field properties do the results of Exercises 21 and 22 appear to validate?

◀ In Exercises 25–34, verify each special case of the Associative Property by showing that the equality holds.

25. $(3 * 5) * 8 = 3 * (5 * 8)$

26. $(3 \circ 5) \circ 8 = 3 \circ (5 \circ 8)$

27. $6 * (7 * 9) = (6 * 7) * 9$

28. $(10 * 4) * 6 = 10 * (4 * 6)$

29. $(6 \circ 8) \circ 4 = 6 \circ (8 \circ 4)$

30. $8 \circ (7 \circ 10) = (8 \circ 7) \circ 10$

31. $10 * (9 * 7) = (10 * 9) * 7$

32. $(8 \circ 9) \circ 10 = 8 \circ (9 \circ 10)$

33. $(9 \circ 9) \circ 6 = 9 \circ (9 \circ 6)$

34. $10 \circ (8 \circ 8) = (10 \circ 8) \circ 8$

35. Prove that each element in B has an inverse with respect to $*$ by indicating the inverse in each case. (Recall that b is the inverse of a with respect to $*$ if $a * b$ is equal to the identity for $*$).

36. Give the inverses with respect to ∘ for each of the elements 1, 2, 3, ..., 10.

◄In Exercises 37–42, verify each special case of the Distributive Property by showing that the equality holds.

37. $6 \circ (8 * 5) = (6 \circ 8) * (6 \circ 5)$

38. $7 \circ (4 * 10) = (7 \circ 4) * (7 \circ 10)$

39. $10 \circ (5 * 6) = (10 \circ 5) * (10 \circ 6)$

40. $9 \circ (9 * 8) = (9 \circ 9) * (9 \circ 8)$

41. $5 \circ (8 * 10) = (5 \circ 8) * (5 \circ 10)$

42. $9 \circ (10 * 7) = (9 \circ 10) * (9 \circ 7)$

43. Prove that $a * b = b * a$ for two cases: (1) $a + b < 11$ and (2) $a + b \geq 11$.

44. Prove that $a \circ b = b \circ a$ for all a and $b \in \{0, 1, 2, 3, \ldots, 10\}$.

45. Consider a set consisting of two elements x and y with operations $+$ and $\cdot$ defined by the following tables:

$+$	x	y
x	x	y
y	y	x

$\cdot$	x	y
x	x	x
y	x	y

That is, $x + x = x, x + y = y, x \cdot x = x, x \cdot y = x$, and so on. Determine by checking on Properties 1 to 6 whether or not this system forms a field.

46. Let $S = \{0, 1, 2\}$ with the operations $+$ and $\cdot$ defined by the following tables:

$+$	0	1	2
0	0	1	2
1	1	2	0
2	2	0	1

$\cdot$	0	1	2
0	0	0	0
1	0	1	2
2	0	2	1

That is, $0 + 1 = 1, 1 + 1 = 2, 0 \cdot 2 = 0, 2 \cdot 1 = 2$, and so on. Does this system form a field? If not, what field properties are not satisfied?

47. Consider the set of numbers $S = \{0, 1, 2, 3\}$ with the operations $+$ and $\cdot$ defined by the tables on the next page. That is, $0 + 1 = 1$, $1 + 3 = 0, 0 \cdot 3 = 0, 3 \cdot 2 = 2$, and so on. Does this system form a field? If not, what field properties are not satisfied?

+	0	1	2	3
0	0	1	2	3
1	1	2	3	0
2	2	3	0	1
3	3	0	1	2

·	0	1	2	3
0	0	0	0	0
1	0	1	2	3
2	0	2	0	2
3	0	3	2	1

48. In all fields, $a \cdot z = z$, where z is the identity element for the operation $+$ and a is any element in F. Complete a proof of this statement by copying and completing the following statements.
    (a) $z = z + z$
    (b) Therefore $a \cdot z = a \cdot (\boxed{?})$.
    (c) But $a \cdot (z + z) = (a \cdot z) + (a \cdot z)$ because of the $\boxed{?}$ Property.
    (d) Thus we know that $a \cdot z = (a \cdot z) + (a \cdot z)$.
    (e) This means that $(a \cdot z)$ is an identity element with respect to the operation $\boxed{?}$.
    (f) Since there is only one identity element z for the operation $+$, it follows that $a \cdot z = \boxed{?}$.

49. **CHALLENGE PROBLEM.** The definition of operation ○ stated that

$$a \circ b = a \cdot b - 11n,$$

where n is the largest integer satisfying the inequality

$$0 \le a \cdot b - 11n < 11.$$

Show that for every a and $b \in B$ there is one and only one integer n which makes the inequality true. Hence prove that the word "largest" is not strictly necessary.

50. **CHALLENGE PROBLEM.** Prove that if $a \cdot b = 0$ for a, $b \in F$, and if $a \neq 0$, then $b = 0$. (*Hint:* Use the Inverse Property with respect to operation $\cdot$ and the results of Exercise 48.)

3.3 ORDERED FIELDS

It was stated in Section 3.1 that the system of rational numbers and the system of real numbers are examples of fields. Both of these systems are, in fact, examples of a more specialized kind of field, one which we call an *ordered field*. An **ordered field** is a field which has the following additional properties.

7. Order Properties

There exists a relation $<$ ("is less than") on F such that for a, b, $c \in F$

(i)   For every $a, b \in F$, one and only one of the following holds:

$$a < b, \qquad a = b, \qquad b < a.$$

(ii) If $a < b$ and $b < c$, then $a < c$.

(iii) If $a < b$, then $a + c < b + c$.

(iv) If $a < b$ and $z < c$, then $ac < bc$. (z is the additive identity of F.)

The Order Properties may seem a little complicated. However, if we consider the rational numbers as an example of an ordered field, the significance of the properties becomes clear.

Property 7(i), often called the **Trichotomy Principle**, simply states that for any two rational numbers either the first is less than the second, or they are equal, or the second is less than the first. The force of the property lies in the fact that *one* of these conditions must hold, but *no more* than one. Property 7(ii) is called the **Transitive Property**. As an illustration from the rational numbers in terms of their customary ordering, consider the numbers 2, 5, and 8. We know that

$$2 < 5, \qquad 5 < 8,$$

and certainly

$$2 < 8.$$

Property 7(iii) is also very simple as applied to rational numbers. We know, for example, that

$$\frac{1}{4} < \frac{1}{2}, \qquad \frac{1}{4} + 5 < \frac{1}{2} + 5, \qquad \text{and} \qquad \frac{1}{4} + (-2) < \frac{1}{2} + (-2).$$

In Property 7(iv) we need to observe carefully the condition $z < c$. For example, $2 < 7$, $0 < 5$, and

$$2 \cdot 5 = 10 < 7 \cdot 5 = 35.$$

The situation when $c < z$, however, is quite different. Thus, for example, we have $-5 < 0$ and do *not* have

$$2(-5) < 7(-5)$$

but rather

$$7(-5) < 2(-5).$$

We shall return to this situation later when we discuss other Properties of Order.

To illustrate further the Order Properties, suppose that we consider a field which is *not* an ordered field. This is the field

$$B = \{0, 1, 2, 3, \ldots , 10; *, \circ\}$$

discussed in Section 3.2 where * corresponds to + and ∘ to ·.

As a preliminary move toward showing that B is not an ordered field we should first determine whether for any a and b in B exactly one of

$$a < b, \qquad a = b, \qquad b < a,$$

is true.

If we consider the natural ordering, as rational numbers, of the elements 0 through 10 in our set, it certainly seems reasonable to assume from experience that the answer is "Yes."

Given, for example, the elements 5 and 7, there does not seem to be any doubt that $5 < 7$. Nor would we hesitate to claim that $8 < 10$, and so on. However, in an ordered field F all of the Order Properties must hold. One of the properties is 7(iii).

If $a < b$ and if c is any element of F, then $a + c < b + c$.

Now try an experiment. If we use the "natural" ordering in the field under investigation, then $6 < 10$ and $2 < 9$. It should then be true, if the above property is to hold, that for any c in B, $6 * c < 10 * c$. But what if we choose 3 as this element c? From our definition of the operation *, we know that

$$6 * 3 = 6 + 3 = 9.$$

However, we also know that

$$10 * 3 = (10 + 3) - 11 = 2.$$

Assembling these two ideas we are faced with the fact that $6 < 10$ but $6 * 3 \not< 10 * 3$ (read "6 * 3 is not less than 10 * 3"), that is,

$$9 \not< 2.$$

By this example we have shown that what seems to be a natural way of defining order in our particular field just does not work. This certainly casts some doubt on the assertion that field B is an ordered field. It seems merely to have a semblance of order. Can we prove, however, that B is *definitely* not an ordered field? That is, can we show that no ordering—"natural" or "unnatural"—is possible?

Consider the elements 0 and 8 in B. By 7(i) either $0 < 8$, or $0 = 8$, or $8 < 0$. Let us start by assuming that

$$0 < 8.$$

By 7(iii), if $0 < 8$, then

$$0 * 3 < 8 * 3, \qquad \text{that is,} \quad 3 < 0.$$

Also, if $0 < 8$, then

$$0 * 0 < 8 * 0.$$

Still further, if $0 < 8$, then

$$0 * 8 < 8 * 8.$$

By 7(ii), since

$$0 * 0 < 8 * 0 \qquad \text{and} \qquad 0 * 8 < 8 * 8,$$

then, since $8 * 0 = 0 * 8$,

$$0 * 0 < 8 * 8 \qquad \text{and hence} \qquad 0 < 5.$$

But if $0 < 5$, then by 7(iv), $0 \circ 5 < 5 \circ 5$; that is, $0 < 3$.

We have shown, then, that the assumption that $0 < 8$ leads to both $0 < 3$ *and* $3 < 0$, which clearly contradicts 7(i).

Now assume that $8 < 0$. Then, by 7(iii),

$$8 * 3 < 0 * 3, \qquad \text{that is,} \quad 0 < 3.$$

From this it follows by 7(iii) that $0 * 3 < 3 * 3$, that is, $3 < 6$. By 7(ii), since $0 < 3$ and $3 < 6$, it follows that $0 < 6$. Since $8 < 0$ and $0 < 3$, it follows by 7(iv) that

$$8 \circ 3 < 0 \circ 3, \qquad \text{that is,} \quad 2 < 0.$$

Now we use $0 < 3$ and $2 < 0$ together with 7(iv) to get

$$2 \circ 3 < 0 \circ 3, \qquad \text{that is,} \quad 6 < 0.$$

Thus from $8 < 0$, we obtained both $6 < 0$ *and* $0 < 6$, again a contradiction of 7(i).

Property 7(i) says that we must have either

$$0 < 8, \qquad 0 = 8, \qquad \text{or} \qquad 8 < 0.$$

But we know that $0 \neq 8$ and have just finished showing that both $0 < 8$ and $8 < 0$ are not possible. Thus we have proved that field B *cannot* be ordered in such a way that it satisfies the Order Properties 7. Hence we conclude that

$$B = \{0, 1, 2, 3, \ldots, 10; *, \circ\}$$

is *not* an ordered field.

Up to now the only symbol used in discussing the order property has been "$<$." It is often more convenient, however, to use the alternate notation $a > b$ (a is greater than b) which, as you know, has the same meaning as $b < a$. You are doubtlessly also familiar with the notation

$$a \leq b \quad (a \text{ is less than or equal to } b)$$

and the equivalent

$$b \geq a \quad (b \text{ is greater than or equal to } a).$$

We also use the word "positive" for elements greater than the additive identity z and "negative" for elements less than z.

From the four Properties of Order that we have stated, other properties can be deduced, properties which are also useful in solving problems. You are asked to prove some of these in the Exercises. In the meantime, as a handy reference, we shall list them here together with a restatement of Properties 7(i), 7(ii), 7(iii), and 7(iv).

We take, as before, a, b, c, d to be elements of F and z to be the additive identity of F.

7(i)   (**Trichotomy**) For every $a, b \in F$, one and only one of the following holds:

$$a < b, \quad a = b, \quad b < a.$$

7(ii)   (**Transitivity**) If $a < b$ and $b < c$, then

$$a < c.$$

7(iii)   If $a < b$, then

$$a + c < b + c.$$

7(iv)   If $a < b$ and $z < c$, then

$$ac < bc.$$

(O–1)   If $a < b$ and $c < z$, then

$$ac > bc.$$

(O-2)   If $a < b$ and $c < d$, then
$$a + c < b + d.$$

(O-3)   If $a > z$, $b > z$, and $a < b$, then
$$a^2 < b^2.$$

(O-4)   If $a > z$, $b > z$, $c > z$, $d > z$, $a < b$, and $c < d$, then
$$ac < bd.$$

(O-5)   If $a > z$ and $b > z$, then
$$a + b > z \quad \text{and} \quad ab > z.$$

(O-6)   If $a > z$ and $b < z$, or if $a < z$ and $b > z$, then
$$ab < z.$$

(O-7)   If $a < z$ and $b < z$, then
$$ab > z.$$

(O-8)   If $a < z$, then $-a > z$ and if $a > z$, then $-a < z$. ($-a$ is the additive inverse of a in F.)

(O-9)   If $a < b$, then there exists an $x \in F$ such that
$$a < x < b.$$

Property O-9, often called the **Density Property**, has far-reaching effects. It tells us, for example, that there is no smallest positive rational number since one can always find a rational number x between 0 and any positive rational number. Note that Property O-9 does *not* hold for the system of natural numbers since there is no natural number x between any natural number n and its successor $n + 1$. (See Exercise 2 of Exercises 2.5.)

We shall sketch a proof of Property O-9 for the field of rational numbers. The property can be proved in a similar manner for any ordered field.

  1. Let a and b be elements in the field of rational numbers such that $a < b$.

  2. By 7(iii),
$$a + a < b + a,$$
that is,
$$2a < b + a.$$

3. By 7(iv),

$$\frac{1}{2}(2a) < \frac{1}{2}(b + a) \qquad \text{since} \qquad 0 < \frac{1}{2}.$$

Thus

$$a < \frac{1}{2}(b + a) = \frac{1}{2}(a + b).$$

4. Again by 7(iii), since $a < b$, we have

$$a + b < b + b,$$

or

$$a + b < 2b.$$

Hence, by 7(iv),

$$\frac{1}{2}(a + b) < b,$$

and we have

$$a < \frac{1}{2}(a + b) < b.$$

5. Finally, if we choose x as $\frac{1}{2}(a + b)$, the property is proved.

The Order Properties form the working basis for the solution of inequalities. We shall illustrate their application using examples from the field, Q, of rational numbers.

Suppose we wish to solve the inequality

$$3x - 5 < x + 11.$$

That is, we want to exhibit in simplest form the set of all rational numbers S such that $r \in S$ implies $3r - 5 < r + 11$. In the following sequence of inequalities state the Order Properties that are used.

$$3x - 5 < x + 11$$
$$3x - 5 + 5 < x + 11 + 5$$
$$3x < x + 16$$
$$3x - x < x - x + 16$$
$$2x < 16$$
$$x < 8.$$

Thus we see that if x is a rational number such that $3x - 5 < x + 11$, then $x < 8$. In other words, the solution set of $3x - 5 < x + 11$ is contained in the set

$$S = \{x : x \in Q \text{ and } x < 8\}.$$

To show that S is actually the solution set of $3x - 5 < x + 11$, we must "reverse" the process. That is, we must show that if $x < 8$, then $3x - 5 < x + 11$.

Assume, then, that

$$x < 8.$$

Then

$$2x < 16$$
$$2x + x < 16 + x$$
$$3x - 5 < 16 + x - 5$$
$$3x - 5 < x + 11.$$

(Give reasons for each step.)

We now see that if x is a rational number, then $3x - 5 < x + 11$ if and only if $x < 8$. Thus the two inequalities $x < 8$ and $3x - 5 < x + 11$ are equivalent, that is, have the same solution set

$$\{x : x \in Q \text{ and } x < 8\}.$$

As a second example consider the inequality

$$13 - 8x < 2x + 33.$$

We have

$$13 + (-13) - 8x < 2x + 33 + (-13)$$
$$- 8x < 2x + 20$$
$$- 8x - 2x < 2x + 20 - 2x$$
$$- 10x < 20.$$

Multiplying by $-\frac{1}{10}$ we have

$$x > -2. \quad \text{(Why?)}$$

By "reversing" the steps as in the first example, we can show that the inequalities $13 - 8x < 2x + 33$ and $x > -2$ are equivalent. Accordingly, the solution set of the inequality $13 - 8x < 2x + 33$ is

$$\{x : x \in Q \text{ and } x > -2\}.$$

(*Note:* In actual practice it is not necessary to go through the "reverse process" if it is clear that all of the operations involved are reversible—as is always the case for the procedure suggested for linear inequalities.)

EXERCISES 3.3

◀ In Exercises 1–10, determine the solution set of each of the given inequalities.

1. $7x + 12 < 3x + 40$
2. $11x - 16 < x + 50$
3. $18 - 7y < y + 82$
4. $25 - 12x < x + 131$
5. $39 - 14r < 54 - r$
6. $10 - \dfrac{1}{2}x < 14 - x$
7. $\dfrac{3}{4}x - 12 < \dfrac{1}{4}x + 10$
8. $0.25x + 1.75 < 50 - 0.5x$
9. $3\dfrac{1}{2}s + 1\dfrac{2}{3} < \dfrac{3}{4}s + 4\dfrac{1}{4}$
10. $13.7z - 21 < 11z + 80.6$

11. Complete the following proof of Property $\mathcal{O}$–1.
    (1) Assume that $a, b, c \in F$ and that $a < b$ and $c < z$. Prove that $ac > bc$ or, equivalently, that $cb < ca$.
    (2) If $c < z$, then $c + (-c) < z + (-c)$. Hence $z < z + (-c)$ because $c + (-c) = \boxed{?}$ and so $z < (-c)$ because $z + (-c) = \boxed{?}$.
    (3) Therefore $a(-c) < b(-c)$ by Property 7($\boxed{?}$).
    (4) It follows that $-(ca) < -(cb)$ since
$$a(-c) = -(ca) \quad \text{and} \quad b(-c) = -(cb).$$
    (5) Then $[-(ca)] + (cb) < -(cb) + (cb)$ by Property 7($\boxed{?}$).
    (6) From this we get
$$(cb) + [-(ca)] < \boxed{?}.$$
    (7) Hence $(cb) + [-(ca)] + (ca) < \boxed{?}$.
    (8) And so $cb < ca$.

12. Prove Property $\mathcal{O}$–2. Given $a < b$ and $c < d$, prove that
$$a + c < b + d.$$
    (*Hint:* If $a < b$, then $a + c < b + c$ and if $c < d$, then $c + b < d + b$.)

13. Prove that if $0 < a$ and $0 < b$, then $0 < a + b$.
    (*Hint:* Use the results of Exercise 12.)

14. Prove that if $0 < a$ and $0 < b$, then $0 < ab$.

15. Prove that $a < b$ if and only if $0 < b + (-a)$.

16. Prove Property ⊙–3.

17. Prove Property ⊙–6.

18. Prove Property ⊙–7. (*Hint:* Use Property ⊙–1.)

19. Prove that in any ordered field F if $a > 0$, then $a^{-1} > 0$, where a^{-1} is the multiplicative inverse of a. (*Hint:* First show that the multiplicative identity of F is positive.)

20. Prove Property ⊙–8. (*Hint:* For the first part, use 7(iii) and the fact that $a + (-a) = 0$.)

◀The Density Property, ⊙–9, states that for any elements a and b in an ordered field F such that $a < b$ there is an element $x \in F$ such that $a < x < b$. Determine such an x for each of the pairs of rational numbers given in Exercises 21–40.

21. 3, 7

22. $\dfrac{1}{3}, \dfrac{1}{2}$

23. $-2, 0$

24. $-4, -3$

25. $-\dfrac{1}{4}, -\dfrac{1}{3}$

26. $-\dfrac{2}{5}, \dfrac{2}{5}$

27. 0.5, 0.6

28. $\dfrac{2}{3}, \dfrac{3}{4}$

29. $-\dfrac{3}{5}, -\dfrac{2}{5}$

30. $\dfrac{3}{7}, \dfrac{3}{5}$

31. $1\dfrac{1}{3}, 1\dfrac{1}{2}$

32. $-0.7, -0.6$

33. 0.25, 0.26

34. $-\dfrac{3}{8}, -\dfrac{1}{4}$

35. $-0.35, -0.34$

36. $\dfrac{1}{12}, \dfrac{1}{11}$

37. $\dfrac{3}{11}, \dfrac{5}{12}$

38. $-\dfrac{5}{12}, -\dfrac{1}{11}$

39. 0.256, 0.255

40. $-\dfrac{2}{3}, -\dfrac{3}{5}$

41. Given two rational numbers $\dfrac{a}{b}$ and $\dfrac{c}{d}$, write a general formula for a rational number $\dfrac{x}{y}$ in terms of a, b, c, and d which will make the inequality

$$\frac{a}{b} < \frac{x}{y} < \frac{c}{d}$$

a true statement.

42. List five rational numbers between $\frac{1}{2}$ and $\frac{1}{4}$.
43. List nine rational numbers between 0.5 and 0.6.
44. List ten rational numbers between $-\frac{1}{4}$ and $-\frac{1}{5}$.
45. Sometimes it is said that the Density Property implies that between any two rational numbers there is an *infinite* number of other rational numbers. State whether or not you agree with this statement and give your reasons.

3.4 A COMPLETE ORDERED FIELD

It has already been remarked that the real numbers are an example of an ordered field. Actually the real numbers have an additional property that is not shared by all ordered fields. Whereas the rational numbers constitute the most familiar example of an ordered field, the real numbers, in addition to being an ordered field, have also the special attribute of being a *complete* ordered field, that is, an ordered field which has along with all the other properties, the *Completeness Property*. A comprehensive understanding of the nature and significance of this property requires careful reading and considerable reflection.

In ordinary usage the adjective "complete" suggests the notion that nothing has been left out. If you own a complete encyclopedia, for example, it is assumed that you have all the volumes. In mathematics, although the term may suggest a similar connotation, its meaning is far more technical and highly specialized.

Before defining the Completeness Property, we shall need to state precisely what we mean when we speak of the "least upper bound" of a given set.

Let F be an ordered field and suppose that S is any subset of F. If there is an element a in F such that for any x in S, $x \leq a$, then we say that a is an **upper bound** of the set S.

For example, take the set $S = \{1, 2, 3, 4, 5\}$ considered as a subset of the field, Q, of rational numbers. Is 7 an upper bound of this set? What about 5? From these two examples we see that an upper bound, in this case 5, can itself be a member of S or, as with 7, it need not be. Name some other upper bounds of this particular set S. What about 1,000,000? What about $\frac{19}{3}$?

As you can see, it is possible to associate with a subset S of the rational numbers another subset B, the set of all upper bounds of S that are in Q. Of course, not all subsets of an ordered set have upper bounds. Q itself has no upper bounds nor does the subset $N = \{1, 2, 3, \ldots\}$ of Q. In such cases, the set B of all upper bounds is the null set.

It may occur that in this set B, the set of all upper bounds of S, there is a smallest element, call it b. If this occurs, that is, if b is less than or equal to every element of B, then we call b the **least upper bound** of S.

Sounds simple enough! But is it?

Once again let $S = \{1, 2, 3, 4, 5\}$. In this case we know that 5 is an upper bound of S. We also know that any rational number less than 5 is not an upper bound of S. Why? Furthermore, we know that all rational numbers greater than 5 are upper bounds of S. From all this we gather that 5 is the least upper bound of S.

Suppose that we look at two other examples. First let T be the subset of Q consisting of all rational numbers less than 10, that is,

$$T = \{x : x \in Q \text{ and } x < 10\}.$$

Is 10 an upper bound of T? Yes, since for all x in T, $x \leq 10$.

Now think carefully. Is 10 the least upper bound of T?

For the sake of argument let us suppose that there is an upper bound of T, call it a, such that $a < 10$. By the Density Property, 0–9, we know that there is a rational number x such that

$$a < x < 10.$$

This means that x is greater than a and x is in T. What does this do to the assumption that a is an upper bound of T?

For a second example choose H to be the set of all rational numbers which are less than or equal to $\frac{1}{2}$, that is,

$$H = \left\{x : x \in Q \text{ and } x \leq \frac{1}{2}\right\}.$$

The rational number $\frac{1}{2}$ is an upper bound of H. It is also in H. Can you show that $\frac{1}{2}$ is the least upper bound of H?

These two examples suggest a way of constructing infinitely many subsets of Q, the field of rational numbers, which have least upper bounds. We can generalize these examples by letting

$$T = \{x : x \in Q \text{ and } x < r\},$$

where r is a specified rational number and

$$H = \{x : x \in Q \text{ and } x \leq r\}.$$

In both instances r is the least upper bound of the given subset.

Now let G be the set of all rational numbers greater than or equal to some given rational number, say 1. Does G have a least upper bound? Does it have any upper bounds for that matter? Since the set of upper bounds for G is empty, there is no least element.

These examples might lead us to the conclusion that if a subset of the rational numbers has any upper bounds at all, it will have a rational number as a least upper bound. The question is: Do *all* subsets of the rational numbers that have upper bounds also have a least upper bound that is a rational number? This is a nontrivial question. You will be asked later to assist in showing that the answer is "No."

Having discussed upper bounds and least upper bounds, we are now in a position to make the following definition.

8. Completeness Property

An ordered field F is **complete** if every nonempty subset of F which has an upper bound has a least upper bound (l.u.b.) in F.

The system of real numbers, which is an ordered field, does have the Completeness Property. The system of rational numbers, also an ordered field, does not. Light will be shed on these two assertions in Section 3.6.

EXERCISES 3.4

◀In Exercises 1–10, give the least upper bounds (if they exist) for each of the given subsets of the field, Q, of rational numbers. State whether or not the indicated l.u.b. is in the subset.

1. $\{2, 4, 6, 8\}$

2. The set of all odd numbers

3. The set of all even numbers less than 99

4. $\{1, 2, 3, \ldots, 10\}$ (that is, the set of integers from 1 to 10)

5. $\{1, 2, 3, \ldots\}$ (that is, the set of all natural numbers)

6. $\{x : x \in Q \text{ and } x \leq 7\}$

7. $\{x : x \in Q \text{ and } x > 4\}$

8. $\left\{x : x \in Q \text{ and } x < \dfrac{1}{2}\right\}$

9. $\left\{1, \dfrac{1}{2}, \dfrac{1}{3}, \dfrac{1}{4}, \ldots\right\}$ (that is, the set of all positive rational numbers with numerator 1)

10. The set of all positive multiples of 5

11-20. A subset S of the rational numbers is said to have a **greatest lower bound** (g.l.b.), x, if every element in S is greater than or equal to x and if x is the largest number for which this is true. For each of the subsets in Exercises 1–10, indicate whether or not a g.l.b. exists and, for the cases where one does exist, name it.

3.5 RATIONAL NUMBERS AND THE COMPLETENESS PROPERTY

In Section 3.4 it was asserted that the rational numbers do *not* have the Completeness Property.

In this section we shall prove, with your help, that a subset of the rational numbers, namely the set S of all positive rational numbers x such that $x^2 < 3$, has rational numbers as upper bounds but no rational number as *least* upper bound. Thus we shall establish the fact that the field of rational numbers is *not* complete.

Study the following argument very carefully, supplying the missing items and reasons.

1. Let S be the set of all positive rational numbers x such that $x^2 < 3$, that is,

$$S = \{x : x \in Q, x > 0, \text{ and } x^2 < 3\}.$$

2. We shall first show that the rational number 2 is an upper bound of S.

3. For, if 2 is not an upper bound, then there is a rational number $y \in S$ with $2 < y$.

4. But if $2 < y$, then $2^2 < y^2$. (Why?) Also since $y \in S$, $y^2 < 3$ by definition of S. By the Transitive Property of order, it follows that $2^2 \boxed{?} 3$, a contradiction. This means that the assumption of statement 3 that 2 is not an upper bound of S must be false.

5. Therefore 2 is an upper bound of S, and so the set B of all $\boxed{?}$ bounds of S is not empty; that is, $B \neq \varnothing$.

6. We shall now show that B does not have a least element and hence show that the field of rational numbers is not complete.

7. Suppose, to the contrary, that B has a least element b, that is, that b is the least upper bound of S.

8. By the Trichotomy Property of order, either $b^2 < 3$, $b^2 = 3$, or $b^2 \boxed{?} 3$.

9. Assume first that $b^2 < 3$. We shall show that under this condition b is *not* an upper bound of S.

10. Let $c = \dfrac{3 - b^2}{3b}$. Since, by assumption, $b^2 < 3$, it follows that $b^2 - 3 < 0$ and so $3 - b^2 > 0$.

11. Because b is an upper bound of S, then $b \geq 1$ since $1 \in S$ due to the fact that $1^2 < 3$.

12. Thus, certainly, $b > 0, \dfrac{1}{3} \cdot \dfrac{1}{b} > 0$, and $\dfrac{3 - b^2}{3b} > 0$ by (10). Hence, since

$$\frac{3 - b^2}{3b} = c,$$

$c \boxed{?} 0$.

13. Since $c = \dfrac{3 - b^2}{3b}$ and $3b > 0$, it follows that $3bc = 3 - b^2$. (Why?) Thus $3bc + b^2 = 3$, $3bc < 3$, and $bc < 1$.

14. Since $b \geq 1$, then $c < 1$ because if $c \geq 1$ and $b \geq 1$, then $bc \geq 1$, and since $c > 0$, $c^2 < c$.

15. We shall now show that $(c + b)^2 < 3$. This means that $(c + b) \in S$ and since $c > 0$, it follows that $b < c + b$. Hence b is *not* an upper bound of S.

16. We show that $(c + b)^2 < 3$ as follows. First,

$$(c + b)^2 = c^2 + 2bc + \boxed{?}.$$

17. From (14) and Order Property 7(iii),

$$c^2 + 2bc + b^2 < c + 2bc + b^2.$$

18. Since, by (11), $b \geq 1$ and, by (12), $c > 0$, it follows that $c \leq bc$. Hence

$$c + 2bc + b^2 \leq bc + 2bc + b^2 = 3bc + b^2$$

and so, by (17),

$$c^2 + 2bc + b^2 < c + bc + b^2 \leq 3bc + b^2$$

and $c^2 + 2bc + b^2 < 3bc + b^2$ by Order Property 7(ii).

19. By (13), $3bc + b^2 = 3$. Therefore, by (18),

$$c^2 + 2bc + b^2 = (c + b)^2 < 3,$$

and statement 15 is verified.

20. For the second case, that is, $b^2 = 3$, it can be proved, as outlined in Exercise 1 of Exercises 3.5, that there is no positive rational number x such that $x^2 = 3$.

21. Consider, then, the last case, that is, $b^2 > 3$. Let

$$d = b - \left(\frac{b^2 - 3}{2b}\right).$$

Since $b^2 > 3$ and $b \geq 1$ (by (11)), then $\dfrac{b^2 - 3}{2b} > 0$. This means that $d < b$ and, since

$$d = \frac{2b^2}{2b} - \left(\frac{b^2 - 3}{2b}\right) = \frac{b^2 + 3}{2b},$$

we know that $d > 0$.

22. But

$$d^2 = \left[b - \left(\frac{b^2 - 3}{2b}\right)\right]^2 = b^2 - 2b\left(\frac{b^2 - 3}{2b}\right) + \left(\frac{b^2 - 3}{2b}\right)^2$$

$$= b^2 - (b^2 - 3) + \left(\frac{b^2 - 3}{2b}\right)^2$$

which is equal to $3 + \left(\dfrac{b^2 - 3}{2b}\right)^2$. Therefore $d^2 \boxed{?} 3$ because $\dfrac{b^2 - 3}{2b} > 0$.

23. This means that d is an upper bound of S. But, also, $d < b$ by (21).

24. Hence b is not the $\boxed{?}$ upper bound.

On the basis of this discussion you can see that if b is the least upper bound of S, then either $b^2 > 3$, or $b^2 < 3$, or $b^2 = 3$. You have seen that if $b^2 > 3$, then b cannot be a least upper bound. If $b^2 < 3$, then b is not even an upper bound. Finally, as you will show in Exercises 3.5, if $b^2 = 3$, then b cannot be a rational number.

Thus we have seen that although some subsets of the field of rational numbers have least upper bounds which are rational numbers, there is at least one subset where the least upper bound is not a rational number. The rational numbers are not complete.

One may think of the set of all real numbers as filling a gap by providing least upper bounds for all subsets of the rational numbers that have upper bounds. In a similar fashion the real numbers which, as you know, also include the rational numbers, provide greatest lower bounds for all subsets of the rational numbers that have lower bounds.

It can also be proved that any subset of the real numbers that has an upper bound has a least upper bound which is also a real number.

Furthermore, although we are not yet in a position to prove this, it can be shown that the field of real numbers is essentially the only field having this least upper bound property. For this reason we can reasonably characterize the system of real numbers as "the one and only complete ordered field."

In Exercise 1 which follows you will show that the positive number x which satisfies the equation $x^2 = 3$ is not a rational number. We designate this irrational number as $\sqrt{3}$. It is the least upper bound of

$$S = \{x : x \in Q, \ x > 0, \text{ and } x^2 < 3\}.$$

Moreover, since $\sqrt{3}$ is a real number, this tells us that the set B of upper bounds of S does have a least element among the real numbers.

EXERCISES 3.5

1. Prove that there is no positive rational number x such that $x^2 = 3$. (*Hint:* Assume that x is a rational number $\dfrac{a}{b}$ with a and b positive integers such that $\dfrac{a^2}{b^2} = 3$. This means that $a^2 = 3b^2$. In the prime factorization of a^2 each factor will appear an even number of times. In the prime factorization of $3b^2$, what about the factor 3?)

2. **CHALLENGE PROBLEM.** Prove that for any natural number n if $\sqrt{n}$ is a rational number, then $\sqrt{n}$ is a natural number.

3.6 SUBFIELDS

In Section 3.5 we characterized the real numbers as a complete ordered field. We also noted that the rational numbers, which may be thought of as a subset of the set of real numbers, constitute an ordered field, although not a complete ordered field. Thus the rational numbers are a **subsystem** of the system of real numbers, having the same two binary operations and all of the field properties. Hence it is customary to designate the system of rational numbers (elements, operations, properties) as a **subfield** of the field of real numbers.

The question arises, "Do the real numbers contain other subfields?" That is, can we pick out other subsets which are also subfields under the same two binary operations? A respectable way to tackle this question is to try an experiment. Suppose we describe a particular subset S as the set of all real numbers which can be written in the form

$$a + b\sqrt{2},$$

where a and b are any rational numbers. Is the system

$$S = \{a + b\sqrt{2} : a \in Q, b \in Q; +, \cdot\}$$

a subfield of the field of real numbers?

Since it is known that the Associative, Commutative, and Distributive Properties hold with respect to addition and multiplication of real numbers, we need only be concerned with closure, identities, and inverses in coming up with an answer.

As an opening, we shall investigate the Closure Property for addition. You can carry the ball from there in the Exercises which follow.

Given two elements in S,

$$a + b\sqrt{2} \quad \text{and} \quad c + d\sqrt{2},$$

we know that

$$(a + b\sqrt{2}) + (c + d\sqrt{2}) = (a + c) + (b\sqrt{2} + d\sqrt{2})$$
$$= (a + c) + (b + d)\sqrt{2}.$$

Since $a + c$ is a rational number and $b + d$ is a rational number, the expression

$$(a + c) + (b + d)\sqrt{2}$$

is in the form prescribed for elements of S, that is, if $a + b\sqrt{2} \in S$ and $c + d\sqrt{2} \in S$, then

$$(a + b\sqrt{2}) + (c + d\sqrt{2}) \in S.$$

Similarly, we must show that the product, $(a + b\sqrt{2})(c + d\sqrt{2})$, of two elements in S can be written as $s + t\sqrt{2}$, where s and t are also rational numbers. Likewise, we have to demonstrate not just that identities and inverses exist but that they too are elements in S, that is, that each can be written in the form $s + t\sqrt{2}$, where s and t are rational numbers. These demonstrations will be carried out in the Exercises which follow.

EXERCISES 3.6

◀In Exercises 1–10, find the sum of the two given elements of S.

1. $3 + 5\sqrt{2}, 6 - 2\sqrt{2}$ **2.** $4 - 3\sqrt{2}, -1 + 7\sqrt{2}$

3. $\frac{1}{2} - \sqrt{2}, \frac{1}{2} + 2\sqrt{2}$ 7. $0.7 + 0.8\sqrt{2}, 0.3 + 0.2\sqrt{2}$

4. $\frac{2}{3} + \frac{1}{2}\sqrt{2}, \frac{1}{3} + \frac{1}{2}\sqrt{2}$ 8. $2.4 - 1.7\sqrt{2}, 1.7 - 2.4\sqrt{2}$

5. $\frac{3}{4} - \frac{1}{4}\sqrt{2}, \frac{1}{4} + \frac{5}{4}\sqrt{2}$ 9. $\frac{3}{5} + \frac{1}{6}\sqrt{2}, \frac{1}{6} + \frac{3}{5}\sqrt{2}$

6. $\frac{1}{5} + \frac{1}{5}\sqrt{2}, \frac{1}{10} - \frac{1}{10}\sqrt{2}$ 10. $\frac{3}{4} - \frac{2}{3}\sqrt{2}, \frac{2}{3} - \frac{3}{4}\sqrt{2}$

11. How would the identity element for addition be put in the form prescribed for elements of S?

12. How would the identity element for multiplication be put in the form prescribed for elements of S?

13. How would any rational number a be put in the form prescribed for elements of S?

14. Does S contain all the rational numbers?

15-24. Determine the products of each of the given pairs of elements in S in Exercises 1–10 and write the results in the form prescribed for elements of S.

25. Find the product

$$(a + b\sqrt{2})(c + d\sqrt{2})$$

and write the result in the prescribed form.

26. Is the set S closed under multiplication?

27. What is the additive inverse of $a + b\sqrt{2}$?

28. To find the multiplicative inverse of an element in S, say, for example, $3 + 4\sqrt{2}$, we must determine a number, $x + y\sqrt{2}$, in S such that

$$(3 + 4\sqrt{2})(x + y\sqrt{2}) = 1 + 0\sqrt{2},$$

since $1 + 0\sqrt{2}$ is the multiplicative identity written in the form prescribed for S. Show that $1 + 0\sqrt{2}$ is the multiplicative identity of S by computing the product

$$(a + b\sqrt{2})(1 + 0\sqrt{2}).$$

29. Find the product

$$(3 + 4\sqrt{2})(x + y\sqrt{2})$$

and write the result in the form prescribed for elements of S.

30. Since

$$(3 + 4\sqrt{2})(x + y\sqrt{2}) = (3x + 8y) + (4x + 3y)\sqrt{2},$$

then we must determine x and y such that

$$(3x + 8y) + (4x + 3y)\sqrt{2} = 1 + 0\sqrt{2}$$

if $x + y\sqrt{2}$ is to be the multiplicative inverse of $3 + 4\sqrt{2}$. Show that $(3x + 8y) + (4x + 3y)\sqrt{2} = 1 + 0\sqrt{2}$ if and only if x and y are such that

$$3x + 8y = 1 \text{ and } 4x + 3y = 0.$$

31. Solve the system of equations of Exercise 30.

32. Using the results of Exercise 31, write the multiplicative inverse of $3 + 4\sqrt{2}$ in the prescribed form.

◀In Exercises 33–45, determine the multiplicative inverses for each of the given numbers and write the results in the prescribed form for elements of S.

33. $2 + 3\sqrt{2}$

34. $3 + 2\sqrt{2}$

35. $4 + 3\sqrt{2}$

36. $3 - 5\sqrt{2}$

37. $5 - 3\sqrt{2}$

38. $3 + 33\sqrt{2}$

39. $\dfrac{1}{2} + \sqrt{2}$

40. $\dfrac{1}{2} - \sqrt{2}$

41. $\dfrac{2}{3} + \dfrac{1}{3}\sqrt{2}$

42. $\dfrac{2}{3} - \dfrac{1}{3}\sqrt{2}$

43. $\dfrac{1}{3} + \dfrac{1}{4}\sqrt{2}$

44. $\dfrac{1}{4} + \dfrac{1}{3}\sqrt{2}$

45. $\dfrac{3}{5} - \dfrac{1}{4}\sqrt{2}$

46. Determine the multiplicative inverse of $3 + 4\sqrt{2}$ by the following alternative method to that described in Exercises 29–31. Since the multiplicative inverse of a real number $x \neq 0$ can be expressed as a reciprocal $\dfrac{1}{x}$, the inverse of $3 + 4\sqrt{2}$ can be written symbolically as $\dfrac{1}{3 + 4\sqrt{2}}$. The job then is to convert this to the proper form, $a + b\sqrt{2}$ where $a, b \in Q$. Multiply both numerator and denominator of $\dfrac{1}{3 + 4\sqrt{2}}$ by $3 - 4\sqrt{2}$ and compare the result with that obtained in Exercise 32.

47.–59. Verify your answers to Exercises 33–45 by the alternative approach of Exercise 46.

60. CHALLENGE PROBLEM. By solving a system of equations for x and y resulting from the equation

$$(a + b\sqrt{2})(x + y\sqrt{2}) = 1 + 0\sqrt{2},$$

show that the multiplicative inverse of $a + b\sqrt{2}$ is given by the formula

$$\left(\frac{a}{a^2 - 2b^2}\right) + \left(\frac{-b}{a^2 - 2b^2}\right)\sqrt{2}.$$

61. **CHALLENGE PROBLEM.** Show that every element of S except $0 + 0\sqrt{2}$ has a multiplicative inverse. (*Hint:* Use the formula

$$\frac{a}{a^2 - 2b^2} + \frac{-b}{a^2 - 2b^2}\sqrt{2}$$

as developed in Exercise 48 and show that if a and b are rational numbers, then $a^2 - 2b^2 \neq 0$ unless a and b are both zero. Recall that $\sqrt{2}$ is an irrational number.)

62. **CHALLENGE PROBLEM.** Show that if $S = \{a + b\sqrt{r} : a, b \in Q; +, \cdot\}$, where r is a fixed positive integer which is not the square of an integer, then S is a subfield of the field of real numbers.

3.7 THE SMALLEST SUBFIELD, THE RATIONAL NUMBERS

In Section 3.6 we examined in some detail the system

$$S = \{a + b\sqrt{2} : a, b \in Q; +, \cdot\}.$$

We can conclude on the basis of text discussion and the Exercises that S is a field—a subfield of the field of real numbers.

Now let us consider more generally the system

$$T = \{a + b\sqrt{r} : a, b \in Q; +, \cdot\}$$

where r is a fixed positive integer that is not the square of an integer (that is, $r \neq s^2$ where s is an integer). The argument to show that T is a field is similar to that used to show that S is a field. Let us run through the argument briefly, omitting details. First, we have

$$(a + b\sqrt{r}) + (c + d\sqrt{r}) = (a + c) + (b + d)\sqrt{r}$$

and

$$(a + b\sqrt{r})(c + d\sqrt{r}) = (ac + bdr) + (ad + bc)\sqrt{r}.$$

Since a, b, c, d, and r are rational numbers, so are $a + c$, $b + d$, $ac + bdr$, and $ad + bc$. Thus T is closed under $+$ and $\cdot$.

The additive and multiplicative identities are, respectively,

$$0 + 0\sqrt{r} \quad \text{and} \quad 1 + 0\sqrt{r}$$

and are in T since 0 and 1 are both rational numbers.

For the additive inverse of $a + b\sqrt{r} \in T$ we have $(-a) + (-b)\sqrt{r} \in T$ since $-a$ and $-b$ are both rational numbers.

Furthermore, by direct multiplication we can show that the multiplicative inverse of $a + b\sqrt{r}$ is

$$\frac{a}{a^2 - b^2 r} + \frac{-b}{a^2 - b^2 r}\sqrt{r}$$

if $a^2 - b^2r \neq 0$. Here, once again, it is clear that $\dfrac{a}{a^2 - b^2r}$ and $\dfrac{-b}{a^2 - b^2r}$ are rational numbers if $a^2 - b^2r \neq 0$ and a and b are rational numbers.

By an argument similar to the one employed in Exercise 49 of Exercises 3.6, it can be shown that the denominator, $a^2 - b^2r$, can be equal to zero if and only if a and b are both zero, that is, if $a + b\sqrt{r}$ is the additive identity. Hence we see that every nonzero element in T has a multiplicative inverse.

As remarked previously, the Associative, Commutative, and Distributive Properties certainly hold in T since they hold for the system of real numbers.

In summary, then, we may assert that T is a subfield of the real numbers.

There are many subfields of the real numbers other than those of this particular form. You will encounter one of these in the Exercises which follow.

We shall not attempt to describe or catalogue all subfields, since this would be a prohibitively long and tedious task. We shall, however, uncover a property which all of these subfields have in common.

In Exercise 14 of Section 3.6 you were asked whether or not the field

$$S = \{a + b\sqrt{2} : a \in Q, b \in Q; +, \cdot\}$$

contained all the rational numbers as a subset. Since any rational number a can be written in the form $a + 0\sqrt{2}$, the answer is assuredly "Yes." This would also be true in general for all subfields T where

$$T = \{a + b\sqrt{r} : a \in Q, b \in Q, r \text{ a fixed integer}; +, \cdot\},$$

since the rational number a can be written as $a + 0\sqrt{r}$.

We now come to a more general question: Do all possible subfields of the real numbers contain the rational numbers as a subfield?

We can arrive at a plausible answer to this question in the following way. First, we know that all fields must contain two unique identity elements. For the real numbers these are 0 and 1, respectively. The Identity Property further assures us that these are the only identity elements for the real numbers. Thus we know, for example, that 0 is the only number which is such that for *every* real number a,

$$a + 0 = a.$$

However, it is conceivable that for a smaller subset of the real numbers, there might be another additive identity that applies to this reduced set. Let x be an element of a subset S and suppose that there is an element $z \in S$ such that

$$x + z = x.$$

But we know that in the field of real numbers

$$x + 0 = x.$$

Hence

$$x + z = x + 0.$$

Now, adding the real number $-x$ to both sides, we get

$$z = 0.$$

This tells us that in every subfield the additive identity is the real number 0.

It can be shown similarly that the multiplicative identity in every subfield must be the real number 1. Furthermore, by the Closure Property of addition we know that if $1 \in F$, it follows that $1 + 1 = 2 \in F$, that $2 + 1 = 3 \in F$, and so on. In a sense, then, this property "generates" all of the natural numbers. Moreover, the field requirements of an additive inverse for each element "guarantee" the inclusion of all negative integers. Finally, since one of the field properties requires a multiplicative inverse for each nonzero number, we are assured the inclusion of all rational numbers as well. For example, since $3 \in F$, $\dfrac{1}{3}$, the multiplicative inverse of 3 is in F and so $2 \cdot \dfrac{1}{3} = \dfrac{2}{3} \in F$. In general, to show $\dfrac{a}{b} \in F$ we consider $\dfrac{1}{b}$, the multiplicative inverse of b, and then $a \cdot \dfrac{1}{b} = \dfrac{a}{b}$.

Thus, by this somewhat informal "constructive" approach, we can surely accept as plausible the assertion that every subfield of the real numbers must contain the rational numbers as a subfield. In this sense, then, we can think of the rational numbers as being the "smallest" subfield of the field of real numbers.

Before leaving, temporarily at least, the subject of subfields it is important to reconsider the Property of Completeness. The particular subfields that we have discussed can all be shown to be ordered fields. Are any of these subfields complete? With respect to the rational numbers it has already been shown that the answer is "No."

But what about the subfield

$$S = \{a + b\sqrt{2} : a \in Q, b \in Q; +, \cdot\}?$$

Consider the subset D of this field consisting of all positive rational numbers a such that

$$a^2 < 6.$$

Just as $\sqrt{3}$ is the l.u.b. of $\{a : a \in Q, a > 0, \text{ and } a^2 < 3\}$, so $\sqrt{6}$ is the l.u.b. of D. Is $\sqrt{6}$ an element in S? Can we find rational numbers a and b such that

$$a + b\sqrt{2} = \sqrt{6}?$$

Suppose that $a + b\sqrt{2} = \sqrt{6}$. Then

$$(a + b\sqrt{2})^2 = (\sqrt{6})^2$$

and

$$a^2 + 2b^2 + 2ab\sqrt{2} = 6 = 6 + 0\sqrt{2}.$$

Therefore

$$a^2 + 2b^2 = 6 \qquad \text{and} \qquad 2ab = 0$$

since if $2ab \neq 0$, then

$$\sqrt{2} = \frac{6 - a^2 - 2b^2}{2ab}$$

would be a rational number, a contradiction. From the field properties and the results of Exercise 50, Section 3.2, we know that if $2ab = 0$, then $a = 0$ or $b = 0$. But $b \neq 0$, for if $b = 0$, then $a = \sqrt{6}$, which contradicts the fact that $\sqrt{6}$ is not a rational number. If $a = 0$, then $2b^2 = 6$ and $b^2 = 3$. Therefore $b = \sqrt{3}$. But $b \neq \sqrt{3}$ because $\sqrt{3}$ is not a rational number. Consequently, there are no rational numbers a and b such that

$$a + b\sqrt{2} = \sqrt{6}.$$

We conclude that the field $S = \{a + b\sqrt{2} : a \in Q, b \in Q; +, \cdot\}$ is not complete.

EXERCISES 3.7

◀ In Exercises 1–10, determine the multiplicative inverse of each number $a + b\sqrt{r}$ and write each inverse in the prescribed form $x + y\sqrt{r}$, where x and y are rational numbers.

1. $1 + 2\sqrt{3}$ 3. $1 - 4\sqrt{5}$

2. $2 + 3\sqrt{3}$ 4. $2 + 5\sqrt{6}$

5. $1 - \sqrt{7}$ **8.** $\dfrac{1}{2} + \sqrt{3}$

6. $6 - \sqrt{5}$ **9.** $\dfrac{1}{3} + \sqrt{5}$

7. $3 + 10\sqrt{11}$ **10.** $\dfrac{2}{3} - \sqrt{7}$

◀For Exercises 11–34, consider the subsystem T of the real numbers defined by

$$T = \{a + b\sqrt[3]{2} + c\sqrt[3]{4} : a \in Q, b \in Q, c \in Q; +, \cdot\}.$$

◀In Exercises 11–15, find the sums of the given pairs of numbers and write each sum as an element of T.

11. $2 + \sqrt[3]{2} + 3\sqrt[3]{4}, \; 1 - 3\sqrt[3]{2} + 5\sqrt[3]{4}$ **14.** $8 - 6\sqrt[3]{4}, \; 1 + \sqrt[3]{2} + \sqrt[3]{4}$

12. $3 - \sqrt[3]{2} + \sqrt[3]{4}, \; 4 - 2\sqrt[3]{2} - \sqrt[3]{4}$

13. $6\sqrt[3]{2} - 5\sqrt[3]{4}, \; 7 + \sqrt[3]{2}$ **15.** $\dfrac{2}{3} + \sqrt[3]{4}, \; \dfrac{1}{2} - \sqrt[3]{2} + \dfrac{1}{2}\sqrt[3]{4}$

◀For Exercises 16–25, write the additive inverses in T of each of the 10 numbers in Exercises 11–15.

◀For Exercises 26–30, form the product of each of the pairs of numbers in Exercises 11–15 and write each of the results as an element of T. It will help to recall that

$$\sqrt[3]{2} \cdot \sqrt[3]{2} = \sqrt[3]{2 \cdot 2} = \sqrt[3]{4}$$

$$\sqrt[3]{2} \cdot \sqrt[3]{4} = \sqrt[3]{2 \cdot 4} = \sqrt[3]{8} = 2$$

$$\sqrt[3]{4} \cdot \sqrt[3]{4} = \sqrt[3]{4 \cdot 4} = \sqrt[3]{16}$$

$$= \sqrt[3]{8} \cdot \sqrt[3]{2} = 2\sqrt[3]{2}.$$

31. **CHALLENGE PROBLEM.** Find the multiplicative inverse of $2 + 3\sqrt[3]{2} + \sqrt[3]{4}$ and write the result as an element of T. (*Hint:* The solution can be obtained by solving a system of three linear equations in three variables in a manner similar to that discussed in Section 3.6.)

32. By means of certain strategies similar to, although more complicated than, those used in Section 3.6 it can be shown that every nonzero real number of the form

$$a + b\sqrt[3]{2} + c\sqrt[3]{4}$$

does have a multiplicative inverse which is an element of T. Write the identities for addition and multiplication in the form $a + b\sqrt[3]{2} + c\sqrt[3]{4} \in T$.

33. Write the product

$$(a + b\sqrt[3]{2} + c\sqrt[3]{4})(d + e\sqrt[3]{2} + f\sqrt[3]{4})$$

as an element of T.

34. CHALLENGE PROBLEM. Write a formula for the multiplicative inverse in T of $a + b\sqrt[3]{2} + c\sqrt[3]{4} \neq 0$.

35. CHALLENGE PROBLEM. Show that if S is a subfield of the real numbers, then the only possible multiplicative identity in S is the real number 1.

3.8   INTEGRAL DOMAINS AND THE INTEGERS

In Section 3.7 it was asserted that the rational numbers are the "smallest" subfield of the field of real numbers. The assertion was first made plausible by a constructive argument which, in effect, showed that in any subfield of the real numbers the required presence of identity elements 0 and 1 along with the Field Properties necessitated the inclusion of all rational numbers. We can also show that if r is any rational number, then $Q' = Q - \{r\}$, the set consisting of all rational numbers except r, is not a field. For if Q' were a field, then the product of any two elements in Q' would also have to be in Q'. (Why?) Now suppose a is an element in Q' which is neither 0 nor 1 and that $r \neq 0$. Then $ra^{-1} \neq r$ and hence $ra^{-1} \in Q'$. But now we have $ra^{-1} \in Q'$, $a \in Q'$, and hence

$$(ra^{-1}) \cdot a = r \cdot (a^{-1} a) = r \cdot 1 = r \in Q',$$

a contradiction.

Finally, if $r = 0$, Q' could not be a field because it would lack an additive identity. Thus we may conclude that if the field Q is reduced by a single element, then the resulting subset is not a field.

Although the field of rational numbers contains no "smaller" subfields, it does contain a subsystem, namely the integers, which might be said to "come close" to being a field. Indeed, it is evident that the set of integers under the operations $+$ and $\cdot$ satisfies all of the Field Properties 1 to 6 with the single exception of the multiplicative inverse property. You are asked to verify this in the forthcoming set of Exercises.

Earlier we noted that the system of rational numbers and the system of real numbers were special cases of a more general class of mathematical systems called fields. In the same manner the integers are a particular example of a more general class of mathematical systems called **integral domains.**

Definition 3.2   An **integral domain** D is a set of at least two elements on which are defined two binary operations with respect to which all the properties of a field hold except, possibly, the multiplicative inverse property, and for which the following Cancellation Property of multiplication holds.

If a, b, and c are any elements in an integral domain such that $c \cdot a = c \cdot b$ and $c \neq z$, the additive identity of D, then

$$a = b.$$

A consequence of the Cancellation Property of multiplication is that if c and a are any elements of an integral domain such that $c \cdot a = 0$, then either $c = 0$ or $a = 0$ (or both). You are asked to prove this in the Exercises.

It is interesting to note that this Cancellation Property also holds for all fields. For if $c \neq z$, then in a field the multiplicative inverse, c^{-1}, exists. As a result, given that

$$c \cdot a = c \cdot b,$$

we can multiply both sides by c^{-1} to obtain the result $a = b$.

Thus we see that the multiplicative inverse property implies the Cancellation Property. The converse, however, is not true. For example, let x and y be integers such that $5x = 5y$. By the Cancellation Property we can conclude that $x = y$. We cannot, on the other hand, conclude that if $5x = 5y$ implies $x = y$, then the system of integers contains the multiplicative inverse of 5. The multiplicative inverse property implies the Cancellation Property, but the Cancellation Property does not imply the existence of multiplicative inverses. It is for this reason that we sometimes refer to the Cancellation Property as being weaker than the multiplicative inverse property.

Since a field has all the properties of an integral domain, every field, including the field of rational numbers, is an integral domain. Thus it is consistent to regard the system of integers as a "subdomain" of the domain of rationals. (We must be careful to distinguish between the use of the word domain here employed as an abbreviation of integral domain and that associated with the definition of a function.)

We shall investigate whether or not the integers contain any proper subsets which are also integral domains. This investigation is taken up in the following Exercises. You are asked to show that the integers satisfy the Order Properties of a field. Hence we can speak of the integers as an *ordered integral domain*.

EXERCISES 3.8

1. If 1 or 0 or both are omitted from the set of integers, will the resulting system be an integral domain? Explain.

2. If 3 is omitted from the set of integers, will the resulting system be an integral domain? Explain.

3. If any integer is omitted from the set of integers, will the resulting system be an integral domain? Explain.

4. Is the set of integers the smallest subdomain of the domain of rational numbers?

5. Is the field

$$F = \{0, 1, 2, 3, \ldots, 10; *, \circ\}$$

discussed in Section 3.2 an integral domain? Explain.

6. Assuming that the Associative, Commutative, and Distributive Properties hold for integers, show that the other integral domain properties hold for the integers, thus establishing the fact that the integers constitute an integral domain.

7. For all fields it was shown that for any $a \in F$, if z is the additive identity, then $a \cdot z = z$. Is this true for all integral domains? Explain.

8. For all integral domains it can be shown that if z is the additive identity and if $a \cdot b = z$, then either a or b must equal z. Copy and complete the following argument to show that this property follows from the Cancellation Property.
   Assume that $a \cdot b = z$ and that $a \neq z$. We wish to show that $b = z$.
   (a) By Exercise 7 we know that $a \cdot z = \boxed{?}$.
   (b) Therefore $a \cdot b = a \cdot z$.
   (c) Since $a \neq z$, by assumption it follows by the $\boxed{?}$ Property that $b = \boxed{?}$.

9. Repeat the argument of Exercise 8 but assume first that $b \neq z$. Then show that $a = z$.

10. Copy and complete the following argument to show that the Cancellation Property can be deduced from the property just proved (sometimes called the zero property of an integral domain).
    Assume that $c \cdot a = c \cdot b$ and that $c \neq z$. We wish to show that $a = b$.
    (a) By assumption, $c \cdot a = c \cdot b$; hence $(c \cdot a) - (c \cdot b) = z$.
    (b) But $(c \cdot a) - (c \cdot b) = c(a - b)$.
    (c) Since $c \neq z$, it follows from Exercises 8 and 9 that $(a - b) = \boxed{?}$.
    (d) Therefore $\boxed{?} = \boxed{?}$.

◀Exercises 11–33 refer to the system $S = \{0, 1, 2, 3, \ldots, 11; *, \circ\}$ with the two binary operations $*$ and $\circ$ defined in an analogous way to those in Section 3.2. That is,

$$a * b = a + b \text{ if } a + b < 12,$$
$$a * b = a + b - 12 \text{ if } a + b \geq 12,$$
$$a \circ b = (a \cdot b) - 12n$$

where n is the unique integer which satisfies the inequality

$$0 \leq [(a \cdot b) - 12n] < 12.$$

◀In Exercises 11–30, determine the following elements of S.

11. $7 * 4$	18. $6 * 6$	25. $7 \circ 7$
12. $3 \circ 2$	19. $6 \circ 6$	26. $9 \circ 9$
13. $5 \circ 6$	20. $8 \circ 8$	27. $7 \circ 6$
14. $11 * 11$	21. $9 * 9$	28. $9 \circ 11$
15. $10 * 11$	22. $10 * 10$	29. $8 \circ 11$
16. $5 * 7$	23. $10 \circ 11$	30. $11 * 1$
17. $8 * 4$	24. $11 \circ 11$	

31. Give the additive inverses of each of the eleven elements of S.

32. (a) If S is a field, then every element except 0 must have a multiplicative inverse. Make a list of the nonzero elements and give in each case the multiplicative inverse, if it exists. Otherwise write "no inverse."
    (b) From the results of (a) can S be a field?

33. (a) Calculate $3 \circ 4$ and $9 \circ 4$.
    (b) From (a) we know that $3 \circ 4 = 9 \circ 4$. If the Property of Cancellation holds, then $3 = 9$. Is S an integral domain?
    (c) From the results of (a) what can you say about the zero property with respect to S?

34. **CHALLENGE PROBLEM.** Show that the integers form an ordered integral domain. (*Hint:* Use the ordering $a < b$ if and only if $0 < b + (-a)$.)

35. **CHALLENGE PROBLEM.** In discussing subfields we considered the subfield of the real numbers

$$S = \{a + b\sqrt{2} : a \in Q, b \in Q; +, \cdot\}$$

and noted that Q is a proper subfield of S, that is, a subfield containing some but not all of the elements of S, and that S is a proper subfield of R, the real numbers.

Is there an integral domain X which contains the integers as a proper subdomain, such that X is a proper subdomain of Q? Give reasons for your decision.

CHAPTER SUMMARY

In this chapter we have examined the system of real numbers as a complete ordered field, that is, as a set of elements F on which are defined two binary operations, $+$ and $\cdot$, such that the following properties hold:

1. CLOSURE with respect to $+$ and $\cdot$.
2. ASSOCIATIVITY with respect to $+$ and $\cdot$.
3. COMMUTATIVITY with respect to $+$ and $\cdot$.
4. Unique ADDITIVE and MULTIPLICATIVE IDENTITIES, z and u, respectively, with $u \neq z$.
5. A unique ADDITIVE INVERSE, $-a$, for each element a and a unique MULTIPLICATIVE INVERSE, a^{-1}, for each element a other than z.
6. DISTRIBUTIVE PROPERTY of multiplication with respect to addition.
7. There exists in F an order relation "$<$" such that for a, b, c, $d \in F$,

   (i) **Trichotomy** For every a, $b \in F$ one and only one of the following holds:
   $$a < b, \qquad a = b, \qquad b < a.$$

   (ii) **Transitivity** If $a < b$ and $b < c$, then $a < c$.
   (iii) If $a < b$, then $a + c < b + c$.
   (iv) If $a < b$ and if $z < c$, then $ac < bc$.

8. COMPLETENESS PROPERTY. Every nonempty subset of the real numbers having an UPPER BOUND has a real number as a LEAST UPPER BOUND.

Other properties of order were deduced from (7). Among these are the following.

(O–1) If $a < b$ and $c < z$, then $ac > bc$.
(O–7) If $a < z$ and $b < z$, then $ab > z$.
(O–8) If $a < z$, then $-a > z$ and if $a > z$, then $-a < z$.
(O–9) If $a < b$, then there exists an $x \in F$ such that $a < x < b$.

We have considered SUBFIELDS of the real numbers, particularly those whose elements are of the form $a + b\sqrt{r}$ with a and b rational numbers and r a fixed integer which is not the square of an integer.

The RATIONAL NUMBERS were characterized as the "smallest" sub-field of the real numbers, that is, every subfield of the real numbers has the field of rational numbers as a subfield. The rational numbers constitute an ordered field with all of the foregoing properties except the Completeness Property.

The INTEGERS were examined as an example of an INTEGRAL DO-MAIN, that is, a system having all of the field properties except possibly for the multiplicative inverse property. Integral domains have a "weaker" property, the CANCELLATION PROPERTY, which states that if a, b, and c are any elements of the integral domain such that $ac = bc$ and $c \neq z$, the additive identity, then $a = b$. Also, in any integral domain it is true that $c \cdot a = 0$ implies $a = 0$ or $c = 0$.

REVIEW EXERCISES

Given the field $B = \{0, 1, 2, 3, \ldots, 10; *, \circ\}$, determine the element in B equal to each of the following with $*$ and $\circ$ defined as in Section 3.2.

1. $4 * 9$ **3.** $4 \circ 7$ **5.** $8 \circ 7$

2. $5 * 8$ **4.** $5 \circ 8$ **6.** $7 \circ 9$

Show that the following are true statements with respect to the field B.

7. $(9 * 4) * 6 = 9 * (4 * 6)$ **9.** $5 \circ (8 * 9) = (5 \circ 8) * (5 \circ 9)$

8. $(9 \circ 4) \circ 6 = 9 \circ (4 \circ 6)$ **10.** $6 \circ (7 * 5) = (6 \circ 7) * (6 \circ 5)$

Find the multiplicative inverse in B of each of the following elements.

11. 8   **12.** 10

In Exercises 13–17, determine the solution set of each of the given inequalities.

13. $6x + 8 < 4x + 20$ **16.** $\dfrac{3}{4}x + 20 < \dfrac{1}{2}x + 21$

14. $12x - 15 < 24 - x$ **17.** $5.7x + 2 < 90 - 2.3x$

15. $\dfrac{1}{2}x + 17 < 31 - \dfrac{3}{2}x$

18. Determine a rational number x such that $\frac{3}{4} < x < \frac{5}{6}$.

19. State the Completeness Property.

20. Give a subset of the rational numbers which has rational numbers as upper bounds but no rational number as a least upper bound.

Given the subfield of the real numbers, $S = \{a + b\sqrt{3} : a \in Q, b \in Q; +, \cdot\}$, determine the following products.

21. $(3 + 4\sqrt{3}) \cdot (5 - 6\sqrt{3})$ **22.** $(1 + 6\sqrt{3}) \cdot (7 - 2\sqrt{3})$

Find the multiplicative inverses of the following elements of the field S of Exercises 21–22.

23. $3 + 4\sqrt{3}$ **24.** $\dfrac{1}{2} - 2\sqrt{3}$ **25.** $7 + \dfrac{2}{3}\sqrt{3}$

26. State the four Order Properties of an ordered field.

27–30. Give four additional Properties of Order which can be deduced from those stated in Exercise 26.

GOING FURTHER: READING AND RESEARCH

In this chapter the main theme has been the concept of a field as a mathematical system or structure. One of the major characteristics of a field is that of having two distinct binary operations. One might ask the question then, "Is there a simpler, perhaps more basic, mathematical structure having but one binary operation?"

The answer is definitely "Yes." Such a structure is not only known in mathematics but also plays a very significant role both in mathematical theory and in a wide variety of applications. This structure is known as a **group.**

The definition of a group, like that of other mathematical systems such as the natural numbers under addition and multiplication, integral domains, fields, and so on, should have a familiar sound.

A group, G, is a set of elements on which is defined a binary operation, $*$, having the following properties.

G–1.   If $x, y \in G$, then $x * y \in G$.

G–2.   If $x, y, z \in G$, then $(x * y) * z = x * (y * z)$.

G–3.   There exists a unique element $e \in G$ such that for any $x \in G$, $x * e = e * x = x$.

G–4.   For every $x \in G$, there exists a unique element $x^{-1} \in G$ such that $x * x^{-1} = x^{-1} * x = e$.

You should have no difficulty identifying these properties by name. The following is an example of a group consisting of four elements. The operation $*$ is defined by the table.

$*$	1	2	3	4
1	1	2	3	4
2	2	4	1	3
3	3	1	4	2
4	4	3	2	1

See if you can identify the operation. What element corresponds to e in the definition? If $x = 2$, what is x^{-1}?

The following is a table for another important group.

*	e	s	t	u	v	w
e	e	s	t	u	v	w
s	s	e	u	t	w	v
t	t	v	e	w	s	u
u	u	w	s	v	e	t
v	v	t	w	e	u	s
w	w	u	v	s	t	e

What is u^{-1}? What is w^{-1}? In this group is it true that for every x and y in the group, $x * y = y * x$, that is, does this group have the Commutative Property? Does the group in the previous example have the Commutative Property? The two examples show that there exist both commutative and noncommutative groups.

Try some experiments with the second group to test Property G–2. For example, it should be true that in this group

$$(u * t) * v = u * (t * v).$$

See if you can construct a table for a group of four elements different from the one you have just seen. (Be sure to test all the properties.)

To learn more about groups and some of the surprising and unusual aspects of group theory, we suggest you do some reading in the following books.

ANDREE, R. V., *Selections from Modern Abstract Algebra*. New York: Holt Rinehart and Winston, 1958.

CARMICHAEL, R. D., *Introduction to the Theory of Groups of Finite Order*. New York: Dover Publications, 1956.

LIEBER, L. R., *Galois and the Theory of Groups*. Lancaster, Pa.: Science Press Printing Co., 1932.

WILDER, R. L., *Introduction to the Foundations of Mathematics*, 2nd ed. New York: John Wiley and Sons, 1965. Pages 165–179.

YARNELLE, J. E., *Finite Mathematical Structures* (Thinking with Mathematics Series). Boston, Mass.: D.C. Heath, 1964.

Erich Hartmann—Magnum

CHAPTER FOUR
COMPLEX NUMBERS

4.1 QUADRATIC EQUATIONS

The solution of a problem involving sequences may lead to a quadratic equation. Suppose, for example, we want to find the first term of an arithmetic sequence that has $S_n = 147$, $t_n = 12$, and $d = \frac{1}{2}$. You should check that the solution of this problem leads to the quadratic equation

$$2t_1^2 - t_1 - 6 = 0.$$

> **Definition 4.1** An equation of the form $ax^2 + bx + c = 0$, where $a, b, c \in R$, $a \neq 0$, is called a **quadratic equation**.

The three commonly used algebraic methods of finding the solution set of a quadratic equation are

(1) factoring,
(2) completing the square, and
(3) by formula.

The first method, **factoring,** is the most convenient way of obtaining the solution set of a quadratic equation (or, as we sometimes say, of "solving the equation") if the polynomial $ax^2 + bx + c$ is easy to factor. It is based on the fact, proved in Section 3.8, that $rs = 0$ (r, $s \in R$) implies $r = 0$ or $s = 0$.

Example 1 Find the solution set of the equation $2t_1^2 - t_1 - 6 = 0$.

Solution:

$$2t_1^2 - t_1 - 6 = (2t_1 + 3)(t_1 - 2) = 0.$$

Hence

$$2t_1 + 3 = 0 \qquad \text{or} \qquad t_1 - 2 = 0.$$

Thus

$$t_1 = -\frac{3}{2} \qquad \text{or} \qquad t_1 = 2.$$

The solution set is $\{-\frac{3}{2}, 2\}$.

The second method, **completing the square,** can be used to find the solution set of a quadratic equation when the quadratic polynomial does not factor readily as well as when it does. The basic idea is that a polynomial of the form $x^2 + bx$ becomes the square of a binomial if we add the term $\left(\dfrac{b}{2}\right)^2$ to it since

$$x^2 + bx + \frac{b^2}{4} = \left(x + \frac{b}{2}\right)^2.$$

We use this technique in the following two examples.

Example 2   Find the solution set of the equation $2t_1^2 - t_1 - 6 = 0$ by completing the square.

Solution:

$$2t_1^2 - t_1 - 6 = 0$$

$$t_1^2 - \frac{1}{2}t_1 = 3$$

$$t_1^2 - \frac{1}{2}t_1 + \left(\frac{1}{2} \cdot \frac{1}{2}\right)^2 = t_1^2 - \frac{1}{2}t_1 + \frac{1}{16} = 3 + \frac{1}{16}$$

$$\left(t_1 - \frac{1}{4}\right)^2 = \frac{49}{16}.$$

We have

$$t_1 - \frac{1}{4} = \frac{7}{4} \quad \text{or} \quad t_1 - \frac{1}{4} = -\frac{7}{4}.$$

Again, we obtain the solution set $\{-\frac{3}{2}, 2\}$.

Example 3   Find the solution set of $2x^2 - 3x - 4 = 0$ by completing the square.

Solution:

$$2x^2 - 3x - 4 = 0$$

$$x^2 - \frac{3}{2}x = \frac{4}{2}$$

$$x^2 - \frac{3}{2}x + \frac{9}{16} = \frac{4}{2} + \frac{9}{16}$$

$$\left(x - \frac{3}{4}\right)^2 = \frac{32 + 9}{16}$$

$$x - \frac{3}{4} = \frac{\sqrt{41}}{4} \quad \text{or} \quad x - \frac{3}{4} = \frac{-\sqrt{41}}{4}$$

$$x = \frac{3 + \sqrt{41}}{4} \quad \text{or} \quad x = \frac{3 - \sqrt{41}}{4}.$$

The solution set is $\left\{ \dfrac{3 + \sqrt{41}}{4}, \dfrac{3 - \sqrt{41}}{4} \right\}$. Sometimes we write

$x_1 = \dfrac{3 + \sqrt{41}}{4}$ and $x_2 = \dfrac{3 - \sqrt{41}}{4}$. We call x_1 and x_2 the **roots**

of the equation $2x^2 - 3x - 4 = 0$.

The third method of finding the solution set of a quadratic equation involves using the *quadratic formula*. The **quadratic formula,** as developed in a first course in algebra, states that the solution set of the quadratic equation $ax^2 + bx + c = 0$ $(a, b, c \in R, a \neq 0)$ over the domain of real numbers is

$$\left\{ \frac{-b + \sqrt{b^2 - 4ac}}{2a}, \frac{-b - \sqrt{b^2 - 4ac}}{2a} \right\},$$

providing that $b^2 - 4ac \geq 0$. This quadratic formula can be established by completing the square and this is the method commonly used in a first course in algebra. In Section 4.6, however, we shall use a different procedure.

Example 4   Find the solution set of $2t_1^2 - t_1 - 6 = 0$ by use of the quadratic formula.

Solution:

$2t_1^2 - t_1 - 6 = 0$

$$a = 2, \quad b = -1, \quad c = -6$$

$$t_1 = \frac{1 + \sqrt{1 - 4(2)(-6)}}{4} = \frac{1 + \sqrt{49}}{4} = \frac{8}{4} = 2$$

or

$$t_1 = \frac{1 - \sqrt{1 - 4(2)(-6)}}{4} = \frac{1 - \sqrt{49}}{4} = \frac{-6}{4} = -\frac{3}{2}.$$

Once again, we obtain the solution set $\{2, -\frac{3}{2}\}$.

Example 5   Find the solution set of $2x^2 - 3x - 4 = 0$ by use of the quadratic formula.

Solution:

$$2x^2 - 3x - 4 = 0$$

$$a = 2, \qquad b = -3, \qquad c = -4$$

$$x_1 = \frac{3 + \sqrt{9 - 4(2)(-4)}}{4} = \frac{3 + \sqrt{41}}{4}$$

$$x_2 = \frac{3 - \sqrt{9 - 4(2)(-4)}}{4} = \frac{3 - \sqrt{41}}{4}.$$

Again, we obtain the solution set $\left\{ \dfrac{3 + \sqrt{41}}{4}, \dfrac{3 - \sqrt{41}}{4} \right\}$.

If the solution set of the equation $ax^2 + bx + c = 0$ is $\{x_1, x_2\}$, then, by definition of solution set, we know that

$$ax_1^2 + bx_1 + c = 0$$

and

$$ax_2^2 + bx_2 + c = 0.$$

Does

$$2\left(\frac{3 + \sqrt{41}}{4} \right)^2 - 3\left(\frac{3 + \sqrt{41}}{4} \right) - 4 = 0?$$

Also, the equation

$$a(x - x_1)(x - x_2) = 0$$

certainly has the solution set $\{x_1, x_2\}$. Conversly, if the solution set of the equation

$$ax^2 + bx + c = 0$$

is $\{x_1, x_2\}$, we know that

$$a(x - x_1)(x - x_2) = ax^2 + bx + c.$$

Does

$$2\left(x - \frac{3 + \sqrt{41}}{4} \right)\left(x - \frac{3 - \sqrt{41}}{4} \right) = 2x^2 - 3x - 4?$$

Example 6   Find the solution set of $x^2 + x + 1 = 0$.

Solution: Since $x^2 + x + 1 = 1x^2 + 1x + 1$, we have $a = 1$, $b = 1$, and $c = 1$. When we attempt to apply the quadratic formula, we find that

$$b^2 - 4ac = 1 - 4 \cdot 1 \cdot 1 = -3.$$

Hence $\sqrt{b^2 - 4ac} = \sqrt{-3}$, which is not a real number since the square of any real number is nonnegative. We conclude therefore that, over the real numbers, the solution set of the equation $x^2 + x + 1 = 0$ is the empty set or, equivalently, that the equation has no roots that are real numbers. In still more abbreviated form we sometimes simply say that the equation has no real roots.

More generally, for a, b, $c \in R$ and $b^2 - 4ac < 0$, the quadratic equation $ax^2 + bx + c = 0$ has no real roots.

The methods we have shown for solving quadratic equations are also useful for solving certain other kinds of equations such as fractional equations and equations involving radicals. When solving fractional equations (equations in which the variable appears in the denominator) or equations involving variables in the radicands of radicals, a check is necessary to assure that the solution set of the final equation is also the solution set of the original equation because the methods we use for solving these two kinds of equations do not always produce a sequence of equivalent equations (equations having the same solution set). We illustrate this point in the next four examples.

Example 7 Find the solution set of

$$x + \frac{1}{x - 2} = 3 + \frac{1}{x - 2}.$$

Solution: To simplify the equation, we multiply each term of the equation by $x - 2$, a step which produces an equivalent equation if $x \neq 2$. We have

$$(x - 2)\left(x + \frac{1}{x - 2}\right) = (x - 2)\left(3 + \frac{1}{x - 2}\right)$$
$$x(x - 2) + 1 = 3(x - 2) + 1$$
$$x^2 - 2x = 3x - 6$$
$$x^2 - 5x + 6 = (x - 3)(x - 2) = 0$$
$$x = 3 \quad \text{or} \quad x = 2.$$

Now we note that if we replace x by 2 in the original equation, we have a zero denominator. Hence 2 is not an element of the solution set of the original equation. If we replace x by 3 in the original equation, however, we have $3 + 1 = 3 + 1$, which is a true sentence. Hence the solution set of the equation

$$1 + \frac{1}{x - 2} = 3 + \frac{1}{x - 2}$$

is $\{3\}$.

Example 8   Find the solution set of the equation

$$\sqrt{3x + 4} = x.$$

Solution: In this case, we square the expressions on both sides of the equal sign, an operation which does not always produce an equivalent equation. Thus, for example, $x = 2$ and $x^2 = 4$ are not equivalent equations (Why not?). Hence when we use this "squaring" procedure it is necessary to check the roots of the final equation obtained by substituting in the original equation.

Thus we have

$$(\sqrt{3x + 4})^2 = x^2$$
$$3x + 4 = x^2$$
$$x^2 - 3x - 4 = (x - 4)(x + 1) = 0$$
$$x = 4 \quad \text{or} \quad x = -1.$$

When we check, we discover that the solution set of the equation $\sqrt{3x + 4} = x$ is $\{4\}$ since replacing x by -1 in $\sqrt{3x + 4} = x$ gives us $\sqrt{1} = -1$, whereas $\sqrt{1} = 1$.

Example 9   Find the solution set of $\dfrac{4}{x + 1} + \dfrac{3}{x} = 2$.

Solution: We have

$$x(x + 1)\left(\frac{4}{x + 1} + \frac{3}{x}\right) = 2[x(x + 1)]$$
$$4x + 3(x + 1) = 2(x^2 + x)$$
$$4x + 3x + 3 = 2x^2 + 2x$$
$$2x^2 - 5x - 3 = (2x + 1)(x - 3) = 0$$
$$x = -\frac{1}{2} \quad \text{or} \quad x = 3.$$

Since neither x nor $x + 1$ is equal to zero when x is replaced by $-\frac{1}{2}$ or 3, we can conclude (barring mistakes in arithmetic!) that the solution set of

$$\frac{4}{x + 1} + \frac{3}{x} = 2$$

is $\{-\frac{1}{2}, 3\}$.

Example 10 Find the solution set of $\sqrt{x + 7} = 2x - 1$.

Solution: We have

$$(\sqrt{x + 7})^2 = (2x - 1)^2$$
$$x + 7 = 4x^2 - 4x + 1$$
$$4x^2 - 5x - 6 = (4x + 3)(x - 2) = 0$$
$$x = -\frac{3}{4} \quad \text{or} \quad x = 2.$$

Here, when we replace x by 2 in the equation

$$\sqrt{x + 7} = 2x - 1,$$

we arrive at the true statement

$$\sqrt{2 + 7} = 2 \cdot 2 - 1$$

since $\sqrt{2 + 7} = 3$ and $2 \cdot 2 - 1 = 3$. But when we replace x by $-\frac{3}{4}$, we obtain

$$\sqrt{-\frac{3}{4} + 7} = 2 \cdot \left(-\frac{3}{4}\right) - 1,$$

which is not a true statement since

$$\sqrt{-\frac{3}{4} + 7} = \sqrt{\frac{25}{4}} = \frac{5}{2} \quad \text{and} \quad 2\left(-\frac{3}{4}\right) - 1 = -\frac{5}{2}.$$

Thus the solution set of $\sqrt{x + 7} = 2x - 1$ is $\{2\}$.

Example 11 Find the solution set of $x^4 - 8x^2 + 16 = 0$.

Solution: We have

$$(x^2)^2 - 8(x^2)^1 + 16 = (x^2 - 4)(x^2 - 4) = 0.$$

Therefore $x^2 - 4 = 0$ and $x = 2$ or $x = -2$. The solution set is $\{2, -2\}$.

EXERCISES 4.1

◀In Exercises 1–25, find, by factoring, the solution set of each equation.

1. $x^2 - 5x - 14 = 0$

2. $2y^2 - 3y - 5 = 0$

3. $6y^2 - y - 12 = 0$

4. $6m^2 - 5m - 4 = 0$

5. $15m^2 - m - 2 = 0$

6. $4t^2 - 21t + 5 = 0$

7. $2t^2 + 7t + 3 = 0$

8. $4 - 12r + 9r^2 = 0$

9. $64r^2 - 48r + 9 = 0$

10. $8s^2 - 8s - 2 = 0$

11. $4x^2 - 20x - 56 = 0$. (See Exercise 1. Compare solution sets.)

12. $6y^2 - 9y - 15 = 0$. (See Exercise 2. Compare solution sets.)

13. $18y^2 - 3y - 36 = 0$. (See Exercise 3. Compare solution sets.)

14. $3t^2 - t = 0$

15. $5x^2 - x = 0$

16. $5x^2 = 0$

17. $5y^2 - 30 = 0$

18. $7s^2 - 5 = 2s$

19. $34y^2 + 17y = 0$

20. $10x^2 + 29x - 21 = 0$

21. $x^2 - (a + b)x + ab = 0$

22. $(2y + 3)(2y - 3) = 9y$

23. $(2y - 1)(y - 1) = 6$

24. $(2 - t)^2 + 4t = 10$

25. $x^2 - ax - x + a = 0$

◀In Exercises 26–30, first form a quadratic equation from the given equation and then solve the resulting quadratic equation by factoring. Be sure to check each element of the solution set of the quadratic equation in the original equation.

26. $\dfrac{15}{x + 4} + \dfrac{6}{x} = 5\dfrac{1}{2}$

27. $\dfrac{3x + 5}{x + 2} = \dfrac{x + 1}{5x + 10}$

28. $\dfrac{6}{y^2 - 4} + \dfrac{1}{y + 2} = \dfrac{3}{7}$

29. $\dfrac{t}{t - 7} - \dfrac{20t}{3t - 4} = \dfrac{8 - t}{3t^2 - 25t + 28}$

30. $\dfrac{2x^2}{x - 1} + \dfrac{6x - 4}{1 - x} = \dfrac{2x + 7}{3}$

◀In Exercises 31–40, complete each of the polynomials in such a way that each is the square of a binomial (that is, "complete the square").

31. $y^2 - 4y + \boxed{?}$

32. $m^2 - 8m + \boxed{?}$

33. $s^2 + 3\dfrac{1}{2}s + \boxed{?}$

34. $t^2 - \dfrac{7}{2}t + \boxed{?}$

35. $a^2 - \dfrac{a}{4} + \boxed{?}$

36. $b^2 - \dfrac{\sqrt{3b}}{4} + \boxed{?}$

37. $x^2 + \boxed{?} + 64$

39. $r^2 - \boxed{?} + \dfrac{3}{4}$

38. $y^2 - \boxed{?} + \dfrac{9}{4}$

40. $s^2 + \boxed{?} + \dfrac{6}{16}$

◀In Exercises 41–50, find the solution set of each equation by the method of completing the square.

41. $x^2 - 4x - 5 = 0$

46. $t^2 - \dfrac{2}{3}t = 1$

42. $y^2 - 4y - 21 = 0$

47. $x^2 - \dfrac{5}{2}x - 5\dfrac{1}{2} = 0$

43. $t^2 - 8t - 1 = 0$

48. $y^2 - \dfrac{7}{2}y = 10$

44. $s^2 + 4s - 4 = 0$

49. $3x^2 - 4x - 6 = 0$

45. $r^2 - 4r = 6$

50. $4y^2 - 5y - 6 = 0$

◀In Exercises 51–60, solve each equation by using the quadratic formula.

51. $2y^2 - 3y - 5 = 0$

52. $8s^2 - 8s - 2 = 0$

53. $10x^2 + 29x - 21 = 0$

54. $3x^2 - 4x - 6 = 0$

55. $4y^2 - 5y - 6 = 0$

56. $6t^2 - 2t - 3 = 0$

57. $5m^2 - 6m + 4 = 0$

58. $4r^2 - 3 = 7r$

59. $2x^2 + 5x - 3 = 0$

60. $3y^2 - y = 8$

◀In Exercises 61–70, find the solution set of each equation.

61. $\dfrac{5}{x - 2} + \dfrac{3}{x - 1} = 4$

62. $\sqrt{x - 7} = x - 6$

63. $\dfrac{4}{y} + \dfrac{2}{y^2 + y} = \dfrac{4y + 1}{y}$

64. $\dfrac{m - 1}{2m + 1} - \dfrac{2m - 3}{m + 3} = \dfrac{3}{2m^2 + 7m + 3}$

65. $2\sqrt{r^2 - 1} = \sqrt{2r - 1}$

66. $\dfrac{4t - 10}{t + 5} - \dfrac{7}{2} = \dfrac{7 - 3t}{t}$

67. $\dfrac{2y^2 - 6}{\sqrt{y}} = 3\sqrt{y}$

68. $\sqrt{y + 2} + 3 = \sqrt{2y + 13}$

69. $\dfrac{x + 2}{x - 2} + \dfrac{x - 2}{x + 2} = \dfrac{8(x - 1)}{4 - x^2}$

70. $\sqrt{9 + 2y} - \sqrt{2y} = \dfrac{5}{\sqrt{9 + 2y}}$

71. Show that if $\sqrt{(x - 5)^2 + y^2} + \sqrt{(x + 5)^2 + y^2} = 12$, then

$$\dfrac{x^2}{36} + \dfrac{y^2}{11} = 1.$$

◀In Exercises 72–86, form a quadratic equation which has the given set as its solution set.

   Example. A quadratic equation with solution set $\{1, -2\}$ is $(x - 1)$ $(x + 2) = 0$ or $x^2 + x - 2 = 0$.

72. $\{3, 4\}$ 80. $\{r, s\}$

73. $\{2, -5\}$ 81. $\{a, b\}$

74. $\{6, 8\}$ 82. $\{1 + \sqrt{2}, 1 - \sqrt{2}\}$

75. $\{-3, -2\}$ 83. $\{-2 + \sqrt{3}, -2 - \sqrt{3}\}$

76. $\left\{4, \dfrac{3}{5}\right\}$ 84. $\{r + \sqrt{s}, r - \sqrt{s}\}$

77. $\left\{3, \dfrac{4}{5}\right\}$ 85. $\left\{\dfrac{2 - \sqrt{3}}{5}, \dfrac{2 + \sqrt{3}}{5}\right\}$

78. $\left\{\dfrac{1}{2}, \dfrac{2}{3}\right\}$ 86. $\left\{\dfrac{3 - \sqrt{5}}{2}, \dfrac{3 + \sqrt{5}}{2}\right\}$

79. $\{0, 4\}$

87. Demonstrate by substitution that $\dfrac{3 + \sqrt{41}}{4}$ is a member of the solution set of the equation $2x^2 - 3x - 4 = 0$.

◀In Exercises 88–96, find the solution set of each equation.

88. $y^4 - 13y^2 + 36 = 0$ 93. $x^4 + 7x^2 + 12 = 0$

89. $t^4 - 5t^2 + 4 = 0$ 94. $36r^4 - 13r^2 + 1 = 0$

90. $r^4 - 8r^2 + 15 = 0$ 95. $(x + 1) - 5\sqrt{x + 1} + 4 = 0$

91. $3\left(\dfrac{1}{m} + 1\right)^2 - \left(\dfrac{1}{m} - 1\right) = 0$ 96. $s^6 - 9s^3 + 8 = 0$

92. $2\left(\dfrac{1}{y} - 1\right)^2 - 3\left(\dfrac{1}{y} - 1\right) - 5 = 0$

97. Prove that if the quadratic equation $x^2 + mx + n = 0$ has the solution set $\{a, b\}$, then $a + b = -m$ and $ab = n$.

98. Derive the quadratic formula by solving the equation $ax^2 + bx + c = 0$ by the method of completing the square. Assume $a, b, c \in R$, $a \neq 0$, and $b^2 - 4ac \geq 0$.

99. Let

$$x_1 = \frac{-b + \sqrt{b^2 - 4ac}}{2a} \quad \text{and} \quad x_2 = \frac{-b - \sqrt{b^2 - 4ac}}{2a}.$$

Show that

$$x_1 + x_2 = -\frac{b}{a} \quad \text{and} \quad x_1 \cdot x_2 = \frac{c}{a}.$$

100. CHALLENGE PROBLEM. Show that the solution set of the equation

$$ax^2 + bx + c = 0$$

is

$$\left\{ \frac{2c}{-b + \sqrt{b^2 - 4ac}}, \frac{2c}{-b - \sqrt{b^2 - 4ac}} \right\}$$

if $a, b, c \in R$, $a \neq 0$, $c \neq 0$, and $b^2 - 4ac \geq 0$.

4.2 A NEW SET OF NUMBERS

As you have seen in Example 6, Section 4.1, if we attempt to apply the quadratic formula to the equation

$$x^2 + x + 1 = 0,$$

we obtain $\sqrt{b^2 - 4ac} = \sqrt{-3}$ and we conclude that the equation does not have a solution in the system of real numbers. Can we develop some number system that might provide us with the solutions for the quadratic equation $x^2 + x + 1 = 0$ and, indeed, with solutions for all quadratic equations? After all, irrational numbers such as $\sqrt{3}$ or π were "invented" to fulfill certain needs.

We are not free, however, to "invent" in just any way at all. What we really want to do is to form an enlargement of the real number system that will not only contain new numbers of the kind we need but also include all the real numbers. Furthermore, we want the operations of addition and multiplication in the enlarged system to have as many properties as possible of those operations in the system of real numbers.

We can approach the problem of such an enlargement either

algebraically or geometrically. At first the geometric approach may not seem fruitful because we have recently concluded that the number line is "complete," that is, without holes or spaces for new numbers. But what about the possibility of associating the numbers of our enlarged system with points in the plane?

Let us first note that, geometrically, we can represent the multiplication of a real number by -1 as a rotation through $180°$. Thus in Figure 4-1, we picture the operation of multiplying 3 by -1. Now,

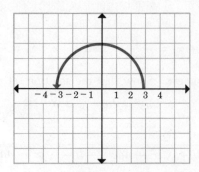

Figure 4-1

if we multiply 3 by -1 once more, we have

$$(-1)(-1)3 = 3.$$

This we illustrate in Figure 4-2. We thus have pictured geometrically

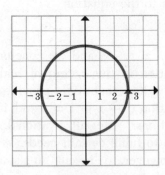

Figure 4-2

the fact that

$$(-1)(-1) = 1.$$

We have said before that there is no real number r such that $r = \sqrt{-1}$, that is, no real number r such that $r^2 = -1$. Symbolically, however, we might consider writing

$$\sqrt{-1}\sqrt{-1} = -1$$

just as $\sqrt{2}\sqrt{2} = 2$. In other words, we have

$$(\sqrt{-1})(\sqrt{-1})3 = (-1)3 = -3.$$

Thus multiplying twice by $\sqrt{-1}$ is equivalent to multiplying once by -1. This suggests that multiplying by $\sqrt{-1}$ corresponds to a 90° rotation as in Figure 4-3 which pictures the operation of multiplying 3 by $\sqrt{-1}$. Thus if point A corresponds to the number 3 on the real

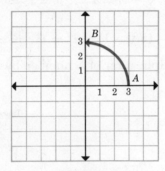

Figure 4-3

number line, then point B corresponds to

$$(\sqrt{-1})3 = 3\sqrt{-1}.$$

What about $3(\sqrt{-1})(\sqrt{-1})$? (See Figure 4-4a.) What about $3(\sqrt{-1})(\sqrt{-1})(\sqrt{-1})$? (See Figure 4-4b.) Can you give a geometric interpretation to the product $(-3)(\sqrt{-1})$?

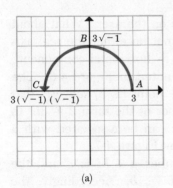

(a)

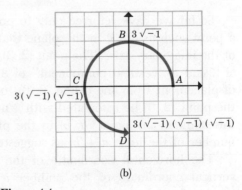

(b)

Figure 4-4

Our discussion suggests that points on the vertical axis, which we call the <u>imaginary axis</u>, correspond to numbers such as $3\sqrt{-1}$, $-2\sqrt{-1}$, $\frac{3}{2}\sqrt{-1}$, and $-\pi\sqrt{-1}$ in the same manner that the real numbers 3, -2, $\frac{3}{2}$, and $-\pi$ correspond to points on the horizontal or <u>real axis</u>. (See Figure 4-5.)

For convenience we shall use the symbol i to represent the newly invented number $\sqrt{-1}$. Thus we write $3i$ for $3\sqrt{-1}$ and $-2i$ for $-2\sqrt{-1}$. Accordingly, we have

$$(\sqrt{-1})(\sqrt{-1}) = i \cdot i = i^2 = -1$$
$$(\sqrt{-1})(\sqrt{-1})(\sqrt{-1}) = i \cdot i \cdot i = i^3 = -1 \cdot i = -i$$

and

$$(\sqrt{-1})(\sqrt{-1})(\sqrt{-1})(\sqrt{-1}) = i^4 = i^2 \cdot i^2 = (-1)(-1) = 1.$$

Do you see that multiplication of a number by i^4 maps the number into itself?

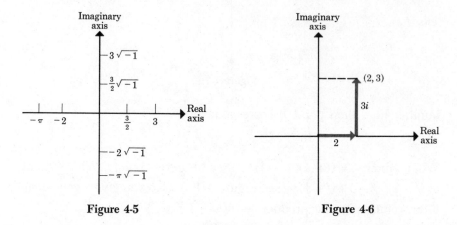

Figure 4-5   Figure 4-6

So far we have referred only to points on the axes. What about a point such as (2, 3) in the plane that is not on either the real axis or the imaginary axis? The point (2, 3) has a "horizontal component" of 2 and a "vertical component" of 3 as noted in Figure 4-6. The displacement from the origin (0, 0) to the point (2, 3) suggests that the point (2, 3) be associated with a number of the form $2 + 3i$. In this manner any point (a, b) in the plane can be associated with a number of the form $a + bi$ as suggested by Figure 4-7.

The horizontal coordinate of the number $a + bi$ is a and the vertical coordinate of the number $a + bi$ is b. Accordingly, the graphs of numbers of the form $a + 0i$ will lie on the horizontal (real)

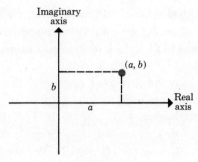

Figure 4-7

axis and the graphs of numbers of the form $0 + bi$ will lie on the vertical (imaginary) axis.

Example. Graph the numbers:
(1) $3 + 0i$, (2) $0 + 4i$, (3) $2 - 3i$, (4) $-3 + 2i$, and (5) $-4 - 3i$.

Solution: See Figure 4-8.

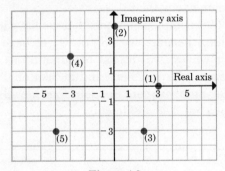

Figure 4-8

A number of the form $a + bi$ where a and b are real numbers is called a **complex number**. The plane that has a complex number assigned to each of its points is called the **complex number plane**. Sometimes the graph is called an Argand diagram in honor of J. R. Argand, one of the first mathematicians to use, in 1806, this method of diagramming complex numbers.

Since each complex number is expressed in the form $a + bi$ where a and b are real numbers, each complex number determines, and is determined by, an ordered pair (a, b) of real numbers. Also, each ordered pair (a, b) of real numbers determines, and is determined by, a point in the coordinate plane. By associating each complex number $a + bi$ with the point (a, b), we establish a one-to-one correspondence between the members of C, the set of complex numbers, and the points in the (x, y)-plane.

Since a and b may be any real numbers, the set C of complex numbers contains numbers such as $3 + 2i$, $4 - 3i$, and $\frac{2}{3} + \frac{5}{2}i$. It also contains numbers such as $\pi + 0i$, $-\frac{3}{2} + 0i$, $1 + 0i$, and $0 + 0i$. We write $\pi + 0i = \pi$, $-\frac{3}{2} + 0i = -\frac{3}{2}$, $1 + 0i = 1$, $0 + 0i = 0$, and so on. In general, $a + 0i = a$ for all $a \in R$. We see, then, that the set of all complex numbers contains the set of real numbers as a subset.

If $a = 0$, we have a number of the form $0 + bi = bi$. Numbers such as $2i$ and $-3i$ are called **pure imaginary** numbers. For any complex number $a + bi$, we call a the **real part** and b the **imaginary part.** We call i the **imaginary unit.** (The use of the word "imaginary" is rather unfortunate implying, as it does, that complex numbers do not "exist." The use of the word, however, is simply an historical legacy from a time when mathematicians were indeed uncertain about the existence of numbers such as $\sqrt{-1}$. Once a geometrical interpretation was developed for the numbers most of the doubt disappeared but the name remained!)

So far the set C of complex numbers that we have invented is rather lacking in usefulness because we do not as yet have any operations. In Section 4.3, however, we discuss the operations of addition and multiplication of complex numbers, and in Section 4.6, we show that the set of complex numbers contains solutions for all quadratic equations with real or complex coefficients.

EXERCISES 4.2

◀In Exercises 1–10, name the equations which have a nonempty solution set if the domain is the set of rational numbers.

1. $3x + 6 = 2$
2. $y^2 - 4 = 0$
3. $t^2 + 3 = 5$
4. $x^2 - 4x - 28 = 0$
5. $x^2 + x - 1 = 0$

6. $5s - 7 = 2$
7. $2x^2 - x = 15$
8. $r^2 + 5 = 0$
9. $3x^2 - 2x + 5 = 0$
10. $4y^2 - 12y + 9 = 0$

◀In Exercises 11–16, an equation is given. For each exercise indicate whether or not it has a nonempty solution set if the domain is N; if it is I; if it is

Q; if it is R. (N = set of natural numbers; I = set of integers; Q = set of rational numbers; R = set of real numbers.)

11. $2x = 3$ **14.** $s + 1 = 1$

12. $y^2 = 3$ **15.** $x^2 + 4 = 0$

13. $t + 3 = 1$ **16.** $r^2 - 3 = 0$

◀ In Exercises 17–20, two number systems are named. Name or describe a property of the first system which is not possessed by the second.

17. Integers, natural numbers

18. Rational numbers, integers

19. Real numbers, rational numbers

20. Complex numbers, real numbers

◀ In Exercises 21–31, express each number in simplest form ($n \in N$).

     Example. $i^3 = i^2 \cdot i = -1 \cdot i = -i$.

21. i^4 **27.** i^{4n+1}

22. i^7 **28.** i^{4n+2}

23. i^8 **29.** i^{4n+3}

24. i^{10} **30.** $1 + 2i - 3i^6 + 7i^2$

25. i^{65} **31.** $i^2 - i^3 - 7i + 5i^5 - i^{60}$

26. i^{4n}

◀ In Exercises 32–43, express each number in the form $a + bi$.

32. 2 **38.** $2 - i^4 + i^5$

33. $5i - i$ **39.** 0

34. $3 + i$ **40.** $16 - 4i^3$

35. π **41.** $\dfrac{3 - i}{2}$

36. $-2i$ **42.** $\dfrac{5 - 7i}{3}$

37. $4 + i^2 + 3i^3$ **43.** $\dfrac{4 - 7i}{2}$

◀ In Exercises 44–55, plot each number in a complex number plane.

44. $4 + 2i$ **47.** $6 + i$

45. $3 - 2i$ **48.** $4i$

46. 0 **49.** $-3i$

50. $-2 - 5i$ **53.** $-4 + i$

51. $-2 + 3i$ **54.** $-\dfrac{10 + 5i}{2}$

52. 3 **55.** $\dfrac{3 - 10i}{2}$

56. CHALLENGE PROBLEM.   Find a geometric interpretation for the sum of two complex numbers.

4.3   OPERATIONS ON COMPLEX NUMBERS

A set of numbers alone is useless; the operations on them make them useful. As you have seen, the set of complex numbers can be developed out of geometric considerations. The operations of addition and multiplication of complex numbers can be developed in the same manner. Rather than do this, however, we shall now give algebraic definitions that are consistent with our geometrical picture of complex numbers and, later on, return to a geometric interpretation of addition and multiplication of complex numbers.

> **Definition 4.2** The system C of complex numbers consists of the set $\{a + bi : a, b \in R\}$ under addition and multiplication where
>
> 1. $a + bi = c + di$ if and only if $a = c$ and $b = d$;
> 2. $(a + bi) + (c + di) = (a + c) + (b + d)i$; and
> 3. $(a + bi)(c + di) = ac + (bc + ad)i + bdi^2$
> $$= (ac - bd) + (bc + ad)i.$$

> Note, in 3, that the first expression, $ac + (bc + ad)i + bdi^2$, for the product $(a + bi)(c + di)$ is exactly what we would obtain by considering $a + bi$ and $c + di$ as polynomials in i. The second equality, $(ac - bd) + (bc + ad)i$, is obtained from the first by using $i^2 = -1$.

Example 1   Find the sum of $2 + 3i$ and $4 + 7i$.

Solution:

$$(2 + 3i) + (4 + 7i) = (2 + 4) + (3 + 7)i$$
$$= 6 + 10i.$$

Example 2 Find the sum of $3 + 0i$ and $6 + 0i$.
Solution:

$$(3 + 0i) + (6 + 0i) = (3 + 6) + (0 + 0)i$$
$$= 9 + 0i.$$

Example 3 Find the sum of $4 + 3i$ and $-3 - 2i$.
Solution:

$$(4 + 3i) + (-3 - 2i) = [4 + (-3)] + [3 + (-2)]i$$
$$= (4 - 3) + (3 - 2)i$$
$$= 1 + 1i = 1 + i.$$

Since the real numbers are closed under addition, a number $(a + c) + (b + d)i$, where $a, b, c, d \in R$, is itself a complex number. Hence the complex numbers are closed under addition.

Each real number a has an additive inverse $(-a)$ such that $a + (-a) = 0$. Accordingly, each complex number $a + bi$ has an additive inverse $(-a) + (-b)i$, which we may also write as $-a - bi$. As in the real number system, we subtract a complex number by adding its additive inverse.

Example 4 Subtract $-2 + 3i$ from $4 + 5i$.
Solution:

$$(4 + 5i) - (-2 + 3i) = (4 + 5i) + (2 - 3i)$$
$$= 6 + 2i.$$

The demonstration that addition of complex numbers is a commutative and associative operation is left to the student to provide as an exercise.

Example 5 Find the product $(3 + 4i)(2 + 3i)$.
Solution:

$$(3 + 4i)(2 + 3i) = (3 \cdot 2 - 4 \cdot 3) + (4 \cdot 2 + 3 \cdot 3)i$$
$$= -6 + 17i.$$

Example 6 Find the product $(3 + 4i)(1 + 0i)$.

Solution:

$$(3 + 4i)(1 + 0i) = (3 \cdot 1 - 4 \cdot 0) + (4 \cdot 1 + 3 \cdot 0)i$$
$$= 3 + 4i.$$

Example 7   Find $(\frac{1}{2} + \frac{3}{2}i)^2$.

Solution:

$$\left(\frac{1}{2} + \frac{3}{2}i\right)\left(\frac{1}{2} + \frac{3}{2}i\right) = \left(\frac{1}{4} - \frac{9}{4}\right) + \left(\frac{3}{4} + \frac{3}{4}\right)i$$
$$= -\frac{8}{4} + \frac{6}{4}i = -2 + \frac{3}{2}i.$$

Example 8   Find $(-2 + \sqrt{3}i)^2$.

Solution:

$$(-2 + \sqrt{3}i)(-2 + \sqrt{3}i) = (4 - 3) + (-2\sqrt{3} - 2\sqrt{3})i$$
$$= 1 - 4\sqrt{3}i.$$

It is worth noting again that we can obtain the same products by treating the factors as polynomials in i and then simplifying by using $i^2 = -1$. Thus, for example, we can write

$$(3 + 4i)(2 + 3i) = 6 + 17i + 12i^2 = 6 + 17i + 12(-1)$$
$$= -6 + 17i.$$

Since the real numbers are closed under multiplication and addition, a number $(ac - bd) + (bc + ad)i$, where $a, b, c, d \in R$, is itself a complex number because $ac - bd$ and $bc + ad$ are real numbers. Hence the complex numbers are closed under the operation of multiplication.

The operation of multiplication of complex numbers is commutative, associative, and distributive with respect to addition as you can demonstrate in the Exercises. The identity for multiplication is $1 + 0i = 1$. Multiplicative inverses are discussed in Section 4.5.

EXERCISES 4.3

◄In Exercises 1–10, find the sum of the two given numbers. Express the sum in the form $a + bi$ $(a, b \in R)$.

1. $3 + 2i, 4 + 5i$ 3. $2 - 3i, -3 + 5i$

2. $-3 + 2i, 4 - 5i$ 4. $4 + i, 1 + 3i$

5. $4, 5i$

6. $3 + 4i, 2i$

7. $6 + i, 7 - i$

8. $6 + 2i, -6 - 2i$

9. $\dfrac{1}{2} + \dfrac{3}{4}i, \dfrac{1}{2} + \dfrac{1}{4}i$

10. $\dfrac{2}{3} - 2i, \dfrac{1}{2} + 4i$

11-20. For Exercises 11–20, subtract the first number given in Exercises 1–10 from the second number. Express the difference in the form $a + bi$ $(a, b \in R)$.

21-30. For Exercises 21–30, find the product of each pair of numbers given in Exercises 1–10. Express each product in the form $a + bi$ $(a, b \in R)$.

◀In Exercises 31–45, write the given expression as a complex number in the form $a + bi$ $(a, b \in R)$.

31. $(1 + i)^3$

32. $(4 + 3i)(4 - 3i)$

33. $(-4 + 3i)(4 + 3i)$

34. $\left(\dfrac{1}{2} - \dfrac{3}{2}i\right)^2$

35. $\left(\dfrac{1}{2} - \dfrac{3}{2}i\right)^3$

36. $\left(\dfrac{1}{2} - \sqrt{2}\,i\right)\left(\sqrt{2} + i\right)$

37. $(2 + 3i)^3$

38. $\left(-\dfrac{1}{2} + \dfrac{\sqrt{3}}{2}i\right)^2$

39. $\left(-\dfrac{1}{2} + \dfrac{\sqrt{3}}{2}i\right)^3$

40. $\left(-\dfrac{1}{2} - \dfrac{\sqrt{3}}{2}i\right)^2$

41. $\left(-\dfrac{1}{2} - \dfrac{\sqrt{3}}{2}i\right)^3$

42. $\left(-\dfrac{1}{2} - \dfrac{\sqrt{3}}{2}i\right)\left(-\dfrac{1}{2} + \dfrac{\sqrt{3}}{2}i\right)$

43. $\left(\dfrac{\sqrt{2}}{2} + \dfrac{\sqrt{2}}{2}i\right)^2$

44. $\left(-\dfrac{\sqrt{2}}{2} + \dfrac{\sqrt{2}}{2}i\right)^2$

45. $\left(\dfrac{\sqrt{3}}{2} + \dfrac{1}{2}i\right)^3$

◀In Exercises 46–57, show that the given complex number is a member of the solution set of the given equation.

46. $x^2 + 1 = 0, i$

47. $x^2 - 2x + 5 = 0, 1 + 2i$

48. $x^2 - 2x + 5 = 0, 1 - 2i$

49. $t^2 - 6t + 10 = 0, 3 + i$

50. $r^2 - 6r + 10, 3 - i$

51. $s^2 + 10s + 34 = 0, -5 + 3i$

52. $t^2 + 10t + 34 = 0, -5 - 3i$

53. $x^2 + 2x + 37 = 0, -1 - 6i$

54. $x^2 + x + 1 = 0$, $-\dfrac{1}{2} - \dfrac{\sqrt{3}}{2}i$ **56.** $x^3 - 1 = 0$, $-\dfrac{1}{2} - \dfrac{\sqrt{3}}{2}i$

55. $2x^2 - x + 2 = 0$, $\dfrac{1}{4} - \dfrac{\sqrt{15}}{4}i$ **57.** $x^3 - x^2 + 2 = 0$, $1 - i$

◀ In Exercises 58–65, plot each number and its additive inverse in a complex plane.

58. $4 + 3i$ **62.** -4
59. $-3 + 2i$ **63.** $0 + 0i$
60. $-5 - 6i$ **64.** $-5 - 4i$
61. $5i$ **65.** $6 - 3i$

66. Determine the real numbers a and b such that
  (a) $(6 - 7i) + (a + bi) = -6 + 5i$.
  (b) $(6 - 7i) - (a + bi) = -3 + 4i$.

67. Express $i^5 + i^2 - i + i^7 - i^3 + i^4$ in the form $a + bi$ ($a, b \in R$).

68. Show that $\dfrac{\sqrt{2}}{2} + \dfrac{\sqrt{2}}{2}i$ is a square root of i.

69. Show that $-\dfrac{1}{2} + \dfrac{\sqrt{3}}{2}i$ is a cube root of 1.

70. Find real numbers x and y such that $(11 + 13i)(x + yi)$ is a real number.

71. Show that $1 + i$ and $1 - i$ are both factors of 2.

72. Factor $x^2 + 1$ using complex numbers.

73. Prove that addition of complex numbers is commutative.

74. Prove that multiplication of complex numbers is commutative.

75. Prove that addition of complex numbers is associative.

76. CHALLENGE PROBLEM. Prove that multiplication of complex numbers is associative.

77. CHALLENGE PROBLEM. Prove that the additive identity for complex numbers is unique.

78. CHALLENGE PROBLEM. Prove that the multiplicative identity for complex numbers is unique.

79. CHALLENGE PROBLEM. Prove that multiplication of complex numbers is distributive with respect to addition.

80. CHALLENGE PROBLEM. Assuming that Exercises 73–79 have been completed, prove that the complex numbers form an integral domain.

4.4   THE SQUARE ROOT OF COMPLEX NUMBERS

We have previously agreed that the symbol "$\sqrt{}$" indicates the positive square root whenever the radicand is a positive real number. Thus $\sqrt{4} = 2$ and $\sqrt{4} \neq -2$. We can write $-\sqrt{4} = -2$. Earlier we agreed that $\sqrt{-1} = i$ and $\sqrt{-1} \neq -i$. We can write, however, $-\sqrt{-1} = -i$.

We have seen that i is a solution of the equation $x^2 + 1 = 0$. A second solution is $-i$ since

$$(-i)^2 + 1 = i^2 + 1 = -1 + 1 = 0.$$

The solution set of the equation $x^2 + 1 = 0$ is, in fact, $\{i, -i\}$. The equation $x^2 + 1 = 0$ or the equivalent equation $x^2 = -1$ is a special case of the equation $x^2 = r$, where $r \in R$ and $r < 0$. But so far we have not defined expressions such as $\sqrt{-4}$ or $\sqrt{-5}$. This we now do.

> **Definition 4.3**   Let a be any real number. If $a \geq 0$, then
>
> 1. $\sqrt{a}$ is the unique nonnegative real number r such that $r^2 = a$ and
> 2. $\sqrt{-a} = i\sqrt{a}$.

Since

$$(i\sqrt{a})^2 = i^2\sqrt{a}\sqrt{a} = -1a = -a,$$

we see that $i\sqrt{a}$ is indeed a square root of $-a$. But $-i\sqrt{a}$ is also a square root of $-a$ since

$$(-i\sqrt{a})^2 = (-i)(-i)\sqrt{a}\sqrt{a} = i^2\sqrt{a}\sqrt{a} = (-1)a = -a.$$

By definition, however, we have $\sqrt{-a} = i\sqrt{a}$ and $\sqrt{-a} \neq -i\sqrt{a}$.

If $a \geq 0$ and $b \geq 0$, then $\sqrt{a}$ and $\sqrt{b}$ are real numbers and $\sqrt{a}\sqrt{b} = \sqrt{ab}$. If $a < 0$ and $b \geq 0$, then we have, by Definition 4.3,

$$\sqrt{a}\sqrt{b} = i\sqrt{-a}\sqrt{b} = i\sqrt{-(ab)}$$

since $-a > 0$ if $a < 0$. Similarly, if $a \geq 0$ and $b < 0$, we also have

$$\sqrt{a}\sqrt{b} = i\sqrt{-(ab)}.$$

Finally, if $a < 0$ and $b < 0$, we have

$$\sqrt{a}\sqrt{b} = (i\sqrt{-a})(i\sqrt{-b}) = i^2\sqrt{(-a)(-b)} = -\sqrt{ab}$$

since $-a > 0$ and $-b > 0$ if $a < 0$ and $b < 0$.

Example 1 $\sqrt{-4} = i\sqrt{4} = 2i.$

Example 2 $\sqrt{-12} = i\sqrt{12} = i\sqrt{4}\sqrt{3} = 2i\sqrt{3}.$

Example 3 $\sqrt{-\dfrac{3}{2}} = i\sqrt{\dfrac{3}{2}} = i\sqrt{\dfrac{6}{4}} = \dfrac{i\sqrt{6}}{2}.$

Example 4 $\sqrt{-3}\sqrt{6} = i\sqrt{3}\sqrt{6} = i\sqrt{18} = 3i\sqrt{2}.$

Example 5 $\sqrt{-2}\sqrt{-3} = (i\sqrt{2})(i\sqrt{3}) = -\sqrt{6}.$

Example 6 $\sqrt{-5}\sqrt{-15} = (i\sqrt{5})(i\sqrt{15})$
$$= -\sqrt{75} = -\sqrt{25}\sqrt{3} = -5\sqrt{3}.$$

Note in Example 6 that $\sqrt{-5}\sqrt{-15} \neq \sqrt{-5\cdot-15} = \sqrt{75}$. The equality $\sqrt{a}\sqrt{b} = \sqrt{ab}$ does not apply in the case when $a < 0$ and $b < 0$.

Each complex number $a + bi$ has two square roots. Suppose, for example, that we wish to find a complex number $x + yi$ with $x, y \in R$ such that $(x + yi)^2 = 5 - 12i$. We have

$$(x + yi)^2 = (x^2 - y^2) + 2xyi.$$

Since

$$(x^2 - y^2) + 2xyi = 5 - 12i,$$

we have, by (1) of Definition 4.2,

(1) $$x^2 - y^2 = 5,$$

and

(2) $$2xy = -12.$$

From (2) we see that $x \neq 0$ and hence

(3) $$y = -\frac{6}{x}.$$

Replacing y in (1) by $-\dfrac{6}{x}$, we obtain

$$x^2 - \frac{36}{x^2} = 5,$$

which gives us, after multiplying each member of the equation by x^2,

(4) $\qquad\qquad\qquad x^4 - 5x^2 - 36 = 0.$

Factoring, we transform (4) to

(5) $\qquad\qquad\qquad (x^2 - 9)(x^2 + 4) = 0.$

Hence

$$x^2 - 9 = 0 \qquad \text{or} \qquad x^2 + 4 = 0.$$

However, since $x \in R$, we cannot have $x^2 + 4 = 0$. If $x^2 - 9 = 0$, we have $x = 3$ or $x = -3$. Using (3), we get

$$x = 3 \text{ and } y = -2 \qquad \text{or} \qquad x = -3 \text{ and } y = 2.$$

Thus it would seem as if the two square roots of $5 + 12i$ were $3 - 2i$ and $-3 + 2i$. Checking, we have

$$(3 - 2i)^2 = (9 - 4) + (-6 - 6)i = 5 - 12i$$

and

$$(-3 + 2i)^2 = (9 - 4) + (-6 - 6)i = 5 - 12i.$$

We conclude that $3 - 2i$ and $-3 + 2i$ are indeed the two square roots of $5 - 12i$. Note that $3 - 2i$ is the negative of $-3 + 2i$ and vice versa.

The procedure we have used for finding the two square roots of $5 - 12i$ can be generalized to show that every nonzero complex number has two square roots which are complex numbers and that each square root is the negative of the other.

In Chapter 14, we shall develop an alternate method of finding square roots of complex numbers which is also applicable to roots of higher order.

Since we are able to find the square roots of a complex number, we can now solve equations of the form $x^2 = z_1$, where $z_1 \in C$, if we take C as the domain of the variable.

Example 7 Find the solutions of $z^2 = 5 - 12i$.

Solution: Since $(3 - 2i)^2 = 5 - 12i$ and $(-3 + 2i)^2 = 5 - 12i$, the solution set of the equation is $\{3 - 2i, -3 + 2i\}$.

EXERCISES 4.4

◀ In Exercises 1–10, write each number in the form ai $(a \in R)$. Express a in simplest form.

1. $\sqrt{-12}$ **5.** $\sqrt{-6}\sqrt{2}$ **8.** $\dfrac{\sqrt{-6}}{\sqrt{2}}$

2. $\sqrt{-8}$ **6.** $\sqrt{-\dfrac{3}{2}}$ **9.** $\dfrac{\sqrt{6}}{\sqrt{-2}}$

3. $\sqrt{-27}$ **7.** $\sqrt{-\dfrac{6}{8}}$ **10.** $\dfrac{\sqrt{-8}}{\sqrt{-2}}i$

4. $\sqrt{-128}$

◀ In Exercises 11–30, perform the indicated operations. Express your answer in the form $a + bi$ where $a, b \in R$.

11. $\sqrt{-4}\sqrt{-5}$ **21.** $\left(\dfrac{1}{2} - \sqrt{-3}\right)^2$

12. $(\sqrt{-3})^2$ **22.** $3\sqrt{-28} + \dfrac{1}{2}\sqrt{-63}$

13. $\sqrt{-8}\sqrt{-2}$ **23.** $\dfrac{3}{\sqrt{-6}}$

14. $-2(\sqrt{-2})^2$ **24.** $4\sqrt{-2\dfrac{1}{2}} - \dfrac{1}{3}\sqrt{-10}$

15. $\sqrt{-9} - \sqrt{-25}$ **25.** $\dfrac{\sqrt{-8}}{\sqrt{-2}}$

16. $\sqrt{-4} - \sqrt{-36} - i$ **26.** $\sqrt{-8} + \sqrt{-2}$

17. $3\sqrt{-100} \div \sqrt{-25}$ **27.** $4\sqrt{-32} - 3\sqrt{-8}$

18. $\sqrt{-\dfrac{3}{2}} - \sqrt{-\dfrac{2}{3}}$ **28.** $\dfrac{4}{3}\sqrt{-\dfrac{3}{2}} + \dfrac{2}{3}\sqrt{-\dfrac{2}{3}}$

19. $(\sqrt{-18})^2$ **29.** $\dfrac{1}{\sqrt{-5}} + \dfrac{2}{5}\sqrt{-5}$

20. $(1 - \sqrt{-2})^2$ **30.** $i\sqrt{-2} + i^3\sqrt{-8} - i^9\sqrt{-18} + i^{11}\sqrt{-\dfrac{1}{2}}$

◀ In Exercises 31–40, find the square roots of each complex number.

31. $4i$ **34.** $8 + 6i$

32. $-9i$ **35.** $15 + 8i$

33. -64 **36.** $3 + 4i$

37. $-5 + 12i$ **39.** $-3 - 4i$

38. $-24 + 70i$ **40.** $-21 - 20i$

◀In Exercises 41–50, find the solution set of each equation ($x, y, z \in C$).

41. $(x + 2)^2 = -9$ **46.** $x^2 = -3 + 4i$

42. $(y - 3)^2 = -25$ **47.** $x^2 = -3 - 4i$

43. $(z - i)^2 = -9$ **48.** $y^2 - 3 = 4i$

44. $(z - 2i)^2 = -36$ **49.** $y^2 + 7 = 24i$

45. $z^2 = -3i$ **50.** $z^2 + 9 = 40i$

4.5 THE FIELD OF COMPLEX NUMBERS

We have considered addition, subtraction, and multiplication of complex numbers. We now consider division. In other number systems we have studied, we defined division in terms of multiplication. We follow the same pattern in the system of complex numbers. Thus, if $a + bi$ and $c + di$ are complex numbers (a and b not both zero), we say that

$$(c + di) \div (a + bi) = x + yi \qquad \text{or} \qquad \frac{c + di}{a + bi} = x + yi$$

if and only if

(1) $(a + bi)(x + yi) = c + di.$

Such a number $x + yi$ with $x, y \in R$ exists and is unique, as we now demonstrate.

From the definition of multiplication we have

$$(a + bi)(x + yi) = (ax - by) + (bx + ay)i.$$

By (1) of Definition 4.2, the right-hand member is equal to $c + di$ if and only if

$$c = ax - by$$

and

$$d = bx + ay.$$

To solve the system of equations

$$c = ax - by$$
$$d = bx + ay$$

we can multiply each member of the first equation by a and each member of the second equation by b. We obtain the system

$$ac = a^2x - aby$$
$$bd = b^2x + aby$$

or, equivalently, by adding the first equation to the second, the system

$$ac = a^2x - aby$$
$$ac + bd = a^2x + b^2x.$$

The second equation above is equivalent to

(2)
$$x = \frac{ac + bd}{a^2 + b^2}$$

since $a^2 + b^2 \neq 0$. (Why?) Similarly, we can show that

(3)
$$y = \frac{ad - bc}{a^2 + b^2}.$$

The check that

$$(a + bi)\left(\frac{ac + bd}{a^2 + b^2} + \frac{ad - bc}{a^2 + b^2}i\right) = c + di$$

is left as an exercise.

The calculations we have just made show us that there is a unique solution, z, to the equation $z_1z = z_2$, where z_1 and z_2 are complex numbers with $z_1 \neq 0$.

Example 1   Let $z_1 = 3 + 4i$ and $z_2 = -2 - 3i$. Find the solution, z, of the equation $z_1z = z_2$.

Solution: Let $z = x + yi$. Then, by Equations (2) and (3), we have

$$x = \frac{3(-2) + 4(-3)}{9 + 16} = \frac{-6 - 12}{25} = \frac{-18}{25}$$

and

$$y = \frac{3(-3) - 4(-2)}{9 + 16} = \frac{-9 + 8}{25} = \frac{-1}{25}.$$

Therefore

$$z = -\frac{18}{25} - \frac{1}{25}i.$$

Check:

$$(3 + 4i)\left(-\frac{18}{25} - \frac{1}{25}i\right) = \left(-\frac{54}{25} + \frac{4}{25}\right) + \left(-\frac{72}{25} - \frac{3}{25}\right)i = -2 - 3i$$

A complex number $x + yi$ is called the multiplicative inverse of the complex number $a + bi$ if

$$(a + bi)(x + yi) = 1.$$

We can find x and y by using Equations (2) and (3). Since $1 = 1 + 0i$, we have $c = 1$ and $d = 0$. Hence

$$x = \frac{a}{a^2 + b^2} \quad \text{and} \quad y = \frac{-b}{a^2 + b^2}.$$

Thus the multiplicative inverse of the complex number $a + bi$ is

$$\frac{a}{a^2 + b^2} - \frac{b}{a^2 + b^2}i$$

and every nonzero complex number has a multiplicative inverse.

It is not at all necessary, however, to use a formula to find the multiplicative inverse of a complex number. Recall that we can "rationalize the denominator" of an expression such as $\dfrac{1}{2 + \sqrt{3}}$, the multiplicative inverse of $2 + \sqrt{3}$, as follows:

$$\frac{1}{2 + \sqrt{3}} \cdot \frac{2 - \sqrt{3}}{2 - \sqrt{3}} = \frac{2 - \sqrt{3}}{2^2 - (\sqrt{3})^2}$$

$$= \frac{2 - \sqrt{3}}{4 - 3} = \frac{2 - \sqrt{3}}{1} = 2 - \sqrt{3}.$$

We can use a similar process with complex numbers. Thus

$$\frac{1}{2 + \sqrt{3}i} \cdot \frac{2 - \sqrt{3}i}{2 - \sqrt{3}i} = \frac{2 - \sqrt{3}i}{2^2 - (\sqrt{3}i)^2} = \frac{2 - \sqrt{3}i}{4 - 3i^2} = \frac{2}{7} - \frac{\sqrt{3}}{7}i.$$

Complex numbers such as $2 + \sqrt{3}i$ and $2 - \sqrt{3}i$, which have the form $a + bi$ and $a - bi$, respectively, are called **conjugates** of each other. We write

$$\overline{a + bi} = a - bi.$$

Thus

$$\overline{2 + \sqrt{3}i} = 2 - \sqrt{3}i$$

and

$$\overline{2 - \sqrt{3}i} = 2 + \sqrt{3}i.$$

What is the conjugate of a complex number, $a = a + 0 \cdot i$, which is also a real number?

It is easy to see that

$$(a + bi)(a - bi) = a^2 + b^2$$

and hence the product of a complex number and its conjugate is a real number. In general, we have

$$\frac{1}{a + bi} \cdot \frac{a - bi}{a - bi} = \frac{a - bi}{a^2 + b^2} = \frac{a}{a^2 + b^2} - \frac{b}{a^2 + b^2}i.$$

Note that $\dfrac{a}{a^2 + b^2} - \dfrac{b}{a^2 + b^2}i$ is the number which we have already shown to be the multiplicative inverse of $a + bi$.

Example 2   Find the multiplicative inverse of $-3 + 7i$.

Solution:

$$\frac{1}{-3 + 7i} \cdot \frac{-3 - 7i}{-3 - 7i} = \frac{-3 - 7i}{(-3)^2 - (7i)^2} = \frac{-3 - 7i}{9 + 49} = -\frac{3}{58} - \frac{7}{58}i.$$

We can also use the complex conjugate to simplify the quotient of two complex numbers. (To "simplify" in this context means to express in the form $a + bi$ which is often referred to as the *standard form* of a complex number.)

Example 3   Express the quotient $\dfrac{3 - 4i}{1 - 3i}$ in the form $a + bi$.

Solution:

$$\frac{3 - 4i}{1 - 3i} \cdot \frac{1 + 3i}{1 + 3i} = \frac{(3 + 12) + (-4 + 9)i}{1 + 9} = \frac{15}{10} + \frac{5}{10}i = \frac{3}{2} + \frac{1}{2}i.$$

Example 4   Simplify $\dfrac{3 - 5i}{4 + 7i}$.

Solution:

$$\frac{3 - 5i}{4 + 7i} \cdot \frac{4 - 7i}{4 - 7i} = \frac{(12 - 35) + (-20 - 21)i}{16 + 49} = -\frac{23}{65} - \frac{41}{65}i.$$

Earlier it has been shown either directly in the text or indicated in the Exercises that in the system of complex numbers:

1. There is closure under addition and multiplication.
2. Addition and multiplication are commutative.
3. Addition and multiplication are associative.
4. Multiplication is distributive over addition.
5. There are unique identity elements for addition and for multiplication.
6. There is a unique additive inverse for each element.

In this section we have demonstrated that every nonzero complex number has a multiplicative inverse. We can therefore conclude that the system of complex numbers forms a field.

Although the system C of complex numbers is a field, it is not an ordered field as is the field of real numbers. We cannot define "greater than" and "less than" for the complex numbers in a manner that will not lead to a contradiction.

To demonstrate this we first repeat, for convenience, the properties of order from Section 3.3 that we shall need in our proof. These are, in terms of C,

7(i)   For every $a, b \in C$ one and only one of the following holds: $a < b, a = b, b < a$.

(O–5) For every $a, b \in C$, if $a > 0$ and $b > 0$, then $a + b > 0$ and $ab > 0$.

(O–8) For every $a \in C$, if $a < 0$, then $-a > 0$ and if $a > 0$, then $-a < 0$.

Recall also that $a < b$ if and only if $b > a$.

Now to establish that the complex numbers are not ordered we first show that $-1 < 0$ as follows:

1. Suppose $-1 > 0$.
2. If $-1 > 0$, then $(-1)(-1) = 1 > 0$. (Property O–5 with $a = b = -1$.)
3. But if $1 > 0$, then $-1 < 0$. (Property O–8).
4. We have shown that the assumption that $-1 > 0$ leads to the statement $-1 < 0$ which contradicts Property 7(i). Hence $-1 > 0$ is false.
5. Since we cannot have $-1 > 0$ and since $-1 \neq 0$, it follows by Property 7(i) that $-1 < 0$.

Now that we have $-1 < 0$ we consider $i \in C$. By 7(i) we must have $i = 0, i < 0$, or $i > 0$. Clearly $i \neq 0$. (Why?) Let us assume that $i < 0$. If $i < 0$, then $-i > 0$ by Property O–8. Since $i^2 = (-i)^2$, it then follows by Property 0–5 that

$$(-i)(-i) = i^2 = -1 > 0$$

which contradicts the result just established that $-1 < 0$. We have one possibility left. Let us assume that $i > 0$. Again by Property 0–5, we have

$$(i)(i) = i^2 = -1 > 0.$$

Since none of the three possibilities, $i = 0$, $i < 0$, $i > 0$ are permissible, we conclude that C is not an ordered field.

EXERCISES 4.5

◀In Exercises 1–15, write the conjugate of each of the following numbers. Find the sum and product of each conjugate pair.

1. $2 + i$
2. $3 + 2i$
3. $2 + 3i$
4. $3 - 5i$
5. $\sqrt{2} - 2i$
6. $-3i$
7. $1 - i$
8. $3 + 8i$

9. $5i$
10. $6 - \pi i$
11. $1 + \sqrt{3}i$
12. $2 - \sqrt{5}i$
13. $\sqrt{3} - i$
14. 7
15. 5

16–30. For Exercises 16–30, write the reciprocal, in the form $a + bi$ where $a, b \in R$, of each of the numbers given in Exercises 1–15.

◀In Exercises 31–35, find the multiplicative inverse of each given number. Express your answer in the form $a + bi$ where $a, b \in R$.

31. $1 - 3i$

32. $2 - 3i$

33. $\dfrac{1}{2} + \dfrac{1}{2}i$

34. $\dfrac{3}{2} - \dfrac{2}{3}i$

35. $\dfrac{3}{5} + \dfrac{4}{5}i$

◀In Exercises 36–45, express each quotient in the form $a + bi$ where $a, b \in R$.

36. $\dfrac{2 + 3i}{i}$

37. $\dfrac{3 + 4i}{5i}$

38. $\dfrac{3}{1 + 2i}$

42. $\dfrac{-3 + 2i}{2 - i}$

39. $\dfrac{-2i}{3 + 4i}$

43. $\dfrac{-4 + 5i}{4 - 7i}$

40. $\dfrac{6 - i}{3 - i}$

44. $\dfrac{-3 - 5i}{3 - 7i}$

41. $\dfrac{1 - i}{1 + i}$

45. $\dfrac{6 - 5i}{\sqrt{2} + i}$

◀In Exercises 46–52, perform the indicated operations. Express each answer in the form $a + bi$ where $a, b \in R$.

46. $\dfrac{1 - i}{1 + i} + \dfrac{2 - i}{-1 + 2i}$

50. $\left(\dfrac{2 - 3i}{1 + i}\right)\left(\dfrac{6 - 3i}{-2 - 5i}\right)$

47. $\dfrac{2 - 3i}{1 - i} + \dfrac{6 - 5i}{-2 - 5i}$

51. $\left(\dfrac{2 + 3i}{2 - i} + \dfrac{4 + i}{6 - i}\right)\left(\dfrac{3 - 2i}{4 + 3i} + \dfrac{1 + 2i}{3 - i}\right)$

48. $\left(\dfrac{1 - 3i}{1 + i}\right)\left(\dfrac{2 - i}{-1 + i}\right)$

52. $\left(\dfrac{1 - i}{1 + i} + \dfrac{5 - 6i}{2 - 3i}\right)\left(\dfrac{4 + 2i}{3i} + \dfrac{5i}{1 - 4i}\right)$

49. $\left(\dfrac{3 + i}{4 - 2i}\right)\left(\dfrac{4 - 3i}{8 - 3i}\right)$

53. Show that the reciprocal of i is equal to the conjugate of i.

◀In Exercises 54–58, solve each equation for x and y.

54. $(2 + 3i)(x + yi) = i$

57. $(4 - 3i)(x + yi) = 4 - 5i$

55. $(3 - i)(x + yi) = 5i$

58. $(2 + i)(x + yi) = 6 - 5i$

56. $(3 - 2i)(x + yi) = 2 - 3i$

59. Show that

$$(a + bi)\left(\frac{ac + bd}{a^2 + b^2} + \frac{ad - bc}{a^2 + b^2}i\right) = c + di.$$

60. **CHALLENGE PROBLEM.** Let z_1 and z_2 be two complex numbers. Prove that
    (a) $\overline{z_1} + \overline{z_2} = \overline{z_1 + z_2}$
    (b) $\overline{z_1} \cdot \overline{z_2} = \overline{z_1 \cdot z_2}$
    (c) $\overline{z_1} \div \overline{z_2} = \overline{z_1 \div z_2}$

4.6 QUADRATIC EQUATIONS OVER C

If r is a positive real number, then there exist two real numbers r_1 and r_2 such that $r_1^2 = r$ and $r_2^2 = r$ and if $r = 0$, then $\sqrt{r} = 0$.

Therefore we can say that real number solutions exist to the equation $ax^2 + bx + c = 0$ for $a, b, c \in R$, $a \neq 0$, and $b^2 - 4ac \geq 0$. The solutions that have been obtained are

$$\frac{-b + \sqrt{b^2 - 4ac}}{2a} \quad \text{and} \quad \frac{-b - \sqrt{b^2 - 4ac}}{2a}.$$

Now suppose that $a, b, c \in R$ as before but that we remove the restriction on $b^2 - 4ac$. We know, from Section 4.4, that complex numbers r_1 and r_2 exist such that $r_1^2 = b^2 - 4ac$ and $r_2^2 = b^2 - 4ac$ when $b^2 - 4ac < 0$.

However, will

$$\frac{-b + \sqrt{b^2 - 4ac}}{2a} \quad \text{and} \quad \frac{-b - \sqrt{b^2 - 4ac}}{2a}$$

still be the two solutions of the equation $ax^2 + bx + c = 0$? That this is so we now demonstrate.

Replacing x in $ax^2 + bx + c$ by $\dfrac{-b + \sqrt{b^2 - 4ac}}{2a}$, we have

$$a\left(\frac{-b + \sqrt{b^2 - 4ac}}{2a}\right)^2 + b\left(\frac{-b + \sqrt{b^2 - 4ac}}{2a}\right) + c =$$

$$a\left[\frac{(-b)^2 - 2b\sqrt{b^2 - 4ac} + (\sqrt{b^2 - 4ac})^2}{(2a)^2}\right]$$

$$+ b\left[\frac{-b + \sqrt{b^2 - 4ac}}{2a}\right] + c$$

$$= a\left(\frac{b^2 - 2b\sqrt{b^2 - 4ac} + b^2 - 4ac}{4a^2}\right) + b\left(\frac{-b + \sqrt{b^2 - 4ac}}{2a}\right) + c$$

(since $(\sqrt{b^2 - 4ac})^2 = b^2 - 4ac$ whether $b^2 - 4ac \geq 0$ or $b^2 - 4ac < 0$)

$$= \frac{2ab^2 - 4a^2c - 2ab\sqrt{b^2 - 4ac} - 2ab^2 + 2ab\sqrt{b^2 - 4ac} + 4a^2c}{4a^2}$$

$$= 0.$$

Hence we have shown that $\dfrac{-b + \sqrt{b^2 - 4ac}}{2a}$ is still a solution to the equation $ax^2 + bx + c = 0$ when $b^2 - 4ac < 0$.

If we replace x by $\dfrac{-b - \sqrt{b^2 - 4ac}}{2a}$ and proceed in the same

manner, we can prove that $\dfrac{-b - \sqrt{b^2 - 4ac}}{2a}$ is also a solution. Thus

we have shown that two solutions over the field of complex numbers of the equation $ax^2 + bx + c = 0$ $(a, b, c \in R, a \neq 0)$ are

$$\frac{-b + \sqrt{b^2 - 4ac}}{2a} \quad \text{and} \quad \frac{-b - \sqrt{b^2 - 4ac}}{2a},$$

when $b^2 - 4ac < 0$ as well as when $b^2 - 4ac > 0$. (What if $b^2 - 4ac = 0$?)

Moreover, these two numbers are the only solutions of the equation $ax^2 + bx + c = 0$ since a quadratic equation can have no more than two solutions. To show this, suppose that there are three distinct numbers x_1, x_2, x_3 that are solutions of the quadratic equation $ax^2 + bx + c = 0$. Then

$$ax_1^2 + bx_1 + c = ax_2^2 + bx_2 + c = ax_3^2 + bx_3 + c = 0.$$

This gives us

$$a(x_1^2 - x_2^2) = b(x_2 - x_1) \quad \text{and} \quad a(x_1^2 - x_3^2) = b(x_3 - x_1)$$

so that

$$a(x_1 - x_2)(x_1 + x_2) = -b(x_1 - x_2) \quad \text{and}$$
$$a(x_1 - x_3)(x_1 + x_3) = -b(x_1 - x_3).$$

Since $x_1 \neq x_2$, $x_1 \neq x_3$, and $a \neq 0$, we have

$$x_1 + x_2 = -\frac{b}{a} = x_1 + x_3$$

so that $x_2 = x_3$, a contradiction. Hence the quadratic equation $ax^2 + bx + c = 0$ $(a, b, c \in R)$ has the solution set

$$\left\{ \frac{-b + \sqrt{b^2 - 4ac}}{2a}, \frac{-b - \sqrt{b^2 - 4ac}}{2a} \right\}$$

over the field of complex numbers.

Example 1 Find the solution set of $x^2 - 2x + 5 = 0$.

Solution: For $x^2 - 2x + 5$, $a = 1$, $b = -2$, $c = 5$. Hence

$$x = \frac{2 + \sqrt{4 - 4(1)(5)}}{2} \quad \text{or} \quad x = \frac{2 - \sqrt{4 - 4(1)(5)}}{2}.$$

We have $x = 1 + 2i$ or $x = 1 - 2i$. The solution set is $\{1 + 2i, 1 - 2i\}$.

Example 2   Find the roots of the equation $3x^2 + x + 1 = 0$.

Solution: For $3x^2 + x + 1$, $a = 3$, $b = 1$, $c = 1$. Hence

$$x = \frac{-1 + \sqrt{1 - 4(3)(1)}}{6} \quad \text{or} \quad x = \frac{-1 - \sqrt{1 - 4(3)(1)}}{6}.$$

The roots are $x_1 = -\dfrac{1}{6} + \dfrac{\sqrt{11}}{6}i$ and $x_2 = -\dfrac{1}{6} - \dfrac{\sqrt{11}}{6}i$.

When a, b, $c \in R$, it is not necessary to find the solution set to determine whether the solutions are real numbers or complex numbers that are not real numbers. We can classify the solutions by examining the number $b^2 - 4ac$, which is called the **discriminant** of the equation $ax^2 + bx + c = 0$. There are three cases.

1. If $b^2 - 4ac > 0$, then $\sqrt{b^2 - 4ac}$ is a positive real number. The two solutions are real numbers.
2. If $b^2 - 4ac = 0$, then $\sqrt{b^2 - 4ac} = 0$. The one solution is a real number.
3. If $b^2 - 4ac < 0$, then $\sqrt{b^2 - 4ac}$ is a pure imaginary number. The two solutions are complex numbers that are not real numbers.

Example 3   Classify the solutions of the equation $2x^2 - x + 1 = 0$.

Solution: For $2x^2 - x + 1$, $a = 2$, $b = -1$, $c = 1$. Hence $b^2 - 4ac = -7$. Thus the solution set consists of two complex numbers that are not real numbers.

Example 4   Classify the solutions of $4t^2 - 36t + 81 = 0$.

Solution: For $4t^2 - 36t + 81 = 0$, $a = 4$, $b = -36$, $c = 81$. Hence $b^2 - 4ac = 0$. Thus the solution set consists of one real number.

If the two solutions of the equation $ax^2 + bx + c = 0$ are complex, nonreal numbers, which is the case when $b^2 - 4ac < 0$, the two solutions will always be of the form

$$\frac{-b}{2a} + \frac{D}{2a}i \quad \text{and} \quad \frac{-b}{2a} - \frac{D}{2a}i,$$

where $D = \sqrt{-(b^2 - 4ac)}$ is a real number. Thus the two complex solutions are complex conjugates

$$r + si \quad \text{and} \quad r - si.$$

Example 5   Find the solutions of the equation $2x^2 - 4x + 3 = 0$.
Solution: When $2x^2 - 4x + 3 = 0$, we have $a = 2$, $b = -4$, and $c = 3$.   Hence

$$x_1 = \frac{4 + \sqrt{16 - 4(2)(3)}}{4} = \frac{4 + \sqrt{-8}}{4} = \frac{4 + \sqrt{8}i}{4} = \frac{4 + 2\sqrt{2}i}{4}$$

$$= 1 + \frac{1}{2}\sqrt{2}i$$

and

$$x_2 = \frac{4 - \sqrt{16 - 4(2)(3)}}{4} = \frac{4 - \sqrt{8}i}{4} = 1 - \frac{1}{2}\sqrt{2}i$$

are the two conjugate roots.

We have shown that the solution set of the equation $ax^2 + bx + c = 0$ for $a, b, c \in R$, $a \neq 0$ is

$$\left\{ \frac{-b + \sqrt{b^2 - 4ac}}{2a}, \frac{-b - \sqrt{b^2 - 4ac}}{2a} \right\}.$$

You may ask if the equation has a nonempty solution set if $a, b, c \in C$. The demonstration that has been given for real numbers a, b, and c is valid when a, b, and c are complex numbers since (1) the system of complex numbers is closed under the operations of addition, subtraction, multiplication, and division (except by 0) and (2) every nonzero complex number has two square roots. In other words, the demonstration for real number coefficients holds for complex number coefficients since if $a, b, c \in C$, then $b^2 - 4ac \in C$ because C is a field. But, as we remarked in Section 4.4, every nonzero complex number has two square roots.   Hence

$$\frac{-b + \sqrt{b^2 - 4ac}}{2a} \in C \quad \text{and} \quad \frac{-b - \sqrt{b^2 - 4ac}}{2a} \in C$$

if $a \neq 0$, where we have not defined which square root is indicated by the symbol "$\sqrt{b^2 - 4ac}$" when $b^2 - 4ac \notin R$; we have only indicated that if one square root of $b^2 - 4ac$ is $\sqrt{b^2 - 4ac}$, the other is $- \sqrt{b^2 - 4ac}$. Every quadratic equation, whether the coefficients are real numbers or, more generally, complex numbers, does have a solution.

As we mentioned in Section 4.1, the derivation of the quadratic formula as commonly presented in a first course in algebra relies on the process of completing the square. A review of this procedure was suggested as Exercise 98 of Exercises 4.1. Here, as an alternative, we use a method that is somewhat similar to that sometimes employed in the derivation of formulas for the solution of cubic (third-degree) and quartic (fourth-degree) equations. In this method, we seek a number m so that the substitution of $y + m$ for x gives an equation of the form $y^2 = r$.

Replacing x by $y + m$ in the equation $ax^2 + bx + c = 0$ (a, b, $c \in C$, $a \neq 0$), we obtain

$$a(y + m)^2 + b(y + m) + c = 0.$$

Multiplying, we have

$$ay^2 + 2amy + am^2 + by + bm + c = 0.$$

Collecting terms, we have

$$(1) \qquad ay^2 + (2am + b)y + (am^2 + bm + c) = 0.$$

To obtain an equation of the form $y^2 = r$, $2am + b$ must be equal to zero. This will be so if $m = -\dfrac{b}{2a}$ since $2a\left(-\dfrac{b}{2a}\right) + b = 0$. Hence we let $m = -\dfrac{b}{2a}$ in (1). We have

$$ay^2 + 0y + a\left(\frac{b^2}{4a^2}\right) - \frac{b^2}{2a} + c = 0$$

or

$$ay^2 = -a\left(\frac{b^2}{4a^2}\right) + \frac{b^2}{2a} - c.$$

Dividing each term by a, we have

$$y^2 = -\frac{b^2}{4a^2} + \frac{b^2}{2a^2} - \frac{c}{a}$$

$$= -\frac{b^2}{4a^2} + \frac{2b^2}{4a^2} - \frac{4ac}{4a^2}$$

$$= \frac{b^2 - 4ac}{4a^2}.$$

Since C contains the square roots of each complex number, we have

$$y = \sqrt{\frac{b^2 - 4ac}{4a^2}} \quad \text{or} \quad y = -\sqrt{\frac{b^2 - 4ac}{4a^2}}.$$

Since

$$x = y + m = y - \frac{b}{2a},$$

we have

$$x = \sqrt{\frac{b^2 - 4ac}{4a^2}} - \frac{b}{2a} \quad \text{or} \quad x = -\sqrt{\frac{b^2 - 4ac}{4a^2}} - \frac{b}{2a}.$$

It thus seems plausible that, just as in the real number case, the solution set of $ax^2 + bx + c = 0$ for a, b, $c \in C$ is

$$\left\{ \frac{-b + \sqrt{b^2 - 4ac}}{2a}, \frac{-b - \sqrt{b^2 - 4ac}}{2a} \right\}$$

where, when $b^2 - 4ac$ is not a nonnegative real number, $\sqrt{b^2 - 4ac}$ is either one of the two complex square roots of $b^2 - 4ac$.

A complete justification for this fact lies in (1) checking that both

$$\frac{-b + \sqrt{b^2 - 4ac}}{2a} \quad \text{and} \quad \frac{-b - \sqrt{b^2 - 4ac}}{2a}$$

do indeed satisfy the equation $ax^2 + bx + c = 0$ (and this, as you should check, can be done exactly as for the case when a, b, $c \in R$) and (2) observing that a quadratic equation cannot have more than two roots. (Here check that our earlier proof of this fact over the real numbers on page 135 also holds over the complex numbers.)

Example 6   Find the solution set of $x^2 + ix + 2 = 0$.

Solution: For $x^2 + ix + 2$, $a = 1$, $b = i$, $c = 2$. Hence

$$x = \frac{-i + \sqrt{i^2 - 4(1)(2)}}{2} = \frac{-i + \sqrt{-9}}{2} = \frac{-i + 3i}{2} = i$$

or

$$x = \frac{-i - \sqrt{i^2 - 4(1)(2)}}{2} = \frac{-i - \sqrt{-9}}{2} = \frac{-i - 3i}{2} = -2i.$$

Checking, we have $i^2 + i \cdot i + 2 = 0$ and $(-2i)^2 + i(-2i) + 2 = 0$. Hence the solution set is $\{i, -2i\}$.

Note that if the coefficients of a quadratic equation are complex, nonreal numbers, the solution set does not necessarily consist of a conjugate pair.

Example 7   Form an equation whose solution set is $\{2 + 3i, -3 + 2i\}$.

Solution: If r_1 and r_2 are the roots of an equation, we know that $(x - r_1)(x - r_2) = 0$. Hence we write

$$[x-(2 + 3i)][x-(-3 + 2i)] = 0.$$

Multiplying, we obtain

$$x^2 + 3x - 2ix - 2x - 6 + 4i - 3ix - 9i - 6 = 0.$$

Collecting terms, we have

$$x^2 + (1 - 5i)x - (12 + 5i) = 0.$$

EXERCISES 4.6

◀ In Exercises 1–5, describe the solution set of each equation. (Do not solve.)

1. $x^2 + 2x + 1 = 0$ 4. $3x^2 - ix + 5 = 0$
2. $3x^2 + 2x - 5 = 0$ 5. $2x^2 + ix - 3 = 0$
3. $3x^2 - 4x + 5 = 0$

◀ In Exercises 6–15, form a quadratic equation that will have the given numbers as the members of its solution set.

6. $4, 5$ 11. $i, -i$

7. $\dfrac{2}{3}, -\dfrac{1}{2}$ 12. $2 + 5i, 2 - 5i$

8. $1 + 2i, 1 - 2i$ 13. $\dfrac{1}{3} + \dfrac{2}{3}i, \dfrac{1}{3} - \dfrac{2}{3}i$

9. $\dfrac{1}{2} - i, \dfrac{1}{2} + i$ 14. $\dfrac{2}{7} + \dfrac{5}{7}i, \dfrac{2}{7} - \dfrac{5}{7}i$

10. 3 15. $-\dfrac{1}{2} + \dfrac{\sqrt{3}}{2}i, -\dfrac{1}{2} - \dfrac{\sqrt{3}}{2}i$

◀ In Exercises 16–30, find the solution set over C of the given equation.

16. $2x^2 + x + 1 = 0$ 17. $3x^2 + 2x + 5 = 0$

18. $2t^2 - t + 1 = 0$

19. $s^2 + 4s + 5 = 0$

20. $y^2 + y + 1 = 0$

21. $3m^2 + 4m + 2 = 0$

22. $6y^2 - 3y = 8$

23. $4x^2 + 5x - 1 = 0$

24. $\dfrac{240}{x + 3} = \dfrac{240}{x} - 4$

25. $\dfrac{x^2 - 1}{x + 1} - 3 = \dfrac{1}{x + 1}$

26. $\dfrac{x(x + 1)}{3} + \dfrac{1 - x}{2} = x^2 + x$

27. $(x + 2)(x - 1) = (x + 1)^2 + x^2 - 5$

28. $\dfrac{x}{2x + 5} + \dfrac{3x}{x - 2} = \dfrac{10}{x - 2}$

29. $\sqrt{x^2 + 4} = x + 1$

30. $\sqrt{x + 8} + \sqrt{x} = 2\sqrt{1 + x}$

31. Find the solution set of $x^2 - x + 1 - i = 0$ over C.

32. Find the solution set of $y^2 + iy + 2 = 0$ over C.

33. Let z_1 and z_2 be complex numbers. If z_1 and z_2 are distinct solutions of the equation $az^2 + bz + c = 0$, show that

$$az^2 + bz + c = a(z - z_1)(z - z_2)$$

for all $z \in C$.

34. CHALLENGE PROBLEM. Find a formula relating the coefficients of a cubic equation in one variable to its solutions r_1, r_2, and r_3.

35. CHALLENGE PROBLEM. Solve the equation $z^4 = -1$ ($z \in C$).

4.7 ARGAND DIAGRAMS

The geometric representation of complex numbers by means of an Argand diagram serves a double purpose. It enables us to interpret statements about complex numbers geometrically and also to express geometric statements in terms of complex numbers.

Recall that the absolute value of a real number a gives us the undirected distance on the real number line from the origin, O, to the point representing a. If A is the point corresponding to the number a, then $|a| = |OA|$. Similarly, then, we define the **absolute value,** $|a + bi|$, of the complex number $a + bi$ to be the measure of the line segment from the origin to the point $P(a, b)$. (See Figure 4-9.) Using the distance formula, we have $|OP| = \sqrt{a^2 + b^2}$. Thus we have

$$|a + bi| = \sqrt{a^2 + b^2}.$$

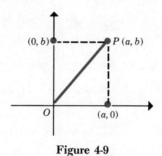

Figure 4-9

Note that if $b = 0$, then $a + bi$ is a real number and we have $|a| = \sqrt{a^2}$, which is consistent with our previous definition of the absolute value of a real number.

Addition of complex numbers can be interpreted geometrically. For example, consider

$$(3 + i) + (1 + 2i) = 4 + 3i.$$

Each of the three complex numbers is graphed in Figure 4-10. If we draw line segments from the origin to the points $3 + i$ and $1 + 2i$ and then from these two points to the point $4 + 3i$, we have a parallelogram as can be easily demonstrated by determining the slopes of the four lines containing the sides of the quadrilateral $ABCD$.

In general, we have $(a + bi) + (c + di) = (a + c) + (b + d)i$, and if we represent these three numbers by the three points $A(a, b)$, $B(c, d)$, and $C(a + c, b + d)$, and draw the four segments $\overline{OA}$, $\overline{AC}$, $\overline{CB}$, and $\overline{OB}$, we have a parallelogram as shown in Figure 4-11.

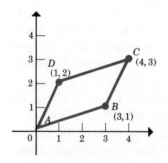

Figure 4-10

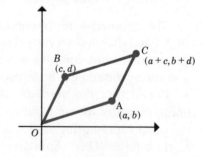

Figure 4-11

To interpret subtraction of complex numbers geometrically, we use the additive inverse. We know that

$$z_1 - z_2 = z_1 + (-z_2) \qquad (z_1, z_2 \in C).$$

Hence we draw a parallelogram using z_1 and $-z_2$ as two vertices. (See Figure 4-12.)

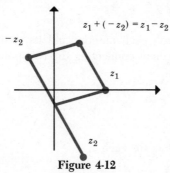

Figure 4-12

The use of the "parallelogram law" for the addition of two complex numbers adds geometrical significance to complex numbers and helps dispel any feelings that these numbers are less "real" than other kinds of numbers.

We could now describe geometric constructions for the product and quotient of complex numbers, but will defer this until Chapter 14 when, with the aid of trigonometry, we can provide a simple and elegant geometric interpretation.

EXERCISES 4.7

◀In Exercises 1–10, graph the given number in the complex number plane.

1. 1

2. -2

3. $-i$

4. $2i$

5. $3 - i$

6. $3 + 5i$

7. $\sqrt{2} - i$

8. $\pi + 2i$

9. $\dfrac{1}{1 + i}$

10. $\dfrac{2 + i}{3 - i}$

11-20. In Exercises 11–20, graph the additive inverse of each number given in Exercises 1–10. Draw a straight line segment connecting the point designating each given number and the point designating its additive inverse.

◀In Exercises 21–25, find $z_1 + z_2$ and $z_1 - z_2$. Illustrate graphically each sum and difference.

21. $z_1 = 1 + i, z_2 = 2 + i$ **24.** $z_1 = -3 + 4i, z_2 = -1 - 3i$

22. $z_1 = 3 + 2i, z_2 = 2 + 3i$ **25.** $z_1 = -2i, z_2 = 2 - 4i$

23. $z_1 = -1 + 2i, z_2 = 2 - i$

◀In Exercises 26–30, graph each number and its conjugate. Then find the midpoint of the line segment connecting the two points.

26. $2 + 2i$ **29.** $3 - 2i$

27. $-2 + 3i$ **30.** $-2i$

28. $-3 - 2i$

◀In Exercises 31–38, find $|z|$ for each z.

31. $z = 3 - 4i$ **35.** $z = i^4 + i^7$

32. $z = 2 + 5i$ **36.** $z = -3 + 2i$

33. $z = 5 - 2i$ **37.** $z = -2i$

34. $z = 1 + i^5$ **38.** $z = 6$

39. Let $z_1 = 4 + 3i$ and $z_2 = -1 + 5i$. Show algebraically and geometrically that

$$|z_1| + |z_2| > |z_1 + z_2|.$$

40. Let $z_1 = -3 + 2i$ and $z_2 = 4 - 3i$. Show algebraically and geometrically that

$$|z_1| + |z_2| > |z_1 + z_2|.$$

41. The statement $|z_1| + |z_2| \geq |z_1 + z_2|$ is known as the triangle inequality. Give a geometric interpretation to $|z_1| + |z_2| \geq |z_1 + z_2|$. Why is it necessary to write "$\geq$" rather than "$>$"?

42. Let $z_1 = 4 + 3i$ and $z_2 = -1 + 5i$. Show algebraically that $|z_1| |z_2| = |z_1 \cdot z_2|$.

43. Show that $z \cdot \overline{z} = |z|^2$ for all complex numbers z.

44. **CHALLENGE PROBLEM.** Prove algebraically that $|z_1| |z_2| = |z_1 \cdot z_2|$ for all complex numbers z_1 and z_2.

45. **CHALLENGE PROBLEM.** Prove that the triangle with vertices $0, 1, z$ $(z \in C)$ is similar to the triangle with vertices $0, z, z^2$. Use the result to describe a geometric construction for z^2.

4.8 REPRESENTATION OF COMPLEX NUMBERS AS ORDERED PAIRS

You are quite familiar with the concept of an ordered pair. Ordered pairs (a, b), for example, are used to represent points in a plane. Frequently, rational numbers, which we generally write in the form $\frac{a}{b}$, are represented as ordered pairs (a, b). Indeed we could have defined the system of complex numbers using the concept of ordered pairs in such a way that it would have been unnecessary to mention i or the square roots of negative numbers. Let us do so now, calling the system C_1 to distinguish it from the system C we have already discussed.

> **Definition 4.4** The system C_1 of complex numbers consists of all ordered pairs (a, b), where a and b are real numbers, together with the following agreements:
>
> 1. $(a, b) = (c, d)$ if and only if $a = c$ and $b = d$.
> 2. $(a, b) + (c, d) = (a + c, b + d)$.
> 3. $(a, b) \cdot (c, d) = (ac - bd, ad + bc)$.

This new system C_1, as defined, has all the properties of the system C we have considered in earlier sections. For example, the real numbers can be pictured as a subset of C_1; this subset together with the operations of addition and multiplication forms a field, and C_1 contains solutions for all equations of the form

$$ax^2 + bx + c = 0,$$

where $a, b, c \in C_1$.

Thus any real number a can be considered as $(a, 0)$ in C_1. If $(a, 0)$ and $(b, 0)$ represent any two real numbers, we have

(1) $$(a, 0) + (b, 0) = (a + b, 0 + 0) = (a + b, 0)$$

and

(2) $$(a, 0) \cdot (b, 0) = (ab - 0 \cdot 0, a \cdot 0 + 0 \cdot b)$$
$$= (ab, 0).$$

Hence the sum and the product of two real numbers are what we expect. Other properties of the real numbers from this point of view can be readily confirmed.

Example 1 Show that $(3, 0)$ is related to a root of the equation $2x + 1 = 7$.

Solution: In C_1 we write $2x + 1 = 7$ as $(2, 0)x + (1, 0) = (7, 0)$. We replace x by $(3, 0)$. We have

$$(2, 0) \cdot (3, 0) + (1, 0) = (6, 0) + (1, 0) = (7, 0)$$

by Definition 4.4.

Not only can the set of real numbers be pictured as a subset of C_1 but also pure imaginary numbers, such as $3i$, $-i$, and $-\frac{2}{3}i$, can be pictured as elements of a subset of C_1. Specifically, any pure imaginary number bi can be considered as $(0, b)$ in C_1. Thus if $(0, a)$ and $(0, b)$ represent any two pure imaginary numbers, we have

$$(3) \qquad (0, a) + (0, b) = (0 + 0, a + b) = (0, a + b)$$

and

$$(4) \qquad (0, a) \cdot (0, b) = (0 \cdot 0 - ab, 0 \cdot b + a \cdot 0) = (-ab, 0).$$

Note that the product of two pure imaginaries is a real number. Most importantly,

$$(0, 1) \cdot (0, 1) = (0, 1)^2 = (-1, 0)$$

corresponding to

$$i \cdot i = i^2 = -1.$$

Example 2 Show that $(0, 1)$ is related to a root of the equation $x^2 + 1 = 0$.

Solution: In C_1, we write $x^2 + 1 = 0$ as

$$(1, 0)x^2 + (1, 0) = (0, 0).$$

We replace x by $(0, 1)$ and obtain, for $(1, 0)x^2 + (1, 0)$,

$$(1, 0) \cdot (0, 1) \cdot (0, 1) + (1, 0).$$

By Definition 4.4, we have

$$(1 \cdot 0 - 0 \cdot 1, 1 \cdot 1 + 0 \cdot 0) \cdot (0, 1) + (1, 0) = (0, 1) \cdot (0, 1) + (1, 0)$$
$$= (-1, 0) + (1, 0).$$

Then we have

$$(-1, 0) + (1, 0) = (0, 0)$$

so that $(0, 1)$ is a solution of the equation

$$(1, 0)x^2 + (1, 0) = (0, 0)$$

just as i is a solution of the equation $x^2 + 1 = 0$.

We can show that the system C_1 forms a field by displaying a proof of each field property. For example, to show that multiplication is commutative, we show that

$$(a, b) \cdot (c, d) = (c, d) \cdot (a, b) \qquad \text{for all } a, b, c, d \in R.$$

Thus we consider

$$(a, b) \cdot (c, d) = (ac - bd, ad + bc)$$

and

$$(c, d) \cdot (a, b) = (ca - db, cb + da).$$

Since addition and multiplication of real numbers are commutative and associative, we have

$$(ac - bd, ad + bc) = (ca - db, cb + da).$$

You can easily determine the additive identity and the multiplicative identity in C_1 and then the additive inverse for any element (a, b) in C_1. The multiplicative inverse for (a, b) is $\left(\dfrac{a}{a^2 + b^2}, \dfrac{-b}{a^2 + b^2}\right)$ if $(a, b) \neq (0, 0)$.

We suggest that the system C_1 contains solutions to all quadratic equations by the following examples.

Example 3   Show that $(2, -1)$ is related to a solution of the equation $x^2 - 4x + 5 = 0$.

Solution: In C_1 we write $x^2 - 4x + 5 = 0$ as

$$x^2 - (4, 0)x + (5, 0) = (0, 0)$$

and check that

$$(2, -1)^2 - (4, 0) \cdot (2, -1) + (5, 0) = (3, -4) - (8, -4) + (5, 0)$$
$$= (0, 0).$$

Example 4   Show that $(0, -2)$ is a solution of $x^2 + (0, 1)x + (2, 0) = (0, 0)$.

Solution: We have

$$(0, -2)^2 + (0, 1) \cdot (0, -2) + (2, 0) = (-4, 0) + (2, 0) + (2, 0)$$
$$= (0, 0).$$

Although we have not investigated all the details, we have seen that there is a significant correspondence between the system C of complex numbers and the system C_1. Not only is there a one-to-one correspondence between the elements of C and C_1 described by

$$a + bi \in C \longleftrightarrow (a, b) \in C_1$$

but also the correspondence is "preserved" under the operations of addition and multiplication. Thus we have

$$2 + 3i \in C \longleftrightarrow (2, 3) \in C_1$$

and

$$1 + 2i \in C \longleftrightarrow (1, 2) \in C_1.$$

The sum $(2 + 3i) + (1 + 2i) = 3 + 5i$ and the sum $(2, 3) + (1, 2) = (3, 5)$. Since

$$3 + 5i \in C \longleftrightarrow (3, 5) \in C_1,$$

we have

$$(2 + 3i) + (1 + 2i) \in C \longleftrightarrow (2, 3) + (1, 2) \in C_1.$$

Also the product $(2 + 3i)(1 + 2i) = -4 + 7i$ and the product $(2, 3) \cdot (1, 2) = (-4, 7)$. Again, since

$$-4 + 7i \in C \longleftrightarrow (-4, 7) \in C_1,$$

we have

$$(2 + 3i)(1 + 2i) \in C \longleftrightarrow (2, 3) \cdot (1, 2) \in C_1.$$

In other words, C and C_1 have identical algebraic structures. Another way of expressing this is to say that C and C_1 are isomorphic.

> **Definition 4.5** Two fields are **isomorphic** if there is a one-to-one correspondence between the elements of the two fields that is preserved under addition and multiplication.

If two fields are isomorphic, their structures are the same; essentially they differ at most in terminology and notation.

EXERCISES 4.8

◀ In Exercises 1–5, express each given complex number as an ordered pair of real numbers. The ordered pair is to be a member of the set of numbers in C_1.

1. $3 + 2i$ 4. $-3i$

2. $-4 - i$ 5. $-\dfrac{1}{2} + \dfrac{3}{2}i$

3. 4

◀ In Exercises 6–10, express each ordered pair as a complex number in standard form.

6. $(3, -2)$ **9.** $\left(5, \dfrac{1}{2}\right)$

7. $(3, 0)$ **10.** $(0, 0)$

8. $(0, \sqrt{2})$

◀ In Exercises 11–15, find the sum and product of each pair of numbers in C_1.

11. $(3, 4), (2, -3)$ **14.** $(0, 2), (3, 0)$

12. $(4, 5), (6, 7)$ **15.** $\left(\dfrac{1}{2}, \dfrac{3}{2}\right), \left(\dfrac{2}{3}, \dfrac{1}{4}\right)$

13. $(-1, 4), (-3, 4)$

◀ In Exercises 16–23, express each of the following products of elements of C_1 in the form (a, b), $a, b \in R$.

16. $(0, 1)^2$ **20.** $(0, 1)^{41}$

17. $(0, 1)^3$ **21.** $(0, 1)^{4n}$ $(n \in N)$

18. $(0, 1)^4$ **22.** $(0, 1)^{4n+1}$ $(n \in N)$

19. $(0, 1)^7$ **23.** $(0, 1)^{4n+2}$ $(n \in N)$

24. In C_1 express in simplest form

$$(0, 1)^3 - (0, 1)^7 + (0, 1)^4 - (0, 1)^6.$$

◀ In Exercises 25–29, give the multiplicative inverse of each of the given numbers in C_1. Demonstrate that the product of each number and its multiplicative inverse is $(1, 0)$.

25. $(2, 3)$ **28.** $\left(2, \dfrac{2}{3}\right)$

26. $(1, 1)$ **29.** $(1, \sqrt{2})$

27. $\left(\dfrac{1}{2}, -1\right)$

◀ In Exercises 30–38, show that the given number is a solution of the given equation. If a coefficient is not indicated, assume it is $(1, 0)$; for example, $x^2 + (1, 0) = (0, 0)$ is

$$(1, 0)x^2 + (1, 0) = (0, 0).$$

30. $(0, 1)$, $x^2 + (1, 0) = (0, 0)$

31. $(1, 2)$, $x^2 - (2, 0)x + (5, 0) = (0, 0)$

32. $(3, 1)$, $y^2 - (6, 0)y + (10, 0) = (0, 0)$

33. $(1, -2)$, $t^2 - (2, 0)t = (-5, 0)$

34. $(3, 0)$, $x^2 - x = (6, 0)$

35. $(2, 0)$, $(2, 0)r^2 + (2, 0) = (5, 0)r$

36. $(0, 1)$, $y^2 + (0, 1)y + (2, 0) = (0, 0)$

37. $\left(\dfrac{3}{2}, \dfrac{\sqrt{11}}{2}\right)$, $m^2 + (-3, 0)m = (-5, 0)$

38. $\left(\dfrac{1}{2}, -\dfrac{\sqrt{3}}{2}\right)$, $x^3 + (1, 0) = (0, 0)$

39. In Exercise 37, one solution of a quadratic equation is given. Find a second solution.

40. In Exercise 38, one solution of a cubic equation is given. Find the other two solutions.

41. Show that addition in C_1 is commutative.

42. Show that the additive identity, $(0, 0)$, in C_1 is unique.

43. Show that C_1 is closed under multiplication.

44. **CHALLENGE PROBLEM.** Show that multiplication in C_1 is associative.

45. **CHALLENGE PROBLEM.** Show that C_1 is a field. (Assume that Exercises 41–44 have been completed.)

4.9   MATRIX REPRESENTATION OF COMPLEX NUMBERS

You are probably familiar with the system of 2×2 matrices and the operations of addition and multiplication that are defined in this system. Thus a 2×2 matrix is an array

$$\begin{pmatrix} a & b \\ c & d \end{pmatrix} \qquad (a, b, c, d \in R)$$

with addition and multiplication defined as follows:

$$\begin{pmatrix} a & b \\ c & d \end{pmatrix} + \begin{pmatrix} e & f \\ g & h \end{pmatrix} = \begin{pmatrix} a + e & b + f \\ c + g & d + h \end{pmatrix}$$

$$\begin{pmatrix} a & b \\ c & d \end{pmatrix} \cdot \begin{pmatrix} e & f \\ g & h \end{pmatrix} = \begin{pmatrix} ae + bg & af + bh \\ ce + dg & cf + dh \end{pmatrix}$$

and equality by

$$\begin{pmatrix} a & b \\ c & d \end{pmatrix} = \begin{pmatrix} e & f \\ g & h \end{pmatrix}$$

if and only if $a = e$, $b = f$, $c = g$, and $d = h$.

Example 1

$$\begin{pmatrix} 2 & 0 \\ -1 & 1 \end{pmatrix} + \begin{pmatrix} 1 & -2 \\ 3 & 4 \end{pmatrix} = \begin{pmatrix} 3 & -2 \\ 2 & 5 \end{pmatrix}$$

$$\begin{pmatrix} 2 & 0 \\ -1 & 1 \end{pmatrix} \cdot \begin{pmatrix} 1 & -2 \\ 3 & 4 \end{pmatrix} = \begin{pmatrix} 2+0 & -4+0 \\ -1+3 & 2+4 \end{pmatrix} = \begin{pmatrix} 2 & -4 \\ 2 & 6 \end{pmatrix}$$

Using 2×2 matrices and the operations of addition and multiplication just defined, a second system can be defined that is also isomorphic to C.

> **Definition 4.6.** The system C_2 consists of all matrices of the form
>
> $$\begin{pmatrix} a & b \\ -b & a \end{pmatrix} \qquad (a, b \in R)$$
>
> together with the operations of addition and multiplication as defined above.

As in C and C_1, the real numbers can be considered as a subset of C_2. The real number a corresponds to the matrix $\begin{pmatrix} a & 0 \\ 0 & a \end{pmatrix}$ in C_2. If $\begin{pmatrix} a & 0 \\ 0 & a \end{pmatrix}$ and $\begin{pmatrix} b & 0 \\ 0 & b \end{pmatrix}$ correspond to the real numbers a and b, respectively, we have

$$\begin{pmatrix} a & 0 \\ 0 & a \end{pmatrix} + \begin{pmatrix} b & 0 \\ 0 & b \end{pmatrix} = \begin{pmatrix} a+b & 0 \\ 0 & a+b \end{pmatrix}$$

and

$$\begin{pmatrix} a & 0 \\ 0 & a \end{pmatrix} \cdot \begin{pmatrix} b & 0 \\ 0 & b \end{pmatrix} = \begin{pmatrix} ab & 0 \\ 0 & ab \end{pmatrix}.$$

Again the set of pure imaginary numbers can be considered a subset of C_2 just as it was considered a subset of C and of C_1. The imaginary number bi corresponds to the matrix $\begin{pmatrix} 0 & b \\ -b & 0 \end{pmatrix}$ in C_2. If $\begin{pmatrix} 0 & a \\ -a & 0 \end{pmatrix}$ and $\begin{pmatrix} 0 & b \\ -b & 0 \end{pmatrix}$ correspond to the imaginary numbers ai and bi, respectively,

we have

$$\begin{pmatrix} 0 & a \\ -a & 0 \end{pmatrix} + \begin{pmatrix} 0 & b \\ -b & 0 \end{pmatrix} = \begin{pmatrix} 0 & a+b \\ -(a+b) & 0 \end{pmatrix}$$

corresponding to $ai + bi = (a+b)i$ and

$$\begin{pmatrix} 0 & a \\ -a & 0 \end{pmatrix} \cdot \begin{pmatrix} 0 & b \\ -b & 0 \end{pmatrix} = \begin{pmatrix} -(ab) & 0 \\ 0 & -(ab) \end{pmatrix}$$

corresponding to $(ai) \cdot (bi) = -(ab)$.

Most importantly, we have

$$\begin{pmatrix} 0 & 1 \\ -1 & 0 \end{pmatrix} \cdot \begin{pmatrix} 0 & 1 \\ -1 & 0 \end{pmatrix} = \begin{pmatrix} -1 & 0 \\ 0 & -1 \end{pmatrix}$$

corresponding to $i \cdot i = -1$.

The matrix

$$\begin{pmatrix} 0 & 0 \\ 0 & 0 \end{pmatrix}$$

is the additive identity for the system since

$$\begin{pmatrix} a & b \\ -b & a \end{pmatrix} + \begin{pmatrix} 0 & 0 \\ 0 & 0 \end{pmatrix} = \begin{pmatrix} a & b \\ -b & a \end{pmatrix}$$

and

$$\begin{pmatrix} 1 & 0 \\ 0 & 1 \end{pmatrix}$$

is the multiplicative identity since

$$\begin{pmatrix} a & b \\ -b & a \end{pmatrix} \cdot \begin{pmatrix} 1 & 0 \\ 0 & 1 \end{pmatrix} = \begin{pmatrix} a+0 & 0+b \\ -b+0 & 0+a \end{pmatrix} = \begin{pmatrix} 1 & 0 \\ 0 & 1 \end{pmatrix} \cdot \begin{pmatrix} a & b \\ -b & a \end{pmatrix}$$

$$= \begin{pmatrix} a & b \\ -b & a \end{pmatrix}.$$

Each matrix $\begin{pmatrix} a & b \\ -b & a \end{pmatrix}$ in C_2 has an additive inverse, $\begin{pmatrix} -a & -b \\ b & -a \end{pmatrix}$, in C_2 since

$$\begin{pmatrix} a & b \\ -b & a \end{pmatrix} + \begin{pmatrix} -a & -b \\ b & -a \end{pmatrix} = \begin{pmatrix} 0 & 0 \\ 0 & 0 \end{pmatrix}.$$

We can now define subtraction in the usual manner. We have

$$\begin{pmatrix} a & b \\ -b & a \end{pmatrix} - \begin{pmatrix} c & d \\ -d & c \end{pmatrix} = \begin{pmatrix} a & b \\ -b & a \end{pmatrix} + \begin{pmatrix} -c & -d \\ d & -c \end{pmatrix}$$

$$= \begin{pmatrix} a - c & b - d \\ -b + d & a - c \end{pmatrix}$$

since $\begin{pmatrix} -c & -d \\ d & -c \end{pmatrix}$ is the additive inverse of $\begin{pmatrix} c & d \\ -d & c \end{pmatrix}$.

We note that the multiplicative inverse of $\begin{pmatrix} a & b \\ -b & a \end{pmatrix}$ is

$$\begin{pmatrix} \dfrac{a}{a^2 + b^2} & \dfrac{-b}{a^2 + b^2} \\ \dfrac{b}{a^2 + b^2} & \dfrac{a}{a^2 + b^2} \end{pmatrix} \qquad \text{(if not both } a \text{ and } b \text{ are zero)}$$

since

$$\begin{pmatrix} a & b \\ -b & a \end{pmatrix} \cdot \begin{pmatrix} \dfrac{a}{a^2 + b^2} & \dfrac{-b}{a^2 + b^2} \\ \dfrac{b}{a^2 + b^2} & \dfrac{a}{a^2 + b^2} \end{pmatrix} = \begin{pmatrix} \dfrac{a^2 + b^2}{a^2 + b^2} & \dfrac{0}{a^2 + b^2} \\ \dfrac{0}{a^2 + b^2} & \dfrac{a^2 + b^2}{a^2 + b^2} \end{pmatrix}$$

$$= \begin{pmatrix} 1 & 0 \\ 0 & 1 \end{pmatrix}.$$

Although not all parts of the proof have been given (more will be asked for in the Exercises), it can be proved that the set of matrices of the form $\begin{pmatrix} a & b \\ -b & a \end{pmatrix}$ $(a, b \in R)$ together with the operations of addition and multiplication form a field.

Other properties of C_2 similar to those of C can also be considered. Recall that the solution set of the equation $x^2 - x - 6 = 0$ is $\{-2, 3\}$.

Example 2   Show that $\begin{pmatrix} -2 & 0 \\ 0 & -2 \end{pmatrix}$ is related to a solution of the equation $x^2 - x - 6 = 0$.

Solution: In C_2, we write

$$X^2 - X - \begin{pmatrix} 6 & 0 \\ 0 & 6 \end{pmatrix} = \begin{pmatrix} 0 & 0 \\ 0 & 0 \end{pmatrix}$$

where X is a matrix. Replacing X by $\begin{pmatrix} -2 & 0 \\ 0 & -2 \end{pmatrix}$, we have

$$\begin{pmatrix} -2 & 0 \\ 0 & -2 \end{pmatrix} \cdot \begin{pmatrix} -2 & 0 \\ 0 & -2 \end{pmatrix} - \begin{pmatrix} -2 & 0 \\ 0 & -2 \end{pmatrix} - \begin{pmatrix} 6 & 0 \\ 0 & 6 \end{pmatrix}$$

$$= \begin{pmatrix} 4 & 0 \\ 0 & 4 \end{pmatrix} - \begin{pmatrix} -2 & 0 \\ 0 & -2 \end{pmatrix} - \begin{pmatrix} 6 & 0 \\ 0 & 6 \end{pmatrix} = \begin{pmatrix} 0 & 0 \\ 0 & 0 \end{pmatrix}.$$

Finally, it can be shown that the system C_2 contains solutions to quadratic equations over C_2. We suggest this by the following examples.

Example 3 Show that $\begin{pmatrix} 2 & -1 \\ 1 & 2 \end{pmatrix}$ is a solution of the equation

$$X^2 - \begin{pmatrix} 4 & 0 \\ 0 & 4 \end{pmatrix} X + \begin{pmatrix} 5 & 0 \\ 0 & 5 \end{pmatrix} = \begin{pmatrix} 0 & 0 \\ 0 & 0 \end{pmatrix},$$

where X is a matrix.

Solution: We have

$$\begin{pmatrix} 2 & -1 \\ 1 & 2 \end{pmatrix} \cdot \begin{pmatrix} 2 & -1 \\ 1 & 2 \end{pmatrix} - \begin{pmatrix} 4 & 0 \\ 0 & 4 \end{pmatrix} \cdot \begin{pmatrix} 2 & -1 \\ 1 & 2 \end{pmatrix} + \begin{pmatrix} 5 & 0 \\ 0 & 5 \end{pmatrix}$$

$$= \begin{pmatrix} 3 & -4 \\ 4 & 3 \end{pmatrix} - \begin{pmatrix} 8 & -4 \\ 4 & 8 \end{pmatrix} + \begin{pmatrix} 5 & 0 \\ 0 & 5 \end{pmatrix} = \begin{pmatrix} 0 & 0 \\ 0 & 0 \end{pmatrix}.$$

Example 4 Show that $\begin{pmatrix} 0 & -2 \\ 2 & 0 \end{pmatrix}$ is a solution of the equation

$$X^2 + \begin{pmatrix} 0 & 1 \\ -1 & 0 \end{pmatrix} X + \begin{pmatrix} 2 & 0 \\ 0 & 2 \end{pmatrix} = \begin{pmatrix} 0 & 0 \\ 0 & 0 \end{pmatrix},$$

where X is a matrix.

Solution: We have

$$\begin{pmatrix} 0 & -2 \\ 2 & 0 \end{pmatrix} \cdot \begin{pmatrix} 0 & -2 \\ 2 & 0 \end{pmatrix} + \begin{pmatrix} 0 & 1 \\ -1 & 0 \end{pmatrix} \cdot \begin{pmatrix} 0 & -2 \\ 2 & 0 \end{pmatrix} + \begin{pmatrix} 2 & 0 \\ 0 & 2 \end{pmatrix}$$

$$= \begin{pmatrix} -4 & 0 \\ 0 & -4 \end{pmatrix} + \begin{pmatrix} 2 & 0 \\ 0 & 2 \end{pmatrix} + \begin{pmatrix} 2 & 0 \\ 0 & 2 \end{pmatrix} = \begin{pmatrix} 0 & 0 \\ 0 & 0 \end{pmatrix}.$$

The correspondence that relates C and C_2 is given by

$$a + bi \in C \longleftrightarrow \begin{pmatrix} a & b \\ -b & a \end{pmatrix} \in C_2,$$

where, in particular, as we noted earlier,

$$i \longleftrightarrow \begin{pmatrix} 0 & 1 \\ -1 & 0 \end{pmatrix}.$$

This correspondence is "preserved" under addition and multiplication. Thus

$$a + bi \in C \longleftrightarrow \begin{pmatrix} a & b \\ -b & a \end{pmatrix} \in C_2$$

and

$$c + di \in C \longleftrightarrow \begin{pmatrix} c & d \\ -d & c \end{pmatrix} \in C_2.$$

Under addition we have

$$(a + bi) + (c + di)$$

$$= (a + c) + (b + d)i \in C \longleftrightarrow \begin{pmatrix} a + c & b + d \\ -(b + d) & a + c \end{pmatrix}$$

$$= \begin{pmatrix} a & b \\ -b & a \end{pmatrix} + \begin{pmatrix} c & d \\ -d & c \end{pmatrix} \in C_2.$$

Under multiplication we have

$$(a + bi)(c + di)$$

$$= (ac - bd) + (bc + ad)i \in C \longleftrightarrow \begin{pmatrix} ac - bd & bc + ad \\ -(bc + ad) & ac - bd \end{pmatrix}$$

$$= \begin{pmatrix} a & b \\ -b & a \end{pmatrix} \cdot \begin{pmatrix} c & d \\ -d & c \end{pmatrix} \in C_2,$$

a generalization of our previous observation that

$$i \cdot i = -1 \longleftrightarrow \begin{pmatrix} -1 & 0 \\ 0 & -1 \end{pmatrix} = \begin{pmatrix} 0 & 1 \\ -1 & 0 \end{pmatrix} \cdot \begin{pmatrix} 0 & 1 \\ -1 & 0 \end{pmatrix}.$$

Hence the field C of complex numbers is isomorphic to the system of matrices of the form $\begin{pmatrix} a & b \\ -b & a \end{pmatrix}$ where $a, b \in R$.

EXERCISES 4.9

◀ In Exercises 1–5, show the correspondent in C of each given matrix in C_2.

1. $\begin{pmatrix} 1 & 2 \\ -2 & 1 \end{pmatrix}$ 4. $\begin{pmatrix} 3 & 0 \\ 0 & 3 \end{pmatrix}$

2. $\begin{pmatrix} 2 & -3 \\ 3 & 2 \end{pmatrix}$ 5. $\begin{pmatrix} 0 & -2 \\ 2 & 0 \end{pmatrix}$

3. $\begin{pmatrix} 4 & 1 \\ -1 & 4 \end{pmatrix}$

◀ In Exercises 6–10, show the correspondent matrix in C_2 of each given number in C.

6. 4 9. $-4 - i$

7. $-3i$ 10. $-\dfrac{1}{2} - \dfrac{3}{2}i$

8. $3 + 2i$

◀ In Exercises 11–15, find the sum of the two given matrices.

11. $\begin{pmatrix} 2 & -1 \\ 1 & 2 \end{pmatrix}, \begin{pmatrix} 3 & 1 \\ -1 & 3 \end{pmatrix}$ 14. $\begin{pmatrix} 3 & 2 \\ -2 & 3 \end{pmatrix}, \begin{pmatrix} 3 & -2 \\ 2 & 3 \end{pmatrix}$

12. $\begin{pmatrix} 1 & -3 \\ 3 & 1 \end{pmatrix}, \begin{pmatrix} 2 & 1 \\ 1 & 2 \end{pmatrix}$ 15. $\begin{pmatrix} \dfrac{1}{2} & \dfrac{1}{3} \\ -\dfrac{1}{3} & \dfrac{1}{2} \end{pmatrix} \begin{pmatrix} \dfrac{1}{2} & -\dfrac{1}{3} \\ \dfrac{1}{3} & \dfrac{1}{2} \end{pmatrix}$

13. $\begin{pmatrix} 6 & 2 \\ -2 & 6 \end{pmatrix}, \begin{pmatrix} 0 & 1 \\ -1 & 0 \end{pmatrix}$

◀ In Exercises 16–20, express each product as a matrix of the form $\begin{pmatrix} a & b \\ -b & a \end{pmatrix}$ for $a, b \in R$.

16. $\begin{pmatrix} 0 & 1 \\ -1 & 0 \end{pmatrix}^2$ 18. $\begin{pmatrix} 0 & 1 \\ -1 & 0 \end{pmatrix}^7$ 20. $\begin{pmatrix} 0 & 1 \\ -1 & 0 \end{pmatrix}^{100}$

17. $\begin{pmatrix} 0 & 1 \\ -1 & 0 \end{pmatrix}^5$ 19. $\begin{pmatrix} 0 & 1 \\ -1 & 0 \end{pmatrix}^9$

◀ In Exercises 21–25, find the product $A \cdot B$.

21. $A = \begin{pmatrix} 2 & -1 \\ 1 & 2 \end{pmatrix}, \quad B = \begin{pmatrix} 3 & 1 \\ -1 & 3 \end{pmatrix}$

22. $A = \begin{pmatrix} 1 & -1 \\ 1 & 1 \end{pmatrix}, \quad B = \begin{pmatrix} 4 & 2 \\ -2 & 4 \end{pmatrix}$

23. $A = \begin{pmatrix} 3 & 2 \\ -2 & 3 \end{pmatrix}$, $B = \begin{pmatrix} 5 & -4 \\ 4 & 5 \end{pmatrix}$

24. $A = \begin{pmatrix} 6 & 2 \\ -2 & 6 \end{pmatrix}$, $B = \begin{pmatrix} 0 & 1 \\ -1 & 0 \end{pmatrix}$

25. $A = \begin{pmatrix} -\dfrac{1}{2} & \dfrac{\sqrt{3}}{2} \\[2ex] -\dfrac{\sqrt{3}}{2} & -\dfrac{1}{2} \end{pmatrix}$, $B = \begin{pmatrix} -\dfrac{1}{2} & \dfrac{\sqrt{3}}{2} \\[2ex] \dfrac{\sqrt{3}}{2} & -\dfrac{1}{2} \end{pmatrix}$

26-30. Find the product $B \cdot A$ using the matrices given in Exercises 21–25.

31. Is it true that matrix multiplication is commutative? Test your conclusion by finding the products $A \cdot B$ and $B \cdot A$ where

$$A = \begin{pmatrix} 1 & -2 \\ 3 & 4 \end{pmatrix} \quad \text{and} \quad B = \begin{pmatrix} 2 & 4 \\ -6 & 1 \end{pmatrix}.$$

◀In Exercises 32–34, give the multiplicative inverse of each matrix. Demonstrate that the product of the given matrix and its multiplicative inverse is the multiplicative identity matrix.

32. $\begin{pmatrix} 2 & 1 \\ -1 & 2 \end{pmatrix}$ **33.** $\begin{pmatrix} 3 & 4 \\ -4 & 3 \end{pmatrix}$ **34.** $\begin{pmatrix} 12 & -5 \\ 5 & 12 \end{pmatrix}$

◀In Exercises 35–39, show that the given matrix is a solution of the given equation ($X \in C_2$).

35. $\begin{pmatrix} 0 & 1 \\ -1 & 0 \end{pmatrix}$, $X^2 + \begin{pmatrix} 1 & 0 \\ 0 & 1 \end{pmatrix} = \begin{pmatrix} 0 & 0 \\ 0 & 0 \end{pmatrix}$

36. $\begin{pmatrix} 3 & 0 \\ 0 & 3 \end{pmatrix}$, $X^2 - X - \begin{pmatrix} 6 & 0 \\ 0 & 6 \end{pmatrix} = \begin{pmatrix} 0 & 0 \\ 0 & 0 \end{pmatrix}$

37. $\begin{pmatrix} 1 & 2 \\ -2 & 1 \end{pmatrix}$, $X^2 - \begin{pmatrix} 2 & 0 \\ 0 & 2 \end{pmatrix}X + \begin{pmatrix} 5 & 0 \\ 0 & 5 \end{pmatrix} = \begin{pmatrix} 0 & 0 \\ 0 & 0 \end{pmatrix}$

38. $\begin{pmatrix} 1 & -2 \\ 2 & 1 \end{pmatrix}$, $X^2 - \begin{pmatrix} 2 & 0 \\ 0 & 2 \end{pmatrix}X + \begin{pmatrix} 5 & 0 \\ 0 & 5 \end{pmatrix} = \begin{pmatrix} 0 & 0 \\ 0 & 0 \end{pmatrix}$

39. $\begin{pmatrix} 1 & -1 \\ 1 & 1 \end{pmatrix}$, $X^3 - X^2 + \begin{pmatrix} 2 & 0 \\ 0 & 2 \end{pmatrix} = \begin{pmatrix} 0 & 0 \\ 0 & 0 \end{pmatrix}$

40. Show that $A = \begin{pmatrix} \dfrac{\sqrt{2}}{2} & \dfrac{\sqrt{2}}{2} \\[2ex] -\dfrac{\sqrt{2}}{2} & \dfrac{\sqrt{2}}{2} \end{pmatrix}$ is a square root of $B = \begin{pmatrix} 0 & 1 \\ -1 & 0 \end{pmatrix}$, that is, show that $A^2 = B$.

41. Show that addition in C_2 is commutative.

42. Show that multiplication in C_2 is commutative.

43. Show that addition in C_2 is associative.

44. CHALLENGE PROBLEM. Find the square root of $\begin{pmatrix} 5 & 12 \\ -12 & 5 \end{pmatrix}$.

45. CHALLENGE PROBLEM. Show that multiplication in C_2 is associative.

46. CHALLENGE PROBLEM. Assuming that Exercises 41–45 have been completed, show that C_2 is a field.

CHAPTER SUMMARY

The system C of COMPLEX NUMBERS consists of a set of numbers of the form $a + bi$, where a and b are real numbers and the agreements that:

1. $a + bi = c + di$ if and only if $a = c$, $b = d$.
2. $(a + bi) + (c + di) = (a + c) + (b + d)i$.
3. $(a + bi)(c + di) = (ac - bd) + (ad + bc)i$.

The system C contains as a subset the set of real numbers. Any real number a is expressed as $a + 0i$. The system C is a field but is not an ordered field.

Every quadratic equation $ax^2 + bx + c = 0$ (a, b, $c \subset C$) has solutions

$$\frac{-b + \sqrt{b^2 - 4ac}}{2a} \quad \text{and} \quad \frac{-b - \sqrt{b^2 - 4ac}}{2a}$$

in C if $a \neq 0$.

Complex numbers can be graphed on the COMPLEX PLANE. Each complex number $a + bi$ determines an ordered pair (a, b), which, in turn, determines a point $P(a, b)$. The undirected distance of a point $P(a, b)$ from the origin is $\sqrt{a^2 + b^2}$. We say that $|a + bi| = \sqrt{a^2 + b^2}$ and call $|a + bi|$ the ABSOLUTE VALUE of the complex number $a + bi$.

The COMPLEX CONJUGATE of a number $a + bi$ is $a - bi$. We write $\overline{a + bi} = a - bi$ and $\overline{a - bi} = a + bi$.

The system of complex numbers can also be defined as a system C_1 of ordered pairs (a, b) of real numbers a and b with the following definitions of addition and multiplication.

$$(a, b) + (c, d) = (a + c, b + d),$$

$$(a, b) \cdot (c, d) = (ac - bd, ad + bc).$$

The two fields C and C_1 are said to be ISOMORPHIC. Two fields are isomorphic if the elements of one can be placed into one-to-one correspondence with the elements of the other and if the correspondence is preserved under addition and multiplication.

The system of 2×2 MATRICES of the form $\begin{pmatrix} a & b \\ -b & a \end{pmatrix}$ where $a, b \in R$ is also isomorphic to the system C of complex numbers.

REVIEW EXERCISES

1. Write the given expression as a complex number of the form $a + bi$ $(a, b \in R)$.

   (a) $i^3 + i^2 - 7i$ (b) $(2 - 3i)^2$

2. Perform the indicated operation and write the product in the form $a + bi$ $(a, b \in R)$.

$$\left(-\frac{1}{2} + \frac{\sqrt{3}}{2}i \right)^3$$

3. Simplify each of the following expressions:

   (a) $\sqrt{-3}\sqrt{-12}$ (b) $\sqrt{-\dfrac{3}{2}} + \sqrt{-\dfrac{2}{3}}$

4. Show that $1 - i$ is a solution of the equation $x^3 - x^2 + 2 = 0$.

5. If $z = 2 - 3i$, find $-z$, $\bar{z}$, $|z|$, $\dfrac{1}{z}$, and $\dfrac{4 + 5i}{z}$. Express each answer in the form $a + bi$ $(a, b \in R)$.

6. Find the solution set of the equation $2x^2 - x + 2 = 0$.

7. Check your answer to Exercise 6.

8. Find three cube roots of 1 by using the equation $x^3 - 1 = 0$. (*Hint:* Factor $x^3 - 1 = 0$ first.)

9. Express $\dfrac{3 - 2i}{1 + 2i}$ in the form $a + bi$ $(a, b \in R)$.

10. Find the multiplicative inverse of $3 - 4i$.

11. Illustrate, by graphing, the sum and difference of $-3 - 4i$ and $-1 - 5i$.

12. One root of a quadratic equation with real coefficients is $3 - 2i$. Find the other root and write a quadratic equation that will have the two numbers as roots.

13. Prove that

$$|a + bi|^2 = (a + bi)\overline{(a + bi)}.$$

14. Determine k so that the equation $x^2 + kx + 3 = 0$ has a single root.

15. Show that $(1, \frac{3}{2})$ is a solution of the equation $(4, 0)x^2 - (8, 0)x + (7,0) = (0, 0)$ over C_1.

16. Show that $\begin{pmatrix} 3 & 2 \\ -2 & 3 \end{pmatrix}$ is a solution of the equation

$$X^2 - \begin{pmatrix} 6 & 0 \\ 0 & 6 \end{pmatrix} X - \begin{pmatrix} -13 & 0 \\ 0 & -13 \end{pmatrix} = \begin{pmatrix} 0 & 0 \\ 0 & 0 \end{pmatrix}$$

over C_2.

17. Find the square root of $-16 + 30i$.

18. Prove that

$$2|z_1|^2 + 2|z_2|^2 = |z_1 + z_2|^2 + |z_1 - z_2|^2$$

for $z_1, z_2 \in C$.

19. Form an equation whose solution set is

$$\{4 + i, 3 - i\}.$$

20. Prove that $\overline{z_1 z_2} = \overline{z_1}\,\overline{z_2}$ for z_1 and $z_2 \in C$.

GOING FURTHER: READING AND RESEARCH

We have seen that it was necessary to "enlarge" the real number system in order to obtain solutions for equations such as $x^2 + 1 = 0$. We know now that the system C of complex numbers does contain solutions for every quadratic equation $ax^2 + bx + c = 0$ where $a, b, c \in C$, and $a \neq 0$.

What about equations such as

$$y^3 - y^2 + y - 1 = 0$$

or

$$3t^5 - 4t^4 + 2t^3 - t^2 + 6t + 4 = 0?$$

Is it necessary to extend the complex number system to have solutions to these equations?

Of course, there is an even more fundamental question. Is it even possible to enlarge the complex number system? When we moved off the real number line and utilized the plane, we were able to develop a new number system. Can we now move off the plane and somehow or other use space to develop a new number system? We have used ordered pairs of real numbers in C_1. What about ordered triples or ordered quadruples?

These and other questions challenged William R. Hamilton, the great Irish mathematician and one of the most precocious geniuses of all time, during the last century. He made some remarkable discoveries. If you are interested in finding out about these, information in the following books will be helpful.

BELL, E. T. *Men Of Mathematics*, (Chapter 19), New York: Simon and Schuster, 1937.

DUBISCH, R. *Introduction To Abstract Algebra*, New York: John Wiley & Sons, 1965.

BELL, E. T. *Mathematics, Queen And Servant Of Science*, New York: McGraw-Hill, 1968.

EVES, H. AND C. W. NEWSOM. *An Introduction To The Foundations And Fundamental Concepts Of Mathematics*, New York: Holt, Rinehart and Winston, Rev. Edition, 1964.

NEWMAN, J. R. *The World Of Mathematics*, Vol. 1, New York: Simon and Schuster, 1968.

CHAPTER FIVE
POLYNOMIAL FUNCTIONS

5.1 POLYNOMIALS

You have already had considerable experience in working with polynomials both in your previous mathematics courses and in the first four chapters of this text. In this section we shall review briefly the operations of addition and multiplication of polynomials, but, more significantly, we shall develop the idea that the system of polynomials under these operations does constitute an integral domain as defined in Chapter 3.

You will recall that the sum of two polynomials is obtained by applying the Distributive, Associative, and Commutative Properties in an appropriate way. For example, if $A = 3x^2 + 4x + 5$ and if $B = 5x^2 + 2x + 7$, then the sum

$$\begin{aligned} A + B &= (3x^2 + 4x + 5) + (5x^2 + 2x + 7) \\ &= (3x^2 + 5x^2) + (4x + 2x) + (5 + 7) \\ &= (3 + 5)x^2 + (4 + 2)x + (5 + 7) \\ &= 8x^2 + 6x + 12. \end{aligned}$$

In practice, of course, it is not necessary to include all these steps. On the other hand, it is important to understand the basic principles behind the process.

As a second example, check that

$$(5y^4 + 6y^2 + 2y - 8) + (y^3 + 3y^2 - 5y + 10)$$
$$= 5y^4 + y^3 + 9y^2 - 3y + 2.$$

Multiplication of polynomials also makes use of the Distributive, Associative, and Commutative Properties in a natural way as illustrated below. To find, for example, the product $A \cdot B$, where $A = x + 2$ and $B = x^2 + 3x + 5$, we can write

$$A \cdot B = (x + 2)(x^2 + 3x + 5).$$

By the Distributive Property this becomes

$$x(x^2 + 3x + 5) + 2(x^2 + 3x + 5)$$
$$= (x^3 + 3x^2 + 5x) + (2x^2 + 6x + 10).$$

Adding, as above, we get

$$A \cdot B = x^3 + 5x^2 + 11x + 10.$$

Frequently, a schematic device is used in the multiplication of polynomials for multiplying $y^3 + 2y^2 + 1$ by $y^2 - 2y + 4$:

$$
\begin{array}{l}
y^3 + 2y^2 \qquad\quad\, + 1 \\
\qquad\;\; y^2 - 2y\; + 4 \\
\hline
y^5 + 2y^4 \qquad\quad + y^2 \\
\quad\; - 2y^4 - 4y^3 \qquad\quad\;\; - 2y \\
\qquad\qquad + 4y^3 + 8y^2 \qquad\quad\; + 4 \\
\hline
y^5 \qquad\qquad\qquad\;\; + 9y^2 - 2y + 4
\end{array}
$$

Observe here the use of the familiar property of exponents which states that

$$y^m \cdot y^n = y^{m+n}$$

when m and n are any natural numbers. You will also note that in the above scheme certain open spaces have been left to accommodate the "missing" powers.

EXERCISES 5.1A (Oral)

◀In Exercises 1–12, find the sum of each of the pairs of polynomials.

1. $3x^2 + 5x - 2,\ 2x^2 - x + 7$
2. $4y^3 - y^2 + 1,\ y - 2$
3. $11x^4 - x^2 + 5x,\ x^5$
4. $t^3 - t^2 + 5t - 2,\ 8$
5. $7z^5 - 2z^2 + 3z - 6,\ 11z^4 + 6z^3$
6. $8s^3 - 2s^2 + 14s - 11,\ 2s^3 - 6s^2 + s + 14$
7. $x^7 - 13x^5 + x^3 - x,\ 3x^6 + x^4 - x^2 + 5$
8. $4y^5 + 12y^3 - y^2 + y,\ 8y^4 + 8y^3 + y^2 - y$
9. $12x^4 - 11x^3 + 21x^2 - x,\ x^5 + 8x^3 - 17x + 4$
10. $x^8 + 7x^4 - x + 12,\ x^7 - 12x^6 + x^5 - 2$
11. $7u^3 + 2u^2 - 6u + 1,\ -7u^3 - 2u^2 + 6u - 1$
12. $x^4 - 13x^3 + 12x^2 - x + 4,\ -x^4 + 13x^3 - 12x^2 + x - 4$

◀In Exercises 13–24, find the product of each of the pairs of polynomials.

13. $x + 2,\ x - 2$
14. $y - 7,\ y + 7$
15. $3z + 12,\ z^2$
16. $x^3,\ 2x^2 - x + 5$
17. $x + 5,\ x - 5$
18. $t - 11,\ t + 11$
19. $s + 10,\ s + 10$
20. $u + 7,\ u - 8.$
21. $2v + 5,\ 3v - 6.$
22. $3y - 1,\ 2y - 3.$
23. $x^2 - 1,\ x^2 + 1.$
24. $x^3 - 1,\ x^3 + 1.$

EXERCISES 5.1B (*Written*)

◄Find the product of each of the following pairs of polynomials.

1. $x^2 + 3x - 6$, $x^2 - 4x + 2$
2. $2x^2 + x - 3$, $x^2 - x + 6$
3. $4y^2 - 7y$, $y^2 - 12$
4. $3s^3 + 6s$, $s - 7$
5. $t^3 + t - 1$, $t^2 - 3t$
6. $5x^2 - 6x + 12$, $6x^2 + 5x - 12$
7. $u^3 + u^2 + u + 1$, $u - 1$
8. $y^4 - y^3 + y^2 - y + 1$, $y + 1$
9. $v^5 + v^4 + v^3 + v^2 + 1$, $v - 1$
10. $x^6 - x^5 + x^4 - x^3 + x^2 - x + 1$, $x + 1$
11. $x^2 - 3x + 9$, $x + 3$.
12. $t^4 - 2t^3 + 4t^2 - 8t + 16$, $t + 2$

The "opposite" process to multiplication is factoring. Study the following examples of factorization of polynomials as a review of this process.

$$2xy^2 - 6x^2 = 2x(y^2 - 3x)$$
$$5ar^2 + 10ar + 15a = 5a(r^2 + 2r + 3)$$
$$4x^2 - 9y^2 = (2x + 3y)(2x - 3y)$$
$$x^4 - 1 = (x^2 + 1)(x^2 - 1) = (x^2 + 1)(x - 1)(x + 1)$$
$$x^2 + 5x + 6 = (x + 3)(x + 2)$$
$$2x^2 - 3x - 5 = (2x - 5)(x + 1)$$
$$4x^2 + 12xy + 9y^2 = (2x + 3y)^2$$
$$ax - ay + bx - by = a(x - y) + b(x - y) = (a + b)(x - y)$$

Note that the numerical coefficients of the polynomials in these examples are all integers. That is, we are considering polynomials and their factorization *over the integers*. Other kinds of polynomials and factorization are possible as, for example,

$$\sqrt{2}\,x + \sqrt{2} = \sqrt{2}(x + 1),$$
$$2x + 1 = 2\left(x + \frac{1}{2}\right),$$
$$x^2 - 3 = (x + \sqrt{3})(x - \sqrt{3}),$$

and

$$x^2 + 1 = (x + i)(x - i).$$

Polynomials and polynomial factorization over the integers, however, are the ones most commonly encountered.

EXERCISES 5.1C (Written)

◀ Factor each of the following polynomials over the integers.

1. $x^2 + 4x + 4$
2. $x^2 - 18x + 81$
3. $x^2 - 64$
4. $81x^2 - 4$
5. $32 - 2y^2$
6. $5s^2 - 125$
7. $4x^2 - 2x - 6$
8. $4x^2 - 10x - 6$
9. $x^2 - xy + xz - yz$
10. $4p + 4q - ap - aq$
11. $16x^4 - 1$

12. $81y^4 - 1$
13. $x^2 + 6x + 9$
14. $y^2 - 20y + 100$
15. $t^2 - 24t + 144$
16. $4x^2 + 12x + 9$
17. $9z^2 - 24z + 16$
18. $u^2 + 7u + 12$
19. $x^2 - 64$
20. $x^2 - 144$
21. $81y^2 - 4$

We now wish to show that the set of polynomials of the type we have been considering is, under addition and multiplication, an integral domain. To accomplish this we shall need a formal definition of the operations $+$ and $\cdot$ on polynomials. Let

$$A = a_0 + a_1 x + a_2 x^2 + \cdots + a_n x^n$$

and

$$B = b_0 + b_1 x + b_2 x^2 + \cdots + b_m x^m$$

where the coefficients

$$a_0, a_1, a_2, \ldots, a_n, b_0, b_1, b_2, \ldots, b_m$$

are elements of some fixed integral domain. Thus they can be rational numbers, or real numbers, or complex numbers as well as integers. They can also be elements from the domain $B = \{0, 1, 2, 3, \ldots, 10, *, \circ\}$ of Section 3.2, a specialized domain as such.

If $n = m$, we define the sum $A + B$ to be

$$(a_0 + b_0) + (a_1 + b_1)x + (a_2 + b_2)x^2 + \cdots + (a_m + b_m)x^m.$$

For example,

$$(1 + 4x + 5x^2) + (2 + 3x + 7x^2) = (1 + 2) + (4 + 3)x + (5 + 7)x^2$$
$$= 3 + 7x + 12x^2.$$

If $n > m$, we define $A + B$ as

$$(a_0 + b_0) + (a_1 + b_1)x + (a_2 + b_2)x^2 + \cdots + (a_m + b_m)x^m$$
$$+ a_{m+1}x^{m+1} + \cdots + a_nx^n.$$

For example,

$$(1 + 4x + 5x^2) + (2 + 3x) = (1 + 2) + (4 + 3)x + 5x^2$$
$$= 3 + 7x + 5x^2.$$

Finally, if $n < m$, $A + B$ can be defined in a similar way, which we leave for the student to do.

The product $A \cdot B$ is defined to be

$$(a_0b_0) + (a_0b_1 + a_1b_0)x + (a_0b_2 + a_1b_1 + a_2b_0)x^2$$
$$+ (a_0b_3 + a_1b_2 + a_2b_1 + a_3b_0)x^3 + \cdots$$
$$+ (a_{n-1}b_m + a_nb_{m-1})x^{(n+m)-1} + a_nb_mx^{n+m},$$

where the sum of the subscripts in each term of the coefficients of x^k is k and where k runs from 0 to $m + n$. (Recall that $x^0 = 1$ if $x \neq 0$.) Thus, for example, for the coefficient,

$$a_0b_3 + a_1b_2 + a_2b_1 + a_3b_0,$$

of x^3 we have

$$0 + 3 = 1 + 2 = 2 + 1 = 3 + 0 = 3.$$

You can see that the formal definition of a sum leads to the same results as the informal definition implied in our examples. To see how the formal definition of multiplication of polynomials relates to the procedures used for multiplication in our examples, examine carefully the following schematic multiplication:

$$
\begin{array}{l}
a_0 \quad + a_1x \quad + a_2x^2 \quad + a_3x^3 \\
\qquad\quad b_0 \quad + b_1x \quad + b_2x^2 \\
\hline
a_0b_0 + a_1b_0x + a_2b_0x^2 + a_3b_0x^3 \\
\qquad\quad a_0b_1x + a_1b_1x^2 + a_2b_1x^3 + a_3b_1x^4 \\
\qquad\qquad\qquad a_0b_2x^2 + a_1b_2x^3 + a_2b_2x^4 + a_3b_2x^5 \\
\hline
a_0b_0 + (a_1b_0 + a_0b_1)x + (a_2b_0 + a_1b_1 + a_0b_2)x^2 \\
\qquad\quad + (a_3b_0 + a_2b_1 + a_1b_2)x^3 + (a_3b_1 + a_2b_2)x^4 + a_3b_2x^5
\end{array}
$$

We now show that under these definitions of addition and multiplication the set P of all polynomials with coefficients from some fixed integral domain D is itself an integral domain.

As a first step we shall need a careful definition of what is meant by the statement that two polynomials are equal. We say that two polynomials are equal if the corresponding coefficients of each of the powers of x are respectively equal. Thus, for example,

$$a_0 + a_1 x + a_2 x^2 + a_3 x^3 = b_0 + b_1 x + b_2 x^2 + b_3 x^3$$

if and only if

$$a_0 = b_0, \qquad a_1 = b_1, \qquad a_2 = b_2, \qquad \text{and} \qquad a_3 = b_3.$$

To proceed with the verification that the set P of all polynomials is an integral domain we would normally begin by showing that the Associative, Commutative, and Distributive Properties hold for both operations. We shall not take the time to do this, however, because the detailed verifications are rather cumbersome and because these properties are so closely related to the corresponding properties of the integral domain D. For example, if we compare $A \cdot B$ with $B \cdot A$, we see that the coefficient of x in $B \cdot A$ is $b_0 a_1 + b_1 a_0$. But this is equal to the coefficient of x, $a_0 b_1 + a_1 b_0$, in $A \cdot B$ because of the fact that both addition and multiplication are commutative operations in D.

With regard to closure, it should be evident from the form of the sum and product in the definitions that the sum and product of two polynomials over D are both polynomials over D since we know that the integral domain D is closed under both addition and multiplication.

We shall now consider the question of identities. Given the polynomial $Z = 0$, where 0 is the additive identity of D, for any polynomial

$$A = a_0 + a_1 x + \cdots + a_n x^n,$$

it is easy to see that

$$A + Z = (a_0 + 0) + a_1 x + \cdots + a_n x^n = A.$$

The polynomial Z, called the **zero polynomial**, is clearly the additive identity of P.

It is equally easy to see that the polynomial $I = 1$, where 1 is the multiplicative identity of D, is the multiplicative identity for P.

For the additive inverse of

$$A = a_0 + a_1 x + \cdots + a_n x^n,$$

we have

$$-A = (-a_0) + (-a_1)x + \cdots + (-a_n)x^n,$$

where $-a_0, -a_1, \ldots, -a_n$ are the additive inverses in D of a_0, $a_1, \ldots, a_n$, respectively. From the definition of addition it is apparent that $A + (-A) = 0$.

The job is virtually done. We have only to investigate the Cancellation Property. That is, we wish to show that if A, B, and C are polynomials over D such that $A \cdot B = A \cdot C$ with $A \neq 0$, then $B = C$. If $B = C = 0$, the conclusion is obvious. Suppose, then, that for

$$A = a_0 + a_1 x + a_2 x^2 + \cdots + a_n x^n \qquad (a_n \neq 0),$$
$$B = b_0 + b_1 x + b_2 x^2 + \cdots + b_m x^m \qquad (b_m \neq 0),$$

and

$$C = c_0 + c_1 x + c_2 x^2 + \cdots + c_r x^r \qquad (c_r \neq 0),$$

we have

$$A \cdot B = a_0 b_0 + (a_0 b_1 + a_1 b_0)x + (a_0 b_2 + a_1 b_1 + a_2 b_0)x^2$$
$$+ (a_0 b_3 + a_1 b_2 + a_2 b_1 + a_3 b_0)x^3 + \cdots$$
$$+ (a_{n-1} b_m + a_n b_{m-1})x^{(n+m)-1} + a_n b_m x^{n+m}$$

equal to

$$A \cdot C = a_0 c_0 + (a_0 c_1 + a_1 c_0)x + (a_0 c_2 + a_1 c_1 + a_2 c_0)x^2$$
$$+ (a_0 c_3 + a_1 c_2 + a_2 c_1 + a_3 c_0)x^3 + \cdots$$
$$+ (a_{n-1} c_r + a_n c_{r-1})x^{(n+r)-1} + a_n c_r x^{n+r}.$$

By definition of equality of polynomials it then follows that

$$a_0 b_0 = a_0 c_0$$
$$a_0 b_1 + a_1 b_0 = a_0 c_1 + a_1 c_0,$$
$$a_0 b_2 + a_1 b_1 + a_2 b_0 = a_0 c_2 + a_1 c_1 + a_2 c_0,$$
$$a_0 b_3 + a_1 b_2 + a_2 b_1 + a_3 b_0 = a_0 c_3 + a_1 c_2 + a_2 c_1 + a_3 c_0,$$
$$\cdots$$
$$a_{n-1} b_m + a_n b_{m-1} = a_{n-1} c_r + a_n c_{r-1},$$
$$a_n b_m = a_n a_r.$$

Now, since the a's, b's, and c's are elements of an integral domain and an integral domain possesses the Cancellation Property, we have

$$a_0 b_0 = a_0 c_0 \qquad \text{implies} \qquad b_0 = c_0 \qquad \text{if } a_0 \neq 0.$$

But then, from $b_0 = c_0$ and

$$a_0 b_1 + a_1 b_0 = a_0 c_1 + a_1 c_0,$$

we get

$$a_0b_1 + a_1b_0 = a_0c_1 + a_1b_0,$$
$$a_0b_1 = a_0c_1,$$

and, finally, $b_1 = c_1$ if $a_0 \neq 0$.

In our discussion thus far we have assumed that $a_0 \neq 0$. But what if $a_0 = 0$? If $a_0 = 0$, we certainly have $a_0b_0 = a_0c_0$. Then our second equality

$$a_0b_1 + a_1b_0 = a_0c_1 + a_1c_0$$

becomes

$$a_1b_0 = a_1c_0$$

and if $a_1 \neq 0$, we again get $b_0 = c_0$ and continue. If, however, we also have $a_1 = 0$, we use our third equality to get

$$a_2b_0 = a_2c_0$$

and get $b_0 = c_0$ if $a_2 \neq 0$ and continue! In other words, we eventually get $b_0 = c_0$ and continue as before unless all of the a's are 0. But not all the a's are 0 since, by hypothesis, $a_n \neq 0$.

The complete details are somewhat cumbersome, but it should seem reasonable that we can, indeed, continue coefficient by coefficient to get

$$b_0 = c_0, \qquad b_1 = c_1, \qquad b_2 = c_2, \ldots, b_m = c_r.$$

Because we must have $b_0 = c_0$, $b_1 = c_1$, $b_2 = c_2$, and so on, if it were true that $m < r$, we would need to have, in this matching of coefficients, $c_r = 0$, a contradiction. Similarly, if it were true that $r < m$, we would have $b_m = 0$, again a contradiction. Hence $m = r$ and $b_m = c_r$ can be written

$$b_m = c_m \qquad \text{or} \qquad b_r = c_r.$$

Thus $B = C$ and we conclude that the Cancellation Property holds for polynomials with coefficients in an integral domain, that is, $A \cdot B = A \cdot C$ and $A \neq 0$ implies $B = C$.

Although not all of the details of the verification have been given, it should seem plausible to you that the set P of polynomials over an integral domain D is indeed an integral domain.

A polynomial

$$a_0 + a_1x + a_2x^2 + \cdots + a_nx^n$$

with $a_n \neq 0$ is said to have **degree** n. Thus the polynomial

$1 + 3x - 5x^2 + x^3$ has degree 3 and the polynomial 2 $(=2x^0)$ has degree 0. We do not, however, assign a degree to the zero polynomial itself.

If A and B are polynomials of degree n and m, respectively, what can you say about the degree of $A \cdot B$? Of $A + B$?

EXERCISES 5.1D

◀ In Exercises 1–12, find the product of each pair of polynomials by using the formal definition, that is, identify $a_0, a_1, \ldots, a_n, b_0, b_1, \ldots, b_m$ and then substitute these values in the formula for the product of polynomials.

1. $3x^2 + 5x - 2, 2x^2 - x + 7$
2. $4y^3 - y^2 + 1, y - 2$
3. $11x^4 - x^2 + 5x, x^5$
4. $t^3 - t^2 + 5t - 2, 8$
5. $7z^5 - 2z^2 + 3z - 6, 11z^4 + 6z^3$
6. $8s^3 - 2s^2 + 14s - 11, 2s^3 - 6s^2 + s + 14$
7. $x^7 - 13x^5 + x^3 - x, 3x^6 + x^4 - x^2 + 5$
8. $4y^5 + 12y^3 - y^2 + y, 8y^4 + 8y^3 + y^2 - y$
9. $12x^4 - 11x^3 + 21x^2 - x, x^5 + 8x^3 - 17x + 4$
10. $x^8 + 7x^4 - x + 12, x^7 - 12x^6 + x^5 - 2$
11. $7u^3 + 2u^2 - 6u + 1, -7u^3 - 2u^2 + 6u - 1$
12. $x^4 - 13x^3 + 12x^2 - x + 4, -x^4 + 13x^3 - 12x^2 + x - 4$

13. **CHALLENGE PROBLEM.** Prove that if A is a polynomial of degree n and B a polynomial of degree m, then $A \cdot B$ is a polynomial of degree $n + m$.

5.2 ZEROS OF POLYNOMIALS

Consider the polynomial $P(x)$ where

$$P(x) = 3x^3 - 5x^2 + 4x - 2.$$

Clearly, this polynomial can be used to define a function P by specifying the domain D since to each number $a \in D$ there corresponds exactly one number, the number

$$P(a) = 3a^3 - 5a^2 + 4a - 2.$$

For example, if the domain is the set of complex numbers, we have

$P(2) = 10$, $P(3) = 46$, $P(0) = -2$, $P(1) = 0$,

$$P(i) = 3i^3 - 5i^2 + 4i - 2 = -3i + 5 + 4i - 2 = i + 3,$$

and so on.

It is appropriate, then, to call

$$P : P(x) = 3x^3 - 5x^2 + 4x - 2$$

a **polynomial function.** The polynomial $P(x)$, then, is the value of the polynomial function P at x.

In this chapter we shall examine polynomial functions with a particular end in view, that of determining the **zeros** of a polynomial function. By this we mean finding the values of x which will make $P(x)$ equal to zero. For example, a zero of the function

$$P : P(x) = 3x^2 - 5x^2 + 4x - 2$$

is 1 since $P(1) = 0$. We also call 1 a zero of the polynomial $3x^2 - 5x^2 + 4x - 2$.

Another description of the same objective is to say that we want to find the solution set of polynomial equations such as

$$3x^3 + 5x^2 - 4x + 2 = 0.$$

You have already had considerable practice in finding the solution sets of quadratic equations. You know that the solution set of the equation

$$ax^2 + bx + c = 0 \qquad (a \neq 0)$$

is

$$\left\{ -\frac{b}{2a} + \frac{\sqrt{b^2 - 4ac}}{2a}, \; -\frac{b}{2a} - \frac{\sqrt{b^2 - 4ac}}{2a} \right\}.$$

Hence the zeros of the polynomial $ax^2 + bx + c$ $(a \neq 0)$ are

$$-\frac{b}{2a} + \frac{\sqrt{b^2 - 4ac}}{2a} \quad \text{and} \quad -\frac{b}{2a} - \frac{\sqrt{b^2 - 4ac}}{2a}.$$

Also, equivalently, we can say that the zeros of the polynomial function

$$P : P(x) = ax^2 + bx + c \qquad (a \neq 0)$$

are

$$-\frac{b}{2a} + \frac{\sqrt{b^2 - 4ac}}{2a} \quad \text{and} \quad -\frac{b}{2a} - \frac{\sqrt{b^2 - 4ac}}{2a}.$$

Frequently, we use factoring to uncover these zeros. Thus, for example, one can quickly find the zeros, 2 and 3, of the polynomial $x^2 - 5x + 6$ by observing that

$$x^2 - 5x + 6 = (x - 2)(x - 3).$$

In this chapter we shall examine the problem of finding zeros of polynomials of degree greater than 2, that is, polynomials of the form

$$P(x) = a_n x^n + a_{n-1} x^{n-1} + \cdots + a_1 x + a_0 \qquad (n > 2)$$

with $a_n \neq 0$, where, for the most part, the coefficients $a_0, a_1, \ldots,$ a_n will be assumed to be integers. Why this assumption? There are basically two reasons. First, the majority of applications of mathematics concerning zeros of polynomials involves polynomials with rational coefficients and, as we shall see, such problems can always be transformed into ones involving polynomials with integral coefficients. Second, the problem of finding zeros of polynomials with nonrational coefficients (either irrational real numbers or complex numbers) is considerably more complicated than are certain aspects, at least, of the problem of finding zeros of polynomials with rational coefficients.

What about the restriction to integral coefficients rather than rational coefficients in general? Consider the equation

$$\frac{1}{3}x^3 + \frac{1}{2}x^2 - \frac{2}{3}x + 5 = 0.$$

It is equivalent to the equation

$$2x^3 + 3x^2 - 4x + 30 = 0$$

since

$$2x^3 + 3x^2 - 4x + 30 = 6\left(\frac{1}{3}x^3 + \frac{1}{2}x^2 - \frac{2}{3}x + 5\right).$$

In general, suppose that we have the equation

$$a_n x^n + a_{n-1} x^{n-1} + \cdots + a_1 x + a_0 = 0$$

with rational coefficients where at least some of them are nonintegral. Then, to produce an equivalent equation with integral coefficients, we have simply to multiply by the LCM of the denominators of the nonintegral coefficients.

Now let us consider the polynomial

$$x^3 - 2x^2 - 5x + 6.$$

We contend that if k is a zero of this polynomial, then $x - k$ is a factor of

$$x^3 - 2x^2 - 5x + 6.$$

Conversely, if $x - k$ is a factor of $x^3 - 2x^2 - 5x + 6$, then k is a zero of the polynomial.

When we say that $x - k$ is a factor of the polynomial $x^3 - 2x^2 - 5x + 6$, we mean, of course, that $x^3 - 2x^2 - 5x + 6$ may be written as

$$(x - k) \cdot P(x),$$

where $P(x)$ is another polynomial.

The statement we have just made lies at the heart of a useful method of finding zeros of polynomials, or, equivalently, of solving polynomial equations. We shall first illustrate the idea as it applies to our example and then proceed to the general case.

If in the polynomial

$$x^3 - 2x^2 - 5x + 6$$

we substitute 3 for x, we get

$$3^3 - 2(3)^2 - 5(3) + 6 = 0.$$

Hence 3 is a zero of the polynomial. It can also be shown by direct multiplication that

$$x^3 - 2x^2 - 5x + 6 = (x - 3)(x^2 + x - 2).$$

Before considering the general case, let us take a look at the division of polynomials. Suppose that we want to divide $x^3 - 3x + 2$ by $x^2 + 4x + 3$. The standard procedure is as follows:

$$
\begin{array}{r}
x - 4 \\
x^2 + 4x + 3 \overline{)\, x^3 \qquad\quad - 3x + 2} \\
\underline{x^3 + 4x^2 + 3x} \\
-4x^2 - 6x + 2 \\
\underline{-4x^2 - 16x - 12} \\
10x + 14
\end{array}
$$

Rewriting our results in multiplicative form, we have

$$x^3 - 3x + 2 = (x^2 + 4x - 3)(x - 4) + (10x + 14),$$

which is of the form

$$f(x) = g(x)q(x) + r(x)$$

where $g(x)$ is the divisor, $q(x)$ is the quotient, and $r(x)$ is the remainder. In our example we note that $r(x) = 10x + 14$ has degree 1 and $1 < 2$, where 2 is the degree of

$$g(x) = x^2 + 4x - 3.$$

As another example, consider $f(x) = x^2 + 4x + 3$ and $g(x) = x + 1$. We have

$$
\begin{array}{r}
x + 3 \\
x + 1 \overline{\smash{\big)}\, x^2 + 4x + 3} \\
\underline{x^2 + x} \\
3x + 3 \\
\underline{3x + 3} \\
0
\end{array}
$$

and so

$$x^2 + 4x + 3 = (x + 1)(x + 3) + 0.$$

Thus the quotient is $x + 3$ and the remainder is 0.

These examples and the general nature of the procedure used for division of polynomials should make plausible the following basic theorem, often called the Division Algorithm Theorem. In this theorem and in the other theorems of this section we consider the polynomials under consideration to have their coefficients in the field of real numbers (and remember that real numbers are a special kind of complex numbers!).

THEOREM 5.1   (Division Algorithm Theorem)

If $f(x)$ and $g(x) \neq 0$ are polynomials, then there exist unique polynomials $q(x)$ and $r(x)$ such that

$$(1) \qquad f(x) = g(x)q(x) + r(x),$$

where either the degree of $r(x)$ is less than the degree of $g(x)$ or $r(x) = 0$.

What is $q(x)$ and what is $r(x)$ if degree of $g(x) >$ degree of $f(x)$?

A proof of this theorem involving the use of the modification of the principle of mathematical induction of Exercise 3, Exercises 2.5, is sketched in one of the Challenge Problems of Exercises 5.2.

It is instructive, by the way, to note the similarity of this theorem to the following theorem about integers.

If a and b are positive integers, then there exist unique integers q and r such that

$$a = bq + r \qquad \text{with} \qquad 0 \leq r < b.$$

Thus, for example, given $a = 14$ and $b = 4$, we have

$$14 = 4 \cdot 3 + 2$$

where $0 \leq 2 < 4$; if $a = 12$ and $b = 4$, we have

$$12 = 4 \cdot 3 + 0$$

where $0 \leq 0 < 4$; and if $a = 4$ and $b = 12$, we have

$$4 = 12 \cdot 0 + 4$$

where $0 \leq 4 < 12$.

A proof of this result can be made using the well-ordering principle (Section 2.5); it is also included as a Challenge Problem in Exercises 5.2.

Now let us return to the problem of finding zeros of polynomials.

Assume that $f(x)$ is a polynomial of degree ≥ 1. Then by Theorem 5.1 we know that

$$(2) \qquad\qquad f(x) = (x - k)q(x) + r(x)$$

where $r(x) = 0$ or the degree of $r(x)$ is less than the degree of $x - k$. But since the degree of $x - k$ is 1, it follows that $r(x)$ must be a polynomial of degree 0, that is, a nonzero integer, or else $r(x)$ must be zero. In any event, $r(x)$ is an integer and we write

$$r(x) = r \in I.$$

(You should recall that the zero polynomial is said to have no degree, but a nonzero integer is a polynomial of degree zero—a subtle but important distinction.)

If we now replace x by k in (2), then (1) becomes

$$f(k) = 0 \cdot q(k) + r = r.$$

This result establishes the Remainder Theorem.

THEOREM 5.2   (Remainder Theorem)

If $f(x)$ is a polynomial and k is any complex number, then $f(k) = r$, where r is the remainder obtained when $f(x)$ is divided by $x - k$.

We shall illustrate the concept by an example. Consider the polynomial

$$f(x) = x^3 - 3x^2 + 5x - 1.$$

Let $x = 2$. We find, then, that

$$f(2) = 2^3 - 3 \cdot 2^2 + 5 \cdot 2 - 1 = 5.$$

Now, dividing $f(x)$ by $x - 2$, we get

$$
\begin{array}{r}
x^2 - x + 3 \\
x - 2 \overline{\smash{\big)}\, x^3 - 3x^2 + 5x - 1} \\
\underline{x^3 - 2x^2} \\
-x^2 + 5x \\
\underline{-x^2 + 2x} \\
3x - 1 \\
\underline{3x - 6} \\
5
\end{array}
$$

Putting this result in the form (1) of Theorem 5.1, we have

$$x^3 - 3x^2 + 5x - 1 = (x - 2)(x^2 - x + 3) + 5$$

where $r = 5$.

We are now ready to substantiate an earlier assertion which we now restate as Theorem 5.3.

THEOREM 5.3   (Factor Theorem)
1. If $x - k$ is a factor of the polynomial $f(x)$, then k is a zero of $f(x)$.
2. If k is a zero of the polynomial $f(x)$, then $x - k$ is a factor of $f(x)$.

Another way of stating Theorem 5.3 is as follows: The number k is a zero of the polynomial $f(x)$ if and only if $x - k$ is a factor of $f(x)$.

To validate (1), we first observe that if $x - k$ is a factor of $f(x)$, then

$$f(x) = (x - k)q(x) + 0.$$

We also know by the Remainder Theorem (Theorem 5.2) that when $f(x)$ is divided by $x - k$, the remainder is equal to $f(k)$. Thus $f(k) = 0$.

To show that (2) is valid we reason as follows: If k is a zero of $f(x)$, then $f(k) = 0$. But $f(k)$ is also equal to the remainder obtained when $f(x)$ is divided by $x - k$. This is equivalent to saying that $x - k$ divides $f(x)$, that is, that $x - k$ is a factor of $f(x)$.

EXERCISES 5.2

◀Exercises 1–10 list two polynomials, the first of which is of the form $x - k$. In each Exercise, (a) divide the second polynomial by the first, (b) substitute k for x in the second polynomial and use the Remainder Theorem to check your division. (*Note:* In a case such as $x + 3$, k is equal to -3.)

1. $x - 3, x^3 - 7x^2 + 2x - 1$
2. $x - 2, x^4 - x^2 + 7$
3. $x + 1, x^5 - 3x^3 + 6x^2 - 5$
4. $x + 3, x^4 + 6x^3 - 7x$
5. $x - 4, x^3 - 8x + 6$

6. $x + 5, 2x^3 + 11x^2 - 14x + 2$
7. $x - 1, 3x^7 - 6x^6 + 5x^5 - 2x + 8$
8. $x - 7, 7x^3 + 3x^2 - x$
9. $x + 8, 5x^3 - 2x + 1$
10. $x - 10, 4x^4 - 3x^2 + 6x - 11$

◀In Exercises 11–20, perform the indicated divisions and express your answers in the form $f(x) = g(x)q(x) + r(x)$.

11. $(6x^2 + x - 2) \div (2x - 1)$
12. $(6x^2 - x - 15) \div (3x - 5)$
13. $(6x^3 + 5x^2 - 8x + 7) \div (2x + 3)$
14. $(2x^3 + 3x^2 - x - 12) \div (x^2 + 3x + 4)$
15. $(6x^3 - 19x^2 + 12x - 5) \div (3x^2 - 2x + 1)$
16. $(x^2 - 2x + 2) \div (2x - 1)$
17. $(7x^2 + 5x - 2) \div (3x - 1)$
18. $(5x^4 - x^3 - x + 1) \div (x^2 - x + 1)$
19. $(x^4 + 2x^3 + x^2 + 3x + 1) \div (x^2 + 2x)$
20. $(2x^4 - x^3 - x + 5) \div (2x^2 - 1)$

21. **CHALLENGE PROBLEM.** Copy and complete the following statements to prove a part of Theorem 5.1. If $f(x)$ and $g(x) \neq 0$ are polynomials such that the degree of $g(x) \leq$ the degree of $f(x)$, then there exist unique polynomials $q(x)$ and $r(x)$ such that

(A) $f(x) = g(x)q(x) + r(x)$

where either the degree of $r(x)$ is less than the degree of $g(x)$ or $r(x) = 0$.

(1) Let $f(x) = a_n x^n + a_{n-1} x^{n-1} + \cdots + a_1 x + a_0$ with $a_n \neq 0$ and $g(x) = b_m x^m + b_{m-1} x^{m-1} + \cdots + b_1 x + b_0$ with $b_m \neq 0$. By hypothesis, we have $n \boxed{?} m$.

(2) If $n = 0$, then $m = 0$ and $f(x) = a_0$ and $g(x) = b_0$. But then we have $a_0 = b_0 \dfrac{a_0}{b_0} + \boxed{?}$ since $g(x) = b_0 \neq 0$. Thus Theorem 5.1 is true when $n = 0$ and hence we now assume that $n \geq 1$.

(3) Define
$$f_1(x) = f(x) - \frac{a_n}{b_m} x^{n-m} g(x).$$

Then $f_1(x)$ is a polynomial of degree less than $\boxed{?}$.

(4) Now let S be the set of all $n \in N$ such that (A) holds for all polynomials $f(x)$ of degree $< n$.

(5) $1 \in S$ since if $n = 1$, $f(x) = a_1 x + a_0$, and $g(x) = b_1 x + b_0$. Then, if $b_1 \neq 0$,

$$a_1 x + a_0 = (b_1 x + b_0)\frac{a_1}{b_1} + \left(a_0 - \frac{a_1 b_0}{\boxed{?}}\right),$$

whereas if $b_1 = 0$, we have $b_0 \neq 0$ since $g(x) \neq 0$ and

$$a_1 x + a_0 = b_0 \left(\frac{a_1}{b_0} x\right) + \boxed{?}.$$

Thus the theorem holds for $n = 1$ and hence $1 \in S$.

(6) By our induction assumption (4) and the results of (3) we know that the theorem holds for $f_1(x)$. Thus there exist polynomials $q_1(x)$ and $r(x)$ such that

$$f_1(x) = g(x)q_1(x) + r(x)$$

where $r(x) = \boxed{?}$ or degree of $r(x) <$ degree of $\boxed{?}$.

(7) From (3) we have

$$f(x) = f_1(x) + \frac{a_n}{b_m} x^{n-m} \boxed{?}$$

$$= [g(x)q_1(x) + \boxed{?}] + \frac{a_n}{b_m} x^{n-m} g(x) \qquad \text{by (6)}$$

$$= \left[\frac{a_n}{b_m} x^{n-m} + q_1(x)\right] g(x) + r(x)$$

$$= q(x)g(x) + r(x)$$

where $q(x) = \boxed{?}$.

(8) Thus the theorem holds for $f(x)$ of degree n and we have shown that $1 \in S$ and that $k \in S$ for $k < n$ implies $n \in S$. By Exercise 3 of Exercises 2.5 it follows that $S = N$.

(9) We have shown that there exist polynomials $q(x)$ and $r(x)$ such that

$$f(x) = g(x)q(x) + \boxed{?}$$

where $r(x) = 0$ or degree of $r(x) <$ degree of $\boxed{?}$.

22. **CHALLENGE PROBLEM.** Copy and complete the following statements to prove that the polynomials $q(x)$ and $r(x)$ of Theorem 5.1 are unique.

(1) Suppose that there is a second pair of polynomials $q'(x)$ and $r'(x)$ such that

$$f(x) = q'(x)g(x) + r'(x).$$

where $r'(x) = \boxed{?}$ or degree of $r'(x) <$ degree of $\boxed{?}$.

(2) Then

$$q'(x)g(x) + r'(x) = q(x)\boxed{?} + r(x)$$

and

$$g(x)[q'(x) - q(x)] = r(x) - \boxed{?}.$$

(3) Now the right-hand side of the last equation is either 0 or of de-

gree $<$ degree of $g(x)$ since degree of $r(x) <$ degree of $g(x)$ and also degree $r'(x) <$ degree of $\boxed{?}$.

(4) But unless $q'(x) - q(x) = 0$ we have a contradiction. (See Challenge Problem 13 of Exercises 5.1D.)

(5) Hence $q'(x) = \boxed{?}$ and

$$g(x) \cdot 0 = 0 = r(x) - r'(x)$$

so that $r'(x) = \boxed{?}$.

23. **CHALLENGE PROBLEM.** Copy and complete the statements to prove the following theorem. Let a and b be positive integers with $a > b$. Then there exist unique integers q and r such that $a = bq + r$ with $0 \leq r < b$.

(1) Let S be the set of all integers of the form $a + bx$, where x is any integer such that $a + bx \geq 0$. We know that when x is any nonnegative integer, then $a + bx \geq 0$; so the set S is not the $\boxed{?}$ set.

(2) Now either all elements in S are positive or else $0 \in S$. In either case, by the well-ordering property, S has a $\boxed{?}$ element r, where $r \geq 0$ and r has the form $a + bx$ for some integer x.

(3) Since r is the least element of S, then $r - b \notin \boxed{?}$.

(4) But since $r = a + bx$, then $r - b = (a + bx) - b = a + b\boxed{?}$. Thus $r - b$, because it has the form

$$a + b(x - 1),$$

is in S if and only if $r - b \geq 0$.

(5) But by (3), $r - b \notin S$. Hence $r - b < 0$, which means that $\boxed{?} < b$.

(6) We have now shown that $0 \leq r < b$. Since $r = a + bx$, if we let $q = -x$, then $x = -q$ so that $r = a - bq$ and $a = \boxed{?} + r$.

(7) To finish the proof we need to show that r and q are unique. Suppose, then, that

$$a = bq + r, \qquad 0 \leq r < b,$$

and also

$$a = bq' + r', \qquad 0 \leq r' < b.$$

We want to show that $q = q'$ and $r = r'$. Now since $bq' + r' = a$ and $bq + r = a$, we know that

$$bq' + r' = bq + r.$$

Hence

$$r' - r = bq - bq' = b(\boxed{?}).$$

(8) Suppose that $r \leq r'$, so that $r' - r \geq 0$. (If $r' < r$, the proof can be completed by simply interchanging r and r' in the remainder of the proof.) But if $r' - r \geq 0$, it follows, since $r' - r \boxed{?} b(q - q')$ by (7), that

$$b(q - q') \geq \boxed{?}$$

and, since b is positive, that $q - q'$ is not negative.

(9) Now, since $r' < b$, it also follows that $r' - r < b$. But by (7) we have $r' - r = b(q - q')$. Therefore

$$b(q - q') < b.$$

Since $b > 0$ we can use Order Property 7(iv) (Section 3.3) to get

$$q - q' < \boxed{?}.$$

(10) However, from (8) we know that $q - q'$ is not negative. Therefore, since there is no positive integer less than 1, it follows that $q - q' = 0$, that is, that $q = \boxed{?}$.

(11) Finally, since $r - r' = b(q - q')$, then $r - r' = \boxed{?}$ if $q = q'$; so $r = \boxed{?}$ and the theorem is proved.

5.3 SYNTHETIC DIVISION

The relationship between k and the factor $x - k$ of a polynomial $f(x)$ enables us to convert the problem of finding zeros to one of determining factors. For example, if we wish to see whether or not 4 is a zero of a given polynomial $f(x)$, we can check to see if $x - 4$ divides $f(x)$.

If one were compelled to use the somewhat cumbersome division procedure illustrated in Section 5.2, this method of checking for zeros would become quite tedious. We shall therefore explore some means of streamlining the process. For example, suppose that we wish to find out if 4 is a zero of $x^4 - 7x^3 + 11x^2 + 7x - 12$. From the preceding discussion we know that 4 is a zero if and only if $x - 4$ divides $x^4 - 7x^3 + 11x^2 + 7x - 12$. Using the conventional division procedure, we have

$$
\begin{array}{r}
x^3 - 3x^2 - x + 3 \\
x - 4 \enclose{longdiv}{x^4 - 7x^3 + 11x^2 + 7x - 12} \\
\underline{x^4 - 4x^3} \\
-3x^3 + 11x^2 \\
\underline{-3x^3 + 12x^2} \\
-x^2 + 7x \\
\underline{-x^2 + 4x} \\
3x - 12 \\
\underline{3x - 12} \\
0
\end{array}
$$

and the answer, at long last, is "Yes."

Business offices, factories, even schools, frequently employ a so-called expediter to make their operations more efficient. Such an expediter will usually suggest that the drinking fountain be placed closer to the workers' desks or that a coffee dispenser be put on the same floor as the workers, and so on.

How might an expediter look at our division operation? First, he might say, "You do not need to write all the x's, just remember where they belong and what the exponents should be. Next, you do not need the stepladder bit and all the 'bringing down' of terms. Finally, if you write 4 instead of -4, the subtraction operation could be replaced by addition."

This is what the result, with a little more simplification, would look like:

$$\underline{4}\begin{array}{|ccccc} 1 & -7 & 11 & 7 & -12 \\ & 4 & -12 & -4 & 12 \\ \hline 1 & -3 & -1 & 3 & 0 \end{array}$$

Study this example carefully. Note that after the "leading" coefficient 1 is placed below the line, it is multiplied by 4 and the result placed under -7. Since

$$-7 + 4 = -3,$$

the -3 is placed in the second column. Multiplying -3 by 4 gives -12. Again, addition gives $11 + (-12) = -1$, and so on.

The 0 on the extreme right is the final remainder, r. The fact that $r = 0$ tells us that $x - 4$ is a factor of

$$f(x) = x^4 - 7x^3 + 11x^2 + 7x - 12.$$

Hence 4 is a zero of $f(x)$.

Compare the "streamlined" operation with the conventional division scheme to verify the following observations.

The quotient by the "long" process is

$$x^3 - 3x^2 - x + 3.$$

The first four numbers written under the line in this streamlined operation, usually called **synthetic division**, are

$$1, -3, -1, 3.$$

By putting appropriate powers of x in their rightful places we can utilize this sequence of numbers to write the quotient. In fact, the synthetic division scheme exhibits the quotient and remainder quite explicitly.

To familiarize yourself with synthetic division, study carefully the following two examples. To divide the same polynomial,

$$f(x) = x^4 - 7x^3 + 11x^2 + 7x - 12,$$

by $x - 3$ and hence to check to see whether 3 is a zero, we write

$$
\begin{array}{r|rrrrr}
3 & 1 & -7 & 11 & 7 & -12 \\
  &   & 3 & -12 & -3 & 12 \\
\hline
  & 1 & -4 & -1 & 4 & 0
\end{array}
$$

Since the remainder is again 0, we have verified that 3 is also a zero of $f(x)$. The quotient in this case is $x^3 - 4x^2 - x + 4$.

To see if -2 is a zero, we would, in the conventional approach, divide by $x + 2$. In the synthetic process we use -2, since $x + 2 = x - (-2)$. The synthetic division operation becomes

$$
\begin{array}{r|rrrrr}
-2 & 1 & -7 & 11 & 7 & -12 \\
   &   & -2 & 18 & -58 & 102 \\
\hline
   & 1 & -9 & 29 & -51 & 90
\end{array}
$$

Evidently -2 is *not* a zero since $r = 90$. However, by the Remainder Theorem, we do know that $f(-2) = 90$. Check that 1 and -1 are also zeros of $f(x)$ and that

$$f(x) = (x - 4)(x - 3)(x - 1)(x + 1).$$

The fact that the remainder can be used to determine functional values will prove useful in the development of some of the theory which we shall be considering later.

If, in a polynomial, the coefficients of one or more powers of x are zero, we must be careful to indicate this in the synthetic division process. For example, to divide $3x^4 - 2x^2 + 3$ by $x - 1$, we write

$$
\begin{array}{r|rrrrr}
1 & 3 & 0 & -2 & 0 & 3 \\
  &   & 3 & 3 & 1 & 1 \\
\hline
  & 3 & 3 & 1 & 1 & 4
\end{array}
$$

so that we conclude that

$$3x^4 - 2x^2 + 3 = (3x^3 + 3x^2 + x + 1)(x - 1) + 4.$$

Another application of synthetic division is as follows. Divide $x^3 - y^3$ by $x - y$.

$$\begin{array}{r|rrrr} y & 1 & 0 & 0 & -y^3 \\ & & y & y^2 & y^3 \\ \hline & 1 & y & y^2 & 0 \end{array}$$

Hence we obtain the factorization

$$x^3 - y^3 = (x - y)(x^2 + xy + y^2).$$

We can use this formula, for example, to write

$$\begin{aligned} 8a^3 - 27b^3 &= (2a)^3 - (3b)^3 \\ &= (2a - 3b)[(2a)^2 + (2a)(3b) + (3b)^2] \\ &= (2a - 3b)(4a^2 + 6ab + 9b^2) \end{aligned}$$

where we replace x by $2a$ and y by $3b$. Check by synthetic division that

$$x^3 + y^3 = (x + y)(x^2 - xy + y^2).$$

We have seen that synthetic division can be very useful in simplifying some polynomial divisions. It does, however, have some limitations! Thus it does not apply at all to the division of a polynomial by a nonlinear polynomial. In fact, it does not apply directly to the division of a polynomial by a linear polynomial $ax + b$ when $a \neq 1$. We can, however, handle division by linear polynomials in general as suggested in the following example.

Divide $3x^3 - 2x^2 + x - 1$ by $2x - 1$. We write

$$\begin{aligned} (3x^3 - 2x^2 + x - 1) &\div (2x - 1) \\ &= \frac{1}{2}(3x^3 - 2x^2 + x - 1) \div \frac{1}{2}(2x - 1) \\ &= \left(\frac{3}{2}x^3 - x^2 + \frac{1}{2}x - \frac{1}{2}\right) \div \left(x - \frac{1}{2}\right) \end{aligned}$$

and then proceed as follows:

$$\begin{array}{r|rrrr} \frac{1}{2} & \frac{3}{2} & -1 & \frac{1}{2} & -\frac{1}{2} \\ & & \frac{3}{4} & -\frac{1}{8} & \frac{3}{16} \\ \hline & \frac{3}{2} & -\frac{1}{4} & \frac{3}{8} & -\frac{5}{16} \end{array}$$

Thus

$$\frac{3}{2}x^3 - x^2 + \frac{1}{2}x - \frac{1}{2} = \left(\frac{3}{2}x^2 - \frac{1}{4}x + \frac{3}{8}\right)\left(x - \frac{1}{2}\right) - \frac{5}{16}$$

and so

$$3x^3 - 2x^2 + x - 1 = \left(\frac{3}{2}x^2 - \frac{1}{4}x + \frac{3}{8}\right)(2x - 1) - \frac{5}{8}.$$

EXERCISES 5.3

◀In Exercises 1–20, use synthetic division to determine the quotient and remainder for each division.

1. $(x^4 - x^3 + 3x^2 - 6x + 1) \div (x - 2)$
2. $(x^4 + 3x^3 - 6x^2 + x - 2) \div (x - 3)$
3. $(x^5 + 3x^4 - 2x^3 + x^2 - 1) \div (x - 1)$
4. $(x^5 + 3x^4 - 2x^3 + x^2 - 1) \div (x + 1)$
5. $(x^5 - x^3 + 2x^2 + 4) \div (x - 4)$
6. $(3x^5 - x^3 + x - 3) \div (x + 2)$
7. $(5x^6 - x^4 + x^2 - 1) \div (x - 1)$
8. $(4x^4 - x^3 + x^2 - 5) \div (x + 4)$
9. $(2x^4 + x^3 + x^2 + 1) \div (x + 5)$
10. $(3x^5 + 3x^2 + 2) \div (x - 2)$
11. $(x^5 - 2x^2 + 1) \div (x + 2)$
12. $(x^7 + 1) \div (x + 1)$
13. $(x^8 + 1) \div (x + 1)$
14. $(x^6 + 64) \div (x - 2)$
15. $(x^6 - 64) \div (x - 2)$
16. $(x^3 - 3x^2 + x - 2) \div (2x - 1)$
17. $(x^3 + 2x^2 - x + 2) \div (2x + 1)$
18. $(x^3 - x^2 + x + 1) \div (3x + 2)$
19. $(x^3 + x^2 - x + 1) \div (3x - 2)$
20. $(2x^3 + x^2 - 1) \div (2x - 1)$

◀In Exercises 21–26, factor the given polynomial.

21. $8x^3 - y^3$
22. $64a^3 + b^3$
23. $x^3 + 27y^3$
24. $125u^3 + 27v^3$
25. $64x^3 - 125y^3$
26. $27a^3 + 8b^3$

27. Show by synthetic division that $x^7 + y^7$ has the factor $x + y$.
28. Simplify the expression

$$\frac{x^7 + 128}{x + 2}.$$

◀ In Exercises 29–30, without multiplying give a suggested answer to the product.

29. $(x + 1)(x^8 - x^7 + x^6 - x^5 + x^4 - x^3 + x^2 - x + 1)$

30. $(x + 2)(x^4 - 2x^3 + 4x^2 - 8x + 16)$

31. CHALLENGE PROBLEM. Use synthetic division to develop a method of factoring $x^k + 1$ where k is an odd positive integer.

5.4 SOLVING EQUATIONS BY SYNTHETIC DIVISION

Since the zeros of a polynomial $f(x)$ are the roots of the equation $f(x) = 0$, we can use synthetic division to test whether or not a given number is a solution of a polynomial equation. We shall now illustrate by example a further extension of the uses of synthetic division in the equation-solving process.

Consider the equation

$$f(x) = x^4 + 2x^3 - 7x^2 - 8x + 12 = 0.$$

We might conjecture that 2 is in the solution set. As a test, then, we have

$$
\begin{array}{r|rrrrr}
2 & 1 & 2 & -7 & -8 & 12 \\
  &   & 2 & 8 & 2 & -12 \\
\hline
  & 1 & 4 & 1 & -6 & 0
\end{array}
$$

By the Remainder Theorem, we know that $r = f(2)$. Since $r = 0$, we have $f(2) = 0$, and hence 2 is a solution of the equation. Thus our conjecture is valid.

As a second conjecture suppose we guess that -3 is also a solution. To check this, we could run through the synthetic division test on the original polynomial. Our previous synthetic process, however, yielded a quotient. This is

$$g(x) = x^3 + 4x^2 + x - 6.$$

In effect, this tells us that

(1) $\begin{aligned} f(x) &= x^4 + 2x^3 - 7x^2 - 8x + 12 \\ &= (x - 2)(x^3 + 4x^2 + x - 6) = (x - 2)g(x). \end{aligned}$

If -3 is a zero of the factor $x^3 + 4x^2 + x - 6$, then -3 certainly is a zero of the given polynomial

$$f(x) = x^4 + 2x^3 - 7x^2 - 8x + 12,$$

since if $g(-3) = 0$, then

$$f(-3) = (-3 - 2)g(-3) = (-5) \cdot 0 = 0.$$

Now suppose we knew that -3 was a zero of $f(x)$. Could we conclude that -3 is also a zero of $g(x)$, that is, that $g(-3) = 0$? If -3 is a zero of $f(x)$, we have, from (1), that

$$(2) \qquad\qquad f(-3) = 0 = (-3 - 2)g(-3).$$

Recall now the "zero" property of any integral domain D (Section 3.8): If a, $b \in D$ and $a \cdot b = 0$, then either $a = 0$ or $b = 0$. Since $-3 - 2 = -5 \neq 0$, it follows from (2) that $g(-3) = 0$.

We conclude that -3 is a zero of $f(x)$ if and only if it is a zero of $g(x)$. Hence we see that a test of -3 as a possible solution of the equation

$$x^4 + 2x^3 - 7x^2 - 8x + 12 = 0$$

can be performed using the so-called **reduced polynomial**, that is, the quotient obtained from the previous division. This test can be exhibited as follows:

$$
\begin{array}{r|rrrr}
-3 & 1 & 4 & 1 & -6 \\
   &   & -3 & -3 & 6 \\
\hline
   & 1 & 1 & -2 & 0
\end{array}
$$

Again a conjecture is validated.

In practice, it is efficient to combine the two tests in the following way:

$$
\begin{array}{r|rrrrr}
2 & 1 & 2 & -7 & -8 & 12 \\
  &   & 2 & 8 & 2 & 12 \\
\hline
  & 1 & 4 & 1 & -6 & 0 \\
-3 & & -3 & -3 & 6 & \\
\hline
  & 1 & 1 & -2 & 0 &
\end{array}
$$

Theoretically, we could continue the process again. However, from the form of the quotient we see that our new "reduced" equation,

$$x^2 + x - 2 = 0,$$

is of degree two. Hence it remains only to solve a quadratic.

By factoring we see that this quadratic equation is equivalent to

$$(x + 2)(x - 1) = 0.$$

Thus the remaining members of our solution set are -2 and 1 and the complete solution set of the equation $x^4 + 2x^3 - 7x^2 - 8x + 12 = 0$ is

$$\{2, -3, -2, 1\}.$$

Up to now we have been using the synthetic division process to test conjectures about possible solutions. Nothing has been said about how one might arrive at such a conjecture. Of all the possible real numbers, how do we select which ones to test as solutions?

Although it is rarely possible to spot a solution by inspection, there are many ways in which one can narrow down the choice. We shall examine some of these techniques. In effect, we shall be discussing methods of "zeroing in on a zero."

Consider, then, the polynomial equation

$$a_n x^n + a_{n-1} x^{n-1} + \cdots + a_1 x + a_0 = 0,$$

where the a's are integers and $a_n \neq 0$. What are the possible solutions?

To deal with this question, we make use of the following theorem.

THEOREM 5.4

Let

$$f(x) = a_n x^n + a_{n-1} x^{n-1} + \cdots + a_1 x + a_0$$

be a polynomial with integral coefficients. If a rational number $\dfrac{t}{s}$ is a solution of the equation $f(x) = 0$ and t and s are integers ($s \neq 0$) whose only common factors are 1 and -1 (that is, $\dfrac{t}{s}$ is in *lowest terms*), then

1. t must be a factor of a_0 and
2. s must be a factor of a_n.

For our example,

$$x^4 + 2x^3 - 7x^2 - 8x + 12 = 0,$$

with the leading coefficient a_n equal to 1, this means that the only possible rational solutions are the factors of 12, that is, $1, 2, 3, 4, 6, 12, -1, -2, -3, -4, -6, -12$. The solution set which we found before, namely, $\{2, -3, -2, 1\}$, is thus consistent with the conclusion drawn by applying the theorem.

To prove the theorem, let us first write

(1) $$a_n x^n + a_{n-1} x^{n-1} + \cdots + a_1 x + a_0 = 0$$

as

(2) $$a_n x^n + a_{n-1} x^{n-1} + \cdots + a_1 x = -a_0.$$

If $\frac{t}{s}$ is a solution, then we know that

(3) $$a_n \left(\frac{t}{s}\right)^n + a_{n-1} \left(\frac{t}{s}\right)^{n-1} + \cdots + a_1 \left(\frac{t}{s}\right) = -a_0.$$

Multiplying each term by s^n changes this to

(4) $$a_n t^n + a_{n-1} t^{n-1} s + \cdots + a_1 t s^{n-1} = -a_0 s^n.$$

By the Distributive Property we can write (4) as

(5) $$t(a_n t^{n-1} + a_{n-1} t^{n-2} s + \cdots + a_1 s^{n-1}) = -a_0 s^n.$$

Our initial job is to show that t is a divisor of a_0. We shall first dispose of the cases when $t = 0, t = 1,$ or $t = -1$. If $t = 0$, then $a_0 = 0$ since, by assumption, $s \neq 0$. Then we have

$$a_0 = 0 = 0 \cdot 0 = t \cdot 0.$$

If $t = 1$ or -1, then t divides all integers including, of course, a_0. Thus the theorem is trivially true for these two cases.

Assume, then, that $t \neq 0, 1,$ or -1. By hypothesis, t has no factors in common with s, hence no factors in common with s^n, since s^n merely duplicates the factors of s n times. Thus we have the following situation: t divides $a_0 s^n$, but t has *no* factors in common (other than 1 and -1) with s^n, that is, t and s^n are relatively prime. A reasonable conclusion, then, is that t must divide a. (A more detailed argument supporting this conclusion is suggested as a Challenge Problem in Exercises 5.4.)

An argument quite similar to this one can be used to prove part 2 of Theorem 5.4 and this is also suggested as a Challenge Problem in Exercises 5.4. Meanwhile let us see how this theorem can be used to good advantage in solving equations.

As an example consider

$$3x^4 - 2x^3 + 6x^2 - x + 8 = 0.$$

What are the possible rational solutions? Any rational number candidate, $\frac{t}{s}$, must be such that t divides 8 and s divides 3. Eligible numerators, then, are $\pm 1, \pm 2, \pm 4,$ and ± 8; eligible denominators are ± 1 and ± 3. Collecting all the rational number candidates, we have

$$C = \left\{\pm 1, \ \pm 2, \ \pm 4, \ \pm 8, \ \pm \frac{1}{3}, \ \pm \frac{2}{3}, \ \pm \frac{4}{3}, \ \pm \frac{8}{3}\right\},$$

sixteen candidates in all counting positive and negative numbers—a rather large number of entries, but not so large when compared with the set of all rational numbers!

As a general procedure we should begin a systematized testing program using each of these possible solutions in turn. Some eliminations are possible at the outset. You can undoubtedly discover some of these yourself.

As a further illustration, we shall apply synthetic division to one of the noninteger candidates. Let us try $\frac{2}{3}$ in the equation

$$3x^4 - 2x^3 + 6x^2 - x + 8 = 0.$$

In the prescribed form we have

$$
\frac{2}{3} \bigg|
\begin{array}{rrrrr}
3 & -2 & 6 & -1 & 8 \\
  & 2 & 0 & 4 & 2 \\
\hline
3 & 0 & 6 & 3 & 10
\end{array}
$$

Since we have a remainder of 10, the test shows that $\frac{2}{3}$ is not a solution. Back to the drawing board!

Suppose that we try $\frac{1}{3}$ as a possible root. Again using synthetic division we have

$$
\frac{1}{3} \bigg|
\begin{array}{rrrrr}
3 & -2 & 6 & -1 & 8 \\
  & 1 & -\dfrac{1}{3} & \dfrac{17}{9} & \dfrac{8}{27} \\
\hline
3 & -1 & \dfrac{17}{3} & \dfrac{8}{9} & \text{(not zero!)}
\end{array}
$$

The rational number $\frac{1}{3}$ has failed the test, but we can certainly profit from this example. Notice that in the third column the number written under the line, $\frac{17}{3}$, is not an integer. From then on matters get progressively worse! The next fraction has denominator 9, and the final entry, the remainder, would have had a denominator of 27 had we computed it. It should be evident from this example and from a little reflection that once a noninteger appears in the quotient line of a synthetic division chart, there is no useful purpose in continuing further.

Here is a final example. The test for $-\frac{2}{3}$, also a candidate at this point, looks like this:

$$
-\frac{2}{3} \begin{array}{|rrrrr} 3 & -2 & 6 & -1 & 8 \\ & -2 & \dfrac{8}{3} & & \\ \hline 3 & -4 & \dfrac{26}{3} & & \end{array}
$$

STOP!

If an entire set of possible rational solutions fails the test, we can, by virtue of the theorem under discussion, conclude that the given equation does not have any *rational* roots. It may, of course (in fact, must!), then have roots which are nonrational real numbers or complex nonreal numbers. Thus, for example, the equation $x^2 - 2 = 0$ has no rational roots (test this statement using Theorem 5.4), but has, of course, the irrational roots $\sqrt{2}$ and $-\sqrt{2}$. Likewise, the equation $x^2 + 1 = 0$ has no rational, in fact no real, roots but has the complex roots i and $-i$.

It is possible that a "reduced" equation may have as a solution a number which has already been identified as a solution of the original equation. For example, the polynomial $x^4 - 4x^3 + x^2 + 12x - 12$ is equal to $(x - 2)(x^3 - 2x^2 - 3x + 6)$; thus, by the factor theorem, it has 2 as a zero. But it also happens that

$$x^3 - 2x^2 - 3x + 6 = (x - 2)(x^2 - 3).$$

Hence 2 is also a zero of the reduced polynomial. The solution set, $\{2, \sqrt{3}, -\sqrt{3}\}$, of the equation

$$x^4 - 4x^3 + x^2 + 12x - 12 = 0$$

has only the three elements 2, $\sqrt{3}$, and $-\sqrt{3}$.

It would have been helpful, however, in this case to "run through" the synthetic division process twice with the number 2 in order to obtain a reduced equation of as small a degree as possible.

In general, if an eligible rational solution passes the first test, it may also be a solution of the reduced equation and should be given a second try. On the other hand, if a number is not a solution of the original equation, it cannot be a solution of any reduced equation. (Why?)

EXERCISES 5.4

◀In Exercises 1–16, find all rational solutions of the given equation. In all instances where the equation "reduces" to a quadratic, complete the solution.

1. $x^3 - 6x^2 + 11x - 6 = 0$
2. $x^3 - 3x^2 + x - 3 = 0$
3. $x^3 + 3x^2 - 2x + 3 = 0$
4. $2x^4 - 9x^3 + 10x^2 - 27x + 12 = 0$
5. $3x^3 - 8x^2 + 11x - 10 = 0$
6. $x^4 + x^3 - 2x^2 + x - 1 = 0$
7. $x^4 - 3x^3 + 4x^2 - 8x - 24 = 0$
8. $6x^4 + 7x^3 + 12x^2 + x - 2 = 0$
9. $9x^3 + 21x^2 - 17x + 3 = 0$
10. $36x^4 - 13x^2 + 1 = 0$
11. $x^4 - x^3 - 7x^2 + x + 6 = 0$
12. $x^3 - 2x^2 + x - 2 = 0$
13. $x^4 - 2x^3 - 13x^2 + 14x + 24 = 0$
14. $2x^3 - 5x^2 - 10x + 25 = 0$
15. $x^4 - 3x^3 - x + 2 = 0$
16. $3x^3 - 2x^2 - 9x + 6 = 0$

◀In Exercises 17–26, list all the possible rational solutions of each of the given equations.

17. $6x^4 - 5x^3 + 10 = 0$
18. $12x^5 - 7 = 0$
19. $8x^6 + 3x - 20 = 0$
20. $2x^5 - 4x^4 + 15 = 0$
21. $21x^5 + 24 = 0$

22. $8x^5 - 3x^4 + 5x^2 - 6 = 0$
23. $9x^6 + 2x^3 - 3x^2 + 8 = 0$
24. $15x^7 - 3x^2 + 2x - 5 = 0$
25. $12x^6 - x^5 + x^4 - 2x^3 + 3x - 15 = 0$
26. $6x^5 + x^4 - 3x^3 + x^2 - 9 = 0$

27. Show that if $a_0 = 0$, then 0 is a solution of the equation

$$a_n x^n + a_{n-1} x^{n-1} + \cdots + a_1 x + a_0 = 0$$

and the remaining roots may be found by solving the equation

$$a_n x^{n-1} + a_{n-1} x^{n-2} + \cdots + a_2 x + a_1 = 0.$$

◀In Exercises 28–33, list, as in Exercises 17–26, all possible rational solutions of each of the given equations.

28. $6x^5 - 10x = 0$ **31.** $11x^3 - 4x = 0$

29. $12x^4 - 5x^3 - 9x = 0$ **32.** $12x^4 - x^2 = 0$

30. $9x^5 - 3x^2 + x = 0$ **33.** $15x^4 + 4x^3 + x^2 = 0$

34. Show that 1 is a solution of the equation

$$a_n x^n + a_{n-1} x^{n-1} + \cdots + a_1 x + a_0 = 0$$

if and only if

$$\sum_{i=0}^{n} a_i = 0.$$

35. Explain the modifications needed in the rule of Exercise 34 to provide a test by inspection for -1 as a solution.

36. Prove that $\sqrt{2}$ is an irrational number by showing that the equation $x^2 - 2 = 0$ has no rational root.

37. Prove that $\sqrt{3}$ is an irrational number by showing that the equation $x^2 - 3 = 0$ has no rational root.

38. CHALLENGE PROBLEM. Generalize the results of Exercises 36 and 37.

39. CHALLENGE PROBLEM. Prove that if t, a_0, and s are integers, t divides $a_0 s^n$ for $n \in N$, and t and s^n are relatively prime, then t divides a_0. (*Hint:* If t divides $a_0 s^n$, then $a_0 s^n = dt$ for some integer d. Then

$$|a_0 s^n| = |a_0||s^n| = |dt| = |d||t|$$

where $|a_0|$, $|s^n|$, $|d|$, and $|t|$ are natural numbers. Now consider the prime factorizations of $|a_0|$, $|s^n|$, and $|t|$. Use the fact that $|t|$ and $|s^n|$ have no prime factors in common to conclude that if p^n (p a prime, $n \in N$) is a factor of t, then p^n is also a factor of a_0.)

40. CHALLENGE PROBLEM. Prove part 2 of Theorem 5.4. (*Hint:* Add $a_0 b^n - a_n t^n$ to both members of Equation (2).)

41. CHALLENGE PROBLEM. Show that an equation of the form

$$a_n x^n + a_{n-1} x^{n-1} + \cdots + a_1 x + a_0 = 0$$

in which all coefficients are positive cannot have positive solutions and that if the a's are alternately positive and negative, that is, $a_n > 0$, $a_{n-1} < 0$, $a_{n-2} > 0$, and so on, then the equation can have no negative solutions.

5.5 A FUNDAMENTAL THEOREM

We have seen that it is possible to find all the rational roots (if any exist) of a polynomial equation in one variable with rational coefficients. (Remember that if some of the coefficients are non-integral, we can produce an equivalent equation by multiplying by the LCM of the denominators of these nonintegral coefficients.) What about polynomial equations with nonrational coefficients and what about nonrational roots?

The subject "theory of equations" is a vast one and we cannot hope to do more than consider a small part of it in this text. Some suggestions for Further Reading and Research are given at the end of the chapter. For now let us note first that a method for approximating the real roots (rational or irrational) of polynomial equations is given in Chapter 6.

Next, what about the possibility of formulas like the quadratic formula for equations of degree higher than 2? Formulas do exist for third degree (cubic) and fourth degree (quartic) equations. These formulas, however, are far more complicated than the quadratic formula and are not very useful for practical purposes. Then, in one of the great mathematical achievements of all time, a youthful French mathematician, Évariste Galois, proved in 1832 that it is impossible to express the zeros of any general polynomial of degree higher than four in terms of powers and roots of its coefficients.

What is there left for us to consider here? For one thing, what about our tacit assumption that polynomial equations with real or complex coefficients always have at least one solution in *C*?

We have certainly encountered equations for which there were no rational roots. A simple example of such an equation is the quadratic equation

$$x^2 - 5 = 0$$

for which the solution set is

$$\{\sqrt{5}, \ -\sqrt{5}\}.$$

Furthermore, you have seen and solved equations with no real number solutions. The classic example of this type is

$$x^2 + 1 = 0$$

with its solution set $\{i, -i\}$. Experience has taught us that if we admit as the solution domain the field of complex numbers (which, as we interpret it, includes all real numbers, and hence all rational numbers), then every equation of the form $f(x) = 0$ where $f(x)$ is a polynomial with coefficients in *C* would seem to have a solution in *C*.

Question: Is this conclusion wishful thinking or is it based on fact?

In 1799 the famous German mathematician Karl Friedrich Gauss came up with a proof of the following spectacular theorem.

THEOREM 5.5

Every equation of the form $f(x) = 0$, where $f(x)$ is a polynomial with coefficients in C, has at least one root in C.

Because of its far-reaching consequences, this famous theorem has been called **The Fundamental Theorem of Algebra.**

Since the integers are complex numbers (of the form $a + 0 \cdot i$ for $a \in I$), the set of polynomials with integers as coefficients that we have been concerned with in this chapter would certainly be covered by the theorem.

There is an extension of the Fundamental Theorem which can be helpful if it is interpreted correctly. This theorem states the following.

> Every equation of the form $f(x) = 0$, where $f(x)$ is a polynomial of degree n with complex coefficients, has exactly n solutions.

To understand and use this theorem properly we have need for a special interpretation of what is meant by an equation having n solutions. To begin with, the equation

$$x^3 - 6x^2 + 11x - 6 = 0$$

can be written as

$$(x - 1)(x - 2)(x - 3) = 0.$$

In this equation the polynomial $f(x)$ has degree 3. According to the Factor Theorem the equation has exactly three solutions, namely 1, 2, 3. So far so good. However, what about the equation

$$x^3 - 6x^2 + 12x - 8 = 0$$

which can be written as

$$(x - 2)(x - 2)(x - 2) = 0?$$

The solution set is a set of exactly one element, that is,

$$\{2\},$$

and yet the associated polynomial is certainly of degree 3.

The apparent contradiction stems from an earlier tradition in the history of the teaching of mathematics in which an equation such as

$$x^3 - 6x^2 + 12x - 8 = 0$$

was interpreted as having a triple solution, that is, three 2's.

In recent times it has been deemed much more reasonable to regard this equation, $x^3 - 6x^2 + 12x - 8 = 0$, as having one solution, namely 2, although the polynomial $x^3 - 6x^2 + 12x - 8$ is indeed the product of three (identical) linear factors so that

$$x^3 - 6x^2 + 12x - 8 = (x - 2)(x - 2)(x - 2).$$

As another example, the equation

$$x^3 - 11x^2 + 35x - 25 = 0,$$

which can be written as

$$(x - 5)(x - 5)(x - 1) = 0,$$

has a solution set of just two elements, that is,

$$\{1, 5\},$$

although the polynomial $x^3 - 11x^2 + 35x - 25$ certainly is the product of three factors, two of which are identical.

The extension of the Fundamental Theorem implies that the equation

$$x^3 - 1 = 0$$

has three solutions. We know that one solution is 1. What are the others?

Using synthetic division we can do the following as a first step:

$$
\underline{1 \rfloor} \quad
\begin{array}{rrrr}
1 & 0 & 0 & -1 \\
  & 1 & 1 & 1 \\
\hline
1 & 1 & 1 & 0
\end{array}
$$

The reduced equation is

$$x^2 + x + 1 = 0.$$

By the quadratic formula the solutions of the reduced equation are

$$-\frac{1}{2} + \frac{\sqrt{3}}{2}i \qquad \text{and} \qquad -\frac{1}{2} - \frac{\sqrt{3}}{2}i.$$

These two complex numbers, known as the complex nonreal cube roots of unity, play an important role in mathematics. They are customarily designated by the special symbols ω and ω^2, respectively.

Synthetic division can also be used to test for complex roots. For example, since

$$
\begin{array}{r|rrrr}
1+i & 1 & -5 & 8 & -6 \\
& & 1+i & -5-3i & 6 \\
\hline
& 1 & -4+i & 3-3i & 0
\end{array}
$$

it follows that $1 + i$ is a root of

$$x^3 - 5x^2 + 8x - 6 = 0,$$

and

$$(x^3 - 5x^2 + 8x - 6) = [x - (1 + i)][x^2 + (-4 + i)x + (3 - 3i)].$$

EXERCISES 5.5

1. Using multiplication in C, as in Chapter 4, show that $\omega \cdot \omega = \omega^2$.
2. Show that $\omega^2 \cdot \omega^2 = \omega$.
3. Show by actual substitution that ω is a solution of $x^2 + x + 1 = 0$.
4. Show that $\omega^2 \cdot \omega = 1$.

◀ In Exercises 5–9, proceed as in the text example to determine the three cube roots of the given number.

5. 8   7. -8 9. -27
6. -1 8. 27

10. Prove that $3 - i$ is a root of $x^3 - 4x^2 - 2x + 20 = 0$.
11. Prove that $2 + 3i$ is a root of $x^3 - 9x^2 + 33x - 65 = 0$.

◀ In Exercises 12–15, form a cubic equation which has as its roots the numbers listed.

12. $2, 1 - 3i, 1 + 3i$ 14. $i, i + 1, 2$
13. $-1, -2 + i, 2 - i$ 15. $-i, 1 - i, 1$

16. **CHALLENGE PROBLEM.** Assume Theorem 5.5 and then prove by induction that if $f(x)$ is a polynomial $a_n x^n + a_1 x^{n-1} + \cdots + a_1 x + a_0$ with coefficients in C, then there exist complex numbers $r_1, r_2, \ldots, r_n$ such that

$$f(x) = a_n(x - r_1)(x - r_2) \cdots (x - r_n).$$

CHAPTER SUMMARY

The polynomial function $f : f(x) = a_n x^n + a_{n-1} x^{n-1} + \cdots + a_1 x + a_0$ with integral coefficients has been examined from the standpoint of determining its ZEROS, that is, solutions of the equation

$$f(x) = 0.$$

Following is a list of the basic theorems considered.

DIVISION ALGORITHM THEOREM. If $f(x)$ and $g(x)$ are polynomials such that the degree of $g(x) \leq$ the degree of $f(x)$, then there exist unique polynomials $q(x)$ and $r(x)$ such that $f(x) = g(x)q(x) + r(x)$ where either the degree of $r(x) <$ the degree of $g(x)$ or $r(x) = 0$.

REMAINDER THEOREM. If $f(x)$ is a polynomial and k is any complex number, then $f(k) = r$ where r is the remainder obtained when $f(x)$ is divided by $x - k$.

FACTOR THEOREM. The number k is a zero of the polynomial $f(x)$ if and only if $x - k$ is a factor of $f(x)$.

THEOREM ON RATIONAL ROOTS. Let $f(x)$ be a polynomial $a_n x^n + a_{n-1} x^{n-1} + \cdots + a_1 x + a_0$, where the a's are integers. If a rational number $\dfrac{t}{s}$ (in lowest terms) is a solution of the equation $a_n x^n + a_{n-1} x^{n-1} + \cdots + a_1 x + a_0$, then

1. t must be a factor of a_0 and
2. s must be a factor of a_n.

FUNDAMENTAL THEOREM OF ALGEBRA. Every equation of the form $f(x) = 0$, where $f(x)$ is a polynomial with complex coefficients, has at least one complex root.

REVIEW EXERCISES

1. Find the sum of the polynomials $5x^4 - 3x^2 + x - 1$, $x^3 + 5x^2 - 3x + 2$, and $7x^4 - x + 6$.
2. Find the product of the polynomials $x^3 - 3x^2 + x - 1$ and $5x^2 - 2x + 3$.

◀In Exercises 3–8, factor the given polynomial over the integers.

3. $2x^2 + 7xy + 6y^2$ 6. $r^4 - r^3 + 4r - 16$
4. $4a^2 - 9b^2$ 7. $16x^4 - 1$
5. $3a^4 - 3a$ 8. $(a - b)^3 - c^3$

9. In the integral domain $B = \{0, 1, 2, 3, \ldots, 10; *, \circ\}$ of Section 3.2, find the sum and the product of the polynomials $2x + 5$ and $3x + 7$ relative to the operations $*$ and $\circ$.
10. Divide $x^4 - 3x^2 + 2x - 5$ by $x + 3$ and then use the Remainder Theorem to check your division.
11. Divide $f(x) = 4x^5 - 3x^4 + x^2 - 2$ by $g(x) = x^2 + x - 2$ and express your answer in the form $f(x) = g(x)q(x) + r(x)$.
12. State the Factor Theorem.

◀In Exercises 13 and 14, use synthetic division to determine the quotient and remainder in each division.

13. $(x^5 - 3x^4 + x^2 - 2) \div (x - 2)$
14. $(2x^5 + x^4 - 3x^3 + 5x - 1) \div (x + 4)$

◀In Exercises 15 and 16, find all the rational roots of the given equation. If the equation "reduces" to a quadratic, complete the solution.

15. $x^3 - 2x^2 + x - 2 = 0$ 16. $x^4 - 4x^3 + x^2 - 3x - 4 = 0$

17. List all the possible rational solutions of the equation

$$5x^3 - 2x^2 - 5x - 10 = 0.$$

18. Find the three cube roots of 125.
19. Form a cubic equation which has -2, $1 + i$, and $1 - i$ as its roots.

GOING FURTHER: READING AND RESEARCH

As we said before, there are all sorts of topics to investigate in the theory of equations and we mention but a few of them here. Let us begin by considering two generalizations of theorems concerning quadratic equations. One is that if the quadratic equation $ax^2 + bx + c = 0$ has real coefficients and a root $c + di$ ($c, d \in R$), it also has the complex conjugate root $c - di$. (See Section 4.6.) It may be shown, more generally, that complex, nonreal roots occur in conjugate pairs, that is, if

$$f(x) = a_n x^n + a_{n-1}x^{n-1} + \cdots + a_1 x + a_0 \qquad (a_n, a_{n-1}, \ldots, a_1, a_0 \in R)$$

and $f(c + di) = 0$ with c, $d \in R$, then, also, $f(c - di) = 0$. To prove this theorem you might begin by forming the product

$$[x - (c + di)][x - (c - di)] = (x - c)^2 + d^2$$

which is a quadratic polynomial with real coefficients. Now divide $f(x)$ by this polynomial to obtain

$$f(x) = [(x - a)^2 + b^2]q(x) + r(x)$$

where $r(x) = 0$ or $r(x) = mx + n$ for m, $n \in R$. Now show that if $r(x) = mx + n$, then $r(a + bi) = 0$ implies $m = n = 0$. (See, for example, M. J. Weiss, *Higher Algebra for the Undergraduate*, revised by R. Dubisch, New York: John Wiley and Sons, 1962.)

Now consider the results of one of the Challenge Problems of Exercises 4.1, namely, that if x_1 and x_2 are roots of the equation $ax^2 + bx + c = 0$, then

$$x_1 + x_2 = -\frac{b}{a} \quad \text{and} \quad x_1 x_2 = \frac{c}{a}.$$

(The coefficients a, b, and c here need not be in R; they can be any complex numbers.)

Can you show that if x_1, x_2, and x_3 are roots of the cubic equation

$$ax^3 + bx^2 + cx + d = 0,$$

then $x_1 + x_2 + x_3 = \frac{b}{a}$, $x_1 x_2 + x_1 x_3 = -\frac{c}{a}$, and $x_1 x_2 x_3 = \frac{d}{a}$? Can you generalize these results to relations between coefficients and roots of the general polynomial equation

$$a_n x^n + a_{n-1} x^{n-1} + \cdots + a_1 x + a_0 = 0$$

with roots $x_1, x_2, \ldots, x_n$? (See, for example, B. E. Meserve, *Fundamental Concepts of Algebra*, Reading, Mass.: Addison-Wesley, 1953.)

The lives of Karl Friedrich Gauss (who proved the Fundamental Theorem of Algebra) and Évariste Galois (who proved the result on the limitations of the possibility of finding formulas for solving polynomial equations) form interesting contrasts. Gauss lived a long and peaceful life and Galois a short and turbulent one. *Men of Mathematics* by E. T. Bell contains a fascinating account of the lives of these two great mathematicians.

Additional references for further reading for this chapter are the following.

ROSE, I. H., *Algebra: An Introduction to Finite Mathematics*, New York: John Wiley and Sons, 1963.

USPENSKY, J. V., *Theory of Equations*, New York: McGraw-Hill, 1948.

CHAPTER SIX
COORDINATE SYSTEMS

6.1 LINEAR GRAPHS

The study of real numbers is assisted by the fact that the real numbers can be put into a one-to-one correspondence with the points on a number line. Of even greater usefulness is the fact that ordered pairs of real numbers can be put into a one-to-one correspondence with points in the plane. It was through this means that the French mathematician René Descartes provided, in the seventeenth century, a link between algebra and geometry known as analytic geometry. The results of this linkage have been far-reaching. As you probably know, Descartes' basic concept involved the establishment of a coordinate system in the plane determined by two mutually perpendicular lines. A plane with such a system is often called a **Cartesian plane** in honor of Descartes or, alternately, a **rectangular coordinate plane**.

This well-known Cartesian coordinate system is illustrated in Figure 6-1, which also illustrates the familiar relationship between ordered pairs of numbers (coordinates) and points in the Cartesian plane. (In our discussion of the coordinate plane and its uses, we shall often write "the point (x, y)" as an abbreviation for "the point with coordinates (x, y)."

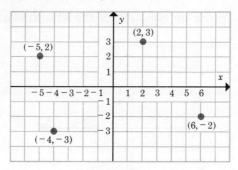

Figure 6-1

To begin with, we shall assume as well established the fact that the solution set of an equation of the form

$$ax + by + c = 0,$$

where a, b, and c are real numbers with a and b not both zero, corresponds in the plane to all points on a line. Actually, we may define a line as a set of points in the plane whose coordinates correspond in one-to-one fashion with the ordered pairs in the solution set of such an equation.

203

It is convenient to regard the equation as the "algebra" part and the line as the "geometry" part of the combine. In this sense, some of our activity will be a kind of lost and found business. Given one partner, find the other, that is, given an equation, determine the geometric configuration or graph and, conversely, given a figure or characterization of it, find the corresponding equation.

As a quick review you are asked to engage in some of this lost and found activity, with respect to lines, in the Exercises which follow. If you have difficulties, do not be discouraged. Further assistance will be furnished in the following sections. Meanwhile, a few examples are given.

Example 1   A line through the origin and the point $(4, 7)$.
Equation: $7x - 4y = 0$.

Example 2   A line parallel to the x-axis and containing the point $(5, -3)$.
Equation: $y = -3$.

Example 3   Write an equation of the line shown below.

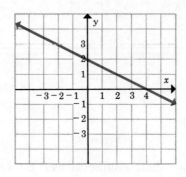

Equation: $x + 2y - 4 = 0$.

EXERCISES 6.1

◀In Exercises 1–10, draw the graph of the line which corresponds to the given equation.

   1. $3x + y - 9 = 0$ 2. $x - 4y + 8 = 0$

3. $x + 6 = 1$ 7. $x = 0$

4. $x = 6$ 8. $y = 0$

5. $3x = 12$ 9. $7x - 8y = 12$

6. $5y + 20 = 0$ 10. $y = 3x$

 In Exercises 11–25, write an equation of the form $ax + by + c = 0$ which corresponds to the indicated line.

11. A line parallel to and three units above the x-axis.

12. A line through the origin and the point $(3, 5)$.

13. A line perpendicular to the x-axis and containing the point $(4, 6)$.

14. A line through the point $(-7, 8)$ and parallel to the y-axis.

15. A line which contains the points $(2, -7)$ and $(0, 0)$.

16. A line through the origin with every x-coordinate (abscissa) of a point on the line three times the corresponding y-coordinate (ordinate).

17. A line through $(0, 0)$ such that the ordinate of each point on the line is five times the corresponding abscissa.

18. A line through all points having an ordinate five more than the abscissa.

19. A line through all points having an ordinate three more than four times the abscissa.

20.
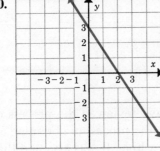

22.
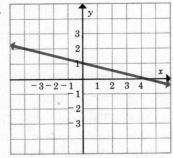

21.
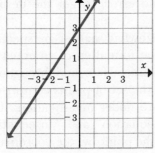

23.
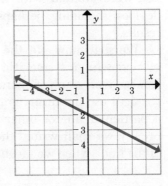

24.

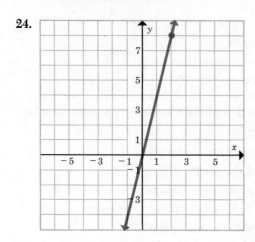

25.

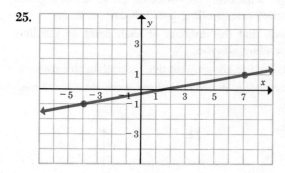

6.2 SLOPE

In the Exercises of Section 6.1 you have reviewed the process of graphing linear equations and, conversely, the process of obtaining an equation from the graph or description of a line. It was assumed that you were familiar with the general concept and could use various devices, including perhaps some trial and error, to get the desired results.

We shall now put the scene into somewhat sharper perspective by reviewing a systematic approach to the formation of an equation from a given graph or characterization of a line. Here, and throughout, we shall use the term "graph" to denote the set of points comprising a particular line or other geometric figure under investigation. For brevity we shall write "the graph of an equation" and "an equation of the line" when we mean "the graph of the points whose coordinates are in the solution set of an equation" and "an equation whose graph is the line," respectively.

The fundamental concept of slope provides a means by which one can proceed systematically from a given graph to the corresponding equation.

In a rectangular coordinate system, the slope of a nonvertical line l, as you may recall, is defined by the formula

$$\frac{y_2 - y_1}{x_2 - x_1}$$

where (x_1, y_1) and (x_2, y_2) are the coordinates of any two distinct points contained in l such that $x_2 \neq x_1$.

For example, the slope of the line containing the points $(-3, 1)$ and $(2, 5)$ is

$$\frac{5 - 1}{2 - (-3)} = \frac{4}{5}.$$

(See Figure 6-2.)

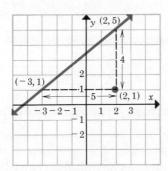

Figure 6-2

By properties of the real numbers we know that

$$\frac{y_2 - y_1}{x_2 - x_1} = \frac{-(y_2 - y_1)}{-(x_2 - x_1)} = \frac{y_1 - y_2}{x_1 - x_2}.$$

Hence the slope of a line does not depend on which of two points is selected for the "first one," that is, (x_1, y_1), and which is selected for the "second," (x_2, y_2). Thus, for example,

$$\frac{5 - 1}{2 - (-3)} = \frac{1 - 5}{-3 - 2} = \frac{4}{5}.$$

Furthermore, the slope of a line is not defined when $x_1 = x_2$, that is, when the line is parallel to the y-axis.

Suppose, then, that we are given a line l identified as the line which contains the two points $(3, 5)$ and $(9, 10)$. How do we proceed to find an equation of l? We note first that the slope of l is

$$\frac{10 - 5}{9 - 3} = \frac{5}{6}.$$

Then, if (x, y) is any point on the line other than the point $(3, 5)$, we know that

(1)
$$\frac{y - 5}{x - 3} = \frac{5}{6}.$$

Essentially, this equation is merely giving in mathematical language the statement that the slope of l as determined by the coordinates $(3, 5)$ and the coordinates (x, y) of any other point on l is equal to $\frac{5}{6}$. Forming an equation, then, can be thought of as the act of translating into mathematical "shorthand" a verbal description of the graph.

An alternative equation related to the same line is

(2)
$$y - 5 = \frac{5}{6}(x - 3).$$

A careful mathematician will note here that Equation (2) is not equivalent to (1) since the solution sets of Equations (1) and (2) are not the same. The point $(3, 5)$ is an element in the solution set of (2) but is not in the solution set of (1). Thus Equation (2) provides a complete description of line l, whereas (1) describes l with the point $(3, 5)$ deleted.

We can write other equations which are equivalent to (2). Two of these are

(3)
$$y = \frac{5}{6}x + 2\frac{1}{2}$$

and

(4)
$$5x - 6y + 15 = 0.$$

You should be able to describe the operations which are used to convert Equation (2) to (3) and to (4).

Form (3) has the advantage of exhibiting the slope, that is, $\frac{5}{6}$, explicitly as the coefficient of x, and the y-coordinate, $2\frac{1}{2}$, of the point where the line intersects the y-axis. The number $2\frac{1}{2}$, which is the value of y when 0 is substituted for x in Equation (3), is usually called the **y-intercept**.

Equation (4) is an example of the **general form**

$$ax + by + c = 0.$$

If we convert the general equation

$$ax + by + c = 0$$

to Form (3), that is, solve for y, we get

$$by = -(ax) - c$$

and so

$$y = -\frac{a}{b}x - \frac{c}{b} \qquad \text{if } b \neq 0.$$

From this we may infer that the line with an equation of the form

$$ax + by + c = 0 \qquad (b \neq 0)$$

has slope $-\dfrac{a}{b}$ and y-intercept $-\dfrac{c}{b}$. What is the situation when $b = 0$?

From the previous discussion we can derive many helpful hints for forming equations of lines from given characterizations. Thus, if we want an equation of the line that contains the point $(2, 7)$ and has a slope of 4, we can use Form (1) and write

$$\frac{y - 7}{x - 2} = 4,$$

from which we can readily obtain the other forms

$$y - 7 = 4(x - 2),$$
$$y = 4x - 1,$$
$$4x - y - 1 = 0,$$

and

$$4x + (-1)y + (-1) = 0.$$

The last equation is, of course, the general form although we also frequently refer to equations such as $4x - y - 1 = 0$ as being in the general form.

More generally, if a line contains the point (s, t) and has slope m, an equation of the line can be written as

$$y - t = m(x - s),$$
$$y = mx + (t - ms),$$

or

$$(-m)x + y + (ms - t) = 0.$$

Finally, if you are given two points, say $(3, 6)$ and $(-2, 4)$, you can compute, usually by inspection, the slope m of the line containing them. In this case

$$m = \frac{6 - 4}{3 - (-2)} = \frac{2}{5}.$$

Then, if you choose the point $(3, 6)$, you can write for an equation of the line

$$y - 6 = \frac{2}{5}(x - 3).$$

Alternatively, if you choose the point $(-2, 4)$, you can write

$$y - 4 = \frac{2}{5}(x + 2),$$

and the end results are the same, that is, the two equations are equivalent. Check this equivalence by converting each equation to the general form $ax + by + c = 0$.

EXERCISES 6.2

◀In Exercises 1–10, write, in the three forms

$$y - t = m(x - s),$$
$$y = mx + d,$$

and

$$ax + by + c = 0,$$

equations of the line determined by the given pair of points. (Recall that m is the slope, d is the y-intercept, and (s, t) are the coordinates of either one of the given points.)

1. $(9, 5)$, $(3, 2)$ **3.** $(0, 0)$, $(5, 10)$

2. $(1, 6)$, $(-1, 4)$ **4.** $(-1, -3)$, $(5, 3)$

5. $(4, -6), (1, -3)$ **8.** $(2, -7), (4, -6)$

6. $(2, 7), (0, 0)$ **9.** $(3, 11), (1, 8)$

7. $(3, 8), (1, 5)$ **10.** $(2, 6), (3, 6)$

◀ In Exercises 11–20, write, in the three forms described for Exercises 1–10, an equation of the line determined by the slope and point given.

11. $m = 1, (-3, 4)$ **16.** $m = -3, (5, -11)$

12. $m = 0, (-8, 6)$ **17.** $m = -\dfrac{1}{2}, \left(2\dfrac{1}{2}, 3\dfrac{3}{4}\right)$

13. $m = \dfrac{3}{4}, (8, -4)$ **18.** $m = -1\dfrac{1}{2}, \left(-\dfrac{3}{4}, 2\right)$

14. $m = 0.7, \left(\dfrac{1}{2}, 1\right)$ **19.** $m = -0.6, (0.3, 1.6)$

15. $m = -\dfrac{3}{4}, \left(\dfrac{2}{3}, -5\right)$ **20.** $m = -\dfrac{5}{6}, \left(-\dfrac{1}{2}, -\dfrac{3}{5}\right)$

◀ In Exercises 21–25, equations are given in general form. From each equation determine the slope and the y-intercept of the line with this equation.

21. $3x + 5y + 11 = 0$ **24.** $\dfrac{2}{3}x - \dfrac{3}{4}y + 1 = 0$

22. $7x - 2y + 10 = 0$ **25.** $0.6x + 0.5y - 1.2 = 0$

23. $10y + 1 = 0$

6.3 FAMILIES OF LINES

In dealing with the general problem of lines and equations, it is often helpful to consider the set of all lines having a common property, for example, the set of all lines passing through a given point or the set of all lines having the same slope. Such sets, usually called families of lines, are illustrated in part in Figures 6-3 and 6-4. (See page 212.)

We can write an equation describing all of the members of the family of lines through the point $(2, 3)$, except the line through $(2, 3)$ parallel to the y-axis, as $y - 3 = m(x - 2)$. Here the letter m which represents the slope and varies with each line is called a parameter of the family. However, this equation does not describe the line through $(2, 3)$ parallel to the y-axis. A separate equation, $x = 2$, is needed to describe this particular member of the family.

To form an equation representing the family of lines all having

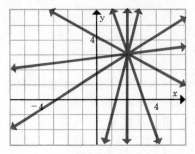

Figure 6-3

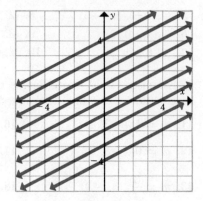

Figure 6-4

slope $\frac{1}{2}$, we choose Form (3) and write

$$y = \frac{1}{2}x + d.$$

Here the parameter d represents the y-intercept, which, of course, varies with each line. In Figure 6-4 the various y-intercepts are -4, -3, -2, -1, 0, 1, 2, 3, and 4.

Note that the set of all y-intercepts for members of the family with an equation of the form $y = \frac{1}{2}x + d$ is the set of all real numbers. Similarly, the family of lines through (2, 3) includes the line $x = 2$ as well as lines having all possible slopes.

You have seen that, in general, lines are determined by two conditions. In finding an equation for a given line it is often useful and instructive to consider one condition at a time. Let us try this.

Example 1   Assume that we want an equation for a line having a slope of $\frac{2}{3}$ and which also passes through the point (6, 7). First, we consider the family of all lines with the given slope. A family equation is

$$y = \frac{2}{3}x + d.$$

Now we wish to select a particular member of the family determined by the second condition, that of containing the point (6, 7). If (6, 7) is to be in the solution set of

$$y = \frac{2}{3}x + d,$$

then a true sentence must be formed when y is replaced by 7 and x by 6. That is,

$$7 = \frac{2}{3}(6) + d$$

must be a true sentence. Evidently the condition is met when $d = 3$ since we know that

$$7 = \frac{2}{3}(6) + 3.$$

Thus we have, as a final equation,

$$y = \frac{2}{3}x + 3.$$

Example 2 Find an equation of the line l containing the points $(1, 8)$ and $(-2, 4)$.

Solution: For the conditions involving two given points we might begin with the family of all lines through one of the points, say $(1, 8)$. An equation for this family is $y - 8 = m(x - 1)$. From here the determination of m could be made by noting that $(-2, 4)$ must also satisfy the equation. Substitution gives us

$$(4 - 8) = m(-2 - 1)$$

or

$$-4 = -3m.$$

From this we see that $m = \frac{4}{3}$, and the unique family member is

$$y - 8 = \frac{4}{3}(x - 1).$$

We write, in abbreviated form, $l : y - 8 = \frac{4}{3}(x - 1)$.

You might try the same problem beginning with the family of lines through $(-2, 4)$. The results should agree.

Suppose, now, that we are asked to write the equation of a line parallel to a given line and passing through a specified point.

What can be said about the equations of parallel lines? In your previous study of mathematics you may have learned that two distinct nonvertical lines are parallel if and only if they have the same slope.

(Distinct vertical lines, that is, lines with equations of the form $x = k_1$ and $x = k_2$ with $k_1 \neq k_2$ are, of course, also parallel.)

If you have not proved this result before, you may be interested in developing such a proof (it really is not too hard) as a Challenge Problem in Exercises 6.3.

Example 3 Find an equation of the line l containing the point $(3, 4)$ and parallel to the line $k : y = 2x + 5$.

Solution: The line k has slope 2 and hence the family of lines parallel to k has slope 2 and the representation

$$y = 2x + d.$$

As in Example 1, we can find a value for d by substituting, in this case, $x = 3$ and $y = 4$ in the equation $y = 2x + d$. Check that this yields $d = -2$ so that $y = 2x - 2$ is an equation of line l.

EXERCISES 6.3

◀In Exercises 1–10, write an equation for the family of lines described.

1. Lines through the point $(2, 5)$.
2. Lines through the origin.
3. Lines having a slope $\dfrac{3}{4}$.
4. Lines parallel to the line $l : 3x - 4y + 7 = 0$.
5. Lines having a slope twice that of the line $l : x - 7y = 12$.
6. Lines parallel to the x-axis.
7. Lines parallel to the y-axis.
8. Lines through the point $\left(\dfrac{1}{3}, \dfrac{1}{8}\right)$.
9. Lines parallel to the line with an equation $\dfrac{1}{2}x + \dfrac{2}{3}y = 5$.
10. Lines through the point $\left(-12, \dfrac{1}{3}\right)$.

◀In Exercises 11–18, write an equation in the general form $ax + by + c = 0$ for the line determined by the two given conditions. (*Note:* You may use any of the approaches illustrated in this section or a method of your own.)

11. The line through the origin parallel to the line containing the points $(-7, 8)$ and $(1, -5)$.

12. The line containing the point (2, 5) and parallel to a line whose slope is $-\dfrac{1}{2}$.

13. The line through the point $(1, -1)$ which is parallel to the line containing the points (2, 1) and $(-8, 6)$.

14. The line through the points $\left(\dfrac{2}{3}, -6\right)$ and $\left(\dfrac{1}{2}, -\dfrac{1}{2}\right)$.

15. The line through (7, 2) which is parallel to the line of Exercise 14.

16. The line through the origin parallel to the line $l : 3x + 2y - 8 = 0$.

17. The line through (2, 5) parallel to the line $l : x + y = 1$.

18. The line through $(-3, 6)$ parallel to the line which contains the points (1, 7) and $(-4, 2)$.

◀ In Exercises 19–23, two conditions for a line are given: the slope and a point on the line. Proceed as in the discussion in the last part of this section to find an equation for the family satisfying the first condition. Then find an equation for the unique line determined by the second condition.

19. Slope $\dfrac{2}{3}$, point (2, 6)

20. Slope $-\dfrac{1}{2}$, point $(1, -3)$

21. Slope -2, point $(4, -8)$

22. Slope $-\dfrac{3}{5}$, point $(11, -4)$

23. Slope 4, point $(-7, 12)$

◀ In Exercises 24–30, for the given pair of equations, state whether the corresponding lines are parallel and distinct, identical, or neither of these.

24. $3x + 8y - 12 = 0$, $4x - 7y + 1 = 0$

25. $7x - 2y + 8 = 0$, $7x - 2y - 8 = 0$

26. $6x + 4y + 7 = 0$, $4x - 6y + 2 = 0$

27. $9x - 8y + 12 = 0$, $3x - 2\dfrac{2}{3}y = -8$

28. $\dfrac{1}{2}x + y = 0$, $2x - y = 30$

29. $x = 5$, $x = 6$

30. $y = 5$, $y = -5$

◀ In Exercises 31–35, draw graphs of each of the given pairs of equations. Use the same axes for both equations of the given pair.

31. $3x + 2y + 8 = 0$, $2x - 3y + 5 = 0$

32. $x - 5y + 2 = 0$, $10x + 2y - 7 = 0$

33. $2x - 6y = 11$, $x + \dfrac{1}{3}y = 11$

34. $\dfrac{1}{2}x + y = 4$, $7x - 3.5y = 1$

35. $y = 6x - 8$, $x + 6y = 4$

36. What relationship between the lines does each of the five pairs of graphs in Exercises 31–35 show?

37–41. Calculate the slopes m_1 and m_2 for each pair of lines in Exercises 31–35 and form the products $m_1 \cdot m_2$.

42. Suggest a possible theorem regarding the relationship of two lines and the products of their respective slopes.

43. On the basis of your conjecture in Exercise 42, write an equation of a line perpendicular to the line $l : 3x + 2y = 7$ and containing the point $(4, 1)$.

44. CHALLENGE PROBLEM. Prove the property that two distinct lines having the same slope are parallel, that is, they do not intersect. (*Hint:* Finding a point where two lines intersect involves the solution of a system of two linear equations

$$ax + by = c$$
$$dx + ey = f.)$$

6.4 DISTANCE

The concept of distance between any two given points plays a prominent part in the study of coordinate systems. Distance can be thought of in two senses: (1) the *directed* distance *from* point A, for example, *to* point B; (2) the idea of distance *between* two points A and B, independent of direction.

In the context of directed distance, we regard the distance from B to A as the additive inverse of the distance from A to B. Thus directed distance can be a negative number, a positive number, or zero.

When we speak of the distance between A and B without use of the word "directed," we consider the distance as a nonnegative number.

To introduce the topic of distance we begin with two points on the x-axis. Let A be the point $(x_1, 0)$ and let B be the point $(x_2, 0)$

which, as an abbreviation, we write as $A(x_1, 0)$ and $B(x_2, 0)$. (See Figure 6-5.)

It seems natural to define the directed distance from A to B as the number $x_2 - x_1$. Accordingly, the directed distance from B to A is $-(x_2 - x_1)$, that is, $x_1 - x_2$.

For $A(-3, 0)$ and $B(5, 0)$, the directed distance from A to B is $5 - (-3) = 8$. The directed distance from B to A is $-3 - 5 = -8$.

For the (undirected) distance between A and B, which we symbolize as AB, we want a nonnegative number independent of direction. Thus it is appropriate for the case where $A = (x_1, 0)$ and $B = (x_2, 0)$ to designate AB as $|x_2 - x_1|$, that is, the absolute value of $x_2 - x_1$. (Recall that $|x| = x$ if $x \geq 0$ and $|x| = -x$ if $x < 0$. Thus $|7| = 7$, $|0| = 0$, and $|-4| = 4$.)

From the designation, $AB = |x_2 - x_1|$, we note that $AB = BA$, since

$$|x_2 - x_1| = |x_1 - x_2|.$$

In the particular case under consideration, that is, $A(-3, 0)$ and $B(5, 0)$, we have

$$AB = |5 - (-3)| = |8| = |-3 - 5| = BA.$$

This distance concept can be readily applied to any two points having the same ordinate. Thus for $A(7, 6)$ and $B(-2, 6)$, we have

$$AB = |-2 - 7| = |9| = |7 - (-2)| = BA.$$

A similar situation holds for points having the same abscissa (x-coordinate). (See Figure 6-6.)

In Figure 6-6, $AB = |2 - (-1)| = |3| = |-1 - 2| = BA$. For $A(-7, -3)$ and $B(-7, -2)$, check that $AB = BA = 1$.

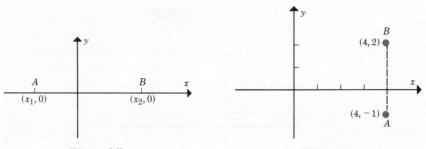

Figure 6-5   Figure 6-6

Here and in what follows we are assuming that measurements on the x- and y-axes are based on the same unit, that is, that the distance from 0 to 1 on the x-axis is equal to the distance from 0 to 1 on the y-axis.

EXERCISES 6.4A

◀In Exercises 1–10, pairs of points $A(x_1, y_1)$ and $B(x_2, y_2)$ are given. Determine in each case the directed distance from A to B, the directed distance from B to A, and the distance AB.

1. $A(3, 5), B(1, 5)$

2. $A(4, -6), B(4, 7)$

3. $A(-2, 1), B(-2, -8)$

4. $A\left(-7, \dfrac{1}{2}\right), B\left(-8, \dfrac{1}{2}\right)$

5. $A\left(\dfrac{3}{4}, 2\right), B\left(\dfrac{3}{4}, -9\right)$

6. $A\left(\dfrac{1}{7}, -\dfrac{3}{4}\right), B\left(\dfrac{1}{7}, -\dfrac{3}{4}\right)$

7. $A\left(\dfrac{1}{4}, \dfrac{1}{8}\right), B\left(\dfrac{1}{2}, \dfrac{1}{8}\right)$

8. $A\left(2\dfrac{2}{3}, 6\right), B\left(1\dfrac{1}{6}, 6\right)$

9. $A(0.2, 1), B(0.8, 1)$

10. $A(1.3, 1.2), B(1.3, -1.2)$

Having considered the notion of distance between points with equal abscissas or equal ordinates, let us now tackle the question of determining the distance between any two points in the plane. For example, given points $A(-2, 1)$ and $B(5, 4)$, how shall we determine AB?

In Figure 6-7, $C(5, 1)$ has been plotted along with $A(-2, 1)$ and $B(5, 4)$ so that a right triangle, ACB, is formed.

From previous considerations we know that $AC = |5 - (-2)| = 7$ and $CB = |4 - 1| = 3$. By the Pythagorean Theorem for right triangles,

$$(AB)^2 = (AC)^2 + (CB)^2.$$

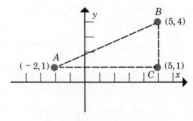

Figure 6-7

Hence in our example

$$(AB)^2 = 7^2 + 3^2,$$

which tells us that

$$AB = \sqrt{49 + 9} = \sqrt{58}.$$

If $A = (x_1, y_1)$ and $B = (x_2, y_2)$, the same sort of analysis as we used in the example above leads to the conclusion that

$$AB = \sqrt{(x_2 - x_1)^2 + (y_2 - y_1)^2}.$$

Using this formula we can see, for example, that the distance AB between $A(-2, 7)$ and $B(3, -5)$ is

$$\sqrt{[3 - (-2)]^2 + (-5 - 7)^2} = \sqrt{25 + 144} = \sqrt{169} = 13.$$

EXERCISES 6.4B

◀In Exercises 1–15, find the distance between the given pair of points.

1. $(2, 1)$, $(-4, 9)$

2. $(3, 6)$, $(1, 15)$

3. $(11, 0)$, $(2, 7)$

4. $(8, 1)$, $(10, 1)$

5. $(-7, -6)$, $(3, 4)$

6. $\left(\frac{3}{4}, \frac{1}{5}\right)$, $\left(-\frac{15}{4}, -\frac{9}{5}\right)$

7. $(0.7, 2)$, $(-0.3, 5)$

8. $(0.2, 1)$, $(-0.3, -1)$

9. $\left(\frac{1}{2}, \frac{3}{4}\right)$, $\left(\frac{1}{2}, \frac{1}{4}\right)$

10. $\left(\frac{5}{6}, \frac{7}{8}\right)$, $\left(\frac{5}{6}, \frac{1}{8}\right)$

11. $\left(\frac{2}{3}, -\frac{1}{2}\right)$, $\left(\frac{1}{4}, -\frac{1}{4}\right)$

12. $(0.7, -0.7)$, $(1.3, -1.6)$

13. $(12, 7)$, $(13, 10)$

14. $\left(3\frac{1}{3}, 2\right)$, $\left(2, 3\frac{1}{3}\right)$

15. $(1.6, -2.7)$, $(2.4, -0.1)$

16. Show that the four points $(2, 3)$, $(7, 3)$, $(2, -5)$, and $(7, -5)$ form the vertices of a rectangle and determine (a) its perimeter, (b) its area, and (c) the length of one of its diagonals.

17. Given the points $A(2, 0)$, $B(7, 3)$, and $C(-2, 4)$, find the perimeter of triangle ABC.

18. A quadrilateral $ABCD$ has vertices $A(-3, 7)$, $B(8, 5)$, $C(7, 0)$, and $D(-3, -1)$. Find the lengths of the diagonals $\overline{AC}$ and $\overline{BD}$.

19. Given a triangle whose vertices are $(8, 5)$, $(0, 0)$, and $(-8, 5)$. Show that the triangle is isosceles.

20. Given a triangle whose vertices are $(1, 1)$, $(9, 1)$, and $(9, 5)$. Use the distance formula and the converse of the Pythagorean Theorem to show that the given triangle is a right triangle.

21. Given the points $A(-3, -2)$ and $B(9, 3)$, determine a third point $C(x, y)$ such that ABC is a right triangle with hypotenuse $\overline{AB}$ and sides parallel to the coordinate axes.

22. Find the perimeter and the area of triangle ABC in Exercise 21.

23. Write an equation whose solution set is the set of all points whose distance from the origin is 5. (*Hint:* Use a formula which gives the distance from any point (x, y) to the point $(0, 0)$.)

24. CHALLENGE PROBLEM. Write an equation whose solution set is the set of all points whose distance from $(2, 3)$ is 7.

25. CHALLENGE PROBLEM. Write an equation whose graph is a circle with radius 10 and center at $(-2, 7)$.

26. CHALLENGE PROBLEM. Write an equation whose graph is a circle with radius r and center at (a, b).

27. CHALLENGE PROBLEM. In the figure below, lines l_1 and l_2 are perpendicular. Use the distance formula, the definition of slope, and the Pythagorean Theorem to show that the product of the slopes of l_1 and l_2 is equal to -1. Thus substantiate the conjecture of Exercise 42 in Section 6.3.

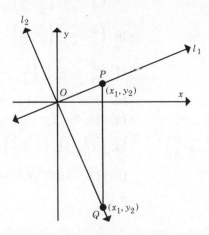

6.5 GRAPHS OF POLYNOMIAL EQUATIONS

In Chapter 5, a discussion of polynomials revealed the fact that the system of polynomials with coefficients in an integral domain is itself an integral domain. We shall now investigate polynomials from a different point of view by taking a look at the graphs of equations of the form

$$y = a_n x^n + a_{n-1} x^{n-1} + \cdots + a_1 x + a_0.$$

For purposes of comparison, let us consider the following sequence of such equations, constructed in ascending order with respect to the highest power of x.

(0)	$y = 2$
(1)	$y = 3x - 1$
(2)	$y = x^2 - x - 12$
(3)	$y = x^3 - 2x^2 - 5x + 6$

You are already familiar with the graph of (0) which involves a polynomial of degree 0 and the graph of (1) involving a polynomial of degree 1.

The graphs of Equations (0), (1), (2), and (3) are shown in Figure 6-8. Equation (0) relating to a polynomial of degree 0, and Equation (1), relating to a polynomial of degree 1, have graphs which are very familiar.

For Equation (2), $y = x^2 - x - 12$, we note that the graph crosses the x-axis at the points $(-3, 0)$ and $(4, 0)$. This tells us that $y = 0$ when $x = -3$ and when $x = 4$.

Thus $\{4, -3\}$ is the solution set of the quadratic equation

$$x^2 - x - 12 = 0.$$

You will note also that the graph crosses the y-axis at the point $(0, -12)$, which is determined, of course, by substituting 0 for x in Equation (2).

A point of special interest is the point $(\frac{1}{2}, -12\frac{1}{4})$ which is called the **minimum point** of the graph.

In general, equations of the form

$$y = ax^2 + bx + c$$

have a graph similar in shape to the one in Figure 6-8c when the coefficient a is positive.

When a is negative, the graph takes on an "inverted" form. For example, the graph of

$$y = -x^2 + 2x + 8$$

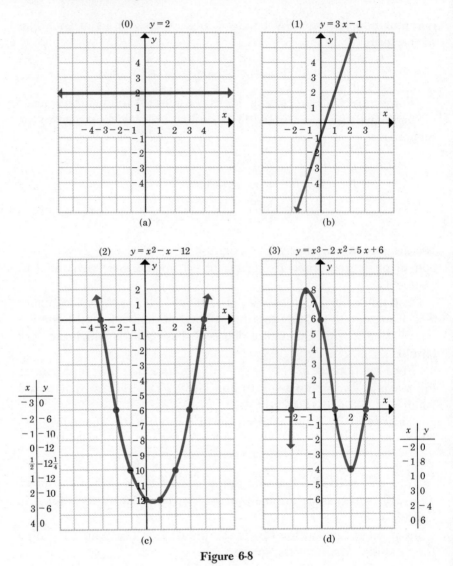

Figure 6-8

is shown in Figure 6-9. In this case $(1, 9)$ is called the **maximum point**. The graph crosses the x-axis at $(-2, 0)$ and $(4, 0)$, and the y-axis at $(0, 8)$.

There is a convenient formula for determining either the minimum point or the maximum point. If x is replaced by $-\dfrac{b}{2a}$ in the equation, the corresponding y-coordinate of the minimum or maximum point is obtained. The proof of this statement is given in Section 7.7. It

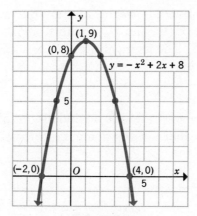

Figure 6-9

should be a challenge to you at this point, however, to see if you can construct your own proof. Meanwhile, it will be helpful to use the formula in the next set of Exercises.

In our first example, $a = 1$ and $b = -1$. Hence $-\dfrac{b}{2a} = \dfrac{1}{2}$, and so $\frac{1}{2}$ is the x-coordinate of the minimum point. When $x = \frac{1}{2}$,

$$y = \left(\frac{1}{2}\right)^2 - \left(\frac{1}{2}\right) - 12 = -12\frac{1}{4},$$

and so $(\frac{1}{2}, -12\frac{1}{4})$ is the minimum point.

In the second example, $y = -x^2 + 2x + 8$, $a = -1$, and $b = 2$. Hence the x-coordinate of the maximum point is $-\dfrac{b}{2a} = 1$. The corresponding y-coordinate is $-1^2 + 2(1) + 8 = 9$ as shown in Figure 6-9.

Before drawing the graphs of equations of the form

$$y = ax^2 + bx + c,$$

it may be expedient to locate certain key points. If the equation

$$ax^2 + bx + c = 0$$

has the real roots, r_1 and r_2, the coordinates $(r_1, 0)$ and $(r_2, 0)$ should be plotted. The point $(0, c)$ will also lie on the graph. These together with the minimum point or maximum point, as discussed, should provide a skeleton framework on which to construct the graph.

You may be tempted, however, to draw the graph by joining the three points $(r_1, 0)$, $(r_2, 0)$, and $(0, c)$ with straight lines as the dashed lines in Figure 6-10. The figure also shows, as a solid line, the correct graph of the equation

$$y = -x^2 + 2x + 8.$$

This illustration demonstrates the advisability of plotting a few additional points when there is any doubt as to the shape of the graph!

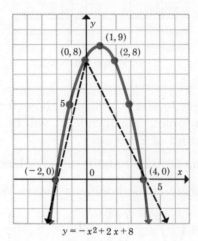

$$y = -x^2 + 2x + 8$$

Figure 6-10

EXERCISES 6.5

◀ In Exercises 1–10, draw a graph of the given equation.

1. $y = x^2 - 4x - 5$
2. $y = x^2 - x - 2$
3. $y = x^2 - 6x + 8$
4. $y = x^2 - 9$
5. $y = -x^2 + 6x - 9$

6. $y = 2x^2 + x - 1$
7. $y = -3x^2 + 5x + 2$
8. $y = 4x^2 - 1$
9. $y = -6x^2 + x + 1$
10. $y = 2x^2 - 7x - 15$

11. Sketch the graph of the equation $y = x^3$.

12. A particle dropped from a point 1000 ft. above the earth, if acted on only by the force of gravity, would fall according to the equation

$$y = 1000 - 16x^2$$

where y represents the distance in feet from the earth at any time and x is the time in seconds from the moment the particle is dropped. Draw the graph of this equation for $0 \leq x \leq 5$. (Use different scales on x- and y-axes.)

13. A rectangular parallelopiped is x ft. wide and is twice as long and 2 ft. higher than it is wide. Write an equation for the total surface area y in terms of x and sketch the graph of this equation.

14. Let y represent the volume of the figure in Exercise 13, write an equation for y in terms of x, and sketch the graph of this equation.

15. The volume of a sphere is given by the formula $y = \frac{4}{3}\pi x^3$, where x is the length of the radius. Using $\pi \approx 3.1$, sketch a graph showing the relation between the length of the radius and the volume for $0 \leq x \leq 3$.

6.6 ABSOLUTE VALUES AND INEQUALITIES

How would one graph the equation $y = |x|$? Recall that

$$|x| = x \qquad \text{if } x \geq 0$$

and

$$|x| = -x \qquad \text{if } x < 0.$$

Hence

$$|x| = \sqrt{x^2},$$

where, as you know, the radical sign $\sqrt{}$ denotes the nonnegative square root. Thus the graph of $y = |x|$ is the same as the graph of

$$y = \sqrt{x^2}.$$

By dividing the domain into two parts, $x < 0$, $x \geq 0$, we see that the graph of $y = \sqrt{x^2}$ appears as in Figure 6-11.

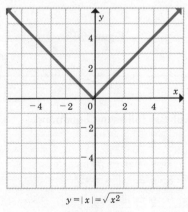

$$y = |x| = \sqrt{x^2}$$

Figure 6-11

The graphs of the equations $y = |x + 2|$, $y = |x| + 2$, and $y = -|x|$ are shown in Figure 6-12.

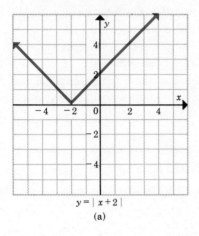

$y = |x + 2|$

(a)

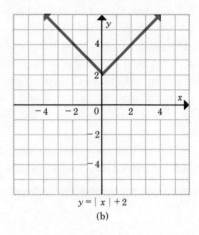

$y = |x| + 2$

(b)

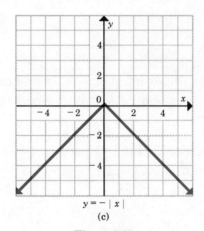

$y = -|x|$

(c)

Figure 6-12

EXERCISES 6.6A

◀Draw graphs of each of the following equations.

1. $y = |x + 3|$ 6. $y = |2x - 4|$
2. $y = |x - 2|$ 7. $y = 4 - |2x|$
3. $y = |2x|$ 8. $y = |x| + |-2|$
4. $y = |x| - 3$ 9. $y = |2 - x|$
5. $y = 3 - |x|$ 10. $y = 2 - |x|$

Much work in mathematics involves the use of inequalities as well as equalities. In your previous studies you have probably been asked to sketch the graphs of inequalities such as

$$3x + 2y < 7$$

and

$$x - 4y \geq 5.$$

We shall review this topic briefly in this section. Such a review will be useful in preparation for some of the activity in Chapter 7.

Consider first the inequality

$$y < x + 2.$$

To determine the set of all pairs (x, y) which satisfy this inequality we first examine the graph of the related equation

$$y = x + 2$$

shown in Figure 6-13.

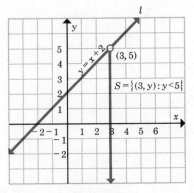

Figure 6-13

Now let us consider a point on the line $l : y = x + 2$, say $(3, 5)$. Next consider the set of all points having x-coordinate 3, that is, the set $T = \{(3, y) : y \in R\}$. Which of these points satisfy the inequality

$$y < x + 2?$$

Clearly the inequality is not satisfied if $x = 3$ and $y \geq 5$. We can see, however, that all ordered pairs of the set

$$S = \{(3, y) : y \in R \text{ and } y < 5\}$$

satisfy the inequality $y < x + 2$. The graph of the set S is also shown in Figure 6-13.

Another point on $l : y = x + 2$ is $(2, 4)$. Consider the set

$$U = \{(2, y) : y \in R \text{ and } y < 4\}.$$

Do all points of U satisfy the inequality $y < x + 2$?
What about the sets

$$V = \{(1, y) : y \in R \text{ and } y < 3\},$$
$$W = \{(-1, y) : y \in R \text{ and } y < 1\},$$

and

$$X = \{(-2, y) : y \in R \text{ and } y < 0\}?$$

The graphs of these sets are shown in Figure 6-14.

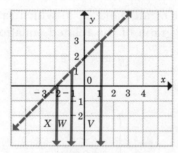

Figure 6-14

Figure 6-15 shows the graphs of other sets similar to the sets S, U, V, W, and X.

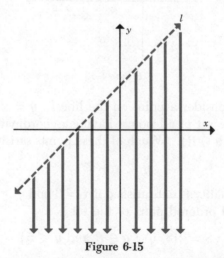

Figure 6-15

From this discussion it should be evident that the complete graph of the inequality $y < x + 2$ consists of the entire region, or half-plane, lying below the line $l : y = x + 2$. This we might indicate by shading as in Figure 6-16.

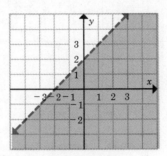

Figure 6-16

However, it is customary, and more convenient, to picture the graph by a set of vertical rays as in Figure 6-15.

Note that the line $l : y = x + 2$ as indicated by the dashed lines in Figures 6-15 and 6-16 is not a part of the graphs. For the inequality

$$y \leq x + 2,$$

however, the line should be included in the graph. It should also be clear that the graph of

$$y > x + 2$$

would lie on the opposite side of the line from the graph of $y < x + 2$.

As a second illustration, consider the inequality

$$x - 3y \leq 4.$$

Using the properties of order, we get

$$-3y \leq -x + 4,$$
$$y \geq \frac{1}{3}x - \frac{4}{3}.$$

(Note that multiplication by $-\frac{1}{3}$ reversed the sense of the inequality.) The related equation is

$$y = \frac{1}{3}x - \frac{4}{3}.$$

The graph of the inequality is shown in Figure 6-17. Since the graph includes all the points whose coordinates satisfy $y = \frac{1}{3}x - \frac{4}{3}$, the line $l : y = \frac{1}{3}x - \frac{4}{3}$ is included in the graph of $y \geq \frac{1}{3}x - \frac{4}{3}$.

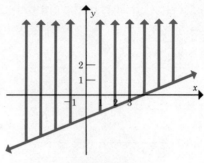

Figure 6-17

A *word of caution:* Note that in its original form the inequality

$$x - 3y \leq 4$$

with its "less than or equal to" symbol might have led us to put the graph of the inequality "below" the line $l : x - 3y = 4$. Our correct use of the Order Properties, however, led to the final form

$$y \geq \frac{1}{3}x - \frac{4}{3}$$

and prevented this mistake.

In considering solution sets of inequalities we have assumed the condition $x, y \in R$. How might we sketch the graph of the set

$$T = \{(x, y) : x + y < 4 \text{ and } x, y \in I\}?$$

Since the coordinates are restricted to the integers, a portion of the graph would appear as in Figure 6-18.

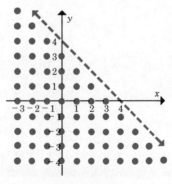

Figure 6-18

EXERCISES 6.6B

◀In Exercises 1–10, sketch the graph of the given set.

1. $S = \{(2, y) : y \in R \text{ and } y \leq 6\}$
2. $S = \{(4, y) : y \in R \text{ and } y > 1\}$
3. $S = \{(-3, y) : y \in R\}$
4. $S = \{(5, y) : y \in R \text{ and } y > -5\}$
5. $S = \{(-4, y) : y \in R \text{ and } y \geq 2\}$
6. $T = \{(x, y) : x + y < 5 \text{ and } x, y \in I\}$
7. $T = \{(x, y) : x < y \text{ and } x, y \in I\}$
8. $T = \{(x, y) : y > x - 2 \text{ and } x, y \in I\}$
9. $T = \{(x, y) : 3x - y > 1 \text{ and } x, y \in I\}$
10. $T = \{(x, y) : 4x + y < 1 \text{ and } x, y \in I\}$

◀In Exercises 11–25, sketch the graph of the given inequality. (*Note:* In Exercises 11–30, we are assuming the condition $x, y \in R$.)

11. $y > x + 6$
12. $y \leq 2x - 1$
13. $x + 3y \leq 6$
14. $3x - y > 0$
15. $4x - 5y \leq 0$
16. $x - \frac{1}{2}y > 2$
17. $\frac{2}{3}x + \frac{1}{2}y \geq 1$
18. $x - y < 8$
19. $x > 6$
20. $y \leq -1$
21. $3y < \frac{1}{2}x - 1$
22. $6y - 2x > 10$
23. $3x \leq 5y - 1$
24. $x + 0.2y < 1.5$
25. $1.4x - 0.7y > 1$

26. **CHALLENGE PROBLEM.** Sketch the graph of $y \geq |x|$.
27. **CHALLENGE PROBLEM.** Sketch the graph of $y < x^2$.
28. **CHALLENGE PROBLEM.** Sketch the graph of $y < |x - 1|$.
29. **CHALLENGE PROBLEM.** Sketch the graph of $y < x^2 - 7x + 12$.
30. **CHALLENGE PROBLEM.** Sketch the graph of $y \leq x^3$.

6.7 APPROXIMATE SOLUTIONS

In discussing the graphs of polynomial equations of degree greater than one, we commented on the points at which the graphs intersected the x-axis, that is, points with zero ordinates. The values of the x-coordinates of such points are the real number solutions of the equations. We have seen how information concerning these roots aids in the construction of the graphs. (See Figures 6-8 and 6-9 in Section 6.5.) Conversely, as we shall see, the graph concept can furnish clues for finding solutions.

In Chapter 5 we saw that if polynomial equations with integral coefficients have rational roots, then these can be found by synthetic division. Irrational solutions of equations of degree higher than two are more difficult to come by.

In this section we shall develop a method for finding approximations to irrational roots as promised in Section 5.5.

Our proposed method of finding an irrational solution to equations of the type under consideration will differ in one very major respect from the procedure for determining solutions which are rational numbers. We shall be finding *approximate* rather than *exact* solutions. In a great many situations, for example in problems involving measurement, one is not able to come up with an exact answer. Usually all that is required is to provide the degree of accuracy that is called for in the given application. Thus if the demands of a problem call for an approximate solution correct to the nearest hundredth, this will be sufficient. There would be no need to struggle for a solution correct to the nearest thousandth.

Our method of finding an approximate solution is, in rough outline, as follows. We first determine, for example, the two consecutive integers between which a sought-for solution lies. Then, thinking in terms of a number line, we find two points closer together than two consecutive integers such that the segment between them still contains the solution, and so on.

The method is perhaps best illustrated by a carefully worked out example. Before starting on the example, however, let us take another look at the graph of the equation $y = x^3 - 2x^2 - 5x + 6$. We saw in Section 6.5 (Figure 6-8d) that the graph crossed the x-axis at the points $(-2, 0)$, $(1, 0)$, and $(3, 0)$ corresponding to the zeros of the polynomial $x^3 - 2x^2 - 5x + 6$. For easy reference, the graph is given again in Figure 6-19.

In general, the graph of an equation of the form

$$y = a_n x^n + a_{n-1} x^{n-1} + \cdots + a_0,$$

with the a's any real numbers, will cross the x-axis at points corre-

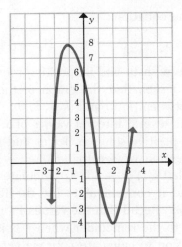

Figure 6-19

sponding to real number solutions of the equation

$$a_n x^n + a_{n-1} x^{n-1} + \cdots + a_0 = 0$$

—a phenomenon which provides the clue to our solution procedure. Let us look, as in Figure 6-20, at a portion of the graph of Figure 6-19 near the point $(3, 0)$. We see that the value of the function for $1 < x < 3$ is negative, whereas the value of the function for $x > 3$ is positive. The significant fact here is that the functional value undergoes a change from negative to positive as x increases from a value slightly less than 3 to a value greater than 3. In the neighborhood of $(1, 0)$, on the other hand, as shown in Figure 6-21, the functional value changes from positive to negative as x increases from a value less than 1 to a value slightly greater than 1.

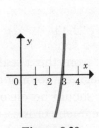

Figure 6-20

Figure 6-21

Consider now the equation

$$f(x) = x^4 + 3x^3 - 5x^2 - 21x - 14 = 0$$

and suppose that we wish to find its positive root. We first find two integers k and $k + 1$ such that the graph of the equation $y = f(x)$ crosses the x-axis between $(k, 0)$ and $(k + 1, 0)$. To do this we need to determine $f(x)$ for several integral values of x.

Since we seek a positive solution, let us find $f(0)$, $f(1)$, $f(2)$, $f(3)$, and so on. By inspection we see that $f(0) = -14$. A fairly simple substitution tells us that $f(1) = -36$. To calculate the other functional values we might find it convenient to use the Remainder Theorem and synthetic division. Recall that the value of a polynomial function f for $x = k$ is equal to the remainder when $f(x)$ is divided by $x - k$. Using synthetic division to calculate $f(2)$, we get

$$\underline{2|}\ \ \begin{array}{ccccc} 1 & 3 & -5 & -21 & -14 \\ & 2 & 10 & 10 & -22 \\ \hline 1 & 5 & 5 & -11 & \boxed{-36} \end{array}.$$

From this we see that $f(2) = -36$. To obtain $f(3)$ we write

$$\underline{3|}\ \ \begin{array}{ccccc} 1 & 3 & -5 & -21 & -14 \\ & 3 & 18 & 39 & 54 \\ \hline 1 & 6 & 13 & 18 & \boxed{40} \end{array}.$$

Hence we see that $f(3) = 40$.

The important consideration here is the fact that $f(2) < 0$, whereas $f(3) > 0$. Thus we may confidently predict that the graph crosses the x-axis somewhere between $(2, 0)$ and $(3, 0)$. In other words, we have located a solution between 2 and 3. Figure 6-22 illustrates the idea behind our next step. The curve represents a small portion of the graph of the equation

$$y = x^4 + 3x^3 - 5x^2 - 21x - 14,$$

where the y-scale has been made different from the x-scale to provide greater detail. We have drawn a line segment joining the points $(2, -36)$ and $(3, 40)$. From the drawing it should be clear that the x-coordinate of the point where this line segment crosses the x-axis furnishes what might be called a first approximation to the solution we are seeking.

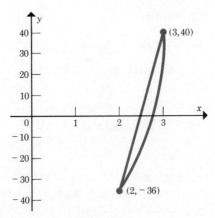

Figure 6-22

We form, then, an equation of the line determined by $(2, -36)$ and $(3, 40)$. This is a familiar operation. The slope is 76. Hence an equation of this line is

$$y + 36 = 76(x - 2).$$

We wish to find the x-coordinate corresponding to a y-coordinate of 0. By substitution we get

$$36 = 76x - 152,$$
$$76x = 188,$$

and

$$x = \frac{188}{76} \approx 2.5.$$

Here, then, is our first approximation rounded to one decimal place. (Recall that the symbol "$\approx$" means "is approximated by.")

Before proceeding to a further refinement, we should first evaluate $f(2.5)$ to see whether it is too large or too small. The graph might furnish a clue, but normally such a detailed graph would not be available.

Again using synthetic division we have

$$
\begin{array}{r|rrrr}
2.5 & 1 & 3 & -5 & -21 & -14 \\
    &   & 2.5 & 13.75 & 21.875 & 2.1875 \\
\hline
    & 1 & 5.5 & 8.75 & 0.875 & -11.8125
\end{array}
$$

and we see that $f(2.5) \approx -12$. Thus, since $f(2.5)$ is negative, we may infer that 2.5 is less than the true solution. Hence the solution lies between 2.5 and 3 as shown in Figure 6-23.

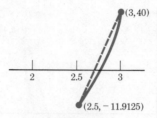

Figure 6-23

For a second approximation, that is, a solution to two decimal places, we repeat the process of considering a line between two points. To do this we might use the two points $(2.5, -12)$ and $(3, 40)$. (See the dashed line in Figure 6-23.) However, in this instance, since $f(2.5) \approx -12$, a rather large negative number, we might suspect that $f(2.6)$ is also a negative number; hence a point with abscissa 2.6 would also lie below the line. To evaluate $f(2.6)$ we write

2.6	1	3	−5	−21	−14
		2.6	14.56	24.856	10.0256
	1	5.6	9.56	3.856	−3.9744

and find that $f(2.6) \approx -4.0$. From the fact that $f(2.6) < 0$ we see that our suspicion was correct. The solution is greater than 2.6, that is, the solution lies between 2.6 and 3.

Does the solution lie between 2.6 and 2.7? A similar check by synthetic division tells us that $f(2.7) = 5.0431 \approx 5.0$, as you may verify. Thus we should use the points $(2.6, -4)$ and $(2.7, 5)$ in the next step.

For a second approximation, then, we can construct another line, the line joining $(2.6, -4)$ and $(2.7, 5)$. The slope of this line is $\dfrac{5 - (-4)}{2.7 - 2.6} = 90$. Hence an equation which describes the line is

$$y + 4 = 90(x - 2.6).$$

(Check this!) If we now let $y = 0$, we obtain $90x = 238$ and hence $x = \frac{238}{90} \approx 2.64$.

The procedure that we have just outlined for finding an approximation to a root of

$$f(x) = x^4 + 3x^3 - 5x^2 - 21x - 14 = 0$$

gives a geometric interpretation to each step in the process. In practice, however, it is more efficient to use merely a succession of synthetic division steps as follows.

Since $f(2) = -36$ and $f(3) = 40$, then $2 < x < 3$ (Figure 6-22). Continuing, we have

$$f(2.5) = -11.912 \qquad \text{so that} \qquad 2.5 < x < 3 \text{ (Figure 6-23)};$$
$$f(2.6) = -3.9744 \qquad \text{so that} \qquad 2.6 < x < 3;$$
$$f(2.7) = 5.0431 \qquad \text{so that} \qquad 2.6 < x < 2.7.$$

A final synthetic division step shows us that $f(2.64) < 0$ and $f(2.65) > 0$ so that we may conclude that a reasonably good approximation for a root is 2.64.

Now that the job of illustrating this approximation process is completed, it is very much in order to indicate that, *in this case,* considerable simplification would have been possible. In investigating the real zeros of any polynomial with integral coefficients it is almost always worthwhile to see first if the polynomial has any rational zeros. Using the theorem on rational roots we see that the only possible rational zeros of $x^4 + 3x^3 - 5x^2 - 21x - 14$ are ± 1, ± 2, ± 7, and ± 14. If we are lucky we might first test -1 and then -2.

$$
\begin{array}{r|rrrrr}
-1 & 1 & 3 & -5 & -21 & -14 \\
   &   & -1 & -2 & 7 & 14 \\
\hline
-2 & 1 & 2 & -7 & -14 & 0 \\
   &   & -2 & 0 & 14 & \\
\hline
   & 1 & 0 & -7 & 0 &
\end{array}
$$

Hence
$$x^4 + 3x^3 - 5x^2 - 21x - 14 = (x + 1)(x + 2)(x^2 - 7)$$

and we see that the positive irrational zero that we were so desperately seeking is actually $\sqrt{7}$ which, to 6 decimal places, is 2.645751. Thus the approximation of 2.64 that we obtained is in error by less than 0.006.

The general method of finding approximate solutions by constructing lines as illustrated above is known as **linear interpolation.**

It is apparent that even an approximation to two decimal places involves a fairly heavy amount of arithmetic. Further refinements to three or four places would obviously make for even rougher going. Fortunately, in this day and age modern computers can be programmed to do the calculations. It is essential, however, that we understand the principles behind the process in order to construct a computer program, that is, tell the computer what to do. Thus even in an age of automation, mathematical know-how is still at a premium. Here, then, is some more useful theory.

You may recall that in first presenting the previous example we asked for *the* positive solution of the equation

$$x^4 + 3x^3 - 5x^2 - 21x - 14 = 0.$$

The use of the definite article "the" implies the existence of one and only one positive solution. How might one have predetermined that such a solution exists without, of course, knowing about the existence of rational roots?

There is a theorem, attributed to the same René Descartes who developed analytic geometry, which can be stated as follows.

THEOREM 6.1

The number of positive real roots of an equation $f(x) = 0$, where $f(x)$ is a polynomial with real coefficients, is either equal to the number v of variations in sign or is equal to $v - e$, where e is a positive even number.

By *variations in sign*, we mean a change in the "sign" (positive or negative) of the coefficients as one proceeds in a "count down" from a_n to a_0. Here are some examples:

$f(x)$	v
$x^4 + 3x^3 - 5x^2 - 21x - 14$	1
$x^5 - 2x^4 + x^3 - x^2 - x + 1$	4
$x^5 + 2$	0
$x^5 - x^2 - 7$	1

(Note, as in the last two examples, that in tabulating the number of variations we ignore the zero coefficients.)

This theorem is quite helpful under certain circumstances. For example, the equation we used to illustrate a method of approximation has one variation and hence one positive real solution. When this is approximated, there is no need to search for other positive roots.

If, for a given polynomial $f(x)$, $v = 2$, the equation $f(x) = 0$ has either two positive solutions or none. If $v = 3$, the equation has either 3 positive solutions or 1.

You will be asked to discuss Descartes' Theorem at greater length in the Exercises.

EXERCISES 6.7

◀ In Exercises 1–10, use Descartes' Theorem to list the possible number of positive solutions for each of the given equations. Give alternatives where they exist. (You need not attempt to solve the equations.)

1. $x^5 + 3x^4 + 6x^3 - x^2 - x - 3 = 0$
2. $x^4 - x^3 + 5x^2 - 6x - 4 = 0$
3. $3x^5 + 4x^4 + 2x^3 + x + 8 = 0$
4. $-8x^5 - 2x^4 - 6x^3 - x - 7 = 0$
5. $x^8 + x^5 - 1 = 0$
6. $x^7 + x^6 - x^5 + x^3 - 1 = 0$
7. $x^{10} - 2 = 0$
8. $x^9 + 1 = 0$
9. $-x^3 - 7 = 0$
10. $x = 0$

◀ In Exercises 11–15, show first that the given equation has no rational solutions. Then find an approximate positive solution for each of the equations by using two linear interpolations. (*Note:* As in the example of this section, you should examine your first (that is, one decimal place) approximation to see if it can be improved on before making the second linear interpolation.)

11. $x^3 + x^2 - 8x - 3 = 0$
12. $x^3 - 7x - 1 = 0$
13. $x^3 + x^2 - 13 = 0$
14. $x^3 - 6x - 11 = 0$
15. $x^3 + x^2 + x - 1 = 0$

◀ In Exercises 16–20, follow the same directions as you did in Exercises 11–15 but instead of linear interpolation use the method of successive synthetic divisions.

16. $x^4 + x^3 - x^2 - 2x - 2 = 0$
17. $x^4 + x^3 - 4x^2 - 5x - 5 = 0$
18. $x^4 + x^3 - 14x^2 - 15x - 15 = 0$
19. $x^5 + x - 23 = 0$
20. $x^5 - x - 1 = 0$

◀ The method of linear interpolation can be used to determine approximate cube roots, fourth roots, etc. For example, the real cube root of 2 is the positive solution of the equation $x^3 - 2 = 0$. In Exercises 21–25, use two linear interpolations to approximate the given root.

21. $\sqrt[3]{3}$ **24.** $\sqrt[4]{5}$

22. $\sqrt[3]{6}$ **25.** $\sqrt[4]{7}$

23. $\sqrt[3]{11}$

26. The volume of a sphere of radius r is $\frac{4}{3}\pi r^3$. A sphere is made of a certain material whose specific gravity is $\frac{2}{3}$, that is, a cubic foot of the material weighs $\frac{2}{3}$ as much as a cubic foot of water. To find the depth d to which a sphere 2 ft. in diameter will sink in water, we proceed as follows. Since $r = 1$, the sphere weighs as much as $\frac{4}{3}\pi \cdot \frac{2}{3}$ cu. ft. of water. The volume of the submerged part of the sphere is $\pi d^2(1 - \frac{1}{3}d)$. This must equal the volume of the displaced water, and so

$$\pi d^2 \left(1 - \frac{1}{3}d\right) = \frac{4}{3}\pi \cdot \frac{2}{3}.$$

Find an approximation to d using two steps according to the method of this section.

27. Determine how far a sphere 2 ft. in diameter will sink in water if the specific gravity of the sphere is $\frac{1}{4}$. (See Exercise 26.)

28. An amount of money P deposited at $r\%$ compounded annually will have a value $P(1 + r)^n$ at the end of n years. If a deposit of $200 is worth $240 at the end of three years, what is the rate, r, of interest to two decimal places?

29. Calculate r as in Exercise 28 for an investment of $100 which in four years is worth $150.

30. The volume of a right circular cylinder is 3π cubic units. The total area of its surface is 10π square units. This means that the radius r of its base is a root of the equation $r^3 - 5r + 3 = 0$. Find the smallest possible value of r to two decimal places.

31. **CHALLENGE PROBLEM.** Give an informal argument to support Descartes' Theorem on the number of positive real roots. (*Hint:* Use synthetic division.)

32. **CHALLENGE PROBLEM.** Develop an analogous theorem to Descartes' Theorem on the possible number of negative real roots of a polynomial equation $f(x) = 0$. (*Hint:* Substitute $-x$ for x in $f(x)$ and use the argument of Exercise 31.)

6.8   ALGEBRAIC PROOFS OF GEOMETRIC THEOREMS

Through the use of graphs and a coordinate system we have seen one aspect of the partnership of geometry and algebra. There is a

further connection which is perhaps even more profound in its impli-
cations.

In your geometry course, you have undoubtedly encountered many
fundamental theorems that had been deduced from a set of geometric
axioms and postulates, that is, geometric properties which are assumed.

One of the far-reaching consequences of the introduction of a
coordinate system is the opportunity it provides for proving geometric
theorems by algebraic means. In this section we shall present an
illustration and let you proceed with a few more. Although we cannot
attempt to cover the ground fully, you should gain from this experience
an appreciation of the enormous potential which this blending of
algebra and geometry unveils.

Before getting into the heart of the matter, however, we need to
examine a few more ideas.

We have already discussed the concept of distance between two
points, which may also be interpreted as the length of a line segment
joining the two points. We now wish to investigate some related ideas.
Given two points P and Q, what are the coordinates of a point M
which lies on the line determined by P and Q and is half the distance
from P to Q? What are the coordinates of a point on the same line
one-third of the distance or any other specified portion of the distance
from P to Q?

Let us first consider an example. Suppose $P = (3, 4)$ and $Q =
(9, 6)$ as in Figure 6-24. Simply by examining this figure what would you
suggest that the coordinates of M might be?

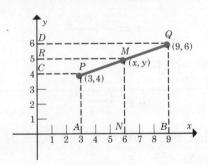

Figure 6-24

From your study of geometry you know that if M is the midpoint
of the segment $\overline{PQ}$, then its projection N on the x-axis is the midpoint
of the segment $\overline{AB}$. Likewise the projection of M on the y-axis is
the midpoint R of the segment $\overline{CD}$. From the graph we can conjecture
that $N = (6, 0)$ and $R = (0, 5)$. Hence it appears that $M = (6, 5)$.

Let us now move from this example to the general case where we are given the points $P(x_1, y_1)$ and $Q(x_2, y_2)$. How do we determine $M(x, y)$, the midpoint of the segment $\overline{PQ}$? From Figure 6-25, we see

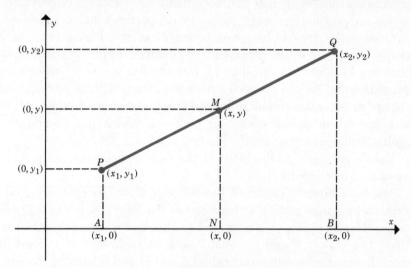

Figure 6-25

that $AN = NB$. How do we determine a point on the x-axis halfway between x_1 and x_2? Since, in Figure 6-25, $x - x_1 = x_2 - x$, it follows that

$$2x = x_1 + x_2$$

or

$$x = \frac{x_1 + x_2}{2}.$$

It seems reasonable also to suggest that

$$y = \frac{y_1 + y_2}{2}.$$

Thus it appears that the coordinates of a point M midway between $P(x_1, y_1)$ and $Q(x_2, y_1)$ are

$$\left(\frac{x_1 + x_2}{2}, \frac{y_1 + y_2}{2} \right).$$

Let us recall, however, that this result was inferred from a particular drawing and some geometric considerations. Furthermore, in arriving at a conjecture we used the fact that $x_2 > x_1$ and $y_2 > y_1$.

We wish now to raise the question as to whether or not $\left(\dfrac{x_1 + x_2}{2}, \dfrac{y_1 + y_2}{2}\right)$ gives us the desired midpoint in all cases. Here is where algebra enters the picture. How can we prove that for any two points, $P(x_1, y_1)$ and $Q(x_2, y_2)$, the midpoint of $\overline{PQ}$ is $M\left(\dfrac{x_1 + x_2}{2}, \dfrac{y_1 + y_2}{2}\right)$? The proof involves two steps. We must show that

(A) M lies on the line l determined by the points P and Q
and
(B) $PM = MQ$.

We can verify (A) by showing that if an equation

$$ax + by + c = 0$$

is satisfied by the coordinates of both P and Q, then it is also satisfied by the coordinates of M.

Suppose, then, that we are given the equation

$$ax + by + c = 0$$

with a and b not both zero. If (x_1, y_1) and (x_2, y_2) are points on the line, then

$$ax_1 + by_1 + c = 0$$

and

$$ax_2 + by_2 + c = 0$$

are true statements. By the property of real numbers which says that if $m = n$ and $r = s$, then $m + r = n + s$, it follows that

$$(ax_1 + by_1 + c) + (ax_2 + by_2 + c) = 0 + 0$$

or

(1) $$a(x_1 + x_2) + b(y_1 + y_2) + 2c = 0.$$

Multiplying each side of (1) by $\frac{1}{2}$, we get

(2) $$a\left(\frac{x_1 + x_2}{2}\right) + b\left(\frac{y_1 + y_2}{2}\right) + c = 0.$$

If (1) is true, then so is (2), by which we see that the coordinates

$$\left(\frac{x_1 + x_2}{2}, \frac{y_1 + y_2}{2}\right)$$

also satisfy the equation $ax + by + c = 0$. Hence point M is indeed on l.

Part (B) can be proved using the fact that

$$PM = \sqrt{\left(\frac{x_1 + x_2}{2} - x_1\right)^2 + \left(\frac{y_1 + y_2}{2} - y_1\right)^2}$$

and a corresponding representation of MQ. You are asked to do this in the Exercises.

The foregoing discussion has shown, in part, how an algebraic proof can be used to substantiate a conjecture based on geometry.

We are now ready to demonstrate some proofs of geometric theorems using coordinates. One well-known theorem in geometry is Theorem 6.2.

THEOREM 6.2

The length of the line segment joining the midpoints of two sides of a triangle is equal to half the length of the third side.

In constructing a proof, it is convenient to locate the triangle in relation to the coordinate axes, as shown in Figure 6-26. This type of orientation reduces the number of symbols needed in the proof without loss of generality.

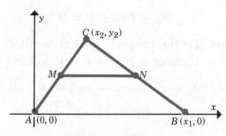

Figure 6-26

For the midpoints M and N we use the midpoint formula to obtain

$$M = \left(\frac{x_2 + 0}{2}, \frac{y_2 + 0}{2}\right) = \left(\frac{x_2}{2}, \frac{y_2}{2}\right)$$

and

$$N = \left(\frac{x_1 + x_2}{2}, \frac{y_2 + 0}{2}\right) = \left(\frac{x_1 + x_2}{2}, \frac{y_2}{2}\right).$$

From Figure 6-26 we see that the length of the third side, AB, is $|x_1|$. Let us now calculate MN. By definition,

$$MN = \sqrt{\left(\frac{x_1 + x_2}{2} - \frac{x_2}{2}\right)^2 + \left(\frac{y_2}{2} - \frac{y_2}{2}\right)^2}$$

which reduces to

$$\sqrt{\left(\frac{x_1}{2}\right)^2} = \frac{1}{2}\sqrt{x_1{}^2}.$$

Since $\frac{1}{2}\sqrt{x_1{}^2} = \frac{1}{2}|x_1|$, we see that $MN = \frac{1}{2}AB$, and our proof is complete.

We shall close this section with an algebraic proof of another geometric theorem. You can give assistance by completing some of the steps.

THEOREM 6.3
The diagonals of a parallelogram bisect each other.

Proof:
1. Locate the parallelogram with respect to the x- and y-axes as shown in Figure 6-27 with all of the coordinates nonnegative.

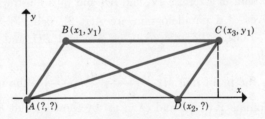

Figure 6-27

2. $A = (\boxed{?}, \boxed{?})$, $B = (x_1, y_1)$, $C = (x_3, y_1)$, and $D = (x_2, \boxed{?})$.
3. The midpoint of the line segment joining B and D is
$$\left(\frac{x_1 + x_2}{2}, \frac{y_1}{2}\right).$$
4. The midpoint of the line segment joining A and C is $(\boxed{?}, \boxed{?})$.
5. Since the figure is, by hypothesis, a parallelogram, we know that $AD = BC$.
6. Since $AD = \boxed{?}$ and $BC = \boxed{?}$, we know that $x_3 - x_1 = x_2$; and hence $x_3 = \boxed{?} + \boxed{?}$.

7. Therefore the coordinates of the two midpoints in steps (3) and (4) are the same and the theorem is proved.

EXERCISES 6.8

◀ In Exercises 1–10, determine the midpoint of the line segment joining the two given points.

1. $(2, 4)$, $(8, 12)$

2. $(1, 7)$, $(9, 13)$

3. $(4, 8)$, $(11, 12)$

4. $(-2, 10)$, $(6, -4)$

5. $(-3, 11)$, $(-8, -11)$

6. $\left(\frac{2}{3}, 1\right)$, $\left(-\frac{1}{4}, 6\right)$

7. $(0.6, 1.4)$, $(3.2, 1.7)$

8. $\left(-\frac{2}{3}, \frac{1}{4}\right)$, $\left(\frac{1}{4}, -\frac{1}{2}\right)$

9. $\left(\frac{3}{5}, \frac{2}{5}\right)$, $\left(-\frac{3}{5}, -\frac{2}{5}\right)$

10. $(1.6, -2.7)$, $(3.8, -2.8)$

11. Find the point of intersection of the diagonals of a parallelogram whose vertices are $(3, 8)$, $(11, 8)$, $(5, 2)$, and $(13, 2)$, respectively.

12. The vertices of a triangle ACB are $A(2, 3)$, $C(5, 11)$, and $B(7, 4)$. Determine the coordinates of the midpoint M of $\overline{AC}$ and the midpoint N of $\overline{BC}$.

13. Using the data of Exercise 12, determine MN.

14. Using the data of Exercise 12, find AB and thus illustrate Theorem 6.2.

15. The vertices of a parallelogram are $A(7, 2)$, $B(10, 7)$, $C(20, 11)$, and $D(17, 6)$. Locate the midpoints of $\overline{AC}$ and $\overline{BD}$ and thus illustrate Theorem 6.3.

16. Complete the proof that $M\left(\dfrac{x_1 + x_2}{2}, \dfrac{y_1 + y_2}{2}\right)$ is the midpoint of the segment joining $P(x_1, y_1)$ and $Q(x_2, y_2)$ by showing that $PM = MQ$.

17. Copy and complete the steps in the following argument: A line segment joining the midpoints of two sides of a triangle is parallel to the third side.

(a) Given the triangle of Figure 6-26, the slope of line $\overleftrightarrow{MN}$ is

$$\frac{\dfrac{y_2}{2} - \dfrac{\boxed{?}}{\boxed{?}}}{\dfrac{x_1 + x_2}{2} - \dfrac{\boxed{?}}{\boxed{?}}} = \frac{\boxed{?}}{\boxed{?}}.$$

(b) The slope of line segment $\overline{AB} = \boxed{?}$.

(c) Therefore the two line segments are $\boxed{?}$.

18. Given the following assertion: A point R on a line segment $\overline{PQ}$ two-thirds of the distance from $P(x_1, y_1)$ to $Q(x_2, y_2)$ has coordinates $(\frac{2}{3}x_2 + \frac{1}{3}x_1, \frac{2}{3}y_2 + \frac{1}{3}y_1)$. Illustrate this assertion for the points $P(1, 4)$ and $Q(10, 16)$. (*Hint:* Find the coordinates of R using the assertion and the given coordinates of P and Q. Now compute PR and PQ.)

19. CHALLENGE PROBLEM. Verify the assertion of Exercise 18 for the general case by showing that $R(\frac{2}{3}x_2 + \frac{1}{3}x_1, \frac{2}{3}y_2 + \frac{1}{3}y_1)$ lies on $\overline{PQ}$ and that $PR = \frac{2}{3}PQ$.

20. CHALLENGE PROBLEM. Prove that the medians of a triangle intersect in a point two-thirds of the distance from each vertex to the midpoint of the side opposite. (*Hint:* See Figure 6-28.)

21. CHALLENGE PROBLEM. Prove the converse of Theorem 6.3; that is, prove that if the diagonals of a quadrilateral bisect each other, then the quadrilateral is a parallelogram. (*Hint:* Using Figure 6-27, show that the coordinates of point C are $(x_1 + x_2, y_1)$.)

22. CHALLENGE PROBLEM. Given the triangle PQR as illustrated in Figure 6-29 where $\overline{RS}$ is perpendicular to the x-axis (that is, $\overline{RS}$ is an altitude of the triangle). Prove that

$$(QR)^2 = (PR)^2 + (PQ)^2 - 2PQ \cdot PS.$$

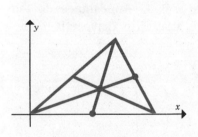

Figure 6-28

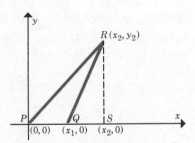

Figure 6-29

6.9   COORDINATES IN THREE DIMENSIONS

Many of the ideas discussed in this chapter relative to a coordinate system in the plane can be extended to a coordinate system in three-space. We shall discuss a few of these briefly. You will have an opportunity to go more deeply into the subject later.

Our rectangular coordinate system in the plane was based on two perpendicular number lines, the x- and y-axes. For a system in three dimensions we use three mutually perpendicular axes as illustrated in Figure 6-30, where the x-, y-, and z-axes are labeled as shown.

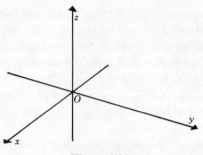

Figure 6-30

We can establish a one-to-one correspondence between points in three-space and ordered triples of real numbers (x, y, z), where z denotes the directed distance (positive, up—negative, down) from the xy-plane, y denotes the distance (positive to the right—negative to the left) from the xz-plane, and x denotes the directed distance (positive, out toward you—negative away from you) from the yz-plane. The general idea is conveyed in Figure 6-31. In this figure, x is the directed distance from 0 to A, which is the same as the distance from B to Q. Hence $x = 5$. The directed distance from O to B is y. Hence $y = 7$. This is the same as the directed distance from A to Q. The directed distance from Q to P, that is, z, is 4. Appearances to the contrary notwithstanding (the figure is drawn in perspective), the parallelogram $OAQB$ is a rectangle.

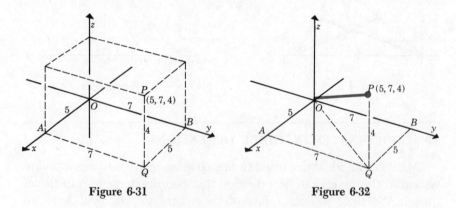

Figure 6-31   Figure 6-32

With the help of Figure 6-32 we can now develop an extension of the distance formula in the plane to a distance formula in three-space. From the Pythagorean Theorem we know that

(1) $$(OQ)^2 = (OA)^2 + (AQ)^2.$$

If we now consider right triangle OQP, we note that OQ is the length of one side, QP is the length of the other side, and OP is the length of the hypotenuse. Applying the Pythagorean Theorem a second time we have

(2) $$(OP)^2 = (OQ)^2 + (QP)^2.$$

If we now use the equality for $(OQ)^2$ from (1) in (2), we get

$$(OP)^2 = (OA)^2 + (AQ)^2 + (QP)^2.$$

Hence, since all distances are nonnegative, we have

$$OP = \sqrt{(OA)^2 + (AQ)^2 + (QP)^2}.$$

Substituting appropriate values from the numerical example in Figure 6-32, we see that in this case

$$OP = \sqrt{5^2 + 7^2 + 4^2} = \sqrt{90}.$$

In general, the formula for the distance d between points $P(x_1, y_1, z_1)$ and $Q(x_2, y_2, z_2)$ is

(3) $$d = \sqrt{(x_2 - x_1)^2 + (y_2 - y_1)^2 + (z_2 - z_1)^2}.$$

This generalization is supported by Figure 6-33. In this figure, $PA = |x_2 - x_1|$, $AB = |y_2 - y_1|$, and $BQ = |z_2 - z_1|$. Thus Formula (3) follows since

$$|x_2 - x_1|^2 = (x_2 - x_1)^2,$$
$$|y_2 - y_1|^2 = (y_2 - y_1)^2,$$

and

$$|z_2 - z_1|^2 = (z_2 - z_1)^2.$$

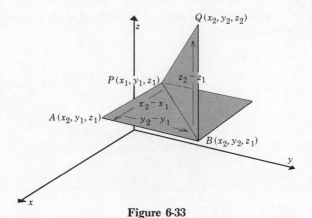

Figure 6-33

From this formula for distance we can move readily to a consideration of a well-known three-dimensional figure, the sphere. Since a sphere can be characterized as the set of all points in three-space which are equidistant from a given point, we can form an equation for this set quite directly.

To find, for example, an equation of a sphere with radius 5 and center at $(1, 3, 6)$, we can write

$$\sqrt{(x - 1)^2 + (y - 3)^2 + (z - 6)^2} = 5,$$

where (x, y, z) is any point on the sphere. From this we form an equivalent equation,

$$(x - 1)^2 + (y - 3)^2 + (z - 6)^2 = 25.$$

Much of our earlier work in this chapter dealt with linear equations of the form $ax + by + c = 0$ which were associated with lines in the plane. We wish now to consider the related idea of equations whose truth sets correspond in three-space to the set of all points in a plane.

To do this we shall need a working definition of what is meant by a plane. The idea of a plane as a "flat surface" without thickness is intuitively familiar. Since it is not easy to use the notion of "flatness" in forming mathematical concepts, however, we shall define a plane more precisely as follows.

A plane in three-space is a nonempty set α of all points such that if (x_1, y_1, z_1) is a fixed point in α and (x, y, z) is any other point in α, then the lines joining (x_1, y_1, z_1) and (x, y, z) must all be perpendicular to a given line, l, containing (x_1, y_1, z_1).

This definition may seem a bit complicated. Figure 6-34, however, should simplify the idea. We shall now use this defining condition to develop an equation for a plane. Let us start with an example and then consider the general case.

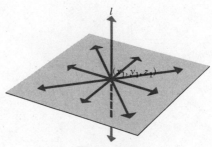

Figure 6-34

Example Determine an equation of a plane, α, which contains the points $(1, 3, 4)$ and is perpendicular to the line joining this point to the origin.

Note first that when we speak of a plane as perpendicular to a given line, we mean that all lines which lie in the plane and intersect the given line are perpendicular to it. (See Figure 6-34.)

Now let (x, y, z) be any point in the plane. We shall require the condition that the line segment joining (x, y, z) to $(1, 3, 4)$ be perpendicular to the line segment joining $(1, 3, 4)$ to $(0, 0, 0)$. (See Figure 6-35.)

If the line segment $\overline{OA}$ is to be perpendicular to $\overline{AB}$, then, by the Pythagorean Theorem,

$$(4) \qquad (OA)^2 + (AB)^2 = (OB)^2.$$

Conversely, if (4) holds, then $\overline{OA}$ is perpendicular to $\overline{AB}$.

Using the distance formula (3), we have

$$(OA)^2 = (1 - 0)^2 + (3 - 0)^2 + (4 - 0)^2,$$
$$(AB)^2 = (x - 1)^2 + (y - 3)^2 + (z - 4)^2,$$
$$(OB)^2 = (x - 0)^2 + (y - 0)^2 + (z - 0)^2.$$

Applying these equalities to Equation (4) and expanding, we get

$$1 + 9 + 16 + x^2 - 2x + 1 + y^2 - 6y + 9 + z^2 - 8z + 16$$
$$= x^2 + y^2 + z^2,$$

which reduces to

$$52 - 2x - 6y - 8z = 0$$

or

$$x + 3y + 4z - 26 = 0.$$

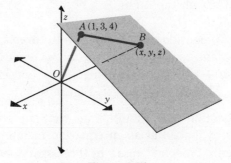

Figure 6-35

This equation, $x + 3y + 4z - 26 = 0$, has the form

(5) $ax + by + cz + d = 0.$

_____

It can be shown, although we shall not attempt to do so here, that any equation of the form (5) with a, b, and c real numbers not all zero is an equation of a plane in three-space and, conversely, that any plane has an equation of this form.

In our example we used as our "perpendicular" line, a line through the origin. This was for convenience in simplifying notation. In practice, the "perpendicular" line can be identified as the line through a point in the plane and any other specified point.

It is interesting to see the analogy between the general equation of a line in the plane, which, as you recall, is

$$ax + by + c = 0$$

and the general equation of a plane in three-space,

$$ax + by + cz + d = 0.$$

If we are working in three-space, it is possible to have equations of a plane which have one or more, but not all, of the coefficients a, b, c equal to zero. The so-called "constant" term may also be zero. Some special cases arise, a few of which follow. You will be asked to identify others.

1. $0x + 0y + z + 0 = 0$.
   (This could, of course, be written $z = 0$, but one must be careful to indicate that this is the equation for a three-dimensional figure.) Equation (1) is an equation of the xy-plane, that is, the set of all points with coordinates of the form $(a, b, 0)$.
2. $ax + by + cz = 0$.
   (A plane through the origin.)
3. $ax + by + d = 0$.
   Equation (3) considered as a relation in three-space is an equation for a plane containing the line whose equation in two-space is $ax + by + d = 0$. The plane (3) is perpendicular to the xy-plane. This is illustrated in Figure 6-36.

In this connection you should note that an equation such as $2x + y - 1 = 0$, if looked at as an equation in the *xy-plane*, would not be considered a "special" case. In three-space, however, such an equation is "special," as indicated.

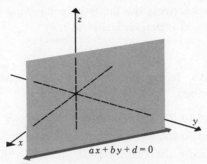

Figure 6-36

EXERCISES 6.9

◀ In Exercises 1–10, determine the distance between the given pair of points.

1. $(1, -2, 3), (2, -1, 5)$

2. $(4, 1, 6), (7, 1, 3)$

3. $(6, -2, -2), (5, -2, -2)$

4. $(4, -6, -2), (1, 7, 11)$

5. $(3, 2, 6), (9, 10, -12)$

6. $(2, -4, 7), (11, -4, -7)$

7. $\left(6, 1, \frac{1}{2}\right), \left(\frac{1}{2}, -\frac{1}{2}, 1\right)$

8. $\left(\frac{1}{4}, 0, 1\right), \left(\frac{1}{3}, \frac{1}{2}, 1\right)$

9. $\left(\frac{1}{2}, 1, \frac{1}{3}\right), (1, 2, 3)$

10. $(0.5, 1.2, 4), (0.1, 1.3, 5)$

◀ In Exercises 11–15, write an equation for the sphere described.

11. Center at $(2, 3, 5)$; radius 4.

12. The set of all points 8 units from the origin.

13. The set of all points 7 units from the point $(2, 0, -3)$.

14. A sphere with radius 12 and center at the point $(2, -4, -6)$.

15. The set of all points 11 units from the point $\left(-2, \frac{1}{2}, \frac{1}{3}\right)$.

◀ In Exercises 16–20, write an equation for each plane.

16. A plane through the point $(1, -2, 3)$ perpendicular to the line which contains this point and the origin.

17. A plane through the point $(3, -3, 4)$ perpendicular to the line containing $(3, -3, 4)$ and $(1, 2, 3)$.

18. A plane through the point $(2, 5, 7)$ perpendicular to the line through $(2, 5, 7)$ and $(2, 5, 0)$.

19. A plane which intersects the line segment joining $(1, 2, 3)$ to $(5, 9, 7)$ at its midpoint and is perpendicular to it.

20. The xz-plane.

◀ CHALLENGE PROBLEMS. In Exercises 21–25, write an equation for each of the special planes described.

21. A plane parallel to the xy-plane and 3 units above it.

22. A plane parallel to the xz-plane and 7 units to its left.

23. A plane containing the line whose equation in the yz-plane is $2y + 3z - 6 = 0$ and which is perpendicular to the yz-plane.

24. A plane through the origin and which contains the z-axis and the line whose equation in the xy-plane is $x = y$.

25. A plane through the origin and which contains the x-axis and the line whose equation in the yz-plane is $y + z = 0$.

CHAPTER SUMMARY

The following properties, definitions, and general concepts of coordinate systems in the plane and in three-space have been examined in this chapter.

1. In the plane every equation of the form $ax + by + c = 0$ with a and b not both zero is the equation of a line and, conversely, an equation for any line in the plane may be written in the form $ax + by + c = 0$.

1a. In three-space, every equation of the form $ax + by + cz + d = 0$ is the equation of a plane. The converse, as in (1), is also true.

2. The distance PQ between two points $P(x_1, y_1)$ and $Q(x_2, y_2)$ in the plane is

$$PQ = \sqrt{(x_2 - x_1)^2 + (y_2 - y_1)^2}.$$

2a. The distance PQ between two points $P(x_1, y_1, z_1)$ and $Q(x_2, y_2, z_2)$ in three-space is

$$PQ = \sqrt{(x_2 - x_1)^2 + (y_2 - y_1)^2 + (z_2 - z_1)^2}.$$

3. In the plane the slope m of the line determined by the points (x_1, y_1) and (x_2, y_2) with $x_2 \neq x_1$ is defined as

$$m = \frac{y_2 - y_1}{x_2 - x_1}.$$

4. The midpoint of a line segment joining the points $P(x_1, y_1)$ and $Q(x_2, y_2)$ has coordinates

$$\left(\frac{x_1 + x_2}{2}, \frac{y_1 + y_2}{2}\right).$$

5. Two distinct nonvertical lines are parallel if and only if they have the same slope. Distinct vertical lines (that is, lines with equations of the form $x = k_1$ and $x = k_2$ with $k_1 \neq k_2$) are also parallel.

In addition to graphs of equations of the form $ax + by + c = 0$, we examined graphs of polynomial equations, graphs of equations involving absolute values, and graphs of linear inequalities. A method of approximating irrational roots of polynomial equations was discussed as well as procedures for obtaining algebraic proofs of certain geometric theorems.

REVIEW EXERCISES

In Exercises 1–6, for the given pair of points write (a) the distance between the two points; (b) the slope of the line joining the two points; and (c) an equation of the line joining the two points.

1. $(3, 8), (2, -7)$ 4. $\left(\frac{1}{3}, 5\right), \left(\frac{1}{2}, -1\right)$

2. $(1, -6), (-4, 11)$ 5. $(0.2, 7), (0.1, 0.5)$

3. $\left(\frac{1}{2}, -3\right), \left(2, \frac{1}{4}\right)$ 6. $\left(1\frac{1}{2}, 3\right), \left(-1\frac{1}{4}, \frac{1}{2}\right)$

In Exercises 7–10, write an equation of the line characterized by the given conditions.

7. A line through $(1, 7)$ with slope $\frac{1}{2}$.

8. A line through the origin parallel to the line $l : 3x - 2y + 8 = 0$.

9. A line through the point $(3, 4)$ and parallel to the line $l : x + y = 6$.

10. A line through the point $(-2, 7)$ and parallel to the y-axis.

In Exercises 11–15, write an equation for the family of lines described.

11. Lines with slope $\frac{3}{4}$.

12. Lines through the origin.

13. Lines through the point $(2, 7)$.

14. Lines whose slopes are not defined.

15. Lines parallel to the line $l : 2x + y - 7 = 0$.

16. Given the points $A(3, 2)$ and $B(-9, -3)$, determine a third point C (x, y) such that ABC is a right triangle with hypotenuse $\overline{AB}$.

17. Draw a graph of the equation $y = x^2 - 4x - 5$ in the plane.

18. Draw a graph of the equation $y = -x^2 - 3x + 4$ in the plane.

In Exercises 19–21, draw a graph of the given inequality. Assume $x, y \in R$.

19. $y > x + 2$ **20.** $y \le 3x - 1$ **21.** $x - 3y < 5$

22. Find an approximate positive solution for the equation $x^3 - 8x - 1 = 0$ using two linear interpolations.

23. Use two linear interpolations to approximate a positive cube root of 2.

In Exercises 24 and 25, determine the midpoint of the line segment joining the two given points.

24. $(3, 8), (-5, 10)$ **25.** $(2, -7), \left(\frac{1}{2}, -17\right)$

26. Determine the distance between the points $(2, 1, 3)$ and $(1, 4, -3)$ in three-space.

27. Write an equation for the plane through $(3, 1, -2)$ and perpendicular to a line containing the points $(3, 1, -2)$ and $(5, 4, 1)$.

28. Write an equation for the sphere with radius 6 and center at the point $(2, 3, -4)$.

29. Write an equation for the plane 8 units above the xy-plane and parallel to it.

30. CHALLENGE PROBLEM. Give an algebraic proof of the following geometric theorem:

The medians to the two equal sides of an isosceles triangle are equal.

GOING FURTHER: READING AND RESEARCH

In the study of lines in the plane the concept of slope played a prominent role. It was possible to construct an equation of a line in a plane if we knew the slope of the line and a point which the line contained.

Is there an analogous concept with respect to lines in three-space? We cannot use slope in the sense of a ratio of "rise to run" in three-space since there is more involved here.

Consider, for example, a line l through the points $P(2, 3, 4)$ and $Q(4, 8, 7)$ as in the following figure. Although we cannot give the slope of l, we can

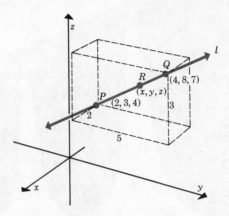

describe its direction. We say that l has three direction numbers, $4 - 2 = 2$, $8 - 3 = 5$, and $7 - 4 = 3$. What are the direction numbers associated with the line through the points $P(-2, 5, 7)$ and $Q(3, -1, 10)$? Can direction numbers be determined for any line if two distinct points on the line are known? Give a formula for the direction numbers d_1, d_2, and d_3 of a line passing through the points $P(x_1, y_1, z_1)$ and $Q(x_2, y_2, z_2)$.

Does this suggest a way of writing an equation for a line in three-space?

Consider the line l in the figure. Let $R(x, y, z)$ be any point on l. How would you describe the direction numbers determined by $R(x, y, z)$ and P $(2, 3, 4)$? Call these D_1, D_2, and D_3. How would you describe the relationship between d_1, d_2, d_3 and D_1, D_2, D_3? It is clear that $d_1 \neq D_1$, $d_2 \neq D_2$, and $d_3 \neq D_3$ if R is not the same point as Q. Think of the idea of proportion. This is the key.

For this and other topics relating to lines and planes in three-space, we suggest that you do some reading in the following books.

DOLCIANI, M. P., E. F. BECKENBACH, et al., *Modern Introductory Analysis* (Chapter 14). New York: Houghton Mifflin, Second Edition, 1967.

MURDOCH, D. C., *Analytic Geometry with an Introduction to Vectors and Matrices*. New York: John Wiley and Sons, 1966.

RINGENBERG, L. A., *College Geometry*. New York: John Wiley and Sons, 1968.

CALLOWAY, J. M. (Editor), *Systems of First Degree Equations in Three Variables*. Supplementary and Enrichment Series, School Mathematics Study Group.

CHAPTER SEVEN
CONIC SECTIONS AND
THEIR GRAPHS

7.1 INTRODUCTION

In your previous work in mathematics, you have studied the graphs of many different equations. One set of equations having particular significance in many areas of mathematics and science is the set of second-degree equations in two variables. Each of these equations is of the form

(1) $$Ax^2 + Bxy + Cy^2 + Dx + Ey + F = 0$$
$$(A, B, C \text{ not all zero}).$$

Except in certain special cases, some of which will be discussed later, every equation of this form has for its graph a **conic section.** Conversely, every conic section in the xy-plane is the graph of an equation of the second degree in x and y.

The name "conic" arises from the relationship of these curves to cones. Very likely you have studied a right circular cone in geometry. Such a cone has a circular base and its vertex is a point on the line that is perpendicular to the plane of the circle at its center as in Figure 7-1a. If this cone is extended without bound in two directions as in Figure 7-1b, a cone of two nappes is formed. The intersection of a cone of two nappes and a plane forms a conic section as shown in Figures 7-2a, b, and c except when the plane passes through the vertex of the cone. In this instance, the intersection is either two lines, one line, or a point as shown in Figure 7-2d, e, and f.

There are several ways to approach the study of the conic sections. The oldest way by far is through the study of the geometry of cones

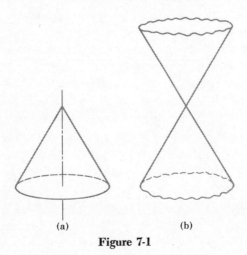

(a)   (b)

Figure 7-1

259

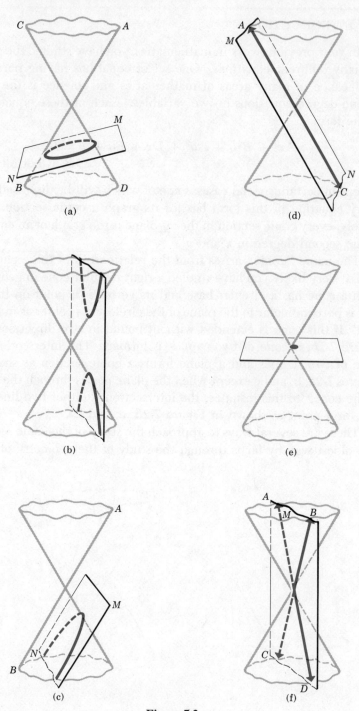

Figure 7-2

as the Greeks did. Apollonius of Perga (about 260–200 B.C.), a geometer who ranks with Euclid and Archimedes, included more than 400 theorems in his comprehensive book on conic sections.

We shall use here an analytic (that is, algebraic) approach that depends on the concept of distance and, in particular, the formula developed in Chapter 6 for the distance, PQ, between two points P (x_1, y_1) and $Q(x_2, y_2)$ in the plane, that is

$$PQ = \sqrt{(x_1 - x_2)^2 + (y_1 - y_2)^2}.$$

Example 1 Find an equation defining the set S of all points in the plane with the property that the distance of each point $P(x, y) \in S$ from the line $l : x = 6$ is always twice the distance of $P(x, y)$ from the point $Q(2, 0)$.

Solution: The line l is perpendicular to the x-axis. (See Figure 7-3.)

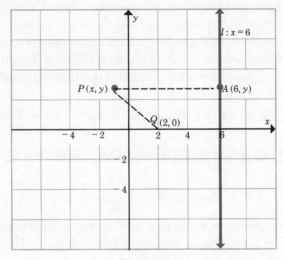

Figure 7-3

We have

$$PA = \sqrt{(x - 6)^2 + (y - y)^2} = \sqrt{(x - 6)^2} = |x - 6|$$

and

$$PQ = \sqrt{(x - 2)^2 + (y - 0)^2} = \sqrt{(x - 2)^2 + y^2}.$$

We have given that

$$2PQ = PA$$

and therefore

(A) $$2\sqrt{(x-2)^2 + y^2} = |x - 6|.$$

We square both sides of Equation (A) and obtain

$$4[(x-2)^2 + y^2] = (x-6)^2.$$

Performing the multiplication, we get

$$4x^2 - 16x + 16 + 4y^2 = x^2 - 12x + 36$$

and, finally,

(B) $$3x^2 + 4y^2 - 4x - 20 = 0,$$

which is an equation of a conic section (actually an ellipse). We must check now to determine if Equation (B) is equivalent to Equation (A).

By a property of the real numbers we know that if $a = b$, then $a^2 = b^2$. Generally speaking, the converse is not true. It is possible to have $a = -5$ and $b = 5$. Then $a^2 = b^2$, but $a \neq b$. However, if we specify that a and b are both nonnegative, then if $a^2 = b^2$, it follows that $a = b$. Now in Equation (A) both the left- and right-hand members *are* nonnegative real numbers for all real numbers x and y. Hence all real numbers which satisfy Equation (B) must satisfy (A). Conversely, all real numbers which satisfy Equation (A) must satisfy (B). Hence

$$S = \{(x, y) : x, y \in R \text{ and } 3x^2 + 4y^2 - 4x - 20 = 0\}.$$

Although, as stated above, every conic section has an equation of the form

(1) $$Ax^2 + Bxy + Cy^2 + Dx + Ey + F = 0,$$

unless stated otherwise we shall consider only such equations that do not have an xy term, that is, equations where $B = 0$. These equations have the form

(2) $Ax^2 + Cy^2 + Dx + Ey + F = 0$ (A and C not both 0).

In Section 7.8 we shall consider briefly some situations in which $B \neq 0$ in (1).

EXERCISES 7.1

1. Find an equation defining the set S of points in the plane with the property that the distance of each point $P(x, y) \in S$ from the point $Q(2, 0)$ is one-third the distance of $P(x, y)$ from the line $l : y = 3$.

2. Find an equation defining the set S of all points in the plane such that the distance of each point $P(x, y) \in S$ from the line $l : y = 8$ is always twice the distance of $P(x, y)$ from the point $Q(0, 3)$.

3. Find an equation defining the set S of all points in the plane such that the difference between the distance of each point $P(x, y) \in S$ from $(0, 5)$ and its distance from $(0, -5)$ is 2.

4. Find an equation defining the set S of all points in the plane such that each point $P(x, y) \in S$ is twice as far from $(4, 0)$ as from the line $l : x = 1$.

5. Find an equation defining the set S of all points in the plane with the property that the distance of each point $P(x, y) \in S$ from the point $(1, 0)$ is one-half the distance of P from the line $l : x = 4$.

◀ In Exercises 6–10, derive an equation of the form

$$Ax^2 + Bxy + Cy^2 + Dx + Ey + F = 0$$

that is equivalent to the given equation.

6. $2(x - 2)^2 - 4(y - 4)^2 = 6$ 9. $5(x + 2)^2 + 5(y + 5)^2 = 18$

7. $\dfrac{(x - 3)^2}{25} - \dfrac{(y - 4)^2}{36} = 1$ 10. $\dfrac{(x - h)^2}{a^2} + \dfrac{(y - k)^2}{b^2} = 1$

8. $2(y - 3) = 5(x - 3)^2$

11. Find an equation defining the set of all points in the plane with the property that the sum of the distance of any point in the set from the point $(4, 0)$ and the distance of the same point in the set from the point $(-4, 0)$ is 10.

12. Let $\angle 1$ in Figure 7-4 be the angle that the intersecting plane forms with the axis of the cone. Let $\angle 2$ be the angle that the cone element $\overleftrightarrow{CD}$

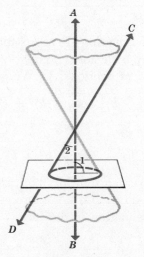

Figure 7-4

makes with the axis $\overleftrightarrow{AB}$. Give conditions on $\angle 1$ and $\angle 2$ for each conic section.

13. **CHALLENGE PROBLEM.** Find an equation defining the set S of all points in the plane with the property that the distance of each point $P(x, y) \in S$ from the point $(5, 0)$ is equal to its distance from the line $x = y$.

7.2   THE CIRCLE

Recall that a circle with center at a point $M(h, k)$ and radius of length r (Figure 7-5) is the set of all points $P(x, y)$ in a plane such that the distance from P to M is equal to r. Thus $MP = r$.

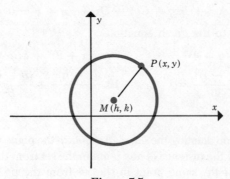

Figure 7-5

We have $MP = \sqrt{(x - h)^2 + (y - k)^2}$ and so

(1) $$\sqrt{(x - h)^2 + (y - k)^2} = r.$$

Note that both sides of Equation (1) represent nonnegative numbers and hence we can square both sides and obtain an equivalent equation as we did in Example 1 of Section 7.1. We get

(2) $$(x - h)^2 + (y - k)^2 = r^2$$

which becomes

(3) $$x^2 + y^2 - 2hx - 2ky + (h^2 + k^2 - r^2) = 0.$$

Since in Equation (3) we can multiply both sides of the equation by any nonzero number, we can also say, more generally, that Equation (3) is equivalent to an equation of the form

(4) $$Ax^2 + Cy^2 + Dx + Ey + F = 0$$

where A and C need not be equal to 1 and yet $A = C \neq 0$. Every circle has an equation of this form.

Is the converse true? That is, is every equation of the form of (4) with $A = C \neq 0$ the equation of a circle? Let us see.

Example 1 Find an equation of a circle whose center is $C(2, 4)$ and the length of whose radius is 5. Write the equation in the form (4).

Solution:

$$\sqrt{(x - 2)^2 + (y - 4)^2} = 5,$$
$$x^2 - 4x + 4 + y^2 - 8y + 16 = 25,$$
(5) $$x^2 + y^2 - 4x - 8y - 5 = 0.$$

Equation (5) has the form of Equation (4) with $A = C = 1$.

Example 2 Describe the set of points whose coordinates satisfy the equation

$$x^2 + y^2 - 8x + 6y - 11 = 0.$$

Solution: First we transform the given equation into the form of Equation (2) by "completing the squares" in the following manner.

$$x^2 + y^2 - 8x + 6y - 11 = 0,$$
$$(x^2 - 8x + \quad) + (y^2 + 6y + \quad) = 11,$$
$$(x^2 - 8x + 16) + (y^2 + 6y + 9) = 11 + 16 + 9,$$
(6) $$(x - 4)^2 + (y + 3)^2 = 36.$$

By comparing Equation (6) with (2), we note that $h = 4$, $k = -3$, and $r = 6$. Hence the set of points whose coordinates satisfy the equation is a circle with radius of length 6 and center at the point $(4, -3)$.

Example 3 Describe the set of points whose coordinates satisfy the equation

$$x^2 + y^2 - 8x + 6y + 25 = 0.$$

Solution: First we transform the given equation into the form of Equation (2) by completing the squares. We have

$$x^2 + y^2 - 8x + 6y + 25 = 0,$$
$$(x^2 - 8x + \quad) + (y^2 + 6y + \quad) = -25,$$
$$(x^2 - 8x + 16) + (y^2 + 6y + 9) = -25 + 16 + 9,$$
$$(x - 4)^2 + (y + 3)^2 = 0.$$

Since $a^2 \geq 0$ for all real numbers a, the last equality is true if and only if $x - 4 = 0$ and $y + 3 = 0$. Hence $x = 4$ and $y = -3$. The only point whose coordinates satisfy the equation is $(4, -3)$. Thus the circle reduces to a point and $r = 0$.

Example 4 Find the solution set of the equation

$$x^2 + y^2 - 14x + 8y + 70 = 0.$$

Solution: Again we use the process of completing the square.

$$x^2 + y^2 - 14x + 8y + 70 = 0,$$
$$(x^2 - 14x + \quad) + (y^2 + 8y + \quad) = -70,$$
$$(x^2 - 14x + 49) + (y^2 + 8y + 16) = -70 + 49 + 16,$$
$$(x - 7)^2 + (y + 4)^2 = -5.$$

There is no ordered pair (x, y) of real numbers that satisfies the last equation since $(x - 7)^2$ and $(y + 4)^2$ are nonnegative for all real numbers x and y. Hence

$$\{(x, y) : x^2 + y^2 - 14x + 8y + 70 = 0\} = \varnothing.$$

The examples suggest that the graph of the solution set of an equation of the form

$$Ax^2 + Cy^2 + Dx + Ey + F = 0 \qquad (A = C \neq 0)$$

may be a circle, a point, or the empty set.

Can you specify the relationship between A, D, E, and F that exists when the solution set is a circle? A point? The empty set?

EXERCISES 7.2

◀In Exercises 1–10, find the length of the radius and the center of the circle with the given equation.

1. $(x - 6)^2 + (x + 2)^2 = 36$
2. $x^2 + (y + 10)^2 = 49$
3. $(x + 11)^2 + y^2 = 64$
4. $x^2 + y^2 - 10x + 9 = 0$
5. $x^2 + y^2 - 4x + 2y + 1 = 0$
6. $x^2 + y^2 + 6(x + y) = -9$
7. $x^2 + y^2 - 14x + 16y + 112 = 0$

8. $x^2 + y^2 - (x + y) = \dfrac{1}{2}$

9. $x^2 + y^2 - \dfrac{2}{3}(x + y) = 3\dfrac{7}{9}$

10. $4(x^2 + y^2) + 4(x + y) = 34$

◀ In Exercises 11–19, find an equation in the form of Equation (4), page 264 of the given set of points.

11. A circle with radius of length 5 and center at (3, 7).

12. A circle with radius of length 12 and center at the origin.

13. A circle with radius of length 7 and center at $(-1, -1)$.

14. The set of all points at a distance of 10 units from the point $(-6, 4)$.

15. The set of all points 11 units from the origin.

16. A circle with center at (2, 1) and passing through the point $(-2, 1)$.

17. A circle with center at $\left(-2\dfrac{1}{2}, -1\dfrac{1}{2}\right)$ and passing through the point (4, 5).

18. A circle with center at $\left(-\dfrac{3}{4}, \dfrac{9}{8}\right)$ and passing through the point $\left(-\dfrac{27}{5}, \dfrac{3}{5}\right)$.

19. A circle (?) with center at the origin and radius of length zero.

◀ In Exercises 20–22, describe the solution set of the given equations.

20. $2x^2 + 2y^2 - 10x + 14y + 18 = 0$

21. $x^2 + y^2 + 5x - 6y + 49 = 0$

22. $5x^2 + 5y^2 - 13y + 7x + 4 = 0$

23. By finding the length of the radius of each circle and the distance between their centers, show that the circles with equations

$$x^2 + y^2 - 4y = 0 \qquad \text{and} \qquad x^2 + y^2 - 4x + 4y + 4 = 0$$

do not have a point in common.

24. CHALLENGE PROBLEM. Show that if the equation

$$x^2 + y^2 + Dx + Ey + F = 0$$

has a solution set containing exactly one ordered pair, then $D^2 + E^2 = 4F$.

25. CHALLENGE PROBLEM. Show that if the equation

$$x^2 + y^2 + Dx + Ey + F = 0$$

has an empty solution set, then $D^2 + E^2 < 4F$.

7.3 THE ELLIPSE

Some of us may occasionally feel we are traveling in circles; all of us in fact are traveling in ellipses since the earth travels in an elliptical path around the sun once a year.

As we did with the circle, we shall begin by defining the ellipse algebraically and then considering its graph.

> **Definition 7.1** An **ellipse** is a set of all points in the plane such that the sum of the distances from each point $P(x, y)$ in the set to two fixed points, F_1 and F_2, is a constant k ($k > F_1F_2$). The two fixed points are called **foci**. (See Figure 7-6.)

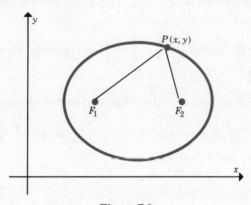

Figure 7-6

By definition, $F_1P + F_2P = k$. The distances F_1P and F_2P are called **focal radii**.

For simplicity, we shall restrict our study of ellipses in this section to those whose foci are points on either the x-axis or the y-axis and which are equally distant from the origin of the Cartesian plane. Let F_1 be the point $(-c, 0)$ and F_2 be the point $(c, 0)$ as in Figure 7-7. We take $a = \dfrac{k}{2}$ so that $k = 2a$ and observe that $2a > 2c > 0$. Using the distance formula and Definition 7.1, we can state that $P(x, y)$ is on the ellipse if and only if

(1) $$\sqrt{(x + c)^2 + y^2} + \sqrt{(x - c)^2 + y^2} = 2a.$$

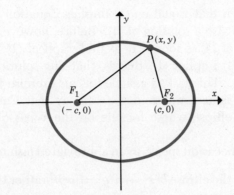

Figure 7-7

Thus
$$\sqrt{(x + c)^2 + y^2} = 2a - \sqrt{(x - c)^2 + y^2}.$$

Squaring each side of this equation, we obtain
$$(x + c)^2 + y^2 = 4a^2 - 4a\sqrt{(x - c)^2 + y^2} + (x - c)^2 + y^2.$$

From this we get
$$(x^2 + 2cx + c^2) + y^2 = 4a^2 - 4a\sqrt{(x - c)^2 + y^2}$$
$$+ (x^2 - 2cx + c^2) + y^2.$$

Simplifying this, we have
$$4cx - 4a^2 = -4a\sqrt{(x - c)^2 + y^2}.$$

Now we divide by 4 and again square both sides to obtain
$$c^2x^2 - 2a^2cx + a^4 = a^2[(x - c)^2 + y^2]$$
and then
$$c^2x^2 - 2a^2cx + a^4 = a^2(x^2 - 2cx + c^2 + y^2).$$

Finally, we get

(2) $$(a^2 - c^2)x^2 + a^2y^2 = a^2(a^2 - c^2).$$

Since $a > c$, we have $a^2 - c^2 > 0$. We let $b^2 = a^2 - c^2$ and then can write (2) as

(3) $$b^2x^2 + a^2y^2 = a^2b^2.$$

If we divide each member of Equation (3) by a^2b^2, we have

(4) $$\frac{x^2}{a^2} + \frac{y^2}{b^2} = 1.$$

We have seen that Equation (1) implies Equation (4), that is, any solution of (1) is also a solution of (4). Before, however, we can assert that the graph of (4) is identical to the graph of (1), we must show that Equation (4) implies (1), that is, that any solution of (4) is also a solution of (1). This task is assigned as an exercise for the student. When this task is accomplished, we can say that Equation (4) is an equation of an ellipse whose foci lie on the x-axis of the Cartesian plane.

We shall on occasion speak, in an abbreviated fashion, of the ellipse $\dfrac{x^2}{a^2} + \dfrac{y^2}{b^2} = 1$ or the ellipse $b^2x^2 + a^2y^2 = a^2b^2$ rather than "the ellipse defined by the equation $\dfrac{x^2}{a^2} + \dfrac{y^2}{b^2} = 1$." Similarly, for brevity we may write, for example, "the line $y = 2x$" for "the line with an equation $y = 2x$" or the line $l : y = 2x$.

Note that the equation

$$\frac{x^2}{a^2} + \frac{y^2}{b^2} = 1,$$

which can be written as $b^2x^2 + a^2y^2 + (-a^2b^2) = 0$, has the form

$$(5) \qquad\qquad Ax^2 + Cy^2 + Dx + Ey + F = 0$$

where $A > 0$, $C > 0$, and $D = E = 0$. (What can be said about F?) Since we can also write (5) in the form

$$(-A)x^2 + (-C)y^2 + (-D)x + (-E)y + (-F) = 0$$

(for example, $3x^2 + 2y^2 + (-5) = 0$ as $(-3)x^2 + (-2)y^2 + 5 = 0$), we can say that when (5) is the equation of an ellipse whose foci are the points $(c, 0)$ and $(-c, 0)$, then $D = E = 0$ and either $A > 0$ and $C > 0$, or $A < 0$ and $C < 0$.

If we replace y by 0 in Equation (4), we can find the **vertices** of the ellipse. Check that these are the points $(-a, 0)$ and $(a, 0)$ (R and S in Figure 7-8). The line segment, $\overline{RS}$, joining the two vertices and containing the foci is called the **major axis** of the ellipse. The midpoint of the major axis (c in Figure 7-8) is called the **center** of the ellipse. Finally, the line segment ($\overline{TU}$ in Figure 7-8) passing through the center of the ellipse, perpendicular to the major axis, and with endpoints lying on the ellipse, is called the **minor axis**. If we replace x by 0 in Equation (4), we find that the coordinates of T and U are $(0, b)$ and $(0, -b)$, respectively.

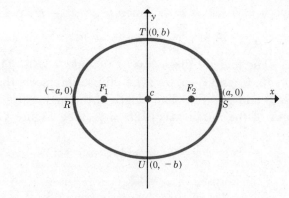

Figure 7-8

If the two foci of an ellipse are points on the y-axis, then the major axis of the ellipse lies along the y-axis. For example, if, as in Figure 7-9, the foci are the points $(0, c)$ and $(0, -c)$ and the sum of the focal radii is $2a$, where $2a > 2c$, then the equation of the ellipse is

$$\frac{x^2}{b^2} + \frac{y^2}{a^2} = 1$$

where, again, $b^2 = a^2 - c^2$. In this case, the vertices of the ellipse are the points $(0, a)$ and $(0, -a)$. The endpoints of the minor axis (the **x-intercepts**) are $(b, 0)$ and $(-b, 0)$.

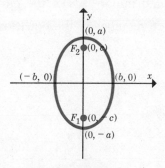

Figure 7-9

Example 1 Find the foci, the vertices, the intercepts, and the lengths of the major and minor axes of the ellipse $16x^2 + 25y^2 = 400$.

Solution: Dividing each member of the equation $16x^2 + 25y^2 = 400$ by 400, we obtain

$$\frac{x^2}{25} + \frac{y^2}{16} = 1 \qquad \text{or} \qquad \frac{x^2}{5^2} + \frac{y^2}{4^2} = 1.$$

We then have $a = 5$ and $b = 4$. Since $b^2 = a^2 - c^2$, we have

$$c^2 = a^2 - b^2 = 25 - 16.$$

Hence $c = 3$. The foci are the points $(3, 0)$ and $(-3, 0)$. The vertices are the points $(5, 0)$ and $(-5, 0)$. The ellipse intersects the y-axis at the points $(0, 4)$ and $(0, -4)$. The length of the major axis is $2a = 10$ and the length of the minor axis is $2b = 8$. (See Figure 7-10.)

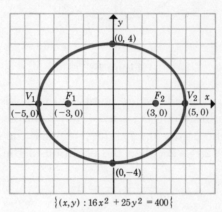

$$\{(x,y) : 16x^2 + 25y^2 = 400\}$$

Figure 7-10

Example 2   Find the foci, the vertices, the intercepts, and the lengths of the major and minor axes of the ellipse $9x^2 + 4y^2 = 36$.

Solution: Dividing each member of the equation $9x^2 + 4y^2 = 36$ by 36, we obtain

$$\frac{x^2}{4} + \frac{y^2}{9} = 1.$$

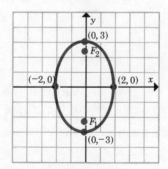

Figure 7-11

Since $9 > 4$, we know that the foci lie on the y-axis. We have $a = 3$, $b = 2$, and $c = \sqrt{9 - 4} = \sqrt{5}$. Thus the foci are the points $(0, \sqrt{5})$ and $(0, -\sqrt{5})$; the vertices are the points $(0, 3)$ and $(0, -3)$; and the x-intercepts are $(2, 0)$ and $(-2, 0)$. The length of the major axis is $2a = 6$ and the length of the minor axis is $2b = 4$. (See Figure 7-11.)

For the ellipse $\dfrac{x^2}{a^2} + \dfrac{y^2}{b^2} = 1$ with the two foci $(c, 0)$ and $(-c, 0)$, we have $k = 2a > 2c > 0$. If $a = b$, however, we have a circle and $c = 0$. Also, if the distance between the two foci is equal to the constant k, then $a = c$ and the ellipse reduces to a line segment. For this reason, a circle and a line segment are sometimes referred to as "limiting" cases of the ellipse.

EXERCISES 7.3A

◀ In Exercises 1–12, find the coordinates of the vertices and of the foci, and the lengths of the major and minor axes of the ellipse whose equation is given.

1. $\dfrac{x^2}{16} + \dfrac{y^2}{9} = 1$ 7. $9x^2 + 36y^2 = 324$

2. $\dfrac{x^2}{36} + \dfrac{y^2}{16} = 1$ 8. $x^2 + 3y^2 = 12$

3. $\dfrac{x^2}{64} + \dfrac{y^2}{9} = 1$ 9. $x^2 = 50 - 2y^2$

4. $\dfrac{x^2}{100} + \dfrac{y^2}{25} = 1$ 10. $8x^2 + 9y^2 = 12$

5. $\dfrac{x^2}{10} + \dfrac{y^2}{5} = 1$ 11. $25x^2 + 4y^2 = 100$

6. $x^2 + 25y^2 = 100$ 12. $36x^2 + 9y^2 = 324$

◀ In Exercises 13–16, use Definition 7.1 to find an equation of the ellipse with foci at points F_1 and F_2 and satisfying the given condition. Point P is on the ellipse.

13. $F_1(2, 0),\ F_2(-2, 0);\ F_1P + F_2P = 6$
14. $F_1(0, 3),\ F_2(0, -3);\ F_1P + F_2P = 8$
15. $F_1(4, 0),\ F_2(-4, 0);\ F_1P + F_2P = 10$
16. $F_1(0, 5),\ F_2(0, -5);\ F_1P + F_2P = 12$

GRAPHING AN ELLIPSE

To sketch the graph of an ellipse,

$$\frac{x^2}{a^2} + \frac{y^2}{b^2} = 1,$$

it is useful to take advantage of symmetry considerations. We say that a curve is **symmetric** with respect to the x-axis if there is a point $(a, -b)$ on the curve corresponding to each point (a, b) on the curve as in Figure 7-12a. Similarly, a curve is symmetric with respect to the y-axis if there is a point $(-a, b)$ on the curve corresponding to each point (a, b) on the curve as in Figure 7-12b.

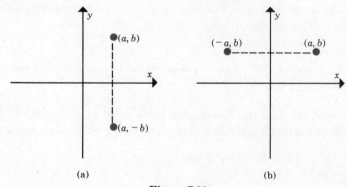

(a)   (b)

Figure 7-12

From these definitions we can see that the graph of an equation is symmetric with respect to the x-axis if an equivalent equation is obtained by replacing (x, y) by $(x, -y)$. Similarly, the graph of an equation is symmetric with respect to the y-axis if an equivalent equation is obtained by replacing (x, y) by $(-x, y)$. Since

$$\frac{x^2}{a^2} + \frac{(-y)^2}{b^2} = 1$$

is equivalent to

$$\frac{x^2}{a^2} + \frac{y^2}{b^2} = 1,$$

the ellipse $\dfrac{x^2}{a^2} + \dfrac{y^2}{b^2} = 1$ is symmetric with respect to the x-axis. Since

$$\frac{(-x)^2}{a^2} + \frac{y^2}{b^2} = 1$$

is equivalent to

$$\frac{x^2}{a^2} + \frac{y^2}{b^2} = 1,$$

the ellipse $\dfrac{x^2}{a^2} + \dfrac{y^2}{b^2} = 1$ is also symmetric with respect to the y-axis.

Example 3 Sketch the ellipse $\dfrac{x^2}{25} + \dfrac{y^2}{16} = 1$.

Solution: From Example 1 we know that the ellipse crosses the x-axis at the points $(5, 0)$ and $(-5, 0)$, and the y-axis at the points $(0, 4)$ and $(0, -4)$. To find a few other points on the curve, we shall use integral values of x, $0 < x < 5$, for easy computation.

Solving the given equation for y, we have

$$y = \frac{\pm \sqrt{400 - 16x^2}}{5}.$$

Now, replacing x by 2, we find $y = \dfrac{\pm \sqrt{336}}{5} \approx \pm 3.7$. Hence the

point $(2, 3.7)$ is approximately on the graph. In a similar manner we find that the point $(4, 2.4)$ is on the graph. The points $(5, 0)$, $(0, 4)$, $(2, 3.7)$, and $(4, 2.4)$ are plotted in Figure 7-13a.

In Figure 7-13b, we have additional points obtained by using symmetry with respect to the x-axis. Now using symmetry with respect to the y-axis we obtain Figure 7-13c. Finally, the graph is completed by connecting the points with a smooth continuous curve as in Figure 7-13d.

Example 4 Sketch the graph of the equation $36x^2 + 9y^2 - 324 = 0$.

Solution: First we transform the given equation into the form of Equation (4) (page 269) by first dividing each term by 324 to get

$$\frac{36x^2}{324} + \frac{9y^2}{324} - 1 = 0$$

from which we obtain

$$\frac{x^2}{9} + \frac{y^2}{36} = 1.$$

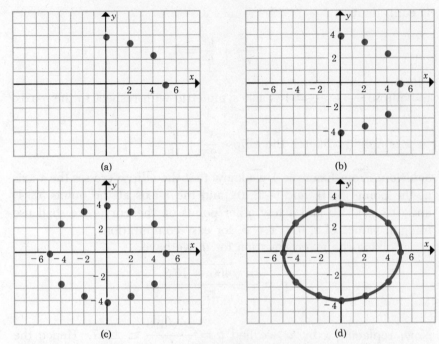

(a)   (b)

(c)   (d)

Figure 7-13

Since $a = 6$, the y-intercepts are 6 and -6. Since $b = 3$, the x-intercepts are 3 and -3. Two other points on the curve are, approximately, $(1, 5.6)$ and $(2, 4.5)$. With the aid of symmetry, we can quickly sketch the curve pictured in Figure 7-14.

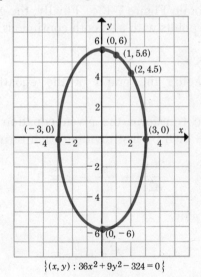

$\{(x, y) : 36x^2 + 9y^2 - 324 = 0\}$

Figure 7-14

Example 5 Discuss the symmetry with respect to the x- and the y-axis of the graph of the equation

$$y = 2x^2 - x + 4.$$

Solution: If we replace (x, y) by $(-x, y)$ in the equation $y = 2x^2 - x + 4$ we obtain

$$y = 2(-x)^2 - (-x) + 4$$

or

$$y = 2x^2 + x + 4,$$

an equation that is not equivalent to $y = 2x^2 - x + 4$. Hence the graph is not symmetric with respect to the y-axis. If we replace (x, y) by $(x, -y)$ we obtain

$$-y = 2x^2 - x + 4$$

or

$$y = -2x^2 + x + 4,$$

again an equation that is not equivalent to $y = 2x^2 - x + 4$. Hence the graph is also not symmetric with respect to the x-axis.

As you know, a circle can be drawn by fixing, with a thumbtack for example, one end of a string at some point, fastening a pencil to the other end, and then rotating the pencil. (See Figure 7-15a.) A similar construction works for an ellipse. The two ends of a string are fixed at two points, F_1 and F_2, and then a pencil is held taut within the loop of the string (which must be longer than the distance between the two points F_1 and F_2). The pencil, when rotated within the stretched loop, will draw an ellipse. (See Figure 7-15b.)

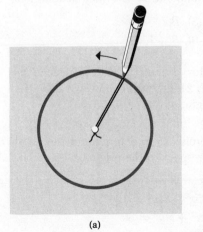

(a)

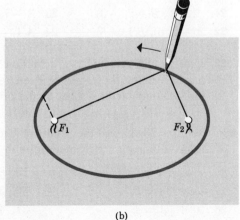

(b)

Figure 7-15

EXERCISES 7.3B

◀In Exercises 1–12, determine if the graph of the given equation is symmetric with respect to the x-axis, or the y-axis, or both.

1. $x + y = 6$

2. $3x - 4 = y$

3. $y = x^2 + 2$

4. $y^2 = x - 3$

5. $x^2 + y^2 = 10$

6. $x^2 + 25y^2 = 100$

7. $y = x^2 + y^2$

8. $x^2 - x = y^2$

9. $xy = 4$

10. $3xy = 7$

11. $4(x - 2)^2 + 9(y + 3)^2 = 36$

12. $16(x + 3)^2 - 9\left(y + \dfrac{3}{2}\right)^2 = 10$

◀In Exercises 13–20, sketch the graph of each equation using the concept of symmetry.

13. $\dfrac{x^2}{16} + \dfrac{y^2}{9} = 1$

14. $\dfrac{x^2}{36} + \dfrac{y^2}{16} = 1$

15. $\dfrac{x^2}{64} + \dfrac{y^2}{9} = 1$

16. $\dfrac{x^2}{100} + \dfrac{y^2}{25} = 1$

17. $x^2 + 25y^2 = 100$

18. $x^2 + 9y^2 = 36$

19. $x^2 + 2y^2 = 50$

20. $x^2 + 4y^2 = 75$

◀In Exercises 21–24, plot the two given points A and B, draw the line segment $\overline{AB}$, and determine by computation the midpoint of $\overline{AB}$. (See Section 6.8.)

21. $A(2, 3)$, $B(-2, -3)$

22. $A(-1, 8)$, $B(1, -8)$

23. $A(-7, 3)$, $B(7, -3)$

24. $A(1, -9)$, $B(-1, 9)$

25. Exercises 21–24 suggest a definition for "symmetry with respect to the origin." Using the results of Exercises 21–24, write your own definition for symmetry with respect to the origin.

◀In Exercises 26–35, use the results of Exercise 25 to determine if the graph of the given equation is symmetric with respect to the origin.

26. $x + y = 0$

27. $3x + 3y = 5$

28. $|x| + |y| = 4$

29. $|x + y| = 4$

30. $25x^2 + 4y^2 = 100$

31. $4x^2 + 9y^2 = 36$

32. $\dfrac{(x-3)^2}{4} + \dfrac{(y+2)^2}{9} = 1$ **34.** $xy = 4$

33. $\dfrac{(x+4)^2}{16} - \dfrac{(y-3)^2}{25} = 1$ **35.** $3xy = 10$

36. Explain why the construction suggested in Figure 7-15b produces an ellipse.

37. CHALLENGE PROBLEM. The statement "If a graph is symmetric with respect to the x-axis and also with respect to the y-axis, then the graph is symmetric to the origin" is either true or false. If true, prove the statement; if false, give a counterexample.

38. CHALLENGE PROBLEM. The statement "If a graph is symmetric with respect to the origin, then it is symmetric with respect to both the x-axis and the y-axis" is either true or false. If true, prove the statement; if false, give a counterexample.

FINDING EQUATIONS OF ELLIPSES

Examples 6, 7, and 8 concerning the ellipse illustrate how an equation of an ellipse can be obtained from information concerning the length of the major axis, the vertices, the foci, or other particulars.

Example 6 Find an equation of the ellipse whose foci are the points $(3\sqrt{3}, 0)$ and $(-3\sqrt{3}, 0)$ and whose major axis has length 12. Sketch the ellipse.

Solution: Since the foci are on the x-axis, the equation has the form $\dfrac{x^2}{a^2} + \dfrac{y^2}{b^2} = 1$. If the length of the major axis is 12, then $a = 6$. One focus is at the point $(3\sqrt{3}, 0)$; hence $c = 3\sqrt{3}$. Since $b^2 = a^2 - c^2$, we have

$$b^2 = 36 - 27 = 9.$$

Hence an equation of the ellipse is

$$\frac{x^2}{36} + \frac{y^2}{9} = 1.$$

By replacing x by 2, we have $y = \sqrt{8} \approx 2.8$. By replacing x by 4, we have $y = \sqrt{5} \approx 2.2$. We use the points $(6, 0)$, $(0, 3)$ $(2, 2.8)$, and $(4, 2.2)$ together with symmetry to sketch the graph as indicated in Figure 7-16.

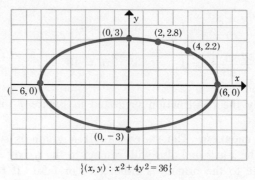

$$\{(x, y) : x^2 + 4y^2 = 36\}$$

Figure 7-16

Example 7   Find an equation of the ellipse the ends of whose minor axis have coordinates $(0, \pm 5)$ and the length of whose major axis is 16.

Solution: The sketch of the ellipse shown in Figure 7-17 tells us that the ellipse has an equation of the form

$$\frac{x^2}{a^2} + \frac{y^2}{b^2} = 1$$

with $a = 8$ and $b = 5$. That is, the ellipse has an equation

$$\frac{x^2}{64} + \frac{y^2}{25} = 1.$$

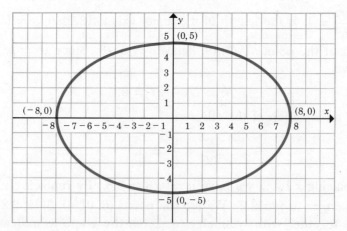

Figure 7-17

Example 8   Find an equation of an ellipse whose minor axis has endpoints (± 5, 0) and whose major axis has length 16.

Solution: Since the minor axis of the ellipse has endpoints (± 5, 0), we know that it lies on the x-axis. The major axis, whose endpoints are (0, ± 8), lies on the y-axis. The foci then lie on the y-axis. This information and the sketch of the ellipse shown in Figure 7-18 tell us that the ellipse has an equation of the form

$$\frac{x^2}{b^2} + \frac{y^2}{a^2} = 1$$

with $a = 8$ and $b = 5$. That is, the ellipse has an equation

$$\frac{x^2}{25} + \frac{y^2}{64} = 1.$$

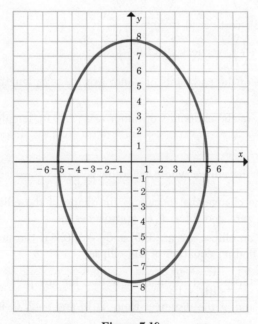

Figure 7-18

EXERCISES 7.3C

◀In Exercises 1–10, sketch the ellipse. Find and label the x- and y-intercepts and the coordinates of the foci.

1. $\dfrac{x^2}{25} + \dfrac{y^2}{9} = 1$ 6. $4x^2 + 32y^2 = 64$

2. $\dfrac{x^2}{36} + \dfrac{y^2}{16} = 1$ 7. $16x^2 + 8y^2 = 48$

3. $\dfrac{x^2}{4} + \dfrac{y^2}{25} = 1$ 8. $15x^2 + 5y^2 = 75$

4. $\dfrac{x^2}{16} + \dfrac{y^2}{64} = 1$ 9. $12x^2 + 8y^2 = 192$

5. $2x^2 + 5y^2 = 50$ 10. $20x^2 + 50y^2 = 125$

◀ In Exercises 11–16, find an equation of the ellipse satisfying the given conditions.

11. Foci $(\pm 2, 0)$, vertices $(\pm 5, 0)$

12. Foci $\left(\pm 3\frac{1}{2}, 0\right)$, vertices $\left(\pm 6\frac{1}{2}, 0\right)$

13. Foci $(0, \pm 3)$, vertices $(0, \pm 5)$

14. Foci $\left(0, \pm 4\frac{1}{2}\right)$, vertices $\left(0, \pm 6\frac{1}{2}\right)$

15. Coordinates of the endpoints of minor axis $(0, \pm 2)$, length of major axis 10.

16. Coordinates of the endpoints of minor axis $(\pm 3, 0)$, length of major axis 10.

17. The foci of the ellipse whose equation is $\dfrac{x^2}{a^2} + \dfrac{y^2}{b^2} = 1$ are F_1 and F_2.

    Describe the change that occurs in the shape of the ellipse as (a) F_1F_2 approaches zero and (b) as F_1F_2 approaches $2a$.

18. The foci of an ellipse are at $(3, 0)$ and $(-3, 0)$. If the sum of the focal radii is 10, find an equation of the ellipse.

19. Given the ellipse $\dfrac{x^2}{16} + \dfrac{y^2}{12} = 1$, show that for any point $P(x, y)$ on the ellipse the sum of the distances from $(2, 0)$ and $(-2, 0)$ is 8.

20. A semielliptic arch over a highway has the equation

$$\dfrac{x^2}{400} + \dfrac{y^2}{225} = 1 \qquad (y > 0).$$

    How much room will there be to spare when a truck, 8 ft. wide and 12 ft. high, drives through the arch down the middle of the highway?

21. **CHALLENGE PROBLEM.** Find an equation of the form $Ax^2 + Cy^2 + Dx + Ey + F = 0$ defining an ellipse whose foci are at the points $(0, 3)$ and $(6, 3)$ and whose major axis has length 10.

22. **CHALLENGE PROBLEM.** Show that Equation (4), page 269,

$$\frac{x^2}{a^2} + \frac{y^2}{b^2} = 1$$

implies Equation (1), page 268,

$$\sqrt{(x + c)^2 + y^2} + \sqrt{(x - c)^2 + y^2} = 2a,$$

where $c^2 = a^2 - b^2$.

23. **CHALLENGE PROBLEM.** Given the ellipse $\dfrac{x^2}{a^2} + \dfrac{y^2}{b^2} = 1$, show that for any point $P(x, y)$ on the ellipse, the sum of the distances from $F_1(c, 0)$ and $F_2(-c, 0)$ is $2a$.

24. **CHALLENGE PROBLEM.** Describe the graph of the inequality

$$\frac{x^2}{9} + \frac{y^2}{16} < 1.$$

Find all integral solutions, that is, find the coordinates of the lattice points

$$\left\{ (x, y) : \frac{x^2}{9} + \frac{y^2}{16} < 1, x \in I, \text{ and } y \in I \right\}.$$

25. **CHALLENGE PROBLEM.** A satellite is placed in an elliptical orbit about the earth so that the North and South poles of the earth lie in the plane of its orbit. Its distance from the North pole plus its distance from the South pole is constant. How high will it be when it passes directly over the North pole if it is 200 miles above the surface of the earth at the moment when it passes through the plane of the equator? Write an equation for its orbit with respect to the center of the earth. (Assume that the diameter of the earth is 8000 miles and that the earth is spherical.)

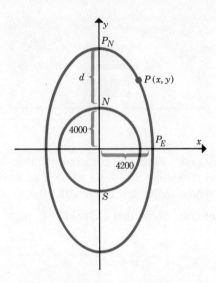

26. **CHALLENGE PROBLEM.** Find the coordinates of four points on the curve defined by the equation $x^2 + 4y^2 = 80$ which are the vertices of a square having diagonals through the origin.

27. **CHALLENGE PROBLEM.** Find an equation of an ellipse which has foci at $(\pm 2\sqrt{5}, 0)$ and passes through $(-3\sqrt{2}, 2\sqrt{2})$.

28. **CHALLENGE PROBLEM.** A rod 12 units long moves so that its endpoints move along two perpendicular lines. Find the path of a point P that is 4 units from one end of the rod.

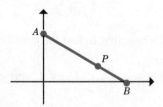

7.4 THE PARABOLA

The graph of the equation $y = x^2$ which you have studied earlier is an example of a curve known as a *parabola*. Figure 7-19 shows examples of various occurrences of parabolas.

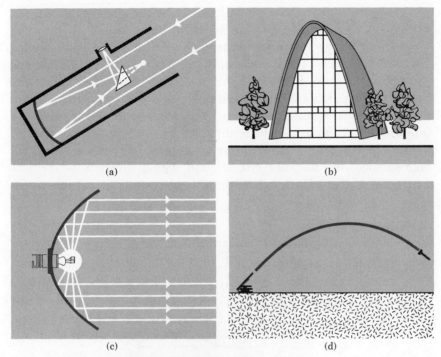

(a)

(b)

(c)

(d)

Figure 7-19

Definition 7.2 A **parabola** is a set of all points P in the plane such that the distance of each point $P(x, y)$ from a fixed point F is equal to the distance of P from a fixed line l. The fixed point is called the **focus**; the fixed line is called the **directrix**. (See Figure 7-20.)

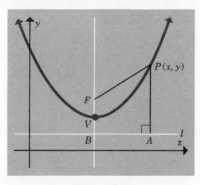

Figure 7-20

In Figure 7-20, we have $PF = PA$, A being the point of intersection of line l with the perpendicular drawn to line l from P.

A line through the focus perpendicular to the directrix is called the **axis of symmetry** of the parabola (or, more simply, the **axis**), and the point V in which the axis intersects the parabola is called the **vertex** of the parabola. Since each point on a parabola is equidistant from focus and directrix, we have $VF = VB$, where B is the point of intersection of the axis and the directrix.

Let us consider a parabola as in Figure 7-21 whose focus, for convenience, is the point $F(0, d)$ and whose directrix is the line $l : y = -d$. Thus the y-axis is the axis of symmetry and the vertex

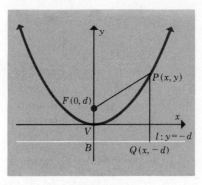

Figure 7-21

of the parabola is the point $(0, 0)$. Let $P(x, y)$ be a point on the parabola. If a perpendicular is drawn from P to line l, the foot of this perpendicular is the point $Q(x, -d)$. By Definition 7.2, $PF = PQ$ and, by the distance formula,

$$PF = \sqrt{x^2 + (y - d)^2} \qquad \text{and} \qquad PQ = \sqrt{(y + d)^2}.$$

Hence $P(x, y)$ is on the parabola if and only if

$$\sqrt{x^2 + (y - d)^2} = \sqrt{(y + d)^2}.$$

This equation is equivalent to

$$x^2 + (y - d)^2 = (y + d)^2$$

or

$$(1) \qquad\qquad\qquad x^2 = 4dy.$$

Hence we conclude that Equation (1) is an equation of the parabola that has the point $(0, d)$ as focus, the line $l : y = -d$ as directrix, and whose vertex is at the point $(0, 0)$. The parabola is symmetric with respect to the y-axis. (Why?)

Example 1 Find the equation of a parabola whose focus is at the point $(0, -d)$ and whose vertex is at the origin. (See Figure 7-22.)

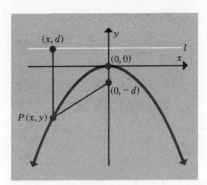

Figure 7-22

Solution: Since the focus is the point $(0, -d)$, the directrix will be the line $l : y = d$. We have

$$\sqrt{x^2 + (y + d)^2} = \sqrt{(y - d)^2}$$

from which, as the student should check, we obtain

$$x^2 = -4dy$$

which can be written as

$$x^2 = 4(-d)y.$$

If the positive direction of the y-axis is, as is customary, drawn upward, then the parabola $x^2 = 4dy$ opens upward if $d > 0$ as is illustrated by Figure 7-21. If $d < 0$, then the parabola $x^2 = 4dy$ opens downward as is illustrated by Figure 7-22.

Figures 7-23a and b illustrate the graphs of $y^2 = 4dx$ for $d > 0$ and $y^2 = 4dx$ for $d < 0$, respectively.

Note that each of the equations, $x^2 = 4dy$, $x^2 = -4dy$, $y^2 = 4dx$, and $y^2 = -4dx$, can be put in the form

(2) $$Ax^2 + Cy^2 + Dx + Ey + F = 0.$$

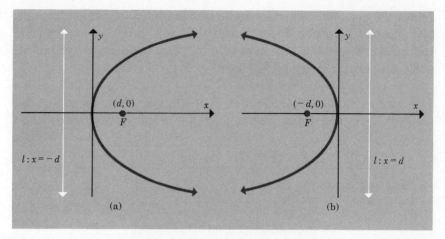

Figure 7-23

We see that when Equation (2) is the equation of a parabola with vertex at the origin and axis either the x-axis or the y-axis, then either $A = E = F = 0$ and C and $D \neq 0$, or $C = D = F = 0$ and A and $E \neq 0$.

Example 2   Find an equation of the parabola whose focus is at the point $(0, \frac{1}{4})$ and whose vertex is at the origin.

Solution: Since $(0, \frac{1}{4})$ is a point on the positive half of the y-axis, the "type" of equation is $x^2 = 4dy$ with $d > 0$. We have $d = \frac{1}{4}$ and an equation of the parabola is $x^2 = y$.

Example 3   Find an equation of the parabola with vertex at the origin and directrix the line $x = -2$.

Solution: Since the directrix lies to the left of the y-axis, the "type" of equation is $y^2 = 4dx$ with $d > 0$. We have $d = 2$ and an equation of the parabola is $y^2 = 8x$.

When you are asked to sketch a parabola, it is necessary to find the coordinates of a few points in order to gauge the size of the "opening". Symmetry, of course, helps in sketching.

Example 4   Find the focus and directrix of the parabola defined by the equation $x^2 = -12y$. Sketch the curve.

Solution: If we compare $x^2 = -12y$ and $x^2 = 4dy$, we have $d = -3$. Hence the focus is the point $(0, -3)$ and the directrix is the line $y = 3$.

Replacing y by -3 in the equation $x^2 = -12y$, we have $x = \pm 6$.
Replacing y by -12 in the equation $x^2 = -12y$, we have $x = \pm 12$.
We use the points $(6, -3)$, $(-6, -3)$, $(12, -12)$, and $(-12, -12)$
to help sketch the graph as indicated in Figure 7-24.

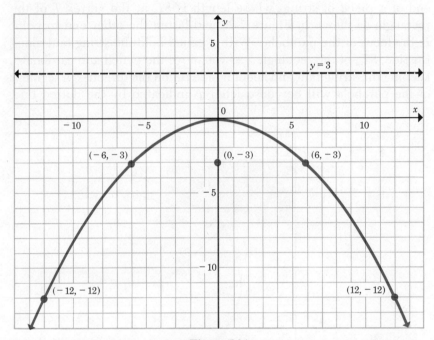

Figure 7-24

The equation $x^2 = 4dy \ (d > 0)$ defines a family of curves as shown
in Figure 7-25. If $d < 0$, what family of curves is defined?

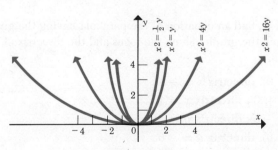

Figure 7-25

Although somewhat harder than constructing a circle or ellipse, there is a way of constructing a parabola by using a piece of string. Draw a line on a piece of paper (l in Figure 7-26) for the directrix of the parabola. Now place a rectangular-shaped piece of cardboard with one side, $\overline{RS}$, on the directrix. With tape fasten one end of a piece of string, whose length is ST, at the vertex T of the rectangle as shown in Figure 7-26 and fasten the other end to point F, the focus of the parabola. With a pencil hold the string taut along the side, $\overline{ST}$, of the rectangle. As you move the pencil and cardboard, keeping the string taut and keeping side $\overline{RS}$ on the directrix, a segment of the parabola will be constructed as indicated by the dotted line.

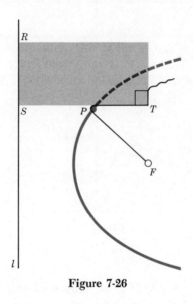

Figure 7-26

EXERCISES 7.4

◀ In Exercises 1–8, find an equation of the parabola having the given specifications and sketch the graph. Show the focus and the directrix of the parabola in your sketch.

1. Focus $(3, 0)$, directrix $x = -3$
2. Focus $(-4, 0)$, directrix $x = 4$
3. Focus $(0, -5)$, directrix $y = 5$
4. Focus $(0, 6)$, directrix $y = -6$
5. Focus $(4, 0)$, vertex $(0, 0)$

6. Focus $(0, -4)$, vertex $(0, 0)$

7. Vertex $(0, 0)$, directrix $y = -\dfrac{5}{2}$

8. Vertex $(0, 0)$, directrix $x = \dfrac{3}{2}$

◀ In Exercises 9–16, find the coordinates of the focus and an equation of the directrix of the parabola defined by the given equation. Sketch a graph of each parabola.

9. $x^2 = -4y$ **13.** $y^2 = -6x$

10. $x^2 + y = 0$ **14.** $3x + 4y^2 = 0$

11. $x^2 = 4y$ **15.** $x^2 = -6y$

12. $2x^2 - 4y = 0$ **16.** $-3x = y^2$

◀ A line segment that is perpendicular to the axis of the parabola at its focus and whose endpoints are on the parabola is called the **latus rectum.** In Exercises 17–20, find the length of the latus rectum of the parabola defined by the given equation.

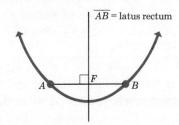

$\overline{AB}$ = latus rectum

17. $y^2 = x$ **19.** $x^2 = -6y$

18. $y^2 = 4x$ **20.** $-3x = y^2$

◀ In Exercises 21–24, use the definition of a parabola to find an equation of the parabola that has the given focus and directrix.

21. Focus $(0, 4)$, directrix $y = 0$

22. Focus $(4, 0)$, directrix $x = 0$

23. Focus $(4, 3)$, directrix $y = 1$

24. Focus $(5, 4)$, directrix $y = 2$

25. Show that the circle whose diameter is the latus rectum of the parabola defined by the equation $x^2 = 2ky$ is tangent to the directrix of the parabola.

26. Explain why the construction shown in Figure 7–26 produces a parabola.

27. **CHALLENGE PROBLEM.** A cable of the Golden Gate suspension bridge is approximately in the shape of a parabola. The supporting towers of the cable are 4200 ft. apart; the cable passes over the supporting tower 746 ft. above the bay; the bridge is 200 ft. above the bay; and the lowest point of the cable is 6 ft. above the roadway. Suppose that the cable shape is exactly that of a parabola. Find the lengths of supporting rods (from the cable to the roadway) at 300-ft. intervals from the center of the bridge to one of the towers.

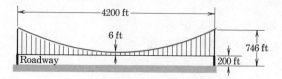

7.5 THE HYPERBOLA

The fourth and last conic section to be studied is the *hyperbola*. Hyperbolas are not as commonly encountered as are circles, ellipses, and parabolas. However, certain kinds of range-finding equipment plot positions using hyperbolas; comets and meteors frequently travel in hyperbolic paths.

> *Definition 7.3* A **hyperbola** is a set of all points in the plane such that the absolute value of the difference of the distances from each point $P(x, y)$ in the set to two fixed points, F_1 and F_2, is a constant. (See Figure 7-27.) The two fixed points are called **foci**.

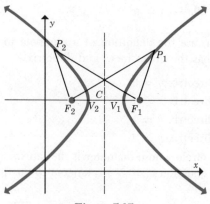

Figure 7-27

If we designate this constant difference by $2a$, we have, as in Figure 7-27,

$$|P_1F_1 - P_1F_2| = 2a$$

and

$$|P_2F_1 - P_2F_2| = 2a.$$

For the moment, let us restrict our study of hyperbolas to those whose foci are on the x-axis, equally distant from the origin. If $P(x, y)$ is any point on a hyperbola whose foci are the two points $F_1(-c, 0)$ and $F_2(c, 0)$ (Figure 7-28) and where $k = 2a < 2c$, then, by Definition 7.3, we have

$$PF_1 - PF_2 = 2a \text{ or}$$
$$PF_1 - PF_2 = -2a.$$

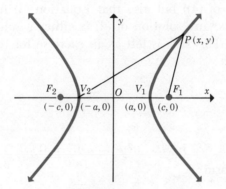

Figure 7-28

Thus we have that $P(x, y)$ is on the hyperbola if and only if

(1) $$\sqrt{(x + c)^2 + y^2} - \sqrt{(x - c)^2 + y^2} = 2a$$

or

(2) $$\sqrt{(x + c)^2 + y^2} - \sqrt{(x - c)^2 + y^2} = -2a.$$

Proceeding as we did in deriving an equation for an ellipse in Section 7.3, we obtain, in both cases (1) and (2), the equation

(3) $$\frac{x^2}{a^2} - \frac{y^2}{c^2 - a^2} = 1$$

and then, since $c^2 - a^2 > 0$, if we let $b^2 = c^2 - a^2$ we obtain

(4) $$\frac{x^2}{a^2} - \frac{y^2}{b^2} = 1.$$

(Note that the relationship, $b^2 = c^2 - a^2$, is *not* the same as for the ellipse. For the hyperbola we have $b^2 = c^2 - a^2$; for the ellipse, $b^2 = a^2 - c^2$.)

Replacing y by 0 in Equation (4) and solving for x, we get $x = a$ and $x = -a$. Hence the points where the hyperbola intersects the line containing the foci, F_1 and F_2, are $(a, 0)$ and $(-a, 0)$. The line containing the foci is called the **transverse axis** and the points of intersection (V_1 and V_2 in Figure 7-28) are called the **vertices**. Note that $V_1V_2 = 2a$. The midpoint, C, of $\overline{V_1V_2}$ is called the **center** of the hyperbola and the line through C perpendicular to $\overline{V_1V_2}$ is called the **conjugate axis**.

A further necessary task is to show that not only does Equation (1) or Equation (2) imply Equation (4) (that is, any solution of (1) or (2) is a solution of (4)) but also that Equation (4) implies Equation (1) or (2) (that is, any solution of (4) is either a solution of (1) or a solution of (2)). This task is left as an exercise for the student.

The equation

$$\frac{x^2}{a^2} - \frac{y^2}{b^2} = 1$$

is equivalent to

$$b^2x^2 - a^2y^2 - a^2b^2 = 0$$

which has the form

(5) $$Ax^2 + Cy^2 + Dx + Ey + F = 0.$$

We see that when Equation (5) is the equation of a hyperbola, with center at the origin and transverse axis the x-axis, then $AC < 0$ (that is, $A > 0$ and $C < 0$ or $A < 0$ and $C > 0$) and $D = E = 0$. (Compare the situation in regard to AC for the ellipse.) What can you say about F?

If the foci lie on the y-axis and the center is at the origin, then the hyperbola has an equation of the form

(6) $$\frac{y^2}{a^2} - \frac{x^2}{b^2} = 1$$

where, again, $b^2 = c^2 - a^2$.

Example 1   Find an equation of the hyperbola whose foci are the points $(0, 5)$ and $(0, -5)$ and whose vertices are the points $(0, 3)$ and $(0, -3)$.

Solution: Since the foci lie on the y-axis, the transverse axis is coincident with the y-axis. Therefore the hyperbola has an equation of the form

$$\frac{y^2}{a^2} - \frac{x^2}{b^2} = 1.$$

We have $a = 3$ and $c = 5$. Thus $b^2 = 25 - 9 = 16$. An equation is

$$\frac{y^2}{9} - \frac{x^2}{16} = 1.$$

Replacing x by 3, we find that the point $(3, \frac{15}{4})$ is on the graph. Since the hyperbola is symmetric with respect to both the x-axis and y-axis (Why?), the points $(3, -\frac{15}{4})$, $(-3, \frac{15}{4})$, and $(-3, -\frac{15}{4})$ are also on the graph. (See Figure 7-29.) We use these four points and the vertices to sketch the graph of the equation

$$\frac{y^2}{9} - \frac{x^2}{16} = 1.$$

Example 2   Sketch the graph of the equation $12x^2 - 4y^2 = 48$.

Solution: The equation $12x^2 - 4y^2 = 48$ is equivalent to

$$\frac{x^2}{4} - \frac{y^2}{12} = 1.$$

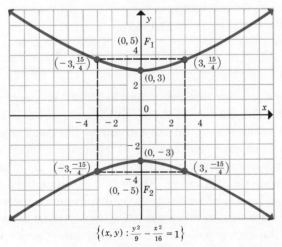

$$\left\{(x, y) : \frac{y^2}{9} - \frac{x^2}{16} = 1\right\}$$

Figure 7-29

This is a hyperbola whose transverse axis is coincident with the x-axis. Since $a^2 = 4$ and $b^2 = 12$, we have $a = 2$ and $b = \sqrt{12}$. Since $c^2 = a^2 + b^2$, we have $c^2 = 4 + 12$ and so $c = 4$. The foci are the points $(4, 0)$ and $(-4, 0)$. The vertices are the points $(-2, 0)$ and $(2, 0)$. If we replace x by 4 in the equation

$$\frac{x^2}{4} - \frac{y^2}{12} = 1,$$

we have $y = \pm 6$. Thus the points $(4, 6)$, $(4, -6)$, $(-4, 6)$, and $(-4, -6)$ are on the graph. Using these points and the vertices, we sketch the graph. (See Figure 7-30.)

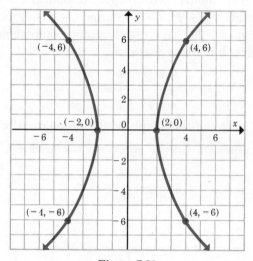

Figure 7-30

If we are graphing the curve $y = \dfrac{1}{x}$ for $x > 0$, we see that y takes on smaller and smaller positive values as larger and larger values are assigned to x. This means, as shown in Figure 7-31, that the distance between the point $A(x, y)$ on the curve and the point $B(x, 0)$ on the x-axis becomes smaller and smaller as x becomes larger and larger.

In fact, for very large values of x, the curve and the x-axis are indistinguishable on a drawing. When the distance between a curve and a straight line approaches zero in this manner, the straight line is called an **asymptote** to the curve. If a curve has an asymptote, then a sketch of the curve will tend to approximate the straight line of the asymptote for large values of x. Thus the drawing of a curve with asymptotes is simplified if the asymptotes are drawn first.

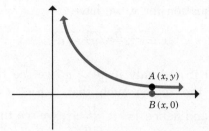

Figure 7-31

If we solve the equation $\dfrac{x^2}{a^2} - \dfrac{y^2}{b^2} = 1$ for y, we obtain

$$y = \pm \frac{b}{a} \sqrt{x^2 - a^2}.$$

Since $x^2 - a^2 \geq 0$ if and only if $|x| \geq a$, we see that no points on the graph lie in the region such that $-a < x < a$. This region, $\{(x, y) : -a < x < a\}$, is shaded in Figure 7-32. Since $\sqrt{x^2 - a^2} < |x|$ and hence $|y| < |\frac{b}{a}x|$, we see that no points on the graph may lie in the region such that $|y| \geq |\frac{b}{a}x|$. This region, which is bounded by the straight lines $y = \frac{b}{a}x$ and $y = -\frac{b}{a}x$, is shaded in Figure 7-33. Since the difference $|x| - |\sqrt{x^2 - a^2}|$ becomes smaller as $|x|$ becomes increasingly larger, we see that the hyperbola is very near the straight lines $y = \frac{b}{a}x$ and $y = -\frac{b}{a}x$ for large $|x|$. The two lines, $y = \frac{b}{a}x$ and $y = -\frac{b}{a}x$, are the asymptotes of the hyperbola

$$\frac{x^2}{a^2} - \frac{y^2}{b^2} = 1.$$

If the foci of the hyperbola are on the y-axis, then the equation of the hyperbola has the form

$$\frac{y^2}{a^2} - \frac{x^2}{b^2} = 1.$$

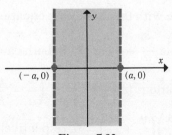

Figure 7-32

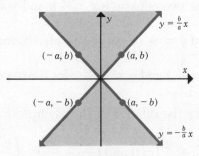

Figure 7-33

If we solve this equation for x, we have

$$x = \pm \frac{b}{a}\sqrt{y^2 - a^2}.$$

Since $y^2 - a^2 \geq 0$ if and only if $|y| \geq a$, we see that no points on the graph lie in the region such that $-a < y < a$. Also since $\sqrt{y^2 - a^2} < |y|$ and hence $|x| < \left|\frac{b}{a}y\right|$, we see that no points may lie in the region such that $|x| \geq \left|\frac{b}{a}y\right|$. This region, which is bounded by the lines $x = \frac{b}{a}y$ and $x = -\frac{b}{a}y$, is shaded in Figure 7-34.

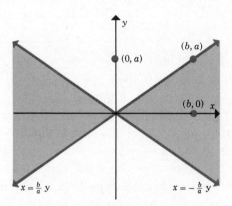

Figure 7-34

Again since the difference $|y| - |\sqrt{y^2 - a^2}|$ becomes smaller as $|y|$ becomes increasingly larger, the lines $x = \frac{b}{a}y$ and $x = -\frac{b}{a}y$ are now the asymptotes to the hyperbola

$$\frac{y^2}{a^2} - \frac{x^2}{b^2} = 1.$$

The two equations $x = \frac{b}{a}y$ and $x = -\frac{b}{a}y$ are equivalent to $y = \frac{a}{b}x$ and $y = -\frac{a}{b}x$, respectively. (Compare with the asymptote equations $y = \frac{b}{a}x$ and $y = -\frac{b}{a}x$ for the hyperbola $\frac{x^2}{a^2} - \frac{y^2}{b^2} = 1$. In both cases, note the relation of the asymptote equations to $\frac{x^2}{a^2} - \frac{y^2}{b^2} = \left(\frac{x}{a} - \frac{y}{b}\right)\left(\frac{x}{a} + \frac{y}{b}\right) = 0$ and $\left(\frac{y^2}{a^2} - \frac{x^2}{b^2}\right) = \left(\frac{y}{a} - \frac{x}{b}\right)\left(\frac{y}{a} + \frac{x}{b}\right) = 0$, respectively.)

Example 3   Sketch the graph of the equation $2x^2 - y^2 = 8$.

Solution: The equation $2x^2 - y^2 = 8$ is equivalent to

$$\frac{x^2}{4} - \frac{y^2}{8} = 1.$$

We have $a = 2$ and $b = \sqrt{8} \approx 2.8$. We plot the four points $(2, 2.8)$, $(2, -2.8)$, $(-2, 2.8)$, and $(-2, -2.8)$, draw the rectangle determined by the four points, draw and extend the diagonals of the rectangle, and sketch the graph as illustrated in Figure 7-35.

Example 4   Sketch the graph of the hyperbola whose foci are the points $(0, \sqrt{34})$ and $(0, -\sqrt{34})$ and whose asymptotes are the two lines $5x + 3y = 0$ and $5x - 3y = 0$.

Solution: Since the foci are the two points $(0, \sqrt{34})$ and $(0, -\sqrt{34})$, we know that the transverse axis lies along the y-axis. The equation $5x - 3y = 0$ is equivalent to $y = \frac{5}{3}x$. Now if we compare $y = \frac{5}{3}x$ to $y = \frac{a}{b}x$, we see that $\frac{a}{b} = \frac{5}{3}$ so that $a = \frac{5b}{3}$. Since $c = \sqrt{34}$ and $b^2 = c^2 - a^2$, we have $b^2 = 34 - \frac{25b^2}{9}$. Hence $\frac{34b^2}{9} = 34$ so that $b = 3$ and $a = \frac{5 \cdot 3}{3} = 5$. Hence we know that an equation of the hyperbola is

$$\frac{y^2}{25} - \frac{x^2}{9} = 1.$$

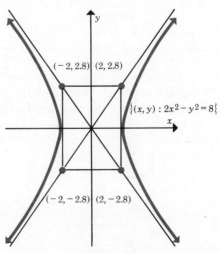

Figure 7-35

We draw a rectangle through the four points $(3, 5)$, $(3, -5)$, $(-3, 5)$, and $(-3, 5)$ and then draw the lines $y = \pm\frac{5x}{3}$. Replacing y by 6 in the equation of the hyperbola, we find that, approximately, the point $(2.0, 6)$ is on the graph. By symmetry we know that $(-2.0, 6)$, $(2.0, -6)$, and $(-2.0, -6)$ are also approximately on the graph. The four points together with the asymptotes and the rectangle enable us to sketch the graph as in Figure 7-36.

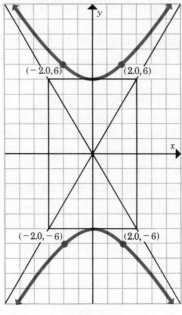

Figure 7-36

EXERCISES 7.5

◀In Exercises 1–12, sketch the hyperbola defined by the given equation, give the coordinates of the vertices and of the foci, and find the equations of the asymptotes.

1. $\dfrac{x^2}{4} - \dfrac{y^2}{4} = 1$

2. $\dfrac{x^2}{9} - \dfrac{y^2}{9} = 1$

3. $\dfrac{y^2}{16} - \dfrac{x^2}{16} = 1$

4. $\dfrac{y^2}{36} - \dfrac{x^2}{36} = 1$

5. $\dfrac{x^2}{9} - \dfrac{y^2}{16} = 1$

6. $\dfrac{x^2}{36} - \dfrac{y^2}{16} = 1$

7. $\dfrac{y^2}{16} - \dfrac{x^2}{25} = 1$ **10.** $9x^2 - 36y^2 = 324$

8. $\dfrac{y^2}{4} - \dfrac{x^2}{25} = 1$ **11.** $4x^2 - 9y^2 = 25$

9. $4x^2 - 25y^2 = 100$ **12.** $9y^2 - 16x^2 = 36$

◀ In Exercises 13–16, find an equation of the hyperbola which has the foci F_1 and F_2, and the given absolute value of the difference of the distances to $P(x, y)$, a point on the hyperbola.

13. $F_1(4, 0)$, $F_2(-4, 0)$; $|PF_1 - PF_2| = 6$
14. $F_1(0, 4)$, $F_2(0, -4)$; $|PF_1 - PF_2| = 6$
15. $F_1(-3\sqrt{2}, 0)$, $F_2(3\sqrt{2}, 0)$; $|PF_1 - PF_2| = 6$
16. $F_1(0, 2\sqrt{3})$, $F_2(0, -2\sqrt{3})$; $|PF_1 - PF_2| = 4$

◀ In Exercises 17–20, find an equation of a hyperbola that fits the given data.

17. Vertices $(\pm 5, 0)$, foci $(\pm 8, 0)$
18. Vertices $(\pm 3, 0)$, foci $(\pm 4, 0)$
19. Asymptotes $3x + 2y = 0$ and $3x - 2y = 0$, foci $(0, \pm 3)$
20. Asymptotes $y = 3x$ and $y = -3x$, foci $(\pm 4, 0)$
21. Show that both Equations (1) and (2) on page 293 imply Equation (4).
22. Show that Equation (4) implies that either Equation (1) holds or (2) holds.
23. Show that the hyperbola with the equation $3x^2 - y^2 = 12$ and the ellipse with the equation $9x^2 + 25y^2 = 225$ have the same foci. Sketch the two curves.
24. CHALLENGE PROBLEM. Find an equation of a hyperbola which has foci at $(0, \pm 2\sqrt{5})$ and passes through the point $(-2\sqrt{6}, 4\sqrt{3})$.
25. CHALLENGE PROBLEM. On a flat plain, the sound of a rifle and that of its bullet striking the target are heard at the same instant. Describe the set of possible locations of the listener.

7.6   TRANSLATION IN THE PLANE

A **translation** is a function that associates each point (x, y) with a point $(x + h, y + k)$ where h and k are real numbers. If (x', y') is the image of point (x, y) under a translation, we have

$$x' = x + h \quad \text{and} \quad y' = y + k.$$

If we have a translation T whose domain is the set of all ordered pairs of real numbers, then each point (x, y) in the plane is mapped into a point (x', y'). (See Figure 7-37.)

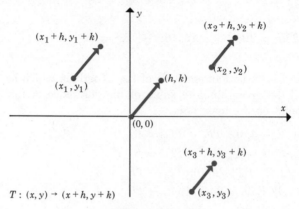

$$T : (x, y) \rightarrow (x + h, y + k)$$

Figure 7-37

A translation "preserves" distances and we say that distance is *invariant* (unchanged) under translation. To prove this, suppose that a translation maps the two points $P(x_1, y_1)$ and $Q(x_2, y_2)$ into $P'(x_1', y_1')$ and $Q'(x_2', y_2')$, respectively. By definition of a translation, we have

$$x_1' = x_1 + h, \qquad x_2' = x_2 + h,$$
$$y_1' = y_1 + k, \qquad y_2' = y_2 + k.$$

Thus

$$
\begin{aligned}
P'Q' &= \sqrt{(x_1' - x_2')^2 + (y_1' - y_2')^2} \\
&= \sqrt{[(x_1 + h) - (x_2 + h)]^2 + [(y_1 + k) - (y_2 + k)]^2} \\
&= \sqrt{(x_1 - x_2)^2 + (y_1 - y_2)^2} = PQ.
\end{aligned}
$$

Since distance is invariant under translation and conic sections are defined in terms of distance, the image of a conic section under translation is another curve identical in shape to that of the given conic section.

In Section 7.2, we learned that a circle whose radius is 2 and whose center is at the point $(3, -4)$ has an equation of the form

(1) $$(x - 3)^2 + (y + 4)^2 = 4.$$

Under the translation given by

$$x' = x - 3$$
$$y' = y + 4$$

we can consider the circle defined by Equation (1) as the graph of the equation

(2) $(x')^2 + (y')^2 = 4$

with respect to the x'-y'-axes shown in Figure 7-38.

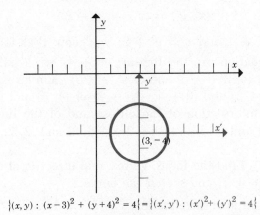

$$\{(x, y) : (x-3)^2 + (y+4)^2 = 4\} = \{(x', y') : (x')^2 + (y')^2 = 4\}$$

Figure 7-38

Similarly, we can see that the graph of the equation

(3) $\dfrac{(x-4)^2}{4} + \dfrac{(y-2)^2}{9} = 1$

is the graph of the equation

(4) $\dfrac{(x')^2}{4} + \dfrac{(y')^2}{9} = 1$

under the translation defined by $x' = x - 4$, $y' = y - 2$ as shown in Figure 7-39.

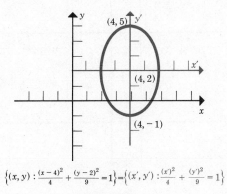

$$\left\{(x, y) : \frac{(x-4)^2}{4} + \frac{(y-2)^2}{9} = 1\right\} = \left\{(x', y') : \frac{(x')^2}{4} + \frac{(y')^2}{9} = 1\right\}$$

Figure 7-39

For the ellipse defined by Equation (4), we know that, with respect to the x'-y'-axes, the foci are $(0, \sqrt{5})$ and $(0, -\sqrt{5})$, the vertices are $(0, 3)$ and $(0, -3)$, and the points of intersection of the ellipse and the x'-axis are $(2, 0)$ and $(-2, 0)$. Since the translation carries (x, y) into

$$(x', y') = (x - 4, y - 2)$$

so that $x = x' + 4$ and $y = y' + 2$, we know that for the ellipse defined by Equation (3), the foci are $(0 + 4, \sqrt{5} + 2)$ and $(0 + 4, -\sqrt{5} + 2)$ or $(4, 2 + \sqrt{5})$ and $(4, 2 - \sqrt{5})$. Similarly, the vertices are $(0 + 4, 3 + 2)$ and $(0 + 4, -3 + 2)$ or $(4, 5)$ and $(4, -1)$, and the points of intersection of the ellipse and of the line $y = 2$ are $(2 + 4, 0 + 2)$ and $(-2 + 4, 0 + 2)$ or $(6, 2)$ and $(2, 2)$.

Example 1   Find the focus, vertex, and directrix of the parabola $(x - 2)^2 = 12(y + 3)$ and sketch the curve.

Solution: The graph of the equation $(x - 2)^2 = 12(y + 3)$ is the image of the graph of the equation $(x')^2 = 12y'$ under the translation defined by $x' = x - 2$ and $y' = y + 3$. Therefore the vertex of the parabola with respect to the x-y-axes is

$$(0 + 2, 0 + (-3)) = (2, -3)$$

and the axis of symmetry is the line $x = 2$. Since $12 = 4d$, $d = 3$. Hence the focus is $(0 + 2, 3 - 3)$ or $(2, 0)$ and the directrix is the line $y = -6$. Figure 7-40 shows a sketch of this parabola.

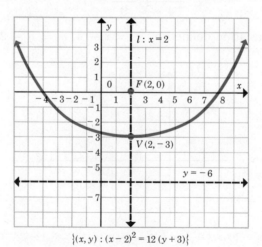

$$\{(x, y) : (x - 2)^2 = 12\,(y + 3)\}$$

Figure 7-40

Our general results concerning translation applied to conic sections can be summarized as follows:

(5) $$\frac{(x-h)^2}{a^2} + \frac{(y-k)^2}{b^2} = 1 \qquad (a > b > 0)$$

is an equation of an ellipse with center at (h, k) and whose major axis is the line $l : y = k$.

(6) $$(x-h)^2 = 4d(y-k)$$

is the equation of a parabola whose vertex is at (h, k) and whose axis of symmetry is the line $l : x = h$.

(7) $$\frac{(x-h)^2}{a^2} - \frac{(y-k)^2}{b^2} = 1$$

is an equation of a hyperbola whose center is at (h, k) and whose transverse axis is the line $l : y = k$.

Any equation of the form

$$Ax^2 + Cy^2 + Dx + Ey + F = 0$$

can be written in a form that simplifies the analysis of its graph. The process of conversion usually depends upon completing the square.

Example 2   Identify the graph of $4x^2 + y^2 - 32x + 6y + 57 = 0$.

Solution:

$$4x^2 - 32x + y^2 + 6y = -57,$$
$$4(x^2 - 8x + \quad) + (y^2 + 6y + \quad) = -57,$$
$$4(x^2 - 8x + 16) + (y^2 + 6y + 9) = -57 + 64 + 9,$$
$$4(x-4)^2 + (y+3)^2 = 16,$$
$$\frac{(x-4)^2}{4} + \frac{(y+3)^2}{16} = 1.$$

The graph is an ellipse whose center is at $(4, -3)$.

Example 3   Identify the graph of $x^2 - 6x - 4y + 1 = 0$.

Solution:

$$x^2 - 6x = 4y - 1,$$
$$x^2 - 6x + 9 = 4y - 1 + 9,$$
$$(x-3)^2 = 4(y+2).$$

The graph is a parabola whose vertex is at $(3, -2)$ and whose axis of symmetry is the line $l : x = 3$.

Example 4   Identify the graph of $9x^2 - 4y^2 + 36x + 8y - 4 = 0$.

Solution:
$$(9x^2 + 36x) - (4y^2 - 8y) = 4,$$
$$9(x^2 + 4x + \quad) - 4(y^2 - 2y + \quad) = 4,$$
$$9(x^2 + 4x + 4) - 4(y^2 - 2y + 1) = 4 + 36 - 4,$$
$$9(x + 2)^2 - 4(y - 1)^2 = 36,$$
$$\frac{(x + 2)^2}{4} - \frac{(y - 1)^2}{9} = 1.$$

The graph is a hyperbola whose center is at $(-2, 1)$.

EXERCISES 7.6

1. Consider the parabola whose equation is $y = x^2 + 2x + 5$.
   (a) Replace x by $x' - 2$ and write the new equation.
   (b) Replace x in the equation $y = x^2 + 2x + 5$ by $x' + 2$ and write the new equation.
   (c) Graph on the same coordinate axes the parabola defined by $y = x^2 + 2x + 5$, the curve defined by the equation obtained in (a), and the curve defined by the equation obtained in (b).

◀ In Exercises 2–7, find an equation of the parabola that satisfies the given conditions. Sketch the curve.

2. Focus $(0, 2)$ and directrix the x-axis
3. Focus $(0, -2)$ and directrix the x-axis
4. Focus $(0, 2)$ and directrix $y = -4$
5. Focus $(-2, 0)$ and directrix $x = 1$
6. Focus $(2d, 0)$ and directrix $x = d$ $(d > 0)$
7. Focus $(2d, d)$ and directrix $x = d$ $(d > 0)$

◀ In Exercises 8–12, sketch each ellipse, labeling the coordinates of the vertices, the foci, and the endpoints of the major and minor axes.

8. $\dfrac{(x - 3)^2}{25} + \dfrac{(y - 5)^2}{9} = 1$ 11. $16x^2 + 9y^2 - 96x + 72y + 144 = 0$

9. $\dfrac{(x + 2)^2}{9} + \dfrac{(y - 1)^2}{16} = 1$ 12. $4x^2 + 9y^2 + 8x - 36y + 4 = 0$

10. $x^2 + 4y^2 + 6x + 9 = 16$

◀In Exercises 13–22, find the coordinates of the vertex, an equation of the axis of symmetry, and sketch the parabola.

13. $y - 3 = x^2$

14. $y + 5 = x^2 + 2x + 1$

15. $y = x^2 - 4x$

16. $y - 3 = 8(x + 4)^2$

17. $y = x^2 + 2x - 3$

18. $y = 2x^2 + 8x - 5$

19. $y = -x^2 + 6x + 7$

20. $y^2 + 2y - 5x + 11 = 0$

21. $x^2 - 2x - y + 8 = 0$

22. $2y^2 + 28y - x + 101 = 0$

23. Find an equation of the ellipse whose vertices are at the points $(5, 2)$ and $(-3, 2)$ and which has one focus at the point $(4, 2)$. Sketch the curve.

24. Find an equation for each of the two ellipses which has a major axis 10 units in length, a minor axis 6 units in length, and whose centers are both at the point $(-2, -1)$.

◀In Exercises 25–29, find the coordinates of the center, the foci, and the vertices of each hyperbola. Give equations of the asymptotes and sketch the curve.

25. $\dfrac{(x + 2)^2}{9} - \dfrac{(y - 5)^2}{4} = 1$

26. $\dfrac{(y - 4)^2}{16} - \dfrac{(x - 1)^2}{1} = 1$

27. $x^2 - y^2 - 2x - 6y - 17 = 0$

28. $4y^2 - x^2 + 12y + 4x + 21 = 0$

29. $9x^2 - 16y^2 - 72x - 96y - 144 = 0$

30. Show that the graph of the equation

$$2x^2 + y^2 - 16x - 6y + 41 = 0$$

is a single point.

31. Show that the solution set of the equation

$$3x^2 + 2y^2 - 18x + 20y + 77 = 0$$

contains just one element.

The following remarks apply to Exercises 32–35. The graph of the equation

$$Ax^2 + Cy^2 + Dx + Ey + F = 0 \qquad \text{(not all } A, C, D, E, F = 0)$$

is

(a) a circle, a point, or ∅ if $A = C \neq 0$.

(b) an ellipse, a point, or ∅ if $A \neq C$ and $AC > 0$.

(c) a parabola, a pair of parallel lines, a line, or ∅ if $AC = 0$.

(d) a hyperbola or a pair of intersecting lines if $AC < 0$.

32. Give equations whose graphs will illustrate each of the possibilities in (a).
33. Give equations whose graphs will illustrate each of the possibilities in (b).
34. CHALLENGE PROBLEM. Give equations whose graphs will illustrate each of the possibilities in (c).
35. CHALLENGE PROBLEM. Give equations whose graphs will illustrate each of the possibilities in (d).

7.7 QUADRATIC FUNCTIONS AND INEQUALITIES

We have discussed the graph of the equation

$$x^2 = 4dy$$

and of the related equation

$$(x - h)^2 = 4d(y - k).$$

Each of these equations is equivalent to an equation of the form

$$y = ax^2 + bx + c \qquad (a \neq 0).$$

Any equation of the form $y = ax^2 + bx + c$ $(a \neq 0)$ defines a quadratic function

$$f : f(x) = ax^2 + bx + c.$$

The equation $y = x^2$ defines a function whose domain is the set of real numbers. The range is the set of nonnegative numbers. The graph of the function, which is very familiar to you, is shown in Figure 7-41. It is a parabola with vertex at $(0, 0)$. Since the vertical axis is the $f(x)$ axis, we see that the minimum value of $f(x)$ is 0. In Chapter 6 you were given an algebraic formula for finding a minimum value. Now we shall establish the validity of this formula.

All functions defined by an equation of the form $y = ax^2$ $(x > 0)$ have a minimum value of 0. If various positive values are assigned

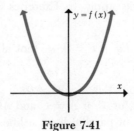

Figure 7-41

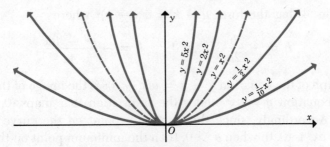

Figure 7-42

to a, we have a family of curves as in Figure 7-42. If various negative values are assigned to a, we have a family of curves as in Figure 7-43. If the parabola opens downward, that is, in the negative direction on the $f(x)$ axis, we can read the maximum value of the function from the graph.

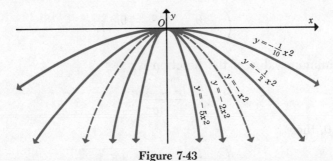

Figure 7-43

Now let us turn to the general quadratic function defined by the equation $y = ax^2 + bx + c$, $a \neq 0$. The equation $y = ax^2 + bx + c$ is equivalent to the equation

$$y + \frac{b^2 - 4ac}{4a} = a\left(x + \frac{b}{2a}\right)^2$$

as we now show. We have

(1) $$y = ax^2 + bx + c,$$

(2) $$y - c = a\left(x^2 + \frac{b}{a}x\right),$$

(3) $$y - c + \frac{b^2}{4a} = a\left(x^2 + \frac{b}{a}x + \frac{b^2}{4a^2}\right),$$

(4) $$y + \frac{b^2 - 4ac}{4a} = a\left(x + \frac{b}{2a}\right)^2.$$

Equation (4) has the form $y - k = a(x - h)^2$ where

$$h = -\frac{b}{2a} \quad \text{and} \quad k = -\frac{b^2 - 4ac}{4a}.$$

The graph of the equation $y - k = a(x - h)^2$ is the image of the graph of the equation $y = ax^2$ under the translation that maps $(0, 0)$ into (h, k). Accordingly, since the minimum point on the curve defined by $y = ax^2$ is $(0, 0)$ when $a > 0$, then the minimum point on the curve defined by

$$y - k = a(x - h)^2$$

is (h, k) when $a > 0$.

Hence, if $a > 0$, the minimum point on the curve defined by

$$y + \frac{b^2 - 4ac}{4a} = a\left(x + \frac{b}{2a}\right)^2$$

is

$$\left(-\frac{b}{2a}, -\frac{b^2 - 4ac}{4a}\right).$$

The minimum value of the function is

$$-\frac{b^2 - 4ac}{4a}.$$

If $a < 0$, then

$$\left(-\frac{b}{2a}, -\frac{b^2 - 4ac}{4a}\right)$$

is the maximum point on the curve and

$$-\frac{b^2 - 4ac}{4a}$$

is the maximum value of the function.

Example 1   Find the maximum (minimum) point on the curve defined by $y = 3x^2 - 6x + 4$.

Solution: We have

$$y - 4 = 3(x^2 - 2x),$$
$$y - 4 + 3 = 3(x^2 - 2x + 1),$$
$$y - 1 = 3(x - 1)^2.$$

Since $a > 0$, the curve opens upward. The minimum point is $(1, 1)$. (See Figure 7-44.)

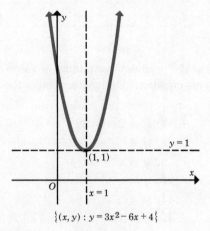

Figure 7-44

Example 2   Find the maximum (minimum) value of the function $F : F(x) = -2x^2 + 5x - 2$.

Solution: We let $y = F(x)$ and have

$$y = -2x^2 + 5x - 2,$$

$$y + 2 = -2\left(x^2 - \frac{5}{2}x\right),$$

$$y + 2 - \frac{25}{8} = -2\left(x^2 - \frac{5}{2}x + \frac{25}{16}\right),$$

$$y - \frac{9}{8} = -2\left(x - \frac{5}{4}\right)^2.$$

Since $a < 0$, the curve opens downward. The maximum point on the curve is $(\frac{5}{4}, \frac{9}{8})$. The maximum value of the function F is $\frac{9}{8}$. (See Figure 7-45.)

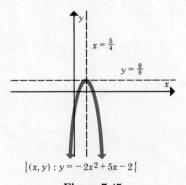

Figure 7-45

EXERCISES 7.7A

◀In Exercises 1–12, find the maximum or minimum value of the function that is described by the given equation. Designate whether the value is a maximum or a minimum.

1. $y = 2x^2 + 8$ 7. $y = 2x^2 + 3x$
2. $y = 3x^2 - 5$ 8. $y = -3x^2 + 2x$
3. $y = -2x^2 - 8$ 9. $y = x^2 + 4x + 7$
4. $y = -3x^2 + 5$ 10. $y = -x^2 + 6x + 4$
5. $y = x^2 + x$ 11. $y = -x^2 + 5x + 3$
6. $y = x^2 + 4x$ 12. $y = x^2 + 7x + 1$

◀In Exercises 13–20, find the maximum or minimum value of the function that is described by the given equation. Designate whether the value is a maximum or a minimum. Sketch a graph and label the axes.

13. $y = x^2 + 7x - 8$ 18. $y = -3x^2 + 6x - 2$
14. $y = x^2 + 5x - 4$ 19. $s = r^2 + 2r - 4$
15. $F(t) = -2t^2 + 4t + 5$ 20. $f(x) = -4x^2 - 8x + 1$
16. $f(s) = 3s^2 - 4s + 2$ 21. $G(x) = 3x^2 - 4x + 5$
17. $g(r) = 2r^2 + r + 1$

Recall that the graph of a line such as $l : 2x + y = 6$ separates the Cartesian plane into two halfplanes. (See Figure 7-46.) The shaded region contains all points in the set $\{(x, y) : 2x + y < 6\}$ and the nonshaded region contains all points in the set $\{(x, y) : 2x + y > 6\}$.

In a similar manner, the graph of a quadratic equation separates the Cartesian plane into two regions. The graph of the equation $y = x^2$ separates the plane in which it lies into two sets of points, $\{(x, y) : y < x^2\}$ and $\{(x, y) : y > x^2\}$. The shaded area in Figure 7-47

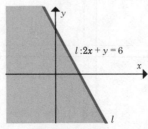

$l : 2x + y = 6$

Figure 7-46

pictures the set $\{(x, y) : y > x^2\}$. The graph of the equation $y = x^2$ is drawn as a dashed line in Figure 7-47 to indicate that it is not part of the graph of the inequality $y > x^2$.

Example 3   Graph the inequality $y \leq x^2 + 3x$.

Solution: First we graph the equation $y = x^2 + 3x$. To do this we first find the equivalent equation

$$y + \frac{9}{4} = \left(x + \frac{3}{2}\right)^2.$$

This equation has as its graph a parabola whose minimum point is $(-\frac{3}{2}, -\frac{9}{4})$. (See Figure 7-48.) To identify the region outside the curve, we determine if $(2, 0)$ is a member of the set $\{(x, y) : y < x^2 + 3x\}$. Since $(2, 0)$ is a member of the set $\{(x, y) : y < x^2 + 3x\}$, we shade the area outside the curve, for this region contains the points of $\{(x, y) : y < x^2 + 3x\}$. We check this picture by noting that $(-1, 0)$ is a member of the set $\{(x, y) : y > x^2 + 3x\}$.

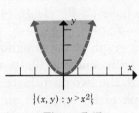

$\{(x, y) : y > x^2\}$

Figure 7-47

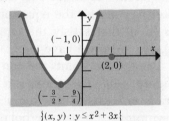

$\{(x, y) : y \leq x^2 + 3x\}$

Figure 7-48

EXERCISES 7.7B

◀In the following Exercises, graph the solution set of each inequality.

1. $3x - 2y > 4$

2. $x + 3y \leq 2$

3. $x > 0$ and $y < 2$

4. $x > -3$ and $y > 2$

5. $y < x^2$

6. $y \geq x^2$

7. $y > x^2 + 4x$

8. $y \leq x^2 - 6x$

9. $y > x^2 + 4x + 3$

10. $y \leq x^2 + 6x - 4$

11. $y \leq 2x^2 + 6x - 3$

12. $y \geq -3x^2 + 6x - 2$

13. $x^2 + y^2 < 4$

14. $x^2 + y^2 \leq 25$

15. $4x^2 + 9y^2 \geq 36$

16. $9x^2 + 16y^2 > 144$

17. $x^2 - y^2 < 4$ 19. $y^2 - 4x \geq 0$

18. $4x^2 - y^2 \geq 9$ 20. $2y^2 - y - 4x > 0$

21. $x^2 - 4y^2 - 6x - 16y + 29 > 0$

22. $2x^2 - y^2 - 16x - 6y + 21 > 0$

23. $3x^2 + 2y^2 - 18x + 20y + 71 < 0$

24. $9x^2 + 5y^2 - 90x + 20y + 150 \geq 0$

25. $16x^2 + 25y^2 - 32x - 150y - 159 \geq 0$

7.8 SPECIAL CASES OF THE CONIC SECTIONS

So far we have been concerned with graphs of equations of the form

$$Ax^2 + Bxy + Cy^2 + Dx + Ey + F = 0,$$

where $B = 0$ but not both A and C are equal to 0. When $B = 0$, the axis of the conic section corresponding to the equation is parallel to one of the coordinate axes. If $B \neq 0$, then the axis of symmetry is not parallel to either coordinate axis.

Let us consider the equation $xy = 4$. We can easily find ordered pairs (x, y) that satisfy the equation. We have $(2, 2)$, $(1, 4)$, $(4, 1)$, $(8, \frac{1}{2})$, etc. Also we have $(-2, -2)$, $(-1, -4)$, $(-4, -1)$, $(-8, -\frac{1}{2})$, etc. Using these points we can construct a graph of the equation as in Figure 7-49. The graph is a hyperbola with center at $(0, 0)$, with its transverse axis on the line $y - x$ and with the two coordinate axes as asymptotes. If a hyperbola is the graph of an equation $xy = k$ $(k \neq 0)$, it is called an **equilateral** or **rectangular** hyperbola. Figure

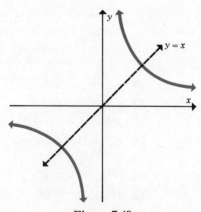

Figure 7-49

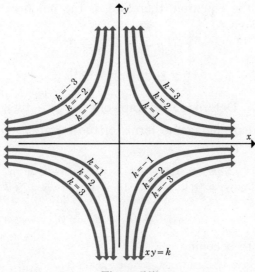

Figure 7-50

7-50 shows the family of graphs of $xy = k$ for $k = 1$, 2, 3, -1, -2, -3. The axis of symmetry of each of these hyperbolas is the line $y = x$ or the line $y = -x$.

As we noted in Section 7.1, a plane cutting a cone may intersect it in a conic section, a point, a line, or two intersecting lines. We call the cases when the intersection is a point, a line, or two intersecting lines, **limiting cases** or **degenerate cases**.

Now let us consider these cases from the algebraic point of view by means of some examples.

Example 1 Determine the graph of $x^2 + 2y^2 = 0$.

Solution: Since $x^2 > 0$ and $y^2 > 0$ for all nonzero real numbers x and y, the only pair of real numbers that satisfies the equation $x^2 + 2y^2 = 0$ is $(0, 0)$. Hence the graph is a single point.

Whenever we have an equation that is equivalent to an equation such as

$$(x + a)^2 - (y + b)^2 = 0,$$

then the graph will be two intersecting lines. Since the expression $(x + a)^2 - (y + b)^2$ factors, we have

$$[(x + a) + (y + b)][(x + a) - (y + b)] = 0.$$

The graph of the equation, therefore, is the union of the graphs of the two equations

$$x + y + a + b = 0$$

and

$$x - y + a - b = 0.$$

Example 2 Determine the graph of $x^2 - y^2 - 6x + 4y + 5 = 0$.

Solution: By rearranging the terms of the equation and completing the squares, we have

$$(x^2 - 6x) - (y^2 - 4y) = -5,$$
$$(x^2 - 6x + 9) - (y^2 - 4y + 4) = -5 + 9 - 4,$$

or

$$(x - 3)^2 - (y - 2)^2 = 0,$$

which factors to become

$$[(x - 3) + (y - 2)][(x - 3) - (y - 2)] = 0$$

or

$$(x + y - 5)(x - y - 1) = 0.$$

The graph, therefore, is the union of the two lines

$$x + y = 5 \qquad \text{and} \qquad x - y = 1.$$

Example 3 Determine the graph of $4x^2 - 9y^2 + 8x + 36y - 32 = 0$.

Solution: We regroup and complete the squares.

$$(4x^2 + 8x) - (9y^2 - 36y) = 32,$$
$$4(x^2 + 2x + 1) - 9(y^2 - 4y + 4) = 32 + 4 - 36,$$
$$4(x + 1)^2 - 9(y - 2)^2 = 0,$$
$$[2(x + 1) + 3(y - 2)][2(x + 1) - 3(y - 2)] = 0.$$

Hence we see that the graph of the equation consists of the two lines

$$l_1 : 2x + 3y - 4 = 0$$

and

$$l_2 : 2x - 3y + 8 = 0,$$

as shown in Figure 7-51.

In the next example we have an equation whose graph is a single line.

Example 4 Determine the graph of $x^2 = 0$. The graph is obviously the graph of $x = 0$, that is, the y-axis.

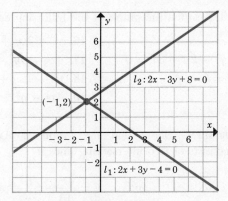

Figure 7-51

Two special cases arise in considering the equations

$$Ax^2 + Bxy + Cy^2 + Dx + Ey + F = 0$$

that do not occur in considering the intersections of a cone and a plane. These are illustrated in our concluding two examples.

Example 5   Determine the graph of $x^2 - 4 = 0$.
Since $x^2 - 4 = (x - 2)(x + 2) = 0$, we see that the graph of $x^2 - 4 = 0$ consists of the two parallel lines $x = 2$ and $x = -2$.

Example 6   Determine the graph of $x^2 + y^2 + 1 = 0$.
Since $x^2 \geq 0$ and $y^2 \geq 0$ for all real numbers x and y, we see that the graph of $x^2 + y^2 + 1 = 0$ is the empty set.

EXERCISES 7.8

◄In Exercises 1–5, sketch the graph of the given equations.

1. $xy = 8$ **4.** $xy + 1 = 0$

2. $xy = -4$ **5.** $xy = \sqrt{15}$

3. $xy - 12 = 0$

◄In Exercises 6–10, identify and sketch the following degenerate conics.

6. $2x^2 + y^2 = 0$ **9.** $xy = 0$

7. $3x^2 - 4y^2 = 0$ **10.** $x^2 - 4y^2 + 2x + 4y = 0$

8. $3x^2 + 4y^2 + 6x + 4y + 5 = 0$

Conic Sections and Their Graphs

◀ In Exercises 11–19, describe the graph of the following equations.

11. $3x^2 + y^2 = 0$

12. $9x^2 - 36x + y^2 - 2y + 37 = 0$

13. $2x^2 + y^2 + 4 = 0$

14. $y^2 = 0$

15. $x^2 - 9 = 0$

16. $x^2 + 2xy + y^2 - 4 = 0$

17. $4x^2 + 9y^2 + 12xy = 9$

18. $y^2 + 12y + 36 = 0$

19. $4x^2 - y^2 + 8x + 4 = 0$

20. **CHALLENGE PROBLEM.** Show that the graph of $Ax^2 + Cy^2 + Dx + Ey + F = 0$ consists of two intersecting lines if

$$AC < 0 \quad \text{and} \quad F = \frac{D^2}{4A} + \frac{E^2}{4C}.$$

CHAPTER SUMMARY

The CONIC SECTIONS are the ELLIPSE (including the circle, a special case of the ellipse), the PARABOLA, and the HYPERBOLA.

It can be shown that every equation of the second degree in x and y

(1) $$Ax^2 + Bxy + Cy^2 + Dx + Ey + F = 0$$

has for its graph a conic section except that in certain limiting cases the graph may be a point, the empty set, a line, two intersecting lines, or two parallel lines. Conversely, every conic section in the Cartesian plane has an equation of the form (1). When $B = 0$, the axes of symmetry of the conic sections determined by (1) are parallel to one of the coordinate axes. The properties of the conic sections are tabulated below.

Circle

$$PC = r$$

$x^2 + y^2 = r^2$
Center: $(0, 0)$, radius of
  length r

$(x - h)^2 + (y - k)^2 = r^2$
Center: (h, k), radius of
  length r

Ellipse

$$PF_1 + PF_2 = 2a, \, a > b, \, b^2 = a^2 - c^2$$

$$\frac{x^2}{a^2} + \frac{y^2}{b^2} = 1 \qquad\qquad \frac{(x-h)^2}{a^2} + \frac{(y-k)^2}{b^2} = 1$$

Center: $(0, 0)$ Center: (h, k)
Vertices: $(\pm a, 0)$ Vertices: $(h \pm a, k)$
Foci: $(\pm c, 0)$ Foci: $(h \pm c, k)$
Length of major axis: $2a$ Length of major axis: $2a$
Length of minor axis: $2b$ Length of minor axis: $2b$

Parabola

$$PF = \text{distance from } P \text{ to directrix}$$

$x^2 = 4dy$ $(x - h)^2 = 4d(y - k)$
Vertex: $(0, 0)$ Vertex: (h, k)
Focus: $(0, d)$ Focus: $(h, d + k)$
Directrix: $y = -d$ Directrix: $y = -d + k$

Hyperbola

$$|PF_1 - PF_2| = 2a, \qquad b^2 = c^2 - a^2$$

$$\frac{x^2}{a^2} - \frac{y^2}{b^2} = 1 \qquad\qquad \frac{(x-h)^2}{a^2} - \frac{(y-k)^2}{b^2} = 1$$

Center: $(0, 0)$ Center: (h, k)
Vertices: $(\pm a, 0)$ Vertices: $(h \pm a, k)$
Foci: $(\pm c, 0)$ Foci: $(h \pm c, k)$

Asymptotes: Asymptotes:

$$y = \pm \frac{b}{a}x \qquad\qquad\qquad y - k = \pm \frac{b}{a}(x - h)$$

An ellipse

$$\frac{x^2}{a^2} + \frac{y^2}{b^2} = 1 \qquad a^2 > b^2$$

has its major axis along the x-axis. An ellipse

$$\frac{x^2}{b^2} + \frac{y^2}{a^2} = 1 \qquad a^2 > b^2$$

has its major axis along the y-axis. A parabola

$$x^2 = 4dy$$

has the y-axis as an axis of symmetry, whereas the parabola

$$y^2 = 4dx$$

has the x-axis as an axis of symmetry. The vertices of the hyperbola

$$\frac{x^2}{a^2} - \frac{y^2}{b^2} = 1$$

lie on the x-axis, whereas the vertices of the hyperbola

$$\frac{y^2}{a^2} - \frac{x^2}{b^2} = 1$$

lie on the y-axis.

A TRANSLATION is a function that maps each point (x, y) in the Cartesian plane into a point (x', y') such that

$$x' = x + h \qquad \text{and} \qquad y' = x + k.$$

Distance is INVARIANT under a translation.

A quadratic function f is defined by $f(x) = ax^2 + bx + c$.
The equation $y = ax^2 + bx + c$ is equivalent to

$$y + \frac{b^2 - 4ac}{4a} = a\left(x + \frac{b}{2a}\right)^2.$$

The maximum (minimum) point on the graph of $y = ax^2 + bx + c$ is

$$\left(\frac{-b}{2a}, \ -\frac{b^2 - 4ac}{4a}\right);$$

the maximum (minimum) value of the function is

$$-\frac{b^2 - 4ac}{4a}.$$

The graph of the quadratic equation $y = ax^2 + bx + c$ divides a Cartesian plane into the region $\{(x, y) : y < ax^2 + bx + c\}$ and a second region, $\{(x, y) : y > ax^2 + bx + c\}$.

The graph of the equation $xy = k$ $(k > 0)$ is an EQUILATERAL hyperbola. The graph lies entirely in the first and third quadrants and both the x- and y-axes are asymptotes. The graph of the equation $xy = k$ $(k < 0)$ lies in the second and fourth quadrants.

REVIEW EXERCISES

◀ In Exercises 1–10, name the graph of each of the equations.

1. $x^2 = 10 - 5y^2$

2. $x^2 + 4x + y^2 = 9$

3. $y^2 = 4x - 1$

4. $36x^2 = 108 + 16y^2$

5. $3y = 4x$

6. $x = \dfrac{5}{y}$

7. $\dfrac{x^2}{2} = y + 6$

8. $(y - 2)^2 = 4x - 5$

9. $5 - 8x^2 + 36y^2 = 0$

10. $4x^2 - 5x = 3y^2 + 4x + 10$

◀ In Exercises 11–20, sketch the curve. Give the coordinates of the focus or foci and of the center and the vertices, if any. Give an equation of the directrix and asymptotes, if any.

11. $x^2 = 8y$

12. $4x^2 + 9y^2 = 36$

13. $9y^2 - 4x^2 = 36$

14. $4x^2 + 4y^2 = 81$

15. $xy = -24$

16. $x^2 = 16y$

17. $4x^2 - 9y^2 = 9$

18. $y^2 = 6x$

19. $(y - 2) = 4(x - 3)^2$

20. $16x^2 + 25y^2 = 225$

◀ In Exercises 21–30, sketch the curve. Label the foci, center, vertices, and asymptotes (if they exist).

21. $x^2 + 3y^2 + 6x + 6 = 0$

22. $x^2 - 4y^2 + 6x + 32y - 59 = 0$

23. $2x^2 - 2x + 3y^2 = \dfrac{19}{6}$

24. $x^2 - y^2 + 3x - y + 8 = 0$

25. $25x^2 + 9y^2 + 150x + 18y + 9 = 0$

26. $25x^2 - 9y^2 + 150x + 18y - 9 = 0$

27. $y^2 + 8x - 6y + 1 = 0$

28. $y^2 - 4y - 2x - 8 = 0$

29. $3x^2 + 3y^2 - 7x + y = 0$

30. $x^2 - 4x + y^2 + 8x = 64$

31. Find the minimum value of the function

$$f : f(x) = x^2 + x - 6.$$

32. Find the maximum value of the function

$$T : T(x) = -x^2 + 12x - 32.$$

33. Find an equation of a circle with center at $(4, -2)$ and passing through the origin.

34. Find an equation of the set of points such that the sum of the distances of any point $P(x, y)$ from $(0, 0)$ and the y-axis is 2.

35. Let $F_1(3, 0)$ and $F_2(-3, 0)$ be two fixed points. Find an equation of the set of points $P(x, y)$ such that

$$(PF_1)^2 - (PF_2)^2 = 12.$$

GOING FURTHER: READING AND RESEARCH

The conic sections, as we have seen, can be defined geometrically using a cutting plane and a cone; they can be defined using the concept of distance as we have done in this chapter. They can also be defined using the concept of *eccentricity*.

To do this we let l be a fixed line and F be a fixed point (not on l) in the Cartesian plane. Let P be a point in the plane, and PR the perpendicular distance from P to the line l. Depending upon what fixed value is assigned to the ratio $\dfrac{PF}{PR}$, called the **eccentricity,** we can have an ellipse, a parabola, or a hyperbola.

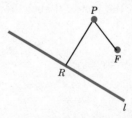

Find an equation defining the set S of points in the plane with the property that the distance of each point $P(x, y) \in S$ from the point $F(2, 0)$ is always twice the distance of P from the line $l : x = 4$.

Find an equation defining the set S of points in the plane with the property that the distance of each point $P(x, y) \in S$ from the point $F(0, 4)$ is always one-half the distance of P from the line $l : y = -3$.

Recall the definition of a parabola. Can you guess how the conic sections are defined using eccentricity?

If you are interested in learning more about this aspect of conic sections, you will find helpful information in each of the books in the following list.

WILSON, W. A. AND J. I. TRACY, *Analytic Geometry*, Boston, Mass: D. C. Heath, 3rd Ed., 1949.
MURDOCH, D. C. *Analytic Geometry*. New York: John Wiley and Sons, 1966.
School Mathematics Study Group. *Analytic Geometry*. Part 2. Pasadena: A. E. Vroman, Inc., 1968.

CHAPTER EIGHT
SYSTEMS OF EQUATIONS

8.1 LINEAR EQUATIONS

Recall that any pair (x, y) of real numbers that satisfies a linear equation

$$ax + by + c = 0 \ (a, b, c \in R)$$

is called a solution of the equation. If we have two linear equations

(1) $$a_1x + b_1y + c_1 = 0$$

and

(2) $$a_2x + b_2y + c_2 = 0,$$

then any pair (x, y) that satisfies both Equations (1) and (2) is called a solution of the system of linear equations. The solution set of a system of two linear equations

$$a_1x + b_1y + c_1 = 0$$

and

$$a_2x + b_2y + c_2 = 0$$

is

$$\{(x, y) : a_1x + b_1y + c_1 = 0\} \cap \{(x, y) : a_2x + b_2y + c_2 = 0\}.$$

Two systems of linear equations that have the same solution set are called equivalent systems. For example, the two systems

$$\begin{array}{ccc} 2x + y = 1 & & x + 3 = 4 \\ x = 1 & \text{and} & y = -1 \end{array}$$

are equivalent since the solution set of both is $\{(1, -1)\}$.

The solution set of a system of two linear equations may be the empty set, an infinite set, or a set with one ordered pair as its sole member.

Example 1   The solution set of the system

$$\begin{array}{c} 3x + 2y - 6 = 0 \\ 6x + 4y + 12 = 0 \end{array}$$

is the empty set. The graph of the equations of the system consists of two parallel lines as shown in Figure 8-1. Such a system is called an inconsistent system.

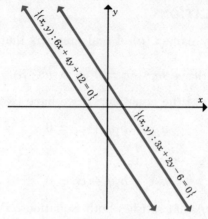

Figure 8-1

Example 2   The solution set of the system

$$3x + 2y - 6 = 0$$
$$6x + 4y - 12 = 0$$

consists of the set

$$\{(x, y) : 3x + 2y - 6 = 0\} = \{(x, y) : 6x + 4y - 12 = 0\}.$$

The graph is one line as shown in Figure 8-2. Such a system is called a **dependent** and **consistent** system.

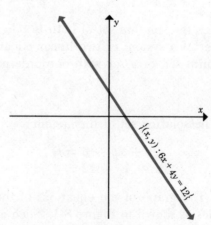

Figure 8-2

Example 3   The solution set of the system

$$2x - 3y = -2$$
$$x + 2y = 13$$

is $\{(5, 4)\}$. The graph of the equations of the system consists of two lines that intersect at the point $(5, 4)$. Such a system is called an **independent** and **consistent** system. (See Figure 8-3.)

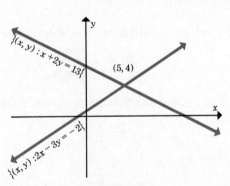

Figure 8-3

Two commonly used algebraic methods of finding the solution set of a linear system are (1) the method of substitution and (2) the method of addition. We will discuss first the method of substitution.

Any solution, (x_0, y_0), of the linear equation $y = m_1 x + b_1$ must have the form $(x, m_1 x + b_1)$. Hence, if (x_0, y_0) is a solution of the system

$$(3) \qquad \begin{aligned} y &= m_1 x + b_1 \\ y &= m_2 x + b_2, \end{aligned}$$

then (x_0, y_0) is also a solution of the system

$$(4) \qquad \begin{aligned} y &= m_1 x + b_1 \\ m_1 x + b_1 &= m_2 x + b_2. \end{aligned}$$

Conversely, if (x_0, y_0) is a solution of (4), then (x_0, y_0) is a solution of (3). Hence (3) and (4) are equivalent systems.

Also, if (x_0, y_0) is a solution of (4) and $m_1 \neq m_2$, then (x_0, y_0) is a solution of

$$(5) \qquad \begin{aligned} y &= m_1 x + b \\ x &= \frac{b_2 - b_1}{m_1 - m_2}. \end{aligned}$$

Conversely, if (x_0, y_0) is a solution of (5), then (x_0, y_0) is a solution of (4). Hence (3), (4), and (5) are equivalent systems if $m_1 \neq m_2$. Since (5) is of the form

$$y = mx + b$$
$$x = a \quad \text{for } a \in R,$$

we can readily find the solution set $\{(a, ma + b)\}$. What is the situation when $m_1 = m_2$?

Example 4   Find by substitution the solution set of the system

$$2x - 3y + 2 = 0$$
$$x + 2y - 13 = 0.$$

Solution: The given system is equivalent to the system

(6)
$$y = \frac{2x + 2}{3}$$

$$y = \frac{13 - x}{2},$$

and (6), by substitution, is equivalent to

(7)
$$y = \frac{2x + 2}{3}$$

$$\frac{2x + 2}{3} = \frac{13 - x}{2},$$

However, (7) is equivalent to

(8)
$$y = \frac{2x + 2}{3}$$

$$x = 5$$

since the solution set of $\dfrac{2x + 2}{3} = \dfrac{13 - x}{2}$ is $\{5\}$. If we replace x by 5 in the equation $y = \dfrac{2x + 2}{3}$, we have $y = 4$. The solution set of (8), therefore, is $\{(5, 4)\}$ and hence the solution set of the given system is $\{(5, 4)\}$.

Example 5   By substitution, solve the system of equations

$$5r - 4s = 22$$
$$r - 3s = 11.$$

Solution: The given system is equivalent to the system

(9)
$$5r - 4s = 22$$
$$r = 11 + 3s$$

and, by substitution, we have the system

(10)
$$5(11 + 3s) - 4s = 22$$
$$r = 11 + 3s.$$

Simplifying (10), we have an equivalent system

$$s = -3$$
$$r = 11 + 3s.$$

Replacing s by -3 in the equation $r = 11 + 3s$, we find $r = 2$. Hence we say that the solution set is $\{(2, -3)\}$ or that the solution is $r = 2$ and $s = -3$.

_____

Now let us consider the second method, that of addition or, more descriptively, *elimination by addition or subtraction.* Clearly, any solution (x_0, y_0) of the system

(11)
$$a_1x + b_1y + c_1 = 0$$
$$a_2x + b_2y + c_2 = 0$$

is also a solution of the system

(12)
$$a_1x + b_1y + c_1 = 0$$
$$(a_2x + b_2y + c_2) + (a_1x + b_1y + c_1) = 0$$

since the second equation of (12) reduces to $0 + 0 = 0$ when x is replaced by x_0 and y is replaced by y_0. Also any solution (x', y') of (12) is a solution of (11) since it follows from (12) that

$$a_2x' + b_2y' + c_2 = 0 \quad \text{if} \quad a_1x' + b_1y' + c_1 = 0.$$

Hence (11) and (12) are equivalent systems. The same kind of argument shows that

$$a_1x + b_1y + c_1 = 0$$
$$a_2x + b_2y + c_2 = 0$$

and

$$a_1x + b_1y + c_1 = 0$$
$$r(a_1x + b_1y + c_1) + s(a_2x + b_2y + c_2) = 0$$

are equivalent systems for all real numbers r and $s \neq 0$.

The equation

$$r(a_1x + b_1y + c_1) + s(a_2x + b_2y + c_2) = 0$$

is called a **linear combination** of the equations

$$a_1x + b_1y + c_1 = 0 \quad \text{and} \quad a_2x + b_2y + c_2 = 0.$$

The essential step in the solution of a system of two equations by "addition" is the formation of a linear combination of the given equations that is itself equivalent to an equation in one variable. The procedure is illustrated in Example 6.

Example 6   Find by addition the solution set of the system

(13)
$$2x - 3y + 2 = 0$$
$$x + 2y - 13 = 0.$$

Solution: System (13) is equivalent to the system

(14)
$$4x - 6y + 4 = 0$$
$$3x + 6y - 39 = 0$$

and (14) is equivalent to the system

(15)
$$(4x - 6y + 4) + (3x + 6y - 39) = 0$$
$$3x + 6y - 39 = 0.$$

When we simplify (15), we have

(16)
$$x = 5$$
$$3x + 6y - 39 = 0.$$

From (16), we obtain $x = 5$ and $y = 4$. Hence the solution set of (13) is $\{(5, 4)\}$.

EXERCISES 8.1

◀In Exercises 1–8, determine whether the two systems of equations are equivalent. Justify your answer.

1. $5x + 4y = 3$ $x = 3$
   $x + y = 0$ and $y = -3$

2. $x = 2$ $x + y = 5$
   $y = 3$ and $x = 2$

3. $y - 4 = 0$ $y = 4$
   $x + y = 3$ and $x + y = 3$

4. $2x + y = 10$ and $4x + 2y = 20$
   $x - y = 6$ $x - y = 6$

5. $2x + 3y = 5$ and $2x + 3y - 5 + 2x - y - 1 = 0$
   $2x - y = 1$ $2x - y = 1$

6. $2x + 3y = 5$ and $2x + 3(2x - 1) = 5$
   $y = 2x - 1$ $y = 2x - 1$

7. $2x - y - 4 = 0$ and $x - y = 1$
   $x - y = 1$ $2(2x - y - 4) - 3(x - y - 1) = 0$

8. $x = x_0$ and $a(x - x_0) + b(y - y_0) = 0$
   $y = y_0$ $c(x - x_0) + d(y - y_0) = 0$

◀In Exercises 9–16, identify, without graphing, each system as being consistent or inconsistent and dependent or independent.

9. $3x - 4y = 6$
   $12x + 3 = 16y$

10. $x + y = 0$
    $6x - 8y = 2$

11. $2x + 5y = 1$
    $4x + 10y = 2$

12. $4x - 3y = 5$
    $10x - 4y = 1$

13. $y = 4x - 3$
    $y = 4x - 5$

14. $x = y - 3$
    $x = 2y - 6$

15. $3x + 4y = 8$
    $12x + 16y = 32$

16. $3x + 4y = 8$
    $3x + 8 = 4y$

17-24. In Exercises 17–24, illustrate each answer to Exercises 9–16 by graphing the equations of the system.

◀In Exercises 25–34, find the solution set of the system by the method of substitution.

25. $x + y = 2$
    $x - y = -1$

26. $x - y = 5$
    $2x + y = 4$

27. $4x - y = 5$
    $-3x + 5y = -21$

28. $8x - 6y = 2$
    $12x + 18y = 18$

29. $\dfrac{r}{8} - \dfrac{s}{4} = 0$

    $\dfrac{8r}{3} - \dfrac{4s}{3} = \dfrac{10}{6}$

30. $\dfrac{8p}{2} - \dfrac{2q}{8} = \dfrac{29}{10}$

    $\dfrac{6p}{5} - \dfrac{3q}{2} = \dfrac{3}{10}$

31. $3r + s = 5$
    $2r + 3s = 7$

32. $4r + 4s = 3$
    $3r - 4s = -23$

33. $m + 2n - 1 = \dfrac{m + n}{5}$

$2m - n = \dfrac{5 - 2m}{6}$

34. $3c - d = 6 + c + 2d$

$3(d - 2) = 2(c + 1) - 10$

◀In Exercises 35–42, find the solution set of the system by the method of addition.

35. $x - y = 4$

$x + y = 7$

39. $\dfrac{8p}{3} - \dfrac{6q}{5} = \dfrac{154}{15}$

$\dfrac{6p}{5} - \dfrac{8q}{3} = \dfrac{176}{45}$

36. $2x - y = 4$

$3x + 2y = 13$

40. $\dfrac{8p}{2} - \dfrac{2q}{8} = \dfrac{29}{10}$

$\dfrac{6p}{5} - \dfrac{3q}{2} = \dfrac{3}{10}$

37. $4r + s = 2$

$6r - 3s = -15$

41. $\dfrac{r + s}{4} + \dfrac{r - s}{2} = 1$

$\dfrac{3r - s}{4} + \dfrac{4r + 20}{11} = 3$

38. $4r - 6s = -2$

$8r + 3s = 6$

42. $\dfrac{3r + 1}{5} = \dfrac{3s + 2}{4}$

$\dfrac{2r - 1}{5} + \dfrac{3s - 2}{4} = 2$

43. Graph, on the same coordinate axes, the equations of each of the following equivalent systems.
   (a) $x = 3$
       $y = 2$
   (b) $x - 3y = -3$
       $x + 3y = 9$
   (c) $x = 3$
       $2(x - 3y + 3) + 3(x + 3y - 9) = 0$
   (d) $3(x - 3y + 3) - (x + 3y - 9) = 0$
       $y = 2$

◀In Exercises 44–54, solve the word problems by setting up a system of two equations in two variables.

44. A man can row downstream 6 miles in 1 hour and return in 2 hours. Find his rate in still water.

45. The sum of the degree measures of the acute angles of an obtuse triangle is 75. If the difference of their measures is 19, find the measures of the three angles of the triangle.

46. Find an equation of the line that passes through the point $(5, 4)$ and the intersection of the lines whose equations are $y = -\frac{1}{4}x + \frac{1}{2}$ and $x + \frac{3}{2}y = -\frac{1}{2}$.

47. If a rectangular field is enlarged by making it 10 yd. wider and 20 yd. longer, the area is increased by 1600 sq. yd. If the field is made smaller by decreasing the length by 10 yd. and the width by 10 yd., the area is decreased by 900 sq. yd. Find the original dimensions of the field.

48. Two fishing boats are 30 miles apart. If they are traveling in the same direction, the faster boat overtakes the slower boat in 8 hours. If they sail toward each other, they meet in 3 hours. What is the speed of each fishing boat? (Assume that the speed of each boat is constant.)

49. One alloy contains three times as much copper as silver, another contains five times as much silver as copper. How much of each alloy must be used to make 14 pounds of an alloy in which there is twice as much copper as silver?

50. If Roy were 4 years older, he would be half as old as his father is now. If Roy were 6 years younger, he would be one-third as old as his father will be 6 years from now. What is Roy's present age?

51. The sum of the digits of a two-digit number is 12. If three times the ten's digit is 1 more than four times the unit's digit, what is the number?

52. A man invested $6000, part at 5% per year and the rest at 4% per year. His total income from both investments is $275 yearly. How much is invested at each rate?

53. A collection of 44 coins has a value of $7.40. If the collection consists of only quarters and dimes, how many of each kind of coin are there?

54. A grocer decided to prepare 10-pound bags of potatoes by mixing two grades of potatoes that sell for 8 cents a pound and 11 cents a pound, respectively. If the 10-pound bag is to sell for 89 cents, how much of each grade of potatoes should the grocer put into each bag?

55. **CHALLENGE PROBLEM.** Prove that the systems (1) and (2) given below are equivalent for $r, s \in R$ and $s \neq 0$.

(1)
$$a_1x + b_1y + c_1 = 0$$
$$a_2x + b_2y + c_2 = 0.$$

(2)
$$a_1x + b_1y + c_1 = 0$$
$$r(a_1x + b_1y + c_1) + s(a_2x + b_2y + c_2) = 0$$

8.2 ONE LINEAR AND ONE SECOND-DEGREE EQUATION

The solution set of a system consisting of one linear equation and one second-degree equation is the set of all ordered pairs of numbers that satisfy both the linear equation and the second-degree equation. For example, if we have a system

$$y = 2x + 3$$
$$x^2 + y^2 = 9,$$

then the solution set is

$$\{(x, y) : y = 2x + 3\} \cap \{(x, y) : x^2 + y^2 = 9\}.$$

To find the solution set of a system that consists of a linear equation and a second-degree equation, we can use the substitution method.

Example 1   Find the solution set of the system

$$y = 2x + 3$$
$$x^2 + y^2 = 9.$$

Solution: An equivalent system is

$$y = 2x + 3$$
$$x^2 + (2x + 3)^2 = 9.$$

Simplifying the second equation of this equivalent system, we have

$$x^2 + (4x^2 + 12x + 9) = 9,$$
$$5x^2 + 12x = 0,$$
$$x(5x + 12) = 0.$$

Hence the given system is equivalent to the system

$$y = 2x + 3$$
$$x(5x + 12) = 0.$$

The equation $x(5x + 12) = 0$ has the solution set $\{0, -\frac{12}{5}\}$. If $x = 0$, then

$$y = 2(0) + 3 = 3;$$

if $x = -\frac{12}{5}$, then

$$y = 2\left(-\frac{12}{5}\right) + 3 = -\frac{9}{5}.$$

Hence the solution set of the system is $\{(0, 3), (-\frac{12}{5}, -\frac{9}{5})\}$.

The graph (Figure 8-4) of the equations of the system shows that the line $y = 2x + 3$ intersects the circle $x^2 + y^2 = 9$ in two points.

Example 2   Find the solution set of the system

$$x^2 - y + 3 = 0$$
$$4x - y - 1 = 0.$$

Solution: The given system is equivalent to the system

$$x^2 - (4x - 1) + 3 = 0$$
$$y = 4x - 1.$$

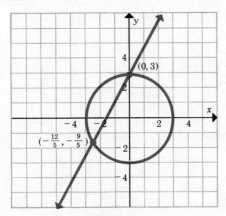

Figure 8-4

Simplifying the first equation of this equivalent system, we have

$$x^2 - 4x + 1 + 3 = 0,$$
$$x^2 - 4x + 4 = 0,$$
$$(x - 2)(x - 2) = 0.$$

Hence the given system is equivalent to

$$y = 4x - 1$$
$$(x - 2)(x - 2) = 0.$$

The equation $(x - 2)(x - 2) = 0$ has the solution set $\{2\}$. If $x = 2$, then $y = 4(2) - 1 = 7$. Hence the solution set of the system is $\{(2, 7)\}$. The graph (Figure 8-5) of the equations of the system shows that

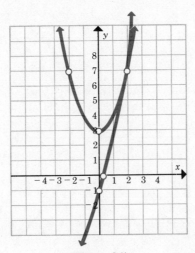

Figure 8-5

the line $y = 4x - 1$ is tangent to the parabola $x^2 - y + 3 = 0$ at $(2, 7)$.

Example 3   Solve the system

$$2x^2 - y^2 = 8$$
$$3x - y = 1.$$

Solution: The given system is equivalent to the system

$$2x^2 - (3x - 1)^2 = 8$$
$$3x - 1 = y.$$

Simplifying the first equation of this equivalent system, we have

$$2x^2 - 9x^2 + 6x - 1 = 8,$$
$$-7x^2 + 6x - 9 = 0,$$
$$7x^2 - 6x + 9 = 0.$$

The equation $7x^2 - 6x + 9 = 0$, as you should check, has no real solutions. Hence the solution set of the system over the field of real numbers is the null set, $\varnothing$.

The graph (Figure 8-6) of the equations of the system shows that the line $3x - y = 1$ and the hyperbola $2x^2 - y^2 = 8$ do not intersect. It is important to note, however, that the system

$$2x^2 - y^2 = 8$$
$$3x - y = 1$$

does have a nonempty solution set over the field of complex numbers.

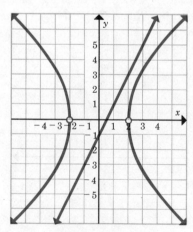

Figure 8-6

If we solve the equation $7x^2 - 6x + 9 = 0$, we find

$$x = \frac{3}{7} + \frac{3\sqrt{6}}{7}i \qquad \text{or} \qquad x = \frac{3}{7} - \frac{3\sqrt{6}}{7}i.$$

By replacing x in the equation $3x - y = 1$ by the two numbers in turn, we find

$$y = \frac{2}{7} + \frac{9\sqrt{6}}{7}i \qquad \text{or} \qquad y = \frac{2}{7} - \frac{9\sqrt{6}}{7}i.$$

A check verifies that the solution set of the system is indeed

$$\left\{ \left(\frac{3}{7} + \frac{3\sqrt{6}}{7}i, \frac{2}{7} + \frac{9\sqrt{6}}{7}i \right), \left(\frac{3}{7} - \frac{3\sqrt{6}}{7}i, \frac{2}{7} - \frac{9\sqrt{6}}{7}i \right) \right\}.$$

Example 4   Solve the system

$$x^2 - y^2 = 0$$
$$x - y = 0.$$

Solution: If we proceed as before, we have

$$x^2 - x^2 = 0$$
$$x - y = 0.$$

Since $x^2 - x^2 = 0$ for every real number x, the solution set of the system consists of all ordered pairs (a, a) where $a \in R$. Since

$$x^2 - y^2 = (x + y)(x - y),$$

the graph of the equation $x^2 - y^2 = 0$ consists of the two intersecting lines, $x - y = 0$ and $x + y = 0$. Since one of the two intersecting lines is coincident with the graph of the second equation in the system, the graph (Figure 8-7) of the equations of the system consists of two intersecting lines.

___

The preceding examples exhibit four different kinds of solution sets for a system consisting of a linear equation and a second-degree equation. Moreover, these are the only kinds of nonempty solution sets for such a system. That is, the solution set over the complex numbers of a system consisting of a linear equation and a second-degree equation may be

(1) a set of consisting of one ordered pair;
(2) a set containing two ordered pairs; or
(3) an infinite set of ordered pairs.

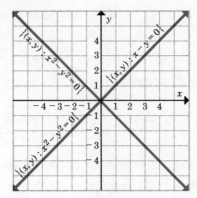

Figure 8-7

Usually we are concerned only with the solution set over the real numbers. This is particularly true if we are studying the graph of the equations in the Cartesian plane. Over the field of real numbers the solution set of a system consisting of a linear equation and a second-degree equation may be

(1) the empty set;
(2) a set consisting of one ordered pair;
(3) a set containing two ordered pairs; or
(4) an infinite set of ordered pairs.

The fourth case can occur only when the conic defined by the second-degree equation is degenerate. (See Section 7.8.)

EXERCISES 8.2

◀ In Exercises 1–18, find the solution set over the field of real numbers of the given system.

1. $x^2 + y^2 = 50$
   $x - y = 0$

2. $x^2 - 4x + 3 = 0$
   $x + 1 = y$

3. $x^2 - y^2 = 9$
   $x + y = 0$

4. $xy = 6$
   $2x - y = 1$

5. $3x^2 - y^2 = 3$
   $2y - x = 8$

6. $xy = 4$
   $x + y = 4$

7. $y = x^2$
   $2x + y = 3$

8. $x^2 = y$
   $2x + y + 1 = 0$

9. $x + y = 11$
   $x^2 + y^2 = 65$

10. $x^2 + y^2 = 25$
    $3x - 4y = 25$

11. $x^2 = y + 4x$
$\quad 2x - y = 5$

12. $x^2 + y = 3$
$\quad 5x + y = 7$

13. $y = x^2 - 4x + 5$
$\quad 2x = y + 4$

14. $25x^2 - 4y^2 = 100$
$\quad x + y = 7$

15. $3x^2 + 2y^2 = 11$
$\quad x - y = 1$

16. $x^2 + y^2 = 50$
$\quad 2x + y = 50$

17. $4x^2 - 3y^2 = 24$
$\quad 3x - 2y = 5$

18. $y = 2x^2 + 5x - 10$
$\quad x - 2y = 7$

19–36. In Exercises 19–36, illustrate your answers to Exercises 1–18 by graphing the equations of the given system.

◀In Exercises 37–41, find the solution set of the given system over the field of real numbers. Sketch the graphs of the equations of the system.

37. $x^2 - 2xy + y^2 = 0$
$\quad x - y = 0$

38. $2x^2 - xy - y^2 = 0$
$\quad 2x + y = 0$

39. $x^2 - y^2 = 0$
$\quad x + y = 0$

40. $x^2 - y^2 + 2x - 4 - y = 3$
$\quad x + y = -3$

41. $x^2 = 6y^2 - xy$
$\quad x + y = -3$

◀In Exercises 42–46, find the solution set of the given system over the field of complex numbers.

42. $2x^2 - y^2 = 8$
$\quad 3x - y = 1$

43. $x^2 = y$
$\quad 2x + y + 2 = 0$

44. $x^2 - y^2 = 75$
$\quad 2x = y$

45. $x^2 + y^2 = 9$
$\quad x + y = 5$

46. $2x + 3y = 2$
$\quad 2x^2 + xy - 2y^2 = 8$

47. Show that one ordered pair in the solution set of the system

$$2x^2 - y^2 = 8$$
$$3x - y = 1$$

is $\left(\dfrac{3}{7} + \dfrac{3\sqrt{6}}{7}i, \ \dfrac{2}{7} + \dfrac{9\sqrt{6}}{7}i \right)$.

◀In Exercises 48–56, solve the problem by first setting up a system of two equations where one equation is linear and the other is of the second degree.

48. Find two numbers whose sum is 10 and the sum of whose squares is 52.

49. The sum of the lengths of the two legs of a right triangle is 68 in. The area of the triangle is 480 sq. in. Find the perimeter of the triangle.

50. The ten's digit of a two-digit number exceeds five times the unit's digit by 1. If the square of the unit's digit is subtracted from the square of the ten's digit, the difference is 35. Find the number.

51. The sum of the perimeters of two squares is 72 in. The sum of the areas of the squares is 234 sq. in. Find the dimensions of each square.

52. The diagonal of a rectangle is two units longer than its base. The height of the rectangle is two units more than one-third the base. What are the dimensions of the rectangle?

53. A school and its parking lot occupy two adjoining square plots of land. The two squares have a total area of 17,725 sq. ft. The length of fencing needed to enclose both is 555 ft. Find the size of the parking lot.

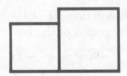

54. If the digits of a two-digit number are interchanged, the number is increased by 36. The product of the digits is 21. Find the number.

55. A rectangular plot of land is enclosed by 200 ft. of fencing. If the plot has an area of 1600 sq. ft., find the dimensions of the plot.

56. The difference between two numbers is $3\sqrt{3}$. The sum of their squares is 87. Find the numbers.

57. CHALLENGE PROBLEM. Find the value(s) of k for which the graph of $y = x + k$ is tangent to the graph of the equation $x^2 + y^2 = 9$.

58. CHALLENGE PROBLEM. A line passing through the point $(0, -5)$ is tangent to the graph of the equation $x^2 = y + 3$. Write an equation of the line.

59. CHALLENGE PROBLEM. Find the solution set of the system

$$\sqrt{x + y} + \sqrt{x - y} = 4$$
$$x + 2y = 13.$$

60. CHALLENGE PROBLEM. Prove that there are four possibilities for the solution set of the system

$$Ax^2 + Bxy + Cy^2 + Dx + Ey + F = 0$$
$$Lx + My + N = 0$$

where not all of A, B, C, D, or E are zero and where M is not zero.

8.3 TWO SECOND-DEGREE EQUATIONS

A system consisting of two second-degree equations is not always solvable by the methods that we have just studied. By using the methods of substitution or linear combinations, we can solve many systems; yet there are other systems that are impossible to solve without general methods for solving cubic (third degree) or quartic (fourth degree) equations in one variable.

The following examples illustrate some of the types of systems for which solution sets can be found by using the methods of this chapter.

Example 1   Find the solution set of the system

$$x^2 + y^2 = 25$$
$$2x^2 - y^2 = 2.$$

Solution: The given system is equivalent to the system

(1)
$$x^2 + (2x^2 - 2) = 25$$
$$2x^2 - 2 = y^2.$$

Simplifying the first equation of (1), we have

$$3x^2 = 27,$$
$$x^2 = 9.$$

Hence the system (1) is equivalent to the system

$$x^2 = 9$$
$$2x^2 - 2 = y^2.$$

The equation $x^2 = 9$ has the solution set $\{3, -3\}$. Replacing x by either 3 or -3 in $2x^2 - 2 = y^2$, we obtain $y = 4$ or $y = -4$. The solution set of the given system therefore is

$$\{(3, 4), (3, -4), (-3, 4), (-3, -4)\}.$$

Figure 8-8 shows the graph of the equations of the system.

Example 2   Solve the system

$$xy = 3$$
$$x^2 + y^2 = 10.$$

Solution: The given system is equivalent to the system

$$\left. \begin{array}{c} y = \dfrac{3}{x} \\[2mm] x^2 + \left(\dfrac{3}{x}\right)^2 = 10 \end{array} \right\} \quad \text{if } x \neq 0.$$

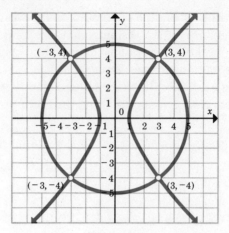

Figure 8-8

We have

$$x^2 + \left(\frac{3}{x}\right)^2 = 10,$$

$$x^4 + 9 = 10x^2,$$

$$x^4 - 10x^2 + 9 = 0.$$

The polynomial $x^4 - 10x^2 + 9$ is of the form $u^2 - 10u + 9$, where $u = x^2$. Since

$$u^2 - 10u + 9 = (u - 9)(u - 1),$$

the equation

$$(x^2 - 9)(x^2 - 1) = 0$$

is equivalent to the equation $x^4 - 10x^2 + 9 = 0$. The given system is therefore equivalent to

$$\left.\begin{array}{r} y = \dfrac{3}{x} \\[2mm] (x^2 - 9)(x^2 - 1) = 0 \end{array}\right\} \quad \text{if } x \neq 0.$$

If any ordered pair (x, y) is to satisfy both equations of the system, we see, from the second equation, that the possible choices for x are $x = 3$, $x = -3$, $x = 1$, and $x = -1$. Replacing x by 3 in the first equation, we have $y = 1$. In a similar manner, we find the four members of the solution set. The solution set is

$$\{(1, 3), (-1, -3), (3, 1), (-3, -1)\}.$$

Figure 8-9 shows the graph of the equations of the system.

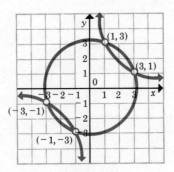

Figure 8-9

Example 3   Solve the system

$$x^2 - y + 4 = 0$$
$$16x^2 + 25y^2 = 400.$$

Solution: The given system is equivalent to

$$x^2 = y - 4$$
$$16(y - 4) + 25y^2 = 400.$$

We have

$$16(y - 4) + 25y^2 = 400,$$
$$16y - 64 + 25y^2 = 400,$$
$$25y^2 + 16y - 464 = 0,$$
$$(25y + 116)(y - 4) = 0.$$

If an ordered pair (x, y) is a solution of the system, we see that $y = 4$ or $y = -\frac{116}{25}$. Replacing y by 4 in the equation $x^2 = y - 4$, we have $x = 0$. Replacing y by $-\frac{116}{25}$ in the same equation, we have $x^2 = -\frac{216}{25}$. Hence there is no real solution corresponding to $y = -\frac{116}{25}$. Over the set of real numbers, the solution set is $\{(0, 4)\}$. Figure 8-10 shows the graph of the equations of the system.

Over the field of complex numbers, the solution set of the system

$$x^2 - y + 4 = 0$$
$$16x^2 + 25y^2 = 400$$

is $\{(0, 4), (\frac{6}{5}\sqrt{6}i, -\frac{116}{25}), (-\frac{6}{5}\sqrt{6}i, -\frac{116}{25})\}$ since

$$x = \frac{6}{5}\sqrt{6}i \qquad \text{or} \qquad x = -\frac{6}{5}\sqrt{6}i$$

if $x^2 = -\frac{216}{25}$. A check verifies that the three pairs of numbers of this set are indeed solutions of the system.

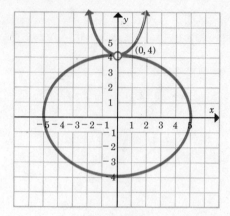

Figure 8-10

Certain systems require the use of both addition and substitution to find the solution set.

Example 4   Find the solution set of the system

$$x^2 + y^2 = 5$$
$$x^2 - xy + y^2 = 7.$$

Solution: First, by addition, we have $(x^2 - xy + y^2) + (-x^2 - y^2) = 7 + (-5)$ and hence obtain the equivalent system

$$x^2 + y^2 = 5$$
$$-xy = 2.$$

Next, by substitution, we form the equivalent system (for $x \neq 0$),

$$x^2 + \left(\frac{2}{-x}\right)^2 = 5$$
$$-xy = 2.$$

We have

$$x^2 + \left(\frac{2}{-x}\right)^2 = 5,$$

$$x^4 + 4 = 5x^2,$$

$$x^4 - 5x^2 + 4 = 0,$$

$$(x^2 - 4)(x^2 - 1) = 0.$$

Hence if any ordered pair is to satisfy both equations of the system, the possible choices for x are $x = 2$, $x = -2$, $x = 1$, and $x = -1$. (And we note that none of these values is 0.) Replacing x in the equation $-xy = 2$ by 2, we get $y = -1$; replacing x by -2, we get $y = 1$; replacing x by 1, we get $y = -2$; and replacing x by -1, we get $y = 2$.

A check shows that the solution set of the system is indeed $\{(2, -1), (-2, 1), (1, -2), \text{ and } (-1, 2)\}$. Figure 8-11 shows the graph of the equations of the given system. (The graph of $x^2 - xy + y^2 = 1$ is an ellipse.)

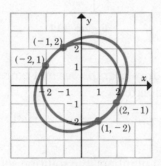

Figure 8-11

Approximate values for the ordered pairs of real numbers in the solution set of a system are often obtained by graphing when the exact solutions for the system cannot be easily obtained algebraically.

Example 5 By graphing, find the solution set of the system

$$x^2 - 4y^2 + 8y - 8 = 0$$
$$x^2 + 9y^2 - 4x - 32 = 0.$$

Solution: The equation $x^2 - 4y^2 + 8y - 8 = 0$ is equivalent to

$$\frac{x^2}{4} - \frac{(y-1)^2}{1} = 1$$

as you should check by going through the process of completing the square (see Section 7.6); the equation $x^2 + 9y^2 - 4x - 32 = 0$, as you should also check, is equivalent to

$$\frac{(x-2)^2}{36} + \frac{y^2}{4} = 1.$$

The solution set will be the coordinates of the points where the hyperbola $\dfrac{x^2}{4} - \dfrac{(y-1)^2}{1} = 1$ and the ellipse $\dfrac{(x-2)^2}{36} + \dfrac{y^2}{4} = 1$ in-

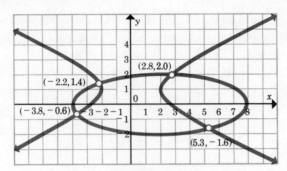

Figure 8-12

tersect. A carefully drawn graph as in Figure 8-12 shows that the solution set is approximately

$$\{(2.8,\ 2.0),\ (-2.2,\ 1.4),\ (-3.8,\ -0.6),\ (5.3,\ -1.6)\}.$$

Over the set of real numbers, the solution set of a system consisting of two second-degree equations may

(1) be the empty set;
(2) have one member;
(3) have two members;
(4) have three members;
(5) have four members; or
(6) in the case of certain degenerate conics be an infinite set.

You have seen an example of case 2 and several examples of case 5. Figure 8-13a should suggest a system of two second-degree equations whose solution set is the empty set; Figure 8-13b a system of two second-degree equations whose solution set has two members; and

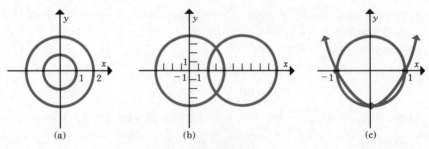

(a)   (b)   (c)

Figure 8-13

Figure 8-13c a system of two second-degree equations whose solution set has three members.

As a simple example of case 6 consider the system

$$x^2 - y^2 = 0$$
$$x^2 + xy - 2y^2 = 0.$$

Since

$$x^2 - y^2 = (x - y)(x + y)$$

and

$$x^2 + xy - 2y^2 = (x - y)(x + 2y) = 0,$$

the graph of the system consists of the lines $x - y = 0$, $x + y = 0$, and $x + 2y = 0$, and any numbers x and y such that $x = y$ satisfy both of the equations $x^2 - y^2 = 0$ and $x^2 + xy - 2y^2 = 0$. The solution set of the system is the infinite set $\{(x, y) : x - y = 0\}$.

EXERCISES 8.3

◀ In Exercises 1–10, find the solution set of each system over the real numbers.

1. $x^2 - y^2 = 4$
   $x^2 + y^2 = 16$

2. $x^2 - y^2 = 12$
   $x^2 - 7y = 2$

3. $x^2 - y^2 = 16$
   $xy = 15$

4. $x^2 + y^2 = 9$
   $2x + 1 = y^2$

5. $4x^2 + 9y^2 = 36$
   $x = y^2$

6. $y = 4x^2$
   $xy = 2$

7. $3x^2 - y^2 + 22 = 0$
   $x^2 + 2y^2 = 107$

8. $x^2 + y^2 = 16$
   $4x^2 - y^2 = 4$

9. $3x^2 - y^2 = 7$
   $2x^2 + 3y^2 = 23$

10. $2x^2 + 3y^2 = 5$
    $y^2 = 3 - 2x$

11–20. In Exercises 11–20, graph the equations of each system in Exercises 1–10.

◀ In Exercises 21–30, find the solution set of each system over the complex numbers.

21. $xy = 6$
    $xy = x^2 + 2$

22. $2x^2 + xy = -10$
    $xy = 2$

23. $2x^2 + 3xy - y^2 = 4$
    $2x^2 + 3xy = 8$

24. $2xy - y^2 = -24$
    $2x^2 + xy = -2$

25. $x^2 - 3xy = -4$
    $x^2 + 9y^2 = 20$

26. $x^2 + 3xy = -5$
    $y^2 - xy = 6$

27. $x^2 + 4y^2 = 20$
    $x^2 - 5xy + 4y^2 = 0$

28. $x^2 + 6xy = 28$
    $8y^2 + xy = 4$

29. $x^2 + y^2 = 13$
    $x^2 - xy + y^2 = 19$

30. $x^2 + xy + y^2 = 4$
    $x^2 + 2y^2 = 12$

31. Show that $(\frac{6}{5}\sqrt{6i}, -\frac{116}{25})$ is a solution of the system

$$x^2 - y + 4 = 0$$
$$16x^2 + 25y^2 = 400.$$

◀In Exercises 32–33, find, by graphing, approximate solutions for the system

32. $4x^2 - 3y^2 = 24$
    $x^2 + y^2 - 4x - 2y = 31.$

33. $x^2 - 2y - 2x + 6 = 0$
    $9x^2 + 25y^2 - 36x - 100y = 89.$

34. Write the system of equations suggested by Figure 8–13a.

35. Write the system of equations suggested by Figure 8–13b.

36. Given that the two curves in Figure 8–13c are a circle and a parabola, write the system of equations suggested by the figure.

37. Graph the system

$$x^2 - y^2 = 0$$
$$x^2 + xy - 2y^2 = 0.$$

◀In Exercises 38–44, solve the problem by first setting up a system of two second-degree equations.

38. The product of two numbers is 96 and the sum of the squares is 292. Find the numbers.

39. The product of the digits of a two-digit number is 24. The sum of the squares of the digits is 73. Find two such numbers.

40. Find two positive numbers, the sum of whose squares exceeds two times their product by 9 and the difference of whose squares exceeds one-half their product by 9.

41. The sum of the squares of the numerator and denominator of a fraction is 65. The sum of the fraction and its reciprocal is $\dfrac{65}{28}$. Find the fraction. (Two solutions)

42. An open-top box is made from a rectangular piece of metal by cutting 2-inch squares from each corner and then folding up the edges. Find the

dimensions of the rectangular piece of metal if the area of the piece of metal is 120 sq. in. and the volume of the box is 96 cu. in.

43. Jane bought a certain number of records for \$16. Two weeks later she found that the price per record had gone up 20 cents and, as a result, bought 4 less for the same money. How many records did she buy the first time?

44. A rectangular-shaped swimming pool with an area of 1200 sq. ft. has a 3-ft. wide tiled area around it. Find the dimensions of the pool if the tiled area is 456 sq. ft.

45. **CHALLENGE PROBLEM.** Show that if (x_0, y_0) is a solution of the system

$$a_1x^2 + b_1x + c_1 = 0$$
$$a_2x^2 + b_2x + c_2 = 0,$$

then (x_0, y_0) is also a solution of the system

$$a_1x^2 + b_1x + c_1 = 0$$
$$r(a_1x^2 + b_1x + c_1) + s(a_2x^2 + b_2x + c_2) = 0$$

for any real numbers r and s such that $s \neq 0$.

8.4 SYSTEMS OF INEQUALITIES

In Section 6.6, we considered solutions of linear inequalities. Here we shall consider solutions of systems of linear inequalities and also solutions of systems consisting of one linear inequality and one quadratic inequality and systems consisting of two quadratic inequalities.

The solution set of a system of two inequalities in two variables is the set of all ordered pairs that are members of the solution sets of both the component inequalities. For example, the solution set of the system

$$2x + 3y < 4$$
$$x - y > 3$$

is the set

$$\{(x, y) : 2x + 3y < 4\} \cap \{(x, y) : x - y > 3\}.$$

One of the most informative means of describing a solution set of a system of two inequalities is by a graph.

Example 1 Find the solution set of the system

$$2x + 3y < 4$$
$$x - y > 3.$$

Solution: Each of the lines corresponding to the equations $2x + 3y = 4$ and $x - y = 3$ is drawn in Figure 8-14. (Dashed lines are used since the two sets

$$\{(x, y) : 2x + 3y = 4\} \qquad \text{and} \qquad \{(x, y) : x - y = 3\}$$

are not part of the solution set of the given system.) In the figure, the region containing all the points whose coordinates satisfy the inequality $2x + 3y < 4$ is marked with vertical rays; the region containing all the points whose coordinates satisfy the inequality $x - y > 3$ is marked with horizontal rays. The solution set of the system is the set of all ordered pairs that are coordinates of the points in the area which is crosshatched. Thus the solution set of the system is indicated by the crosshatched area.

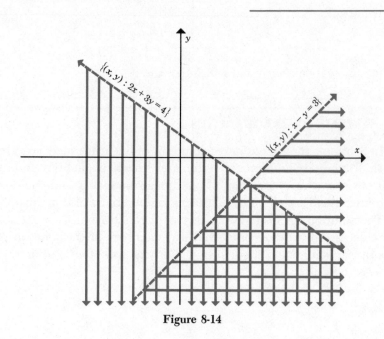

Figure 8-14

A system consisting of one linear inequality and one quadratic inequality is solved similarly.

Example 2   Solve, by graphing, the system

$$x^2 + y^2 \le 5$$
$$4x - 5y > -4.$$

Solution: We draw the circle defined by the equation $x^2 + y^2 = 5$, making a "solid" curve since points on the circle are part of the graph of the inequality $x^2 + y^2 \leq 5$. Since $0 < 5$, $(0, 0)$ is a member of the solution set of the inequality $x^2 + y^2 < 5$. We mark the area within the circle with horizontal lines as shown in Figure 8-15. The line defined by the equation $4x - 5y = -4$ is drawn as a dashed line (Why?) and the region below the line is marked with vertical rays (Why?). The solution set of the system is indicated by the cross-hatching.

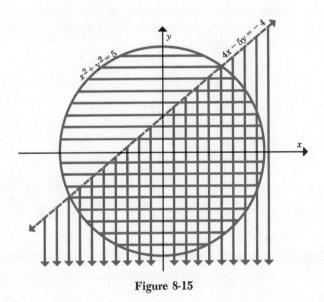

Figure 8-15

Solution sets of systems of two second-degree inequalities are graphed in a similar way.

Example 3 By graphing find the solution set of the system

$$x^2 + y^2 \leq 64$$
$$4x^2 - y^2 > 16.$$

Solution: The circle $x^2 + y^2 = 64$ and the hyperbola $4x^2 - y^2 = 16$ are graphed, the hyperbola being drawn as a dashed curve since it is not part of the graph of the system. We note that the point $(0, 0)$ is part of the graph of the inequality $x^2 + y^2 \leq 64$ and not part

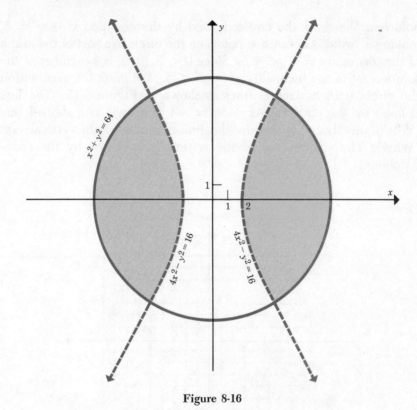

Figure 8-16

of the graph of the inequality $4x^2 - y^2 > 16$. Accordingly, we know that the graph of the given system is the shaded area in Figure 8-16.

A system may consist of more than two inequalities.

Example 4 Find, by graphing, the solution set of the system

$$y < 2x^2$$
$$4x^2 + 9y^2 \leq 36$$
$$y \geq 0.$$

Solution: We graph the parabola $y = 2x^2$, the ellipse $4x^2 + 9y^2 = 36$, and the line $y = 0$. The graph of the solution set is the region below the parabola, inside or on the ellipse, and above or on the x-axis as indicated in Figure 8-17.

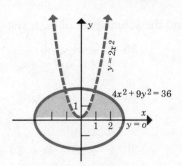

Figure 8-17

EXERCISES 8.4

◄In Exercises 1–20, find the solution set of each system by graphing.

1. $x + y \leq 4$
   $x - y > 4$

2. $3x - y \leq 6$
   $x + 2y > 3$

3. $x \geq 2$
   $y < 3$

4. $x^2 < 9$
   $y > 1$

5. $y^2 \leq 9$
   $x < 0$

6. $x^2 < 4$
   $y^2 \leq 4$

7. $x + y > 2$
   $x^2 + y^2 \leq 25$

8. $4x^2 + 9y^2 \geq 36$
   $y < 2$

9. $x^2 + y^2 < 16$
   $x + 2y > 2$

10. $4x^2 - 9y^2 \leq 36$
    $x - 3y \geq 1$

11. $x^2 > 9$
    $9 < y^2$

12. $x^2 + y^2 < 4$
    $x^2 - y^2 \geq 4$

13. $x^2 + y > 4$
    $x^2 + 4y^2 < 16$

14. $x^2 + 16y^2 \geq 16$
    $x^2 + 4y^2 \leq 16$

15. $y^2 - x^2 \geq 8$
    $y^2 < 9x$

16. $x^2 + y^2 \geq 13$
    $xy < 6$

17. $3x^2 + 2y^2 \leq 44$
    $x^2 \leq y$

18. $y^2 < 4x$
    $2x + y < 4$

19. $4x^2 + 3y^2 \geq 192$
    $3y^2 - x^2 < 12$

20. $y = -x^2 + 8x$
    $y = x^2 - 12x + 32$

◄In Exercises 21–24, indicate the solution set of each system by graphing.

21. $x^2 + y^2 < 16$
    $x + 2y - 2 \leq 0$

22. $y < x^2 + 2x - 3$
    $2x - y - 1 \leq 0$

23. $y < x^2 - 2x + 3$
    $2x - y - 1 > 0$

24. $x^2 + y^2 \leq 25$
    $x^2 + y^2 \geq 25$

◀ In Exercises 25–30, find the solution set of each system by graphing.

25. $x < 2$
$y > -3$
$y - x \leq 0$

26. $y^2 \leq 4x$
$x^2 \geq 4y$
$x - y \leq 0$

27. $x^2 + y^2 \geq 9$
$9x^2 + 16y^2 \leq 144$
$x \geq 0$

28. $x^2 + y^2 < 36$
$16x^2 + 4y^2 \leq 64$
$|x| < 2$

29. $4x^2 + 9y^2 \geq 36$
$2x^2 > y$
$y > 0$

30. $x^2 + y^2 - 4x - 2y + 1 \leq 0$
$y \leq x^2 - 4x + 4$
$y \leq 2$

8.5 SYSTEMS OF LINEAR EQUATIONS

Mathematicians who apply mathematics in engineering and similar technologies frequently encounter an array of linear equations such as

$$2x + 3y + 4z + 3t = 5$$
$$x + 2y - z + t = 1$$
$$-x - y + 2z - t = 5.$$

The array consists of 3 linear equations in 4 variables. It is not unusual to have many such equations in many variables, say, for example, 25 equations in 25 variables.

To solve a system of 25 equations in 25 variables by our familiar methods using pencil and paper is a formidable task. Happily a procedure has been developed that is readily adapted to electronic computors.

The procedure, in a very systematic way, converts a set of equations such as

$$2x - 2y + 2z = 4$$
$$x - 2y - 2z = -1$$
$$2x + y + 3z = 1$$

to a set of equivalent equations of the form

$$x + 0 + 0 = 1$$
$$0 + y + 0 = 2$$
$$0 + 0 + z = -1$$

which exhibits the solution set $\{1, 2, -1\}$ explicitly. Although the procedure may seem rather mechanical, which indeed it is, it has great power since it can be easily applied to large systems.

Let us start by illustrating the procedure for solving the foregoing system of 3 linear equations:

$$
\begin{align}
2x - 2y + 2z &= -4 \tag{1} \\
x - 2y - 2z &= -1 \tag{2} \\
2x + y + 3z &= 1. \tag{3}
\end{align}
$$

(I)

We proceed as follows: (a) multiply Equation (1) by $\frac{1}{2}$ to obtain Equation (1a) of the system II given below; multiply Equation (1) by $-\frac{1}{2}$ and add this new equation to Equation (2) to obtain Equation (2a) in II; multiply Equation (1) by -1 and add to Equation (3) to obtain Equation (3a) in II. This gives the following system:

$$
\begin{align}
x - y + z &= -2 \tag{1a} \\
0 - y - 3z &= 1 \tag{2a} \\
0 + 3y + z &= 5. \tag{3a}
\end{align}
$$

(II)

(Recall that adding the equation "$C = D$" to the equation "$A = B$" means to form the equation "$A + C = B + D$".)

The second step is similar to the first. Retain Equation (1a) as Equation (1b) in the system III; multiply Equation (2a) by -1 to obtain Equation (2b) in III; multiply Equation (2a) by 3 and add the new equation to Equation (3a) to obtain Equation (3b) in III. We have

$$
\begin{align}
x - y + z &= -2 \tag{1b} \\
0 + y + 3z &= -1 \tag{2b} \\
0 + 0 - 8z &= 8. \tag{3b}
\end{align}
$$

(III)

In the third step we multiply Equation (3b) by $-\frac{1}{8}$ to obtain Equation (3c) in the system IV; multiply Equation (3b) by $\frac{3}{8}$ and add to Equation (2b) to obtain Equation (2c) in IV; multiply Equation (3b) by $\frac{1}{8}$ and add to Equation (1b) to obtain Equation (1c) in IV. We obtain

$$
\begin{align}
x - y + 0 &= -1 \tag{1c} \\
0 + y + 0 &= 2 \tag{2c} \\
0 + 0 + z &= -1. \tag{3c}
\end{align}
$$

(IV)

The last step is brief. Retain Equations (3c) and (2c) and add Equation (2c) to Equation (1c). We get

$$
\begin{align}
x + 0 + 0 &= 1 \\
0 + y + 0 &= 2 \\
0 + 0 + z &= -1
\end{align}
$$

(V)

or, more familiarly,

$$x = 1$$
$$y = 2$$
$$z = -1.$$

In each step, the only two operations used have been

(1) multiplying each member of an equation by a nonzero number; and

(2) replacing an equation by another equation obtained by adding to it a multiple of one of the other equations. Since the properties of equality for real numbers assure us that for all real numbers A, B, C, and D,

> $A = B$ and $C = D$ implies $A + C = B + D$ and, conversely, $A + C = B + D$ and $C = D$ implies $A = B$

and also that, for all real numbers A, B, and C,

> $A = B$ implies $A \cdot C = B \cdot C$ and $A \cdot C = B \cdot C$ implies $A = B$ if $C \neq 0$,

it follows that operations (1) and (2) above yield a system of equations equivalent to the original system. Thus system V is equivalent to system I.

Example 1   Find the solution set of the system

$$x + y = 1$$
$$2x + 4y = 10.$$

Solution: The operations that are used to obtain the next equivalent system are indicated in parentheses below the system.

(I)
$$\begin{aligned} x + y &= 1 &\qquad (1)\\ 2x + 4y &= 10 &\qquad (2)\\ (E_1, \; -2E_1 + E_2) \end{aligned}$$

(II)
$$\begin{aligned} x + y &= 1 &\qquad (1a)\\ 0 + 2y &= 8 &\qquad (2a) \end{aligned}$$

The notation $(E_1, \; -2E_1 + E_2)$ shows how we proceed from system I to system II. E_1 shows that we retain Equation (1) as Equation (1a).

Next, $-2E_1 + E_2$ shows that we multiply Equation (1) by -2 and add the resulting equation to Equation (2) to obtain Equation (2a).

We repeat system II for convenience.

$$x + y = 1 \qquad \text{(1a)}$$

(II)

$$0 + 2y = 8 \qquad \text{(2a)}$$

$$\left(E_1, \frac{1}{2}E_2\right)$$

(III)

$$x + y = 1 \qquad \text{(1b)}$$

$$0 + y = 4 \qquad \text{(2b)}$$

The notation $(E_1, \frac{1}{2}E_2)$ shows how we go from system II to system III, E_1 indicates that we retain Equation (1a) unchanged as Equation (1b), $\frac{1}{2}E_2$ indicates that we form Equation (2b) by multiplying Equation (2a) by $\frac{1}{2}$.

Now we repeat system III for convenience.

(III)

$$x + y = 1 \qquad \text{(1b)}$$

$$0 + y = 4 \qquad \text{(2b)}$$

$$(E_1 - E_2, E_2)$$

(IV)

$$x + 0 = -3 \qquad \text{(1c)}$$

$$0 + y = 4 \qquad \text{(2c)}$$

The notation $(E_1 - E_2, E_2)$ shows that we form Equation (1c) of system IV by subtracting Equation (2b) from Equation (1b) and that we retain Equation (2b) of system III as Equation (2c) of system IV.

From system IV we observe that the solution set of IV and therefore also of I is $\{(-3, 4)\}$.

A third operation in addition to the two we have been using is sometimes necessary because the leading coefficient of the first equation must not be 0 as we start our procedure. For example, in solving the system

$$y + z = 4$$
$$x - y + z = 1$$
$$x + 2y + z = 3,$$

the first step is to interchange the first two equations. This gives

$$x - y + z = 1$$
$$0 + y + z = 4$$
$$x + 2y + z = 3,$$

and then we proceed as in the first example. (Alternatively, we could interchange the first and third equations.)

Example 2   Find the solution set of the system

$$x + y - z = 2$$
$$2x + y + z = -1$$
$$3x - y - z = 6.$$

Solution: The operations that are used to obtain the next equivalent system are again indicated by the notation in the parentheses below each system. For example, the notation $(E_1, -2E_1 + E_2, -3E_1 + E_3)$ indicates that we obtain II from I by

(1) retaining (1) as (1a).
(2) adding -2 times (1) to (2) to get (2a).
(3) adding -3 times (1) to (3) to get (3a).

(I)
$$\begin{aligned} x + y - z &= 2 & (1)\\ 2x + y + z &= -1 & (2)\\ 3x - y - z &= 6 & (3) \end{aligned}$$
$$\left(E_1, \ -2E_1 + E_2, \ -3E_1 + E_3\right)$$

(II)
$$\begin{aligned} x + y - z &= 2 & (1a)\\ 0 - y + 3z &= -5 & (2a)\\ 0 - 4y + 2z &= 0 & (3a) \end{aligned}$$
$$\left(E_1, \ (-1)E_2, \ -4E_2 + E_3\right)$$

(III)
$$\begin{aligned} x + y - z &= 2 & (1b)\\ 0 + y - 3z &= 5 & (2b)\\ 0 + 0 - 10z &= 20 & (3b) \end{aligned}$$
$$\left(-\frac{1}{10}E_3 + E_1, \ -\frac{3}{10}E_3 + E_2, \ -\frac{1}{10}E_3\right)$$

(IV)
$$\begin{aligned} x + y + 0 &= 0 & (1c)\\ 0 + y + 0 &= -1 & (2c)\\ 0 + 0 + z &= -2 & (3c) \end{aligned}$$
$$\left((-1)E_2 + E_1, \ E_2, \ E_3\right)$$

$$x + 0 + 0 = 1 \tag{1d}$$
(V) $$\qquad 0 + y + 0 = -1 \tag{2d}$$
$$0 + 0 + z = -2 \tag{3d}$$

Hence the solution set is $\{(1, -1, -2)\}$.

_____

The strategy of the procedure we have been using is to produce, in a very systematic way, a system equivalent to the given one for which the solution set is self-evident. To describe this procedure concisely in the case of 3 equations in 3 variables we can look at just the coefficients of the variables formed in the rectangular array

$$
\begin{array}{ccc l}
a & b & c & (ax + by + cz) \\
d & e & f & (dx + ey + fz) \\
g & h & i & (gx + hy + iz).
\end{array}
$$

Our procedure then transforms a to 1 (if a were 0, we would have first interchanged the first equation with the second or third). Then it transforms d and g to 0 so that we have

$$
\begin{array}{ccc}
1 & b' & c' \\
0 & e' & f' \\
0 & h' & i'.
\end{array}
$$

Now it transforms e' to 1 (again, if e' were 0, an interchange of the second equation with, in this case, the third one would be necessary). Then h' is transformed to 0, resulting in

$$
\begin{array}{ccc}
1 & b' & c' \\
0 & 1 & f'' \\
0 & 0 & i''.
\end{array}
$$

Now i'' is transformed to 1 and then f'' and c' to 0, resulting in

$$
\begin{array}{ccc}
1 & b' & 0 \\
0 & 1 & 0 \\
0 & 0 & 1.
\end{array}
$$

Finally, b' is transformed to 0, resulting in

$$
\begin{array}{ccc}
1 & 0 & 0 \\
0 & 1 & 0 \\
0 & 0 & 1.
\end{array}
$$

EXERCISES 8.5A

◀ Solve each system of equations using the procedures just described.

1. $2x + 3y = 13$
$2x - y = 1$

6. $2x - y + 3z = 3$
$x - 3y + z = 2$
$3x + 2y - 2z = 11$

2. $4x - y = 0$
$6x + 5y = 13$

7. $x + 2y + 3z = 3$
$2x - y - 2z = -8$
$3x + 4y + 5z = 1$

3. $x + y + z = 2$
$x - y - z = 0$
$x + y - z = -2$

8. $x + 3y + z = 0$
$x + 4z = -2$
$-6y + z = 1$

4. $x + 2y + 3z = 1$
$2x - y - z = -2$
$3x - 2y + z = 3$

9. $x - y + z = 2$
$x + y + z = 0$
$3x + 2y + 4z = 1$

5. $2x - y + 3z = 14$
$3x - 2y - z = 5$
$4x - 3y - 2z = -1$

10. $x + y - z = 4$
$2x - y + 3z = -3$
$x - y + 2z = -4$

The rectangular array that results from exhibiting the coefficients of the variables may have reminded you of the arrays of Chapter 4 which we called 2 × 2 matrices. In general, as you may know, an array of mn numbers in m rows and n columns is called an **m** × **n** matrix. Thus, for example,

$$\begin{pmatrix} 2 & 3 & -1 & 4 \\ 1 & 0 & 3 & 2 \\ 2 & -1 & 5 & 1 \end{pmatrix}$$

is a 3 × 4 matrix.

If we agree always to write the linear equations of any system with the same order of variables, we can denote any system by a matrix. Each of the systems, for example, that we have considered so far can be expressed in matrix notation. Thus:

$$\begin{matrix} x + y = 1 \\ 2x + 4y = 10 \end{matrix} \quad \longleftrightarrow \quad \begin{pmatrix} 1 & 1 & 1 \\ 2 & 4 & 10 \end{pmatrix},$$

$$\begin{matrix} 2x - 2y + 2z = -4 \\ x - 2y + 2z = -1 \\ 2x - y + 3z = 1 \end{matrix} \quad \longleftrightarrow \quad \begin{pmatrix} 2 & -2 & 2 & -4 \\ 1 & -2 & 2 & -1 \\ 2 & -1 & 3 & 1 \end{pmatrix},$$

$$
\begin{aligned}
y + z &= 4 \\
x - y + z &= 1 \\
x + 2y + z &= 3
\end{aligned}
\qquad \longleftrightarrow \qquad
\begin{pmatrix}
0 & 1 & 1 & 4 \\
1 & -1 & 1 & 1 \\
1 & 2 & 1 & 3
\end{pmatrix},
$$

(Note that if a variable of an equation if missing, it is necessary to use a zero in the matrix notation.)
and

$$
\begin{aligned}
x + y - z &= 2 \\
2x + y + z &= -1 \\
3x - y - z &= 6
\end{aligned}
\qquad \longleftrightarrow \qquad
\begin{pmatrix}
1 & 1 & -1 & 2 \\
2 & 1 & 1 & -1 \\
3 & -1 & -1 & 6
\end{pmatrix}
$$

Example 3 Display the system

$$
\begin{aligned}
x + y &= 3 \\
x - z &= 0 \\
y + z &= -1
\end{aligned}
$$

in matrix form.

Solution:

$$
\begin{pmatrix}
1 & 1 & 0 & 3 \\
1 & 0 & -1 & 0 \\
0 & 1 & 1 & -1
\end{pmatrix}.
$$

The operations we shall use on a matrix will correspond to the operations on a system of equations. Thus the three algebraic operations we have used in the previous examples are paralleled by three matrix *row operations*.

Definition 8.1 The three row operations,

1. Interchange of any two rows,
2. Multiplication of all elements of any row by a non-zero number,
3. Addition of an arbitrary multiple of any row to any other row,

are called **elementary row operations** on a matrix. Two matrices are said to be **row equivalent** if and only if each can be transformed into the other by means of elementary row operations.

Here, of course, the third row operation means that we add corresponding elements of each row. For example, adding twice the first row to the second row in the matrix

$$\begin{pmatrix} 1 & 2 & -3 \\ 1 & 0 & 4 \\ 3 & -1 & 1 \end{pmatrix}$$

means to form

$$\begin{pmatrix} 1 & 2 & -3 \\ 1 + 2(1) & 0 + 2(2) & 4 + 2(-3) \\ 3 & -1 & 1 \end{pmatrix} = \begin{pmatrix} 1 & 2 & -3 \\ 3 & 4 & -2 \\ 3 & -1 & 1 \end{pmatrix}.$$

Using row operations, we can now indicate more quickly and precisely the procedure we have been using to solve systems of linear equations.

Arrows will be used to denote equivalent matrices and the operations that are used to obtain equivalent matrices will be indicated by the notation above the arrows. Thus, for example, $(R_1, \frac{1}{2}R_1 + R_2)$ will mean that we take a matrix of two rows and write a matrix which is row equivalent to it by writing the first row again but replacing the second row by "$\frac{1}{2}$ the first row added to the second row." In the following examples, check each step carefully.

Example 4   Solve the system

$$2x + 3y = -4$$
$$x - 2y = 5$$

using matrix notation.

Solution:

$$\begin{pmatrix} 2 & 3 & -4 \\ 1 & -2 & 5 \end{pmatrix} \xrightarrow{\left(\frac{1}{2}R_1, \ -\frac{1}{2}R_1 + R_2\right)} \begin{pmatrix} 1 & \frac{3}{2} & -2 \\ 0 & -\frac{7}{2} & 7 \end{pmatrix}$$

$$\xrightarrow{\left(\frac{3}{7}R_2 + R_1, \ -\frac{2}{7}R_2\right)} \begin{pmatrix} 1 & 0 & 1 \\ 0 & 1 & -2 \end{pmatrix}.$$

The last matrix represents the system

$$x + 0 = 1$$
$$0 + y = -2.$$

Hence the solution set is $\{(1, -2)\}$.

Example 5 Using matrix notation, solve the system

$$
\begin{aligned}
2x - 2y + 2z &= -4 \\
x - 2y - 2z &= -1 \\
2x + y + 3z &= 1.
\end{aligned}
$$

Solution:

$$
\begin{pmatrix}
2 & -2 & 2 & -4 \\
1 & -2 & -2 & -1 \\
2 & 1 & 3 & 1
\end{pmatrix}
$$

$$
\xrightarrow{\left(\frac{1}{2}R_1,\ -\frac{1}{2}R_1 + R_2,\ (-1)R_1 + R_3\right)}
\begin{pmatrix}
1 & -1 & 1 & -2 \\
0 & -1 & -3 & 1 \\
0 & 3 & 1 & 5
\end{pmatrix}
$$

$$
\xrightarrow{\left(R_1 + (-1)R_2,\ (-1)R_2,\ 3R_2 + R_3\right)}
\begin{pmatrix}
1 & 0 & 4 & -3 \\
0 & 1 & 3 & -1 \\
0 & 0 & -8 & 8
\end{pmatrix}
$$

$$
\xrightarrow{\left(\frac{4}{8}R_3 + R_1,\ \frac{3}{8}R_3 + R_2,\ -\frac{1}{8}R_3\right)}
\begin{pmatrix}
1 & 0 & 0 & 1 \\
0 & 1 & 0 & 2 \\
0 & 0 & 1 & -1
\end{pmatrix}.
$$

The last matrix represents the system

$$
\begin{aligned}
x + 0 + 0 &= 1 \\
0 + y + 0 &= 2 \\
0 + 0 + z &= -1.
\end{aligned}
$$

Hence the solution set is $\{(1, 2, -1)\}$.

A distinct advantage of the matrix process is that it always provides us with the solution set of a system even when the solution set is the null set or consists of more than one element.

Example 6 Find the solution set of the system

$$
\begin{aligned}
x + y + z &= 4 \\
2x - y + z &= 6 \\
4x - 5y + z &= 10.
\end{aligned}
$$

Solution:

$$\begin{pmatrix} 1 & 1 & 1 & 4 \\ 2 & -1 & 1 & 6 \\ 4 & -5 & 1 & 10 \end{pmatrix}$$

$$\underrightarrow{\left(R_1,\ -2R_1 + R_2,\ -4R_1 + R_3\right)} \quad \begin{pmatrix} 1 & 1 & 1 & 4 \\ 0 & -3 & -1 & -2 \\ 0 & -9 & -3 & -6 \end{pmatrix}$$

$$\underrightarrow{\left(R_1 + \tfrac{1}{3}R_2,\ -\tfrac{1}{3}R_2,\ -3R_2 + R_3\right)} \quad \begin{pmatrix} 1 & 0 & \tfrac{2}{3} & \tfrac{10}{3} \\ 0 & 1 & \tfrac{1}{3} & \tfrac{2}{3} \\ 0 & 0 & 0 & 0 \end{pmatrix}.$$

The last matrix represents the system

$$x + \frac{2}{3}z = \frac{10}{3}$$

$$y + \frac{1}{3}z = \frac{2}{3}$$

$$0 = 0.$$

The equation $0 = 0$ places no restrictions on the system and can be disregarded. Hence we have two equations,

$$x = \frac{10}{3} - \frac{2}{3}z \text{ and}$$

$$y = \frac{2}{3} - \frac{1}{3}z.$$

Whatever value is given to z, this value and the corresponding values for x and y determined by these equations satisfy the original system. A few solutions are shown in the following table.

z	x	y
-1	4	1
-4	6	2
11	-4	-3

The solution set is $\left\{\left(\dfrac{10 - 2z}{3}, \dfrac{2 - z}{3}, z\right) : z \in R\right\}$.

Example 7 Find the solution set of the system

$$x + 2y - 2z = 5$$
$$3x - y - z = -2$$
$$2x - 3y + z = 1.$$

Solution:

$$\begin{pmatrix} 1 & 2 & -2 & 5 \\ 3 & -1 & -1 & -2 \\ 2 & -3 & 1 & 1 \end{pmatrix}$$

$$\underrightarrow{\left(R_1, -3R_1 + R_2, -2R_1 + R_3\right)} \quad \begin{pmatrix} 1 & 2 & -2 & 5 \\ 0 & -7 & 5 & -17 \\ 0 & -7 & 5 & -9 \end{pmatrix}$$

$$\underrightarrow{\left(R_1 + \tfrac{2}{7}R_2, -\tfrac{1}{7}R_2, (-1)R_2 + R_3\right)} \quad \begin{pmatrix} 1 & 0 & -\tfrac{4}{7} & \tfrac{1}{7} \\ 0 & 1 & -\tfrac{5}{7} & \tfrac{17}{7} \\ 0 & 0 & 0 & 8 \end{pmatrix}.$$

The last row of the matrix represents the equation $0 = 8$. Since $0 \neq 8$, it follows that the solution set for the system is $\varnothing$.

Sometimes an interchange of rows is necessary as for the system

$$y + z = 2$$
$$x - y + z = 1$$
$$x + 2y - 3z = 4$$

where the leading coefficient in the first equation is 0. The matrix is

$$\begin{pmatrix} 0 & 1 & 1 & 2 \\ 1 & -1 & 1 & 1 \\ 1 & 2 & -3 & 4 \end{pmatrix}$$

which we first transform to

$$\begin{pmatrix} 1 & -1 & 1 & 1 \\ 0 & 1 & 1 & 2 \\ 1 & 2 & -3 & 4 \end{pmatrix}$$

by interchanging the first and second rows.

Schematically (omitting any needed interchange of rows because of zero coefficients), the procedure in the case of three equations in three variables can be pictured as follows:

$$\begin{pmatrix} a & b & c & d \\ e & f & g & h \\ i & j & k & l \end{pmatrix} \longrightarrow \begin{pmatrix} 1 & b' & c' & d' \\ 0 & f' & g' & h' \\ 0 & j' & k' & l' \end{pmatrix}$$

$$\longrightarrow \begin{pmatrix} 1 & 0 & c'' & d'' \\ 0 & 1 & g'' & h'' \\ 0 & 0 & k'' & l'' \end{pmatrix} \longrightarrow \begin{pmatrix} 1 & 0 & 0 & d''' \\ 0 & 1 & 0 & h''' \\ 0 & 0 & 1 & l''' \end{pmatrix}$$

The steps indicated show a unique solution (d''', h''', l'''). If there is an infinite number of solutions as in Example 6, then one or two of the last rows will consist entirely of zeros and if the solution set is empty as in Example 7, the last row will be of the form $(0 \ \ 0 \ \ 0 \ \ l''')$ with $l''' \neq 0$. (Compare the outline of our procedure given earlier on page 359.)

This method of solving linear systems by using matrix representation can be easily applied to systems of equations that are much more complex than those we have considered. Furthermore, as we remarked before, the systematic nature of the process lends itself particularly well to programming in electronic computors.

EXERCISES 8.5B

◀ In Exercises 1–10, which are the same as those in Exercises 8.5A, solve using matrix notation.

1. $2x + 3y = 13$
$2x - y = 1$

2. $4x - y = 0$
$6x + 5y = 13$

3. $x + y + z = 2$
$x - y - z = 0$
$x + y - z = -2$

4. $x + 2y + 3z = 1$
$2x - y - z = -2$
$3x - 2y + z = 3$

5. $2x - y + 3z = 14$
$3x - 2y - z = 5$
$4x - 3y - 2z = -1$

6. $2x - y + 3z = 3$
$x - 3y + z = 2$
$3x + 2y - 2z = 11$

7. $x + 2y + 3z = 3$
$2x - y - 2z = -8$
$3x + 4y + 5z = 1$

8. $x + 3y + z = 0$
$x + 4z = -2$
$-6y + z = 1$

9. $x - y + z = 2$
$\quad x + y + z = 0$
$\quad 3x + 2y + 4z = 1$

10. $x + y - z = 4$
$\quad 2x - y + 3z = -3$
$\quad x - y + 2z = -4$

◀In Exercises 11–17, find the solution set of the following systems using matrix notation.

11. $x + 2y - z = 5$
$\quad x + y + 2z = 11$
$\quad x + y + 3z = 14$

12. $x + 2y - z = -1$
$\quad 2x + 2y - 3z = -1$
$\quad 4x - y + 2z = 11$

13. $4x - y + z = 6$
$\quad 3x + 2y - 4z = 2$
$\quad 7x + y - 3z = 5$

14. $x + y + z = 2$
$\quad 2x + 2y + 2z = 5$
$\quad x - y + z = 7$

15. $x - 2y - 3z = 2$
$\quad x - 4y - 13z = 14$
$\quad 3x - 5y - 4z = 0$

16. $x + 2y - z = 3$
$\quad x - y + z = 4$
$\quad 4x - y + 2z = 14$

17. $x + 2y - z = 3$
$\quad 4x - y + 2z = 15$
$\quad x + y - z = 2$

18. Solve the following system:

$$x + y + z - w = 1$$
$$x - y + 3z + 2w = 2$$
$$2x + y + 3z + w = -2$$
$$x - 2y + z + 3w = 10.$$

19. Solve the following system:

$$x + y - z + w = 0$$
$$2x - y + 2z + w = -5$$
$$3x + y + z - 3w = 13$$
$$x - 2y + 2z - 2w = 1.$$

20. Solve the following system:

$$\frac{3}{x} + \frac{2}{y} + \frac{1}{z} = 5$$

$$\frac{2}{x} + \frac{3}{y} - \frac{2}{z} = -1$$

$$\frac{4}{x} + \frac{1}{y} + \frac{3}{z} = 10.$$

(*Hint:* Let $u = \dfrac{1}{x}$, $v = \dfrac{1}{y}$, $w = \dfrac{1}{z}$, and rewrite the equations as $3u + 2v + w = 5$, and so on.)

21. Solve the following system:

$$\frac{1}{x} + \frac{1}{y} + \frac{1}{z} = 2$$

$$\frac{2}{x} - \frac{3}{y} - \frac{1}{z} = 11$$

$$\frac{1}{x} + \frac{2}{y} + \frac{3}{z} = 2$$

(*Hint:* See Exercise 20.)

22. Solve each of the following systems using matrix notation:

(a) $2x + 3y = -4$
$\quad\ 4x + 6y = 8$

(b) $2x + 3y = 4$
$\quad\ 4x + 6y = 8$

23. Give a geometric interpretation of the systems in Exercise 22.

24. Give a geometric interpretation of the system in Example 5, page 363, of the text. (*Hint:* See Section 6.9.)

25. Give a geometric interpretation of the system in Example 6, page 363, of the text. (*Hint:* See Section 6.9.)

26. The sum of the digits of a three-digit number is 7. The hundred's digit is 2 less than twice the sum of the ten's digit and the unit's digit. Five times the unit's digit plus twice the sum of the hundred's digit and ten's digit is 20. Find the number.

27. Maria has a collection of pennies, nickels, and dimes in her purse. The number of coins is 12 and the collection amounts to 55 cents. If the number of pennies and nickels together is three times the number of dimes, how many of each kind of coin does she have?

28. Find an equation of the circle which contains the points $(3, -7)$, $(-2, -2)$, and $(-1, -5)$.

29. Three trucks are hauling gravel. The first day, when 78 cu. yd. were hauled, the first truck hauled 4 loads, the second truck hauled 3 loads, and the third truck hauled 5 loads. On the second day, 81 cu. yd. were hauled by the trucks in 5, 4, and 4 loads, respectively. On the third day, 69 cu. yd. were hauled in 3, 5, and 3 loads, respectively. Assuming that each truck always hauled its capacity, find the capacity of each truck.

30. John has $30,000 divided among three investments. His total income from the three investments is $1350 per year. The income from his first investment is $150 more than the sum of his income from the other two investments. If his investments yield 6%, 4%, and 3%, respectively, how much has he invested in each?

31. CHALLENGE PROBLEM. Example 6 of the text, page 363, displays a dependent system. Show that any one of the three equations may be expressed as a linear combination of the other two equations.

CHAPTER SUMMARY

Two equations are equivalent if and only if they have the same solution set. Similarly, two systems of equations are equivalent if and only if they have the same solution set. The solution set of a system is the intersection of the solution sets of the component equations.

For all $r, s \in R$ (r and s not both 0)

$$r(a_1x + b_1y + c_1) + s(a_2x + b_2y + c_2) = 0$$

is called a LINEAR COMBINATION of the two equations

$$a_1x + b_1y + c_1 = 0 \qquad \text{and} \qquad a_2x + b_2y + c_2 = 0.$$

A system of two linear equations in two variables can be solved by SUBSTITUTION or ADDITION.

A system of one linear equation and one second-degree equation can be solved by substitution.

A system of two second-degree equations cannot always be solved unless general methods of solving cubic and quartic equations are available. Certain systems, however, can be solved using either substitution, addition, or both.

A system of m linear equations in n variables can be represented in MATRIX FORM. Using ROW OPERATIONS, any (3×4) matrix can be reduced to a matrix of the form

$$\begin{pmatrix} 1 & 0 & 0 & a \\ 0 & 1 & 0 & b \\ 0 & 0 & 1 & c \end{pmatrix} \qquad (a, b, c \in R)$$

if the system has a unique solution, to the form

$$\begin{pmatrix} 1 & 0 & 0 & a \\ 0 & 1 & 0 & b \\ 0 & 0 & 0 & 0 \end{pmatrix}$$

if the system has more than one solution, and to the form

$$\begin{pmatrix} 1 & 0 & 0 & a \\ 0 & 1 & 0 & b \\ 0 & 0 & 0 & c \end{pmatrix} \qquad (c \neq 0)$$

if the system has no solution.

REVIEW EXERCISES

1. Find the solution set of each of the following systems. Graph the equations of the system.

   (a) $x - y = -1$ (b) $x - y = -1$ (c) $x - y = -1$
       $2x + 4y = 19$ $x - y = 6$ $3y - 3x = 3$

2. Determine if $(3, 2)$ is an element of the solution set of the system
   $$2x - 3y = 0$$
   $$x + y - 5 = 0$$
   $$5x - 3y - 9 = 0.$$

   Sketch the graph of the system. Does the graph verify that $(3, 2)$ is an element of the system?

3. Determine which of the following systems are consistent. (Do not solve the system.) If the system is consistent, determine whether it is also dependent.

   (a) $x + y = 1$ (c) $y = 3x + 2$

   $\frac{1}{2}x = 2 - \frac{1}{2}y$ $y = 3x + 5$

   (b) $y = 2x - 1$ (d) $12x = 13y + 8$

   $x - \frac{1}{2}y = \frac{1}{2}$ $4x - 2 = 5y$

4. (a) Form, using linear combinations, two systems equivalent to
   $$x + y = 5$$
   $$2x - y = 1.$$

   (b) Draw the graphs of the equations of the two equivalent systems on the same coordinate axes.

   (c) Next form a linear combination that has no term in y after simplification. Graph the two linear combinations on the same coordinate axes used in (b).

5. Find the solution set of the system
   $$2y^2 + xy = 5$$
   $$x + 4y = 7.$$

6. Solve the system
   $$x^2 + 4y^2 = 25$$
   $$3x + 8y = 25.$$

   Sketch the graph of the system.

7. Find the solution set of the system
   $$x^2 - y^2 = 0$$
   $$x + y = 0.$$

   Give a geometric interpretation of the system.

8. Show geometrically, by sketching the various possibilities, the possible solution sets of a system whose individual components are an equation of a hyperbola and an equation of a parabola.

9. Find the solution set of the system

$$4x^2 + y^2 = 100$$
$$y^2 = 4x + 20.$$

Sketch the graph of the system.

10. Find the solution set of the system

$$x^2 - y^2 = 41$$
$$2x^2 - y^2 = 7.$$

Sketch the graph of the system.

11. Find, by graphing, the solution set of the system

$$x^2 - y^2 \le 9$$
$$x^2 + 2y^2 \ge 4.$$

12. Find, by graphing, the solution set of the system

$$x^2 + y^2 - 4x - 2y \le -1$$
$$y - 4 \le x^2 - 4x.$$

13. Using matrix notation, find the solution set of the system

$$x + 3y - 4z = 17$$
$$-2x - y + 2z = -4$$
$$-x + 2y - 3z = 11.$$

Give a geometric interpretation of the system.

14. Using matrix notation, find the solution set of the system

$$x + 5y - z = 3$$
$$2x - 3y + z = 3$$
$$x + 18y - 4z = 6.$$

Give a geometric interpretation of the system.

15. Show that the equations in Exercise 14 are dependent by expressing one as a linear combination of the other two.

16. Find over the field of complex numbers, the solution set of the system

$$y = x^2$$
$$y = x - 4.$$

17. Find over the field of complex numbers, the solution set of the system

$$ix + (2 - i)y + 6i = 0$$
$$x - iy = 0.$$

GOING FURTHER: READING AND RESEARCH

In Chapter 4, we defined equality, addition, and multiplication for 2×2 matrices. Let us look again at the definition of multiplication. By definition,

$$\begin{pmatrix} a & b \\ c & d \end{pmatrix}\begin{pmatrix} e & f \\ g & h \end{pmatrix} = \begin{pmatrix} ae + bg & af + bh \\ ce + dg & cf + dh \end{pmatrix}.$$

Similarly, we define

$$\begin{pmatrix} a & b \\ c & d \end{pmatrix}\begin{pmatrix} x \\ y \end{pmatrix} = \begin{pmatrix} ax + by \\ cx + dy \end{pmatrix}.$$

Thus

$$\begin{pmatrix} 1 & 0 \\ 0 & 1 \end{pmatrix}\begin{pmatrix} x \\ y \end{pmatrix} = \begin{pmatrix} 1x + 0 \\ 0 + 1y \end{pmatrix} = \begin{pmatrix} x \\ y \end{pmatrix}.$$

Check that, for any 2×2 matrix

$$\begin{pmatrix} a & b \\ c & d \end{pmatrix},$$

we have

$$\begin{pmatrix} 1 & 0 \\ 0 & 1 \end{pmatrix}\begin{pmatrix} a & b \\ c & d \end{pmatrix} = \begin{pmatrix} a & b \\ c & d \end{pmatrix}\begin{pmatrix} 1 & 0 \\ 0 & 1 \end{pmatrix} = \begin{pmatrix} a & b \\ c & d \end{pmatrix}.$$

That is, check that the matrix

$$\begin{pmatrix} 1 & 0 \\ 0 & 1 \end{pmatrix}$$

is a multiplicative identity in the system of 2×2 matrices.

Does every 2×2 matrix have a multiplicative inverse? That is, given

$$\begin{pmatrix} a & b \\ c & d \end{pmatrix},$$

can we always find a 2×2 matrix

$$\begin{pmatrix} a' & b' \\ c' & d' \end{pmatrix}$$

such that

$$\begin{pmatrix} a' & b' \\ c' & d' \end{pmatrix}\begin{pmatrix} a & b \\ c & d \end{pmatrix} = \begin{pmatrix} a & b \\ c & d \end{pmatrix}\begin{pmatrix} a' & b' \\ c' & d' \end{pmatrix} = \begin{pmatrix} 1 & 0 \\ 0 & 1 \end{pmatrix}?$$

What must be the conditions on a, b, c, and d for a multiplicative inverse to exist?

Multiplicative inverses do exist for some matrices. You can check that

(a) $\begin{pmatrix} 2 & 7 \\ 1 & 6 \end{pmatrix} \begin{pmatrix} \dfrac{6}{5} & \dfrac{-7}{5} \\ \dfrac{-1}{5} & \dfrac{2}{5} \end{pmatrix} = \begin{pmatrix} \dfrac{6}{5} & \dfrac{-7}{5} \\ \dfrac{-1}{5} & \dfrac{2}{5} \end{pmatrix} \begin{pmatrix} 2 & 7 \\ 1 & 6 \end{pmatrix} = \begin{pmatrix} 1 & 0 \\ 0 & 1 \end{pmatrix}$

and

(b) $\begin{pmatrix} 3 & 5 \\ 2 & 4 \end{pmatrix} \begin{pmatrix} \dfrac{4}{2} & \dfrac{-5}{2} \\ -\dfrac{2}{2} & \dfrac{3}{2} \end{pmatrix} = \begin{pmatrix} \dfrac{4}{2} & \dfrac{-5}{2} \\ \dfrac{-2}{2} & \dfrac{3}{2} \end{pmatrix} \begin{pmatrix} 3 & 5 \\ 2 & 4 \end{pmatrix} = \begin{pmatrix} 1 & 0 \\ 0 & 1 \end{pmatrix}.$

Try to find a formula for finding the multiplicative inverse of a 2×2 matrix when such an inverse exists.

Multiplicative inverses of matrices can be related to solutions of systems of linear equations. Consider, for example, the system

$$2x + 7y = 1$$
$$x + 6y = 2.$$

In matrix notation we can write

$$\begin{pmatrix} 2 & 7 \\ 1 & 6 \end{pmatrix} \begin{pmatrix} x \\ y \end{pmatrix} = \begin{pmatrix} 1 \\ 2 \end{pmatrix}.$$

If we now multiply each member of the above equation by the multiplicative inverse of

$$\begin{pmatrix} 2 & 7 \\ 1 & 6 \end{pmatrix},$$

we have, assuming that matrix multiplication is associative,

$$\begin{pmatrix} \dfrac{6}{5} & -\dfrac{7}{5} \\ -\dfrac{1}{5} & \dfrac{2}{5} \end{pmatrix} \begin{pmatrix} 2 & 7 \\ 1 & 6 \end{pmatrix} \begin{pmatrix} x \\ y \end{pmatrix} = \begin{pmatrix} \dfrac{6}{5} & -\dfrac{7}{5} \\ -\dfrac{1}{5} & \dfrac{2}{5} \end{pmatrix} \begin{pmatrix} 1 \\ 2 \end{pmatrix}$$

from which we obtain

$$\begin{pmatrix} 1 & 0 \\ 0 & 1 \end{pmatrix} \begin{pmatrix} x \\ y \end{pmatrix} = \begin{pmatrix} -\dfrac{8}{5} \\ \dfrac{3}{5} \end{pmatrix}$$

or

$$\begin{pmatrix} x \\ y \end{pmatrix} = \begin{pmatrix} -\dfrac{8}{5} \\ \dfrac{3}{5} \end{pmatrix}$$

and hence the solution set of the given system of equations is $\{(-\tfrac{8}{5}, \tfrac{3}{5})\}$.

Using the multiplicative inverse of

$$\begin{pmatrix} 3 & 5 \\ 2 & 4 \end{pmatrix},$$

can you apply this procedure to the system

$$3x + 5y = -1$$
$$2x + 4y = 3?$$

Each of the foregoing equations and statements about 2×2 matrices can be matched with similar equations and statements about 3×3 matrices. Try to generalize these results to 3×3 matrices and systems of three linear equations in three variables.

If you are interested in learning more about matrices generally, you will find the following books helpful.

School Mathematics Study Group, *Introduction to Matrix Algebra*, Pasadena: A. E. Vroman, Inc., 1965.

NERING, E. D. *Linear Algebra and Matrix Theory*, New York: John Wiley and Sons, 1963.

DAVIS, P. J. *The Mathematics of Matrices*, Boston: Blaisdell, 2nd Ed., 1965.

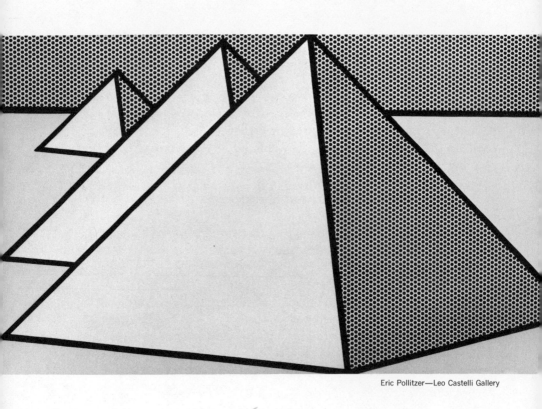

Eric Pollitzer—Leo Castelli Gallery

CHAPTER NINE

ARRANGEMENTS, SUBSETS, AND THE BINOMIAL THEOREM

9.1 A PROBLEM OF COMMUNICATION

In earlier times displaying signal flags was one of the principal means of communicating from one ship to another. Even in our electronic age this custom still persists on ceremonial occasions.

Let us consider a problem. Suppose that we have four signal flags each of a distinctive color or design and suppose that a message consists of a display of these four flags in a certain order. Rearrange the order and you have a different message.

Question: How many different messages can be sent out in this manner?

For convenience let us assume that the flags are colored red, white, blue, and green and we identify them by r, w, b, and g, respectively. One somewhat unscientific way of answering the question would be to list all the possibilities and then count them.

An alternative approach might be preferable, however. Let us consider the 4 positions and number them 1, 2, 3, 4. In choosing a message, first a color is selected for position number 1. There are 4 choices: r, w, b, or g. After a choice has been made for the first spot, 3 possibilities remain for the second. To fill the first 2 positions, then, we have these choices:

> with r in position number 1: rw, rb, rg;
> with w in position number 1: wr, wb, wg;
> with b in position number 1: br, bw, bg;

and finally

> with g in position number 1: gr, gw, gb.

The total number of choices for the first 2 spots is clearly 12, that is, $4 \cdot 3$.

Now, associated with any one of these 12 choices for the first 2 spots are 2 remaining colors for position number 3. For example, with rw we can have rwb or rwg. There are also 2 choices for spot number 3 associated with any of the other 11 choices for 1 and 2. We see, then, that the total number of options for positions 1, 2, and 3 is $12 \cdot 2$, that is, $4 \cdot 3 \cdot 2$. Given any 3 flags in these positions, there is obviously only 1 choice left for the number 4 spot.

It seems reasonable, then, to deduce that the total number of messages can be calculated as

$$4 \cdot 3 \cdot 2 \cdot 1 = 24.$$

Although the factor 1 does not affect the product, it is included so that all positions may be represented.

You can see that if we increase the number of flags to 5, then the number of choices for the first position would be 5, for the second, 4, and so forth. Hence the total number of messages in this case is

$$5 \cdot 4 \cdot 3 \cdot 2 \cdot 1 = 120.$$

To illustrate the idea further, suppose that the flag situation is altered as follows. We now suppose that there are 4 flags of each color so that messages such as *rrrr*, or *rrbb*, or *wggg*, etc., are possible. Under these conditions how many different messages consisting of 4 flags can be dispatched? The difference here is that once a choice is made for spot number 1, there are still 4 available choices for spot number 2; similarly for positions 3 and 4. Hence the number of different messages is

$$4 \cdot 4 \cdot 4 \cdot 4 = 256.$$

These signal flag problems illustrate a fundamental mathematical principle that has a wide variety of applications. The principle is intuitively quite reasonable.

Let $A_1, A_2, \ldots, A_k$ be a sequence of k situations in which choices need to be made. (For example, in the flag arrangement problem we have just considered, A_1 would be the situation in which we choose the first flag, A_2 the situation in which we choose the second flag, A_3 the situation in which we choose the third flag, and A_4 the situation in which we choose the fourth flag.) If N_1 is the number of different choices (decisions) to be made with respect to A_1, if N_2 is the number of choices with respect to A_2, $\ldots$, and if N_k is the number of choices with respect to A_k, then the total number of possible sequences of choices, one for each of the k situations, is

$$N_1 \cdot N_2 \cdot N_3 \cdot \ldots \cdot N_k.$$

(In our first example, $N_1 = 4$, $N_2 = 3$, $N_3 = 2$, and $N_4 = 1$; in our second example, $N_1 = 5$, $N_2 = 4$, $N_3 = 3$, $N_4 = 2$, and $N_5 = 1$; and in our third example, $N_1 = N_2 = N_3 = N_4 = 4$.) We shall call this basic result the principle of choice (*PC* for short).

As a further illustration of *PC*, consider the following question.

How many three-digit numerals can be formed using the digits 1 through 9

(1) if no digit can be repeated?

(2) if the same digit can be used more than once?

To answer question (1) we see that there are 9 possible choices for the unit's digit. However, after a unit's digit is chosen, only 8 choices remain for the ten's digit, and then, after a ten's digit is picked, there are 7 options for the hundred's digit.

The answer to (1) then is

$$9 \cdot 8 \cdot 7 = 504,$$

that is, 504 different numerals. In this case, $N_1 = 9$, $N_2 = 8$, and $N_3 = 7$.

In question (2) with repetitions permissible we have 9 choices for each place. Hence the total number of possible numerals in this instance is

$$9 \cdot 9 \cdot 9 = 729.$$

Here $N_1 = N_2 = N_3 = 9$.

The principle of choice can be applied in still another context. Suppose that you have a set S consisting of, say, 8 elements. How would you calculate the total number of possible subsets of S?

Recall that T is a subset of S, that is, $T \subseteq S$, if every element in T is also in S. Under this definition we regard S as a subset of itself. We may also infer from this definition that the empty or null set, $\varnothing$, is a subset of every set. Hence in counting the number of subsets of S we must include both S and $\varnothing$ in the tally.

In determining the number of subsets of a set of 8 elements we might consider making a list. Let the elements of S be

$$\{a, b, c, d, e, f, g, h\}.$$

We could begin with $\varnothing$, then list the subsets of one element, that is, $\{a\}$, $\{b\}$, $\{c\}$, and so on, then the subsets of two elements, that is $\{a, b\}$, $\{a, c\}$, $\{b, c\}$, etc. A long job, especially since the number of subsets turns out to be 256, a fact which we shall presently verify! How, then, could we have reached the conclusion that the answer is 256 without the "long count"?

Consider the manner in which any particular subset might have been chosen. Take, for example, the subset

$$B = \{b, d, f, g\}.$$

We can imagine a "decision-making" process in which we scan the complete set S from a to h deciding for each element in turn whether it is to be included or not. In the case of the subset

$$B = \{b, d, f, g\},$$

the results of the decisions would be

a	b	c	d	e	f	g	h
No	Yes	No	Yes	No	Yes	Yes	No

There are 8 decisions in all, each involving 2 choices.

A similar tally on the subset $\{a, b, c\}$ would be

a	b	c	d	e	f	g	h
Yes	Yes	Yes	No	No	No	No	No

It appears, then, that the total number of subsets is equal to the number of possible tabulations of the type listed.

With this in mind we are now ready for an application of *PC* as follows. Any given tabulation may be thought of as a sequence of 8 decisions, where the number of choices for each decision in the sequence is 2, that is, Yes or No. Hence $N_i = 2$ for $i = 1, 2, \ldots, 8$. Using *PC* we may conclude that the total number of tabulations is

$$2 \cdot 2 \cdot 2 \cdot 2 \cdot 2 \cdot 2 \cdot 2 \cdot 2 = 2^8 = 256.$$

Hence S has 256 subsets, as previously predicted.

We should note, in passing, that the tabulation corresponding to the null set $\varnothing$ is

a	b	c	d	e	f	g	h
No	No	No	No	No	No	No	No

whereas for the set S itself we have

a	b	c	d	e	f	g	h
Yes	Yes	Yes	Yes	Yes	Yes	Yes	Yes

From the foregoing discussion we should now be in a position to consider the general problem of finding the number of subsets of a set of n elements.

Since the choice for any element to be or not to be in a subset is still a two-choice decision, it follows that $N_i = 2$ in all cases. Hence a set of n elements has 2^n subsets.

Possibly a simpler way of associating a subset with a decision process would be merely to use the symbol 1 for inclusion and 0 for exclusion. For example, if

$$S = \{a, b, c, d, e, f, g, h\},$$

a subset

$$A = \{c, d, e, h\}$$

could be associated with the notation

$$0 \quad 0 \quad 1 \quad 1 \quad 1 \quad 0 \quad 0 \quad 1.$$

For $B = \{b, d, f, g\}$ we would have

$$0 \quad 1 \quad 0 \quad 1 \quad 0 \quad 1 \quad 1 \quad 0$$

and for $\varnothing$

$$0 \quad 0 \quad 0 \quad 0 \quad 0 \quad 0 \quad 0 \quad 0.$$

If you have had experience with the binary (base 2) system of numeration, you will recognize the above as examples of binary numerals. It is pertinent to note that with 8 positions we can write 256 $(= 2^8)$ different binary numerals.

It is also possible to prove by induction that 2^n is the total number of subsets of a set of n elements. We now outline such a proof for you to complete. (Before proceeding further, perhaps you should quickly review Section 2.3.)

To begin the proof we first let

$S = \{n : n \in N$ and $2^n =$ the number of subsets of a set of n elements$\}$.

1. Is $1 \in S$? Yes, since a set with one element, say a, has 2 subsets, $\{a\}$ and $\boxed{?}$, and $2^1 = \boxed{?}$.
2. Assume, now, that $k \in S$, and consider a set T with k elements. By assumption, T has $\boxed{?}$ subsets.
3. Let T' be the set which contains all the elements of T and one additional element s, that is, T' has $\boxed{?}$ elements.
4. All subsets of T are certainly subsets of T'. There are $\boxed{?}$ of these.
5. Now to each of these subsets of T adjoin the element s. How many new subsets of T' containing s will this give us? (Note that $\{s\}$ can be thought of as $\varnothing \cup \{s\}$.)
6. Thus T' has exactly 2^k subsets that do *not* contain s and exactly

2^k subsets that *do* contain s. In all, T' has $2^k + 2^k = 2(\boxed{?})$ subsets.

7. But $2(2^k) = 2^{\boxed{?}}$.

8. We have shown that if $k \in S$, then $k + 1 \in S$; hence S is an $\boxed{?}$ set and our proof is complete.

Although our induction proofs have been restricted to the set of natural numbers, the above formula, 2^n for the number of subsets, does apply when $n = 0$. We have $2^0 = 1$ and note that the empty set $\varnothing$ has exactly one subset, namely itself.

EXERCISES 9.1

1. How many subsets does a set of 10 elements have?
2. How many subsets does a set of 12 elements have?
3. How many different arrangements can be made with the letters in the word SIGNAL?
4. How many different arrangements can be made with the letters in the word REGAL?
5. How many different messages can be sent using 10 different flags if each message consists of 3 flags in a certain order?
6. What would the answer to Exercise 5 have been if the message-sending equipment had included 30 flags, 3 of each color?
7. Consider again a collection of 10 different flags. This time a message consists of any display of one or more flags without regard to order. What is the total number of messages in this situation?
8. How many different seating arrangements are possible for 8 students in a row of 8 chairs?
9. How many different seating arrangements are possible for 8 students if the row contains 9 chairs? (Note that one seating arrangement is considered to be different from another even if the difference is only in the location of the empty seat.)
10. Two dice are thrown. Each die can show a number from 1 to 6. What is the total number of possible outcomes?
11. If 3 dice are thrown, what is the total number of possible outcomes?
12. An experiment consists of throwing a die, then drawing a card at random from a deck of 52 cards, and, finally, tossing a coin. How many different outcomes could this experiment have?
13. On the menu of a restaurant there are 7 choices of appetizers, 3 different soups, 10 main dishes, and 5 desserts. How many different four-course dinners can one choose?

14. Assume that on the restaurant menu of Exercise 13 there is a less expensive three-course dinner which stipulates that the dinner includes either appetizer or soup, but not both. How many different three-course dinners can be ordered?

15. How many different five-digit numerals can be formed from the numerals 1 to 7 if no digit is to be repeated?

16. How many different four-digit numerals can be formed from the numerals 1 to 5 if any digit may be used as often as you wish?

17. How many arrangements of three letters can be formed from the alphabet when (a) repetitions are not permitted and (b) repetitions are permitted?

18. How many different ten-place binary numerals can be written?

19. A number written in base five uses the symbols 0, 1, 2, 3, 4. How many different four-place base five numerals can be written?

20. There are 7 horses in a race. In how many different ways can the winning combinations of first place, second place, and third place occur?

21. How many different arrangements can be made of the letters in the word PROBLEMS?

22. CHALLENGE PROBLEM. In how many ways can 10 dinner guests be seated around a circular table if a difference in seating depends only on who is next to whom? (*Hint:* A rearrangement in which each person moves one seat to the right or left is not considered a change.)

23. CHALLENGE PROBLEM. Determine the number of different arrangements of the letters in the word SUBSET.

24. CHALLENGE PROBLEM. How many different arrangements can be made with the letters in the word MATHEMATICS?

25. CHALLENGE PROBLEM. How many different subsets of 4 elements each can be formed from a set of 10 elements?

26. CHALLENGE PROBLEM. How many different subsets can be formed from a set of 10 elements if each subset is to contain an odd number of elements?

27. CHALLENGE PROBLEM. Tom, Dick, and Harry play 10 games. A typical outcome might be that Tom wins 3 games, Dick 5 games, and Harry 2 games. How many different outcomes are possible if ties are not considered?

9.2 ARRANGEMENTS AND SUBSETS

In this section we consider a few more variations on the general theme of "how to count without counting." We also systematize the procedures by the introduction of two very useful functions.

First, however, a word on notation. To indicate a product of the form $1 \cdot 2 \cdot 3 \cdot \ldots \cdot n$ we use the symbol $n!$ (read "n factorial").

Examples

$$5! = 1 \cdot 2 \cdot 3 \cdot 4 \cdot 5, \qquad 7! = 1 \cdot 2 \cdot 3 \cdot 4 \cdot 5 \cdot 6 \cdot 7,$$

and

$$10! = 1 \cdot 2 \cdot 3 \cdot 4 \cdot 5 \cdot 6 \cdot 7 \cdot 8 \cdot 9 \cdot 10.$$

We can also write 5!, for example, as

$$5 \cdot 4 \cdot 3 \cdot 2 \cdot 1.$$

As you have seen, however, many problems involve products that do not have as factors all the natural numbers from a given natural number down to 1. The number of possible arrangements of 10 different letters taken three at a time, for example, is $10 \cdot 9 \cdot 8$. How can we express a product using factorial symbols?

Since $10 \cdot 9 \cdot 8 = \dfrac{10 \cdot 9 \cdot 8 \cdot 7 \cdot 6 \cdot 5 \cdot 4 \cdot 3 \cdot 2 \cdot 1}{7 \cdot 6 \cdot 5 \cdot 4 \cdot 3 \cdot 2 \cdot 1}$,

we can write

$$10 \cdot 9 \cdot 8 = \frac{10!}{7!}.$$

In general, if we wish to express a product of the form

$$n(n - 1)(n - 2) \ldots$$

where the number of factors is r, we write

$$\frac{n!}{(n - r)!} \qquad \text{if } r < n.$$

For example, $12 \cdot 11 \cdot 10 \cdot 9$, where $n = 12$, $r = 4$, and $n - r = 8$, can be written as

$$\frac{12!}{8!}.$$

What about the expression

$$\frac{n!}{(n - r)!}$$

when $r = n$? If we consider a product of n factors of the form $n(n - 1)(n - 2) \ldots$, we have precisely

$$n(n - 1)(n - 2) \cdot \ldots \cdot 2 \cdot 1 = n!.$$

Thus for $r = n$ we would like to have

$$\frac{n!}{(n-r)!} = n!.$$

This is very easy to achieve. We simply define $0! = 1$ and then have, for $n = r$,

$$\frac{n!}{(n-r)!} = \frac{n!}{0!} = \frac{n!}{1} = n!.$$

EXERCISES 9.2A

1. Evaluate 5!

2. Evaluate $\dfrac{6!}{3!}$

3. Evaluate $6! \cdot 3!$

4. Evaluate $\dfrac{13!}{10!}$

5. Evaluate $\dfrac{6!}{3! \cdot 2!}$

6. Evaluate $\dfrac{10!}{2!}$

7. Evaluate $\dfrac{20!}{17!}$

8. Evaluate $\dfrac{8!}{2^8}$

9. Evaluate $\dfrac{15!}{7! \cdot 8!}$

10. Evaluate $\dfrac{12!}{3! \cdot 4! \cdot 5!}$

11. If a product $n(n-1)(n-2) \ldots$ is to contain exactly r factors where $r < n$, what is the rth factor?

12. How many factors does the expression $(20)(20-1)(20-2) \ldots (20-12)$ represent?

Let us now consider a type of arrangement problem with a different twist. We have seen that the number of ways in which 4 flags can be arranged is

$$4 \cdot 3 \cdot 2 \cdot 1 = 4!.$$

In general, the number of different arrangements of n distinct objects in $n!$.

Thus, the number of possible arrangements of the letters in the word SIGNAL is 6!. However, the number of different arrangements of the letters in the word SUBSET (Exercise 23 of Exercises 9.1) is not 6! since an interchange of the two S's does not produce a different arrangement.

To deal with the general case of duplicated letters (or any other objects), let us look at the following example. Suppose that we want the number of different arrangements of the letters *rrrbg*. We might regard this as an arrangement of 5 flags, 3 of which are red, 1 of which is blue, and 1 green. We shall begin by imagining that the red flags are of different shades and hence distinct, and will indicate this by subscripts, $r_1r_2r_3gb$. Under this condition we would have $5! = 120$ different arrangements. Let us now consider a partial listing of these arrangements as follows:

$r_1r_2r_3gb$	$r_1r_2r_3bg$	$br_1r_2r_3g$
$r_1r_3r_2gb$	$r_1r_3r_2bg$	$br_1r_3r_2g$
$r_2r_1r_3gb$	$r_2r_1r_3bg$	$br_2r_1r_3g$
$r_2r_3r_1gb$	$r_2r_3r_1bg$	$br_2r_3r_1g$
$r_3r_1r_2gb$	$r_3r_1r_2bg$	$br_3r_1r_2g$
$r_3r_2r_1gb$	$r_3r_2r_1bg$	$br_3r_2r_1g$

In this partial list the 18 different arrangements are grouped in sets of 6. The complete list constructed in this pattern would contain 20 such sets of 6, where in each set the positions of the *g* and *b* remain fixed and the *r*'s are rearranged in all possible ways.

If we now remove the subscripts, the items in each set of 6 become identical. Thus the total number of different arrangements reduces to

$$\frac{1}{6}(5!) = \frac{1}{6}(120) = 20.$$

The reason for grouping 6 items in each subset is to emphasize the fact that the 3 letters (red flags) r_1, r_2, and r_3 can be arranged in $3! = 6$ ways. This then suggests that a formula for the number of arrangements of *rrrbg* be written as

$$\frac{5!}{3!}$$

which, as you see, is equal to 20. From this we may now infer that the number of different arrangements of the letters in SUBSET is equal to

$$\frac{6!}{2!} = 360.$$

This idea can be extended to arrangements involving more than one set of duplications. For example, the number of arrangements of the letters *rrrbbg* is

$$\frac{6!}{3!2!} = 60.$$

More generally, we can see that the number of different arrangements of n objects, of which there are x copies of one kind, y copies of another, and so on, is given by the formula

$$\frac{n!}{x!y! \cdots}.$$

As an example of this formula, there are

$$\frac{11!}{4!4!2!} = 34,650$$

different arrangements of the letters in the word MISSISSIPPI. Here we have 4 S's, 4 I's, 2 P's, and one lonely M. To include M, we could have written the answer as

$$\frac{11!}{4!4!2!1!}.$$

With this in mind we can state, with greater generality, that the number of different arrangements of n objects is given by the formula

$$\frac{n!}{n_1!n_2! \cdots n_k!}$$

where n_i represents the number of objects of the ith kind and

$$n_1 + n_2 + \cdots + n_k = n.$$

In this light we might regard a situation involving n distinct objects as a special case in which $n_i = 1$ for all i.

In the examples which we have just been discussing, we presumed the inclusion of *all* the given letters or objects in each arrangement. Let us now return to the general problem of counting the number of arrangements of n distinct objects taken r at a time where $r \leq n$.

We may regard the general counting device for this as a function, P, of the two variables n and r, where n and r are nonnegative integers with $n \geq r$. Thus P can be defined as

$$P : P(n, r) = \frac{n!}{(n - r)!}.$$

(The letter P stems from the word *permutations* which has been traditionally used for "arrangements.")

Examples

$$P(10,\ 3) = \frac{10!}{7!} = 10 \cdot 9 \cdot 8 = 720.$$

$$P(7,\ 4) = \frac{7!}{3!} = 7 \cdot 6 \cdot 5 \cdot 4 = 840.$$

Earlier we discussed the problem of counting the total number of subsets of a given set. A related problem is that of determining the number of subsets, all of which are of a particular size. For example, how many subsets of 4 elements each can be constructed from a set of 10 elements? (See Challenge Problem 25 of Exercises 9.1.)

The problem is related to that of finding the number of arrangements of 10 distinct objects taken 4 at a time, but with a *major* difference. Recall that a formula for the number of different arrangements in this case is

$$P(10,\ 4) = 10 \cdot 9 \cdot 8 \cdot 7 = 5040.$$

In the arrangement situation, however, we regarded the order in which the objects appeared as significant. For example, *abcd*, *bcad*, *bdac*, etc., are considered to be distinct entities. From each set of 4 letters we could form $4! = 24$ different arrangements.

On the other hand, if we are considering a subset such as $\{a,\ b,\ c,\ d\}$ selected from a set, say, of 10 letters,

$$\{a,\ b,\ c,\ d,\ e,\ f,\ g,\ h,\ i,\ j\},$$

we consider $\{a,\ b,\ c,\ d\}$, $\{a,\ c,\ d,\ b\}$, $\{b,\ c,\ d,\ a\}$, etc., as identical subsets regardless of the order in which the elements are listed.

Let us now compare the two situations. If we were obliged to list all of the 4-letter arrangements that could be formed from 10 letters on the one hand and the number of subsets of 4 letters each from a set of 10 letters on the other, the two lists, with many abbreviations, might look like this.

	Arrangements	Subsets	
4!, that is, 24 arrangements	abcd cdab dbac ⋮ bacd	$\{a,\ b,\ c,\ d\}$	1 subset

	Arrangements		Subsets	
24 arrangements	$\left\{\begin{array}{l} bcde \\ \vdots \\ ebcd \end{array}\right.$		$\{b, c, d, e\}$	1 subset
24 arrangements	$\left\{\begin{array}{l} ghij \\ \vdots \\ jhgi \end{array}\right.$		$\{g, h, i, j\}$	1 subset

and so on. If we completed the list, we would eventually end up with 5040 different arrangements because we have already noted that

$$P(10, 4) = 5040.$$

The question is: How many different subsets would we have? It should be apparent from the above scheme that there are 24 arrangements for each subset. Hence the number of different subsets corresponding to 5040 arrangements would be

$$\frac{5040}{24} = \frac{5040}{4!} = 210.$$

Since $5040 = P(10, 4)$, the number of subsets in this particular example could be represented as

$$\frac{P(10, 4)}{4!}.$$

In general, the number of subsets of r elements that can be formed from a set of n elements is given by the formula

$$\frac{P(n, r)}{r!} = \frac{n!}{r!(n - r)!}.$$

This formula too can be thought of as a function of the two variables n and r. It is customary to label the function $C(n, r)$ since the number of subsets of r elements which can be formed from a set of n elements has been traditionally called the number of **combinations** of n objects taken r at a time.

Thus we have

$$C(n, r) = \frac{P(n, r)}{r!}.$$

For example, the number of subsets of 3 elements which can be formed from a set of 8 objects is

$$C(8,\ 3) = \frac{P(8,\ 3)}{3!} = \frac{8!}{3!5!} = \frac{8\cdot 7\cdot 6}{1\cdot 2\cdot 3} = 56.$$

A widely used alternative notation for $C(n,\ r)$ is

$$\binom{n}{r}.$$

Examples

$$\binom{9}{2} = \frac{9\cdot 8}{1\cdot 2} = 36;$$

$$\binom{11}{3} = \frac{11\cdot 10\cdot 9}{1\cdot 2\cdot 3} = 165.$$

Note: It is the custom in many texts to write $\binom{n}{r}$ as $\dfrac{n!}{(n-r)!r!}$.

It should be easy to see that this is equal to $C\,(n,\ r)$.

So far the discussions in this chapter have been aimed at providing a battery of convenient shortcuts for counting—counting that involved numbers of sufficient magnitude to merit special devices.

There are always risks involved in presenting the many specialized formulas such as we have done. One risk is the development of a notion that either $P(n,\ r)$, $C(n,\ r)$, 2^n, or n^r can be applied to all counting situations. Another is the risk of confusion as to which formula is appropriate for which problem.

By way of summary, the following illustrations point up some of the signs for which you should be watching.

Let us consider five somewhat related questions.

1. Given the digits 1, 2, . . . , 9, how many four-digit numerals can be formed if no digit is repeated?
2. How many different four-digit numerals can be constructed with the digits 1, 2, . . . , 9 if digits may be repeated?
3. How many different four-digit numerals can be formed with the digits 1, 2, . . . , 9 if no digit is repeated and if no two numerals having the same set of digits are allowed (for example, not both of 2143 and 3124 are allowed)?

4. How many different numerals of 1, 2, 3, 4, . . . up to 9 digits can be formed from the digits 1, 2, 3, . . . , 9 if no digit is repeated in any numeral and if no two numerals can have exactly the same set of digits (for example, not both of 234 and 432 are allowed)?
5. How many different four-digit numerals can be formed from the digits 0, 1, 2, . . . , 9 if repetitions are allowed?

For question 1 we are clearly concerned with arrangements, although the word is not specifically mentioned since, for example, 1234 is obviously a different numeral from 4132. Thus we see that $P(n, r)$ is appropriate and the required answer is

$$P(9, 4) = 9 \cdot 8 \cdot 7 \cdot 6 = 3024.$$

Question 2 poses a different problem. Permission to repeat digits allows for numbers such as 1133, 2225, 7777, and so on. For such a problem neither $P(n, r)$ nor $C(n, r)$ is appropriate. We need instead the principle of choice. For the first digit we have 9 choices. We also have 9 choices for each of the other three digits. Thus the answer in this case is

$$9 \cdot 9 \cdot 9 \cdot 9 = 9^4 = 6561.$$

Question 3, since the conditions permit the inclusion of one and only one of the numerals 1234, 1432, 2431, and so on, is concerned essentially with eliminating all but one of the arrangements of any four specific digits. Thus we have a situation comparable with that of determining the number of subsets of 4 elements from a set of 9. Hence $C(n, r)$ is appropriate and we have for question 3,

$$C(9, 4) = \binom{9}{4} = \frac{9 \cdot 8 \cdot 7 \cdot 6}{1 \cdot 2 \cdot 3 \cdot 4} = 126.$$

Question 4 is similar to 3 in that by eliminating rearrangements we are talking essentially about subsets. In this instance, however, subsets of any size from 1 through 9 elements are included. Thus we have the set of all subsets except for $\varnothing$. The appropriate formula, then, is $2^n - 1$ or, in this case,

$$2^9 - 1 = 512 - 1 = 511.$$

Question 5 is an example of a problem for which none of the formulas $P(n, r)$, $C(n, r)$, 2^n, or n^r applies precisely because of the special aspect of the digit 0 whereby expressions such as 0125, 0036,

0007, and so on, are not considered as four-digit numerals. A moment's reflection on this question, however, tells us that we are essentially considering the numerals representing all numbers from 1000 to 9999 inclusive. Hence the answer is 9000. An alternative to this, using *PC*, would give us $9 \cdot 10 \cdot 10 \cdot 10$ since there are 10 choices for every place except the first one on the left. It is reassuring to see that both answers are the same.

At this point a quick review of the basic concepts and formulas examined thus far in this chapter seems in order.

1. The number of different *arrangements* of n distinct objects taken r at a time is

$$P(n, r) = \frac{n!}{(n-r)!}.$$

2. The number of subsets (combinations) of r elements that can be formed from a set of n elements is

$$C(n, r) = \binom{n}{r} = \frac{P(n, r)}{r!} = \frac{n!}{(n-r)!r!}.$$

3. The number of arrangements of n objects taken r at a time if each object can be repeated as many as r times is n^r.

4. The number of arrangements of n objects of which n_1 are of one kind, n_2 of another, etc., is

$$\frac{n!}{n_1!n_2! \ldots}.$$

EXERCISES 9.2B

◀Calculate each of the following:

1. $\binom{8}{4}$ 5. $\binom{12}{5}$

2. $P(10, 3)$ 6. $\binom{12}{7}$

3. $C(11, 4)$ 7. $P(11, 11)$

4. $C(11, 7)$ 8. $\binom{15}{1}$

9. $\dbinom{11}{5}$ **15.** $\dbinom{25}{25}$

10. $P(12, 7)$ **16.** $\dbinom{20}{18}$

11. $\dbinom{13}{10}$ **17.** $\dbinom{20}{2}$

12. $\dbinom{13}{3}$ **18.** $\dbinom{50}{4}$

13. $\dbinom{20}{15}$ **19.** $\dbinom{50}{48}$

14. $\dbinom{15}{14}$ **20.** $\dbinom{52}{13}$

21. How many subsets of 5 elements each can be formed from a set of 14 elements?

22. How many committees of 4 can be selected from a group of 12 citizens?

23. How many five-digit numerals can be formed from the digits 1, 2, 3, 4, 5, 6, and 7 if no digit may be repeated?

24. How many five-digit numerals can be formed from the digits 1, 2, 3, 4, 5, 6, and 7 if the digits may be repeated?

25. How many different committees of 5 persons can be chosen from 8 boys and 10 girls if each committee is to contain exactly 3 boys and 2 girls? (*Hint:* Calculate the male contingent, that is, 3 boys, and the female contingent, 2 girls, separately. Then use *PC.*)

26. How many different committees of 5 persons can be chosen from 8 boys and 10 girls if there must be 4 girls and 1 boy on each committee?

27. Change the conditions of Exercise 25 to the requirement that there be at least 3 boys on each committee. (*Hint:* Choices include 3 boys, 4 boys, or 5 boys.)

28. How many different football teams of 11 can be selected from a squad of 20 if any one of the 20 can play any position?

29. Vary the conditions of Exercise 28 by having 4 quarterbacks on the squad of 20 who can play no other position, whereas none of the other 16 is a potential quarterback.

30. There are 12 seats in a particular row in a theatre. How many different ways can 8 persons occupy these seats?

31. Determine the number of different arrangements of the letters in the word PARALLEL.

32. How many different arrangements can be made of the letters in the word ALGEBRAICALLY?

33. There are 12 teams in a league. How many games must be played so that each team will play every other team exactly twice?

34. It takes at least $\frac{2}{3}$ of the states in the United States to ratify an amendment to the constitution. Thus a minimum number of 34 states out of 50 is required. How many different combinations of 34 states are possible?

35. In how many different ways can a total of 10 be made with a throw of 3 dice?

36. Using nickels, dimes, and quarters, how many different collections of change can be used to make 50 cents?

37. The game of poker involves, among other things, dealing out a hand of 5 cards from a deck of 52. How many different poker hands are there?

38. **CHALLENGE PROBLEM.** An ordinary deck of cards contains 4 suits of 13 denominations each. A pair consists of two of any particular denomination, for example, two 3's or two kings. How many poker hands are there that have no pairs?

39. **CHALLENGE PROBLEM.** Prove that $\binom{n}{r} = \binom{n}{n-r}$ where $r \leq n$. Use $\binom{n}{r} = \dfrac{n!}{(n-r)!r!}$.

40. **CHALLENGE PROBLEM.** How many different ways can 8 persons occupy 12 seats in a row if the 8 persons consist of 4 men and 4 women and if it is required that between any 2 men there shall be a woman and between any 2 women there should be a man? (Note that it is not required that all people have adjacent seats.)

9.3 SOME PROBABILITY APPLICATIONS

Each of 20 capsules in a bowl encloses a slip of paper marked with a different number from the set of integers

$$S = \{1, 2, \ldots, 20\}.$$

A blindfolded person draws a capsule out of the bowl. Under such conditions we agree that any particular number is equally likely to be drawn as any other number. What are the chances that the number drawn will be less than 4? Since there are 3 numbers less than 4, and 20 numbers in all, it seems eminently reasonable to say that the chances are 3 in 20.

What are the chances of getting a number greater than 7? There are 13 numbers greater than 7 and 20 numbers in all, and so it seems reasonable to say that the chances are 13 in 20.

It is customary to put answers like these in the form of a ratio and say that the **probability** of drawing a number less than 4 is $\frac{3}{20}$.

Similarly, the probability of drawing a number greater than 7 is $\frac{13}{20}$.

Suppose that a bowl contains 12 marbles, of which 5 are black and 7 are white. If a marble is drawn at random, the probability that it is black is $\frac{5}{12}$. What is the probability that it will be white?

We can generalize these examples as follows:

> If a set S (usually called a **sample space**) contains n equally likely outcomes of an experiment and if A is a subset of S (usually called an **event**) containing m outcomes, then the **probability** of an outcome occurring which is in the set A is $\frac{m}{n}$.

Using P to stand for the probability function, $N(A)$ to represent the number of elements in any set A, and $N(S)$ to represent the number of elements in the sample space S, we can rewrite this statement as

$$P(A) = \frac{N(A)}{N(S)}.$$

If, in the capsule situation, A is the event that the number drawn is a multiple of 3, then, since there are 6 multiples of 3 between 1 and 20,

$$P(A) = \frac{6}{20} = \frac{3}{10}.$$

Often the numbers $N(A)$ and $N(S)$ are formidably large and, in such cases, it becomes expedient to utilize the counting devices discussed in Sections 9.1 and 9.2 when calculating a probability.

EXERCISES 9.3

Suppose that capsules in a bowl contain slips of paper each marked with a different arrangement of four letters from the word FORMIDABLE and that all such four-letter arrangements are included. Determine, in Exercises 1–5, the respective probabilities associated with each of the following events. The capsule drawn contains:

1. No vowel.
2. The letter F.
3. The letters A and B.

4. The letters F, O, and R.

5. The letter M or the letter D, but not both.

6. What is the probability of throwing a total of 5 with 2 dice?

7. What is the probability of drawing a heart out of an ordinary deck of cards?

8. A committee (see Exercise 25 of Exercises 9.2B) of 5 is chosen from 8 boys and 10 girls. What is the probability that the chosen committee will consist entirely of girls? (*Hint:* Let S be the set of all possible committees of 5, A the set of all possible female committees.)

9. What is the probability that in the situation described in Exercise 8 the chosen committee will be all male?

10. Refer to Exercise 8 and give the probability that the chosen committee will contain 3 boys and 2 girls.

11. A set contains 12 elements. A person is asked to select any subset at random. What is the probability that the set selected will contain 3 elements?

12. What is the probability that the randomly chosen set of Exercise 11 will contain more than 10 elements?

13. Show that if A is any event in a sample space S, then the probability of A not happening is equal to $1 - P(A)$. (*Hint:* Assume that A contains m elements and S contains n elements. How many elements of S are not in A?)

14. Refer to Exercises 8 and 13 and determine the probability that the chosen committee of 5 will not contain just girls.

15. What is the probability that the committee chosen in Exercise 8 will not be all male?

16. Six people consisting of 2 adults and 4 children have selected 6 adjacent seats in a movie house. What is the probability that there is one adult on each end?

17. Given the conditions of Exercise 16, what is the probability that the 2 adults are exactly in the middle?

18. Determine the probability that the total on a throw of 2 dice is not 7.

19. What is the probability that a card drawn from a deck is not a spade?

20. A bowl contains 7 white marbles, 8 red marbles, and 10 blue marbles. What is the probability that a marble drawn at random is not red?

21. A group contains 12 Democrats, 8 Republicans, and 5 Independents. Seven people are selected at random. What is the probability that 4 will be Democrats and 3 will be Republicans?

22. Given the conditions of Exercise 21, find the probability that the committee contains no Democrats.

23. **CHALLENGE PROBLEM.** An experiment consists of tossing a single die, then drawing a card from a deck. What is the probability that both of the resulting numbers will be the same? (*Hint:* $N(S)$ can be calculated using PC. $N(A)$ can be counted.)

24. **CHALLENGE PROBLEM.** Find the probability that if 5 cards (a poker hand) are drawn from a deck, they will be all of one suit. There are 4 suits with 13 cards in each suit. (*Hint:* The number of spade hands, for example, is $C(13, 5)$.)

25. **CHALLENGE PROBLEM.** A rapid system of transmitting a signal for a number uses a binary notation indicated by lights. With a row of 6 lights, for example, one can signal the number 25 as follows:

Here the colored circles indicate that the light is on. If buttons lighting the lights are activated at random (for example, a baboon plays with the machine), what are the probabilities of each of the following outcomes?
(a) The signal shows the number 63.
(b) The signal shows a number greater than 50.
(c) The signal shows a number less than 20.
(d) The signal shows a positive even number.
(e) The signal shows a number which is a positive multiple of 5.

9.4 THE BINOMIAL THEOREM

As we indicated earlier, the principle of choice enters the stage in many variations of role. You will recall its first appearance as an aid in computing the number of messages that could be sent out with four signal flags. We shall now see how it offers assistance in the problem of finding the nth power of a binomial.

The job of computing the square of a binomial is a familiar and relatively simple one. Any student of elementary algebra knows, for example, that

$$(x + y)^2 = x^2 + 2xy + y^2.$$

You may also have made use of the fact that

$$(x + y)^3 = x^3 + 3x^2y + 3xy^2 + y^3$$

or, even further, the fact that

$$(x + y)^4 = x^4 + 4x^3y + 6x^2 y^2 + 4xy^3 + y^4.$$

In all these cases the results can be easily attained by reiterated multiplication. In many applications, however, we are called upon to find the nth power of a binomial where n is quite large. If, for example, we were asked to expand $(x + y)^{12}$, we would certainly not want to use ordinary multiplication if we knew of some mathematical theories or strategies that could offer convenient shortcuts. To investigate some of these theories and strategies, let us begin by exploring some examples.

To find the product

$$(a + b)(c + d),$$

we apply the Distributive Property as follows:

$$(a + b)(c + d) = a(c + d) + b(c + d)$$
$$= ac + ad + bc + bd.$$

The result is a sum of the products ac, ad, bc, and bd, each product obtained by selecting a term from the first binomial and multiplying it by a term in the second binomial until all the possibilities have been exhausted.

For the product

$$(x + y)^2 = (x + y)(x + y)$$

we can do the same thing, that is,

$$(x + y)(x + y) = (x \cdot x) + (x \cdot y) + (y \cdot x) + (y \cdot y),$$

and for

$$(x + y)^3 = (x + y)(x + y)(x + y)$$

we can form products of three factors each by selecting again a term from each binomial until all options are used. This gives us initially

$$(x \cdot x \cdot x) + (y \cdot x \cdot x) + (x \cdot y \cdot x) + (x \cdot x \cdot y)$$
$$+ (y \cdot y \cdot x) + (y \cdot x \cdot y) + (x \cdot y \cdot y) + (y \cdot y \cdot y).$$

In both cases, as you know, some of the terms can be combined to produce a more compact form. Thus

$$(x \cdot x) + (x \cdot y) + (y \cdot x) + (y \cdot y) = x^2 + 2xy + y^2,$$

and

$$(x \cdot x \cdot x) + (y \cdot x \cdot x) + (x \cdot y \cdot x) + (x \cdot x \cdot y) + (y \cdot y \cdot x)$$
$$+ (y \cdot x \cdot y) + (x \cdot y \cdot y) + (y \cdot y \cdot y) = x^3 + 3x^2y + 3xy^2 + y^3.$$

For the product

$$(x + y)^5 = (x + y)(x + y)(x + y)(x + y)(x + y)$$

the same principle applies, although the selections are more numerous. Here again we make a series of choices, one letter from each binomial until all the options are used. From an earlier discussion we know that the total number of possibilities is $2^5 = 32$ since there are two "decisions" to make for each binomial and there are five binomials.

Spread out to its full length, then, the expanded product for $(x + y)^5$ would consist of a series of 32 terms. However, as we shall see, many of these terms can be combined.

Let's begin, then, by exhibiting a few choices. We might start by showing a preference for x's, that is, choose an x from each binomial. This would give us as our first product,

$$x \cdot x \cdot x \cdot x \cdot x = x^5.$$

Next, we might allow one y to enter the picture. Clearly, there are several ways of doing this. Here they are:

$$y \cdot x \cdot x \cdot x \cdot x, \qquad x \cdot y \cdot x \cdot x \cdot x, \qquad x \cdot x \cdot y \cdot x \cdot x,$$
$$x \cdot x \cdot x \cdot y \cdot x, \qquad x \cdot x \cdot x \cdot x \cdot y.$$

Since we know by the Commutative Property of multiplication that all of these products are equal, we can condense the exhibit by writing the sum of all of these five terms as $5x^4y$. Thus far, then, we have $x^5 + 5x^4y$. It seems orderly now to make selections involving 2 y's. Before we proceed, however, let's take stock. How many choices are there for terms involving 3 x's and 2 y's? Rather than count them, let's utilize the mathematical ideas we have been discussing earlier in the chapter. What we are really asking here is the number of ways of arranging the letters $xxxyy$, that is, of arranging 5 objects of which 3 are alike and 2 are alike. From the discussion in Section 9.2 we recall that a formula for this is $\dfrac{n!}{n_1!n_2!}$, which, in our case, where $n = 5$, $n_1 = 3$, $n_2 = 2$, becomes

$$\frac{5!}{3!2!} = 10.$$

There is an alternative way of interpreting the problem of arrangements of $xxxyy$. If we number the positions as follows,

$$
\begin{array}{ccccc}
x & x & x & y & y \\
1 & 2 & 3 & 4 & 5,
\end{array}
$$

we can think of the arrangements in terms of the positions occupied by the x's since the y's merely fill in the unoccupied spaces. In this arrangement the x's are in positions 1, 2, 3. In the arrangement

$$x \quad y \quad y \quad x \quad x$$

the x's occupy positions 1, 4, 5. How many choices of position are there for 3 x's or, equivalently, in how many ways can you choose 3 numbers from the set $\{1, 2, 3, 4, 5\}$? No difficulty here! How many subsets of 3 elements can you form from a set of 5 elements? The answer is certainly

$$C(5, 3) = \binom{5}{3} = 10.$$

We seem to have come up with two different ways to obtain an answer for the same problem. There is no conflict involved, however. The number of arrangements of n objects consisting of two kinds, n_1 of the first kind and n_2 of the second, is given by the formula

$$\frac{n!}{n_1! n_2!}$$

where $n_1 + n_2 = n$.

On the other hand, the number of subsets of n_1 elements that can be formed from a set of n elements is

$$C(n, n_1) = \frac{P(n, n_1)}{n_1!}.$$

But we know that

$$P(n, n_1) = \frac{n!}{(n - n_1)!}.$$

Thus

$$\frac{P(n, n_1)}{n_1!} = \frac{n!}{n_1!(n - n_1)!}.$$

Since $n_1 + n_2 = n$ and hence $n - n_1 = n_2$, we see that

$$\frac{n!}{n_1!(n - n_1)!} = \frac{n!}{n_1! n_2!}.$$

Thus we have

$$C(n, n_1) = \binom{n}{n_1} = \frac{n!}{n_1! n_2!}$$

if $n_1 + n_2 = n$.

We can conclude, then, that the number of arrangements of n objects of two kinds, n_1 of the first and n_2 of the second, is equal to the number of subsets of n_1 elements that can be formed from a set of n elements.

In the binomial product,

$$(x + y)^5 = (x + y)(x + y)(x + y)(x + y)(x + y),$$

we see that we have 32 products of 6 different types. These types contain 5, 4, 3, 2, 1, and 0 x's, respectively.

To condense the 32 terms into compact form we need, finally, to know how many of each type there are. But this has already been shown to depend on the number of x's in each type. For example, we can see that the number of terms containing two x's is

$$C(5, 2) = \binom{5}{2} = \frac{5!}{2!3!} = \frac{5 \cdot 4}{2 \cdot 1} = 10.$$

Thus we can write $(x + y)^5$ as

$$\binom{5}{5}x^5 + \binom{5}{4}x^4y + \binom{5}{3}x^3y^2 + \binom{5}{2}x^2y^3 + \binom{5}{1}xy^4 + \binom{5}{0}y^5.$$

Here the coefficient $\binom{5}{4}$, for example, is the number of subsets of 4 elements which can be formed from a set of 5 elements. It follows that in each term the i in the symbol $\binom{5}{i}$ is equal to the corresponding exponent of x.

A word, perhaps, is in order about the form of the final coefficient, $\binom{5}{0}$. Conceptually, this offers no problem since we may interpret it as the number of subsets with no elements. There being one and only one empty set $\varnothing$ it is clear that $\binom{5}{0} = 1$. Also, however, the formula

$$\binom{n}{r} = \frac{n!}{r!(n - r)!}$$

gives us

$$\frac{n!}{0!(n - 0)!} = \frac{1}{0!}$$

when $r = 0$. Now we have already, in Section 9.2, defined $0! = 1$.

Hence, for $r = 0$ we have

$$\binom{n}{r} = \frac{1}{1} = 1$$

as desired.

On the basis of the foregoing discussion, it seems reasonable to generalize these results to the expansion of $(x + y)^n$, where n is any natural number. At this time, however, we shall not prove our conjecture. This proof will be undertaken in Section 9.6. Meanwhile let us accept as highly plausible the general formula, known as the **Binomial Theorem,**

$$(1) \quad (x + y)^n = \binom{n}{n}x^n + \binom{n}{n-1}x^{n-1}y + \binom{n}{n-2}x^{n-2}y^2 + \cdots$$

$$+ \binom{n}{1}xy^{n-1} + \binom{n}{0}y^n.$$

For example,

$$(x + y)^{12} = \binom{12}{12}x^{12} + \binom{12}{11}x^{11}y + \binom{12}{10}x^{10}y^2 + \cdots$$

$$+ \binom{12}{1}xy^{11} + \binom{12}{0}y^{12}.$$

Using Σ notation we can write (1) in compact form as

$$(x + y)^n = \sum_{i=0}^{n} \binom{n}{n-i}x^{n-i}y^i.$$

The symbol on the right may seem complicated at first. However, if we take a numerical example, such as the one above where $n = 12$ and allow i to range over the whole numbers from 0 to 12, the pattern should emerge. Thus

$$\sum_{i=0}^{12} \binom{12}{12-i}x^{12-i}y^i = \binom{12}{12-0}x^{12-0}y^0 + \binom{12}{12-1}x^{12-1}y$$

$$+ \binom{12}{12-2}x^{12-2}y^2 + \cdots,$$

which is consistent with the result previously obtained.

Since the coefficients $\binom{n}{r}$ play such a prominent role in the process of binomial expansion, it seems in order to look for possible shortcuts in computing such numbers. The way can often be shortened by using the relation

$$\binom{n}{r} = \binom{n}{n-r}.$$

The proof of this relation (which you may already have made in Challenge Problem 39 of Exercises 9.2B) rests on the fact that

$$\binom{n}{r} = \frac{n!}{r!(n-r)!}.$$

Likewise

$$\binom{n}{n-r} = \frac{n!}{(n-r)![n-(n-r)]!} = \frac{n!}{(n-r)!r!}.$$

Since, clearly, $\dfrac{n!}{(n-r)!r!} = \dfrac{n!}{r!(n-r)!}$, the proof is complete.

Using this relationship, we see that a term such as $\binom{12}{9}$ can be computed more easily by noting that $\binom{12}{9} = \binom{12}{3}$. Similarly, $\binom{10}{8} = \binom{10}{2}$, $\binom{15}{11} = \binom{15}{4}$, and so on.

The relationship, $\binom{n}{r} = \binom{n}{n-r}$, enables us to write the Binomial Theorem in two forms:

(1) $(x + y)^n = \binom{n}{n}x^n + \binom{n}{n-1}x^{n-1}y + \binom{n}{n-2}x^{n-2}y^2 + \cdots$

$$+ \binom{n}{1}xy^{n-1} + \binom{n}{0}y^n$$

and

(2) $(x + y)^n = \binom{n}{0}x^n + \binom{n}{1}x^{n-1}y + \binom{n}{2}x^{n-2}y^2 + \cdots$

$$+ \binom{n}{n-1}xy^{n-1} + \binom{n}{n}y^n.$$

In Σ notation these two forms are

$$(x + y)^n = \sum_{i=0}^{n} \binom{n}{n-i} x^{n-i} y^i \quad \text{and} \quad (x + y)^n = \sum_{i=0}^{n} \binom{n}{i} x^{n-i} y^i,$$

respectively.

As an illustration of the use of form (2) let us expand $(x + 1)^8$. We have

$(x + 1)^8$

$$= \sum_{i=0}^{8} \binom{8}{i} x^{8-i} 1^i$$

$$= \binom{8}{0} x^8 + \binom{8}{1} x^7 \cdot 1 + \binom{8}{2} x^6 \cdot 1^2 + \binom{8}{3} x^5 \cdot 1^3 + \binom{8}{4} x^4 \cdot 1^4$$

$$+ \binom{8}{5} x^3 \cdot 1^5 + \binom{8}{6} x^2 \cdot 1^6 + \binom{8}{7} x \cdot 1^7 + \binom{8}{8} \cdot 1^8$$

$$= x^8 + 8x^7 + 28x^6 + 56x^5 + 70x^4 + 56x^3 + 28x^2 + 8x + 1.$$

Note how the relationship $\binom{n}{r} = \binom{n}{n-r}$ also provides a kind of symmetry in the array of coefficients.

EXERCISES 9.4A

◀ In Exercises 1–5, write the expansions of the given binomials.

1. $(x + y)^5$ 4. $(y + 1)^9$
2. $(a + b)^6$ 5. $(p + q)^{10}$
3. $(x + 1)^7$

6. Write in Σ notation the expansion of $(a + b)^{12}$ using form (1).
7. Give the first four terms of the expansion of $(1 + x)^{15}$.
8. Give the fifth term of $(y + 1)^{18}$.
9. Give the tenth term of $(r + s)^{12}$.
10. Give the seventh term of $(x + 1)^{13}$.
11. Give the fourth term of $(m + n)^{20}$.
12. Write the last three terms of $(x + y)^{18}$.
13. Write in Σ notation the expansion of $(p + q)^{11}$ using form (2).
14. Write the first four terms of $(x + 1)^{14}$.

15. Write the last four terms of $(1 + y)^{20}$.

16. Give the term of $(x + y)^{12}$ which has y^5 as a factor.

17. Give the two terms of $(a + b)^{11}$ which have 330 as coefficient.

18. Give the terms of $(x + y)^{14}$ which have 1001 as coefficient.

19. Give the first three terms of $(a + 1)^{25}$.

20. The binomial $(1 + 1)^7$, if expanded formally, could be written as

$$\binom{7}{0} + \binom{7}{1} + \binom{7}{2} + \cdots + \binom{7}{7}.$$

The terms $\binom{7}{2}, \binom{7}{3}$, etc., give the number of subsets with 2 elements, 3 elements, etc., which can be formed from a set of 7 elements. Discuss how this idea can be used to determine the total number of subsets of a given set.

The Binomial Theorem provides a formal rule for finding the terms in a binomial expansion. As is often the case, however, a formal rule, although essential mathematically, does not always provide the most efficient means of computation. Let us examine once more the expansion, in form (2), of $(x + y)^n$:

$$(x + y)^n = \binom{n}{0}x^n + \binom{n}{1}x^{n-1}y + \binom{n}{2}x^{n-2}y^2 + \binom{n}{3}x^{n-3}y^3 + \cdots$$

If we write this as

$$x^n + \frac{n}{1}x^{n-1}y + \frac{n(n-1)}{1 \cdot 2}x^{n-2}y^2 + \frac{n(n-1)(n-2)}{1 \cdot 2 \cdot 3}x^{n-3}y^3$$
$$+ \frac{n(n-1)(n-2)(n-3)}{1 \cdot 2 \cdot 3 \cdot 4}x^{n-4}y^4 + \cdots,$$

we can see a recursive relationship between any given coefficient and the one immediately following. Consider, for example, the second, third, and fourth terms,

$$\frac{n}{1}x^{n-1}y, \quad \frac{n(n-1)}{1 \cdot 2}x^{n-2}y^2, \quad \text{and} \quad \frac{n(n-1)(n-2)}{1 \cdot 2 \cdot 3}x^{n-3}y^3.$$

The coefficient, $\frac{n}{1}$, of the second term becomes the next coefficient, $\frac{n(n-1)}{2}$, when multiplied by $\frac{n-1}{2}$. If we now multiply

$\dfrac{n(n-1)}{1 \cdot 2}$ by $\dfrac{n-2}{3}$, we get the next coefficient. What must we multiply

$$\frac{n(n-1)(n-2)}{1 \cdot 2 \cdot 3}$$

by to get the coefficient,

$$\frac{n(n-1)(n-2)(n-3)}{1 \cdot 2 \cdot 3 \cdot 4},$$

in the following (fifth) term?

What is the coefficient of the term that follows

$$\frac{n(n-1)(n-2)(n-3)(n-4)(n-5)}{1 \cdot 2 \cdot 3 \cdot 4 \cdot 5 \cdot 6} x^{n-6}y^6?$$

This could be obtained by multiplying by $n-6$ and dividing by 7. But $n-6$ is the exponent of x, and 7 is one more than the exponent of y.

Consider the following two consecutive terms in the expansion of $(x+y)^8$:

$$56x^5y^3 \quad \text{and} \quad 70x^4y^4.$$

To obtain the coefficient 70, we multiply 56 by 5 and divide by 4 since 5 is the exponent of x, and 4 is one more than the exponent of y in the term $56x^5y^3$. (Note that $\dfrac{56 \cdot 5}{4} = 70$.)

What is the coefficient of the term that follows $210x^6y^4$ in the expansion of $(x+y)^{10}$? From previous considerations this should be

$$\frac{210 \cdot 6}{5} = 252.$$

We know that $\dbinom{10}{4} = 210$. Check to see if $\dfrac{210 \cdot 6}{5} = \dbinom{10}{5}$.

EXERCISES 9.4B

◀ In each of Exercises 1–10 a particular term of an indicated expansion is given. Find the next term.

1. $(x + y)^7$, $35x^4y^3$ **6.** $(x + y)^{14}$, $2002x^9y^5$

2. $(a + b)^{10}$, $120a^7b^3$ **7.** $(x + 1)^8$, $56x^3$

3. $(r + s)^{12}$, $495r^4s^8$ **8.** $(a + 1)^{10}$, $252a^5$

4. $(x + y)^{11}$, $165x^8y^3$ **9.** $(1 + z)^{12}$, $220z^3$

5. $(a + b)^{15}$, $105x^{13}y^2$ **10.** $(1 + b)^{20}$, $1140b^3$

9.5 MORE ON THE BINOMIAL THEOREM

Thus far we have confined our attention to the expansion of simple binomials such as $x + y$ and $x + 1$. Certain modifications take place when the terms x and/or y are replaced by more complicated expressions.

To find $(a - b)^5$, for example, we can use form (2) with x replaced by a and y replaced by $(-b)$. In this case you must remember that $(-b)^i = -b^i$ if i is odd, and $(-b)^i = b^i$ when i is even.

To find the first four terms in the expansion of $(a - 3b)^7$, we can write

$$(a - 3b)^7$$
$$= \binom{7}{0}a^7 + \binom{7}{1}a^6(-3b) + \binom{7}{2}a^5(-3b)^2 + \binom{7}{3}a^4(-3b)^3 + \cdots$$
$$= a^7 + 7a^6(-3b) + 21a^5(-3b)^2 + 35a^4(-3b)^3 + \cdots$$
$$= a^7 - 21a^6b + 189a^5b^2 - 945a^4b^3 + \cdots .$$

To determine a specified term of any given binomial expansion, such as the fifth term of $(2x - 5)^{10}$, without carrying out the entire expansion, we can proceed as follows. Form (2) tells us that the fifth term of $(x + y)^n$ is $\binom{n}{4}x^{n-4}y^4$. We now replace n by 10, x by $2x$, and y by -5. By these substitutions we get

$$\binom{10}{4}(2x)^6(-5)^4 = 210(2x)^6(-5)^4.$$

Finally, since $(2x)^6 = 64x^6$ and $(-5)^4 = 625$, we have

$$\binom{10}{4}(2x)^6(-5)^4 = 210 \cdot 64 \cdot 625x^6 = 8,400,000x^6.$$

(This may have seemed like a big job, but it surely does not compare with the task of finding the fifth term of $(2x - 5)^{10}$ by repeated multiplications!)

A word of advice may be timely at this point. Consider the expansion

$$(x + y)^7 = \binom{7}{0}x^7 + \binom{7}{1}x^6y + \binom{7}{2}x^5y^2 + \binom{7}{3}x^4y^3 + \binom{7}{4}x^3y^4$$

$$+ \binom{7}{5}x^2y^5 + \binom{7}{6}xy^6 + \binom{7}{7}y^7.$$

If, for example, we were asked to give the fifth term, we might be inclined to associate the fifth term with the number 5 in some way. However, if we look at the actual expansion, we see that the fifth term is

$$\binom{7}{4}x^3y^4$$

where the number 5 is not much in evidence. The catch is the fact that the exponent of y is always one *less* than the number of the term in which it appears. This, of course, is due to the fact that the first term, x^7, does not have a y factor. Thus y^1 is in the second term, y^2 is in the third, and so forth. Generally, the following rule-of-thumb may be helpful in determining the kth term of $(x + y)^n$. First think of $k - 1$ as the exponent of y. Then in the coefficient $\binom{n}{i}$, i is equal to the exponent of y, that is, $i = k - 1$ also. For the exponent of x we note that the sum of the exponents of x and y is always equal to n. Hence we write, for the kth term,

$$\binom{n}{k - 1}x^{[n-(k-1)]}y^{k-1}.$$

For example, the ninth term of the expansion of $(x + y)^{12}$ is

$$\binom{12}{8}x^4y^8$$

and the fourth term of $(x + y)^{10}$ is

$$\binom{10}{3}x^7y^3.$$

The Binomial Theorem can be used to determine approximate values of powers of a number such as 1.01. For example, if we wish to approximate $(1.01)^5$, we can write

$$(1.01)^5 = (1 + 0.01)^5$$

and use form (2). This gives us, if we use only the first four terms,

$$(1 + 0.01)^5 \approx \binom{5}{0} + \binom{5}{1}(0.01) + \binom{5}{2}(0.01)^2 + \binom{5}{3}(0.01)^3$$
$$= 1 + 5(0.01) + 10(0.0001) + 10(0.000001)$$
$$= 1.05101.$$

Since $(0.01)^n$ becomes very small as n increases, the first four terms of the binomial expansion of $(1 + 0.01)^5$ furnish a fairly good approximation to $(1.01)^5$. For an expression such as $(0.98)^5$ we write $(1 - 0.02)^5$. (In such a case we must remember that $(-0.02)^i$ is negative when i is odd.)

EXERCISES 9.5

◀ In Exercises 1–10, find the first three terms of each expansion.

1. $(2x - y)^8$
2. $(x - 2y)^8$
3. $(a - b)^9$
4. $(m + n)^{11}$
5. $(m^2 - 1)^7$

6. $(1 - x^2)^8$
7. $(2x + y)^7$
8. $(x + 2y)^{10}$
9. $(3x^2 + y)^8$
10. $(x - y)^{12}$

◀ In Exercises 11–25, expand each binomial and simplify each term of the expansion.

11. $(3x - 2y)^4$
12. $(2a - b)^5$
13. $(m - n)^5$
14. $(r - s)^7$

15. $(u - 3v)^6$

16. $(2x - 3y)^4$

17. $(m - 2n)^5$
18. $(m^2 - n)^6$

19. $(2m - 1)^8$
20. $(x^2 - 1)^5$
21. $(x^2 - x)^4$
22. $(x - 2y)^6$

23. $\left(x - \dfrac{y}{2}\right)^6$

24. $\left(\dfrac{x}{3} - y^2\right)^4$

25. $(2x^2 - 3y^2)^5$

◀ In Exercises 26–40, find and simplify the designated term.

26. 3rd term of $(x + y)^6$
27. 4th term of $(a - b)^8$
28. 5th term of $(a - 2b)^6$

29. 8th term of $(2x - y)^{10}$
30. 3rd term of $(r - s)^{12}$
31. 6th term of $(m^3 - n^2)^9$

32. 7th term of $(a^2 - b)^9$

33. 5th term of $\left(\dfrac{3}{x} - \dfrac{x}{3}\right)^8$

34. 4th term of $\left(x + \dfrac{y}{2}\right)^7$

35. 4th term of $\left(\dfrac{1}{x} - x^2\right)^6$

36. 6th term of $\left(\dfrac{r}{s} + s\right)^{10}$

37. 3rd term of $(3r^2 - s)^5$

38. 5th term of $(x^2 - y^2)^9$

39. Middle term of $(2x - y)^8$

40. Middle term of $(a - b^2)^{12}$

◀ In Exercises 41–49, find an approximation for the indicated power by using the first four terms of a binomial expansion.

41. $(1.02)^4$

42. $(0.98)^6$

43. $(1.01)^6$

44. $(0.97)^4$

45. $(1.02)^6$

46. $(0.99)^5$

47. $(0.98)^7$

48. $(1.03)^5$

49. $(1.01)^8$

50. CHALLENGE PROBLEM. Compute

(a)
$$\binom{8}{5}, \binom{8}{4}, \text{ and } \binom{9}{5}.$$

Also compute

(b)
$$\binom{10}{4}, \binom{10}{3}, \text{ and } \binom{11}{4}.$$

On the basis of the results of (a) and (b) make a conjecture regarding the following equality:

$$\binom{n}{r} + \binom{n}{r-1} = \binom{?}{?}$$

9.6 THE BINOMIAL THEOREM AND INDUCTION

In Section 9.4 the formula

(1) $(x + y)^n$
$$= \binom{n}{n}x^n + \binom{n}{n-1}x^{n-1}y + \cdots + \binom{n}{1}xy^{n-1} + \binom{n}{0}y^n$$

was inferred by considering the cases when $n = 1, 2, 3, 4,$ or 5.

We now present the major part of a proof by induction that (1) holds for any natural number n. A few details, however, are left as an exercise for the student. Let us begin in the usual way by letting S be the subset of the set of all natural numbers for which (1) holds.

We wish to show that $S = N$. To do this, as you recall, we must first show that $1 \in S$. This can readily be done by observing that

(2) $$(x + y)^1 = \binom{1}{1}x^1 + \binom{1}{0}y^1$$

since $\binom{1}{1} = \binom{1}{0} = 1$. Our next task is to show that the set S is inductive. To do this we must show that the assumption that $k \in S$ leads to the conclusion that $k + 1$ is also in S.

The assumption $k \in S$ says that

(3) $$(x + y)^k = \binom{k}{k}x^k + \binom{k}{k-1}x^{k-1}y + \cdots + \binom{k}{1}xy^{k-1} + \binom{k}{0}y^k.$$

Since we know that

$$(x + y)(x + y)^k = (x + y)^{k+1},$$

it seems like a good idea to multiply both sides of Equation (3) by $(x + y)$. This gives us $(x + y)^{k+1}$ on the left, which is what we want. What happens on the right? The product

$$(x + y)\left[\binom{k}{k}x^k + \binom{k}{k-1}x^{k-1}y + \binom{k}{k-2}x^{k-2}y^2 + \cdots \right.$$
$$\left. + \binom{k}{1}xy^{k-1} + \binom{k}{0}y^k \right]$$

becomes, by the Distributive Property,

$$x\left[\binom{k}{k}x^k + \binom{k}{k-1}x^{k-1}y + \binom{k}{k-2}x^{k-2}y^2 + \cdots \right.$$
$$\left. + \binom{k}{1}xy^{k-1} + \binom{k}{0}y^k \right]$$
$$+ y\left[\binom{k}{k}x^k + \binom{k}{k-1}x^{k-1}y + \binom{k}{k-2}x^{k-2}y^2 + \cdots \right.$$
$$\left. + \binom{k}{1}xy^{k-1} + \binom{k}{0}y^k \right].$$

This gives us

$$\left[\binom{k}{k}x^{k+1} + \binom{k}{k-1}x^k y + \binom{k}{k-2}x^{k-1}y^2 + \cdots\right.$$
$$\left. + \binom{k}{1}x^2 y^{k-1} + \binom{k}{0}xy^k\right]$$

$$+ \left[\binom{k}{k}x^k y + \binom{k}{k-1}x^{k-1}y^2 + \binom{k}{k-2}x^{k-2}y^3 + \cdots\right.$$
$$\left. + \binom{k}{1}xy^k + \binom{k}{0}y^{k+1}\right].$$

Adding

$$\binom{k}{k-1}x^k y \text{ and } \binom{k}{k}x^k y, \ \binom{k}{k-2}x^{k-1}y^2 \text{ and } \binom{k}{k-1}x^{k-1}y^2,$$

etc., we obtain, by the Distributive Property,

$$(4) \quad \binom{k}{k}x^{k+1} + \left[\binom{k}{k-1} + \binom{k}{k}\right]x^k y + \left[\binom{k}{k-2} + \binom{k}{k-1}\right]x^{k-1}y^2$$
$$+ \left[\binom{k}{k-3} + \binom{k}{k-2}\right]x^{k-2}y^3 + \cdots + \left[\binom{k}{0} + \binom{k}{1}\right]xy^k + \binom{k}{0}y^{k+1}.$$

Before proceeding further we should perhaps form a picture of what the formula should look like if it is to validate the conclusion that $k + 1 \in S$. On the basis of form (1) we should have

$$(5) \quad (x + y)^{k+1} = \binom{k+1}{k+1}x^{k+1} + \binom{k+1}{k}x^k y + \binom{k+1}{k-1}x^{k-1}y^2 + \cdots$$
$$+ \binom{k+1}{1}xy^k + \binom{k+1}{0}y^{k+1}$$

if $k + 1 \in S$. Our job, then, is to show that (4) is identical with the right-hand side of (5). To do this let us first dispose of the first and last terms of (4). Since

$$\binom{k}{k} = \binom{k+1}{k+1} = \binom{k}{0} = \binom{k+1}{0} = 1$$

because

$$\binom{n}{n} = \binom{n}{0} = 1 \qquad \text{for all } n \in N,$$

it follows that the first and last terms of (4) are satisfactory.

There remains, then, the task of showing that

$$\binom{k}{k} + \binom{k}{k-1} = \binom{k+1}{k},$$

$$\binom{k}{k-1} + \binom{k}{k-2} = \binom{k+1}{k-1},$$

and so forth, which means, in general, that we need to show that

$$\binom{n}{r} + \binom{n}{r-1} = \binom{n+1}{r}$$

for all natural numbers n and r with $r \leq n$. (See Exercise 50 of Exercises 9.5.)

The proof of this result is quite straightforward. You will be given an opportunity to construct a proof in the Exercises. It should be clear that when this final step is achieved, Formula (1) (that is, the Binomial Theorem) will have been proved by induction to hold for all natural numbers.

The formula

$$\binom{n}{r} = \frac{n!}{r!(n-r)!}$$

appearing in the Binomial Theorem was first used to compute the number of subsets of r elements in a set of n elements. Thus far it has been meaningful only when n is a natural number and r is a whole number. (Remember that we have $\binom{n}{0} = 1$ since we defined $0! = 1$. See page 385.)

If we now consider the expression

$$\binom{n}{r} = \frac{n!}{r!(n-r)!} = \frac{n(n-1)\cdots(n-r+1)(n-r)!}{r!(n-r)!}$$
$$= \frac{n(n-1)(n-2)\cdots(n-r+1)}{r!}$$

apart from the subset context, then we can evaluate $\binom{n}{r}$ when n is any nonzero rational number and r is a natural number. For example, we can write

$$\binom{-3}{4} = \frac{(-3)(-4)(-5)(-6)}{4 \cdot 3 \cdot 2 \cdot 1} = 15$$

when $n = -3$, $r = 4$, and $n - r + 1 = -3 - 4 + 1 = -6$;

$$\binom{-2}{5} = \frac{(-2)(-3)(-4)(-5)(-6)}{5 \cdot 4 \cdot 3 \cdot 2 \cdot 1} = -6$$

when $n = -2$, $r = 5$, and $n - r + 1 = -2 - 5 + 1 = -6$; and

$$\binom{\frac{1}{2}}{4} = \frac{(\frac{1}{2})(-\frac{1}{2})(-\frac{3}{2})(-\frac{5}{2})}{4 \cdot 3 \cdot 2 \cdot 1} = -\frac{5}{128}$$

when $n = \frac{1}{2}$, $r = 4$, and

$$n - r + 1 = \frac{1}{2} - 4 + 1 = -\frac{5}{2}.$$

Such expressions play a part in the development of mathematical formulas that have various useful applications.

EXERCISES 9.6

1. Copy and complete the following proof that

$$\binom{n}{r} + \binom{n}{r-1} = \binom{n+1}{r}.$$

(1) $$\binom{n}{r} = \frac{n(n-1)\cdots(n-r+2)(n-r+1)}{1 \cdot 2 \cdot 3 \cdot \ldots \cdot (r-1)r}$$

(2) $$\binom{n}{r-1} = \frac{n(n-1)\cdots(n-r+3)(\boxed{?})}{1 \cdot 2 \cdot 3 \cdot \ldots \cdot (r-2)(\boxed{?})}$$

(3) $$\binom{n+1}{r} = \frac{(n+1)(n)\cdots(n-r+3)(\boxed{?})}{1 \cdot 2 \cdot 3 \cdot \ldots \cdot (r-1)(\boxed{?})}$$

(4) To form the sum of $\binom{n}{r}$ and $\binom{n}{r-1}$ in (1) and (2) we use the common denominator $1 \cdot 2 \cdot 3 \cdot \ldots \cdot r$. Thus

$$\binom{n}{r} + \binom{n}{r-1}$$

$$= \frac{n(n-1)\cdots(n-r+2)(n-r+1)}{1 \cdot 2 \cdot 3 \cdot \ldots \cdot r}$$

$$+ \frac{n(n-1)\cdots(n-r+3)(n-r+2)}{1 \cdot 2 \cdot 3 \cdot \ldots \cdot (r-1)} \cdot \frac{r}{r}$$

$$= \frac{[n(n-1)\cdots(n-r+2)(n-r+1)] + [n(n-1)\cdots(n-r+2)(\boxed{?})]}{1 \cdot 2 \cdot 3 \cdot \ldots \cdot r}.$$

(5) Both expressions in brackets have the common factors $n(n-1)\cdots$ $(n-r+2)$; hence the numerator can be written as

$$[n(n-1)\cdots(n-r+2)][(n-r+1)+r].$$

The expression within the brackets on the right in (5) is equal to $\boxed{?}$.
(6) Hence this numerator is equal to

$$(n+1)(n)(n-1)\cdots(n-r+2).$$

(7) Thus $\binom{n}{r}+\binom{n}{r-1}$ is equal to

$$\frac{(n+1)(n)(n-1)\cdots(n-r+2)}{1\cdot2\cdot3\cdot\ \ldots\ \cdot\boxed{?}}$$

which, by (3), is equal to $\left(\dfrac{\boxed{?}}{\boxed{?}}\right)$. Hence the demonstration is complete.

◀In Exercises 2–8, evaluate the given expression.

2. $\binom{\frac{1}{2}}{5}$ 6. $\binom{-\frac{3}{4}}{5}$

3. $\binom{\frac{1}{4}}{3}$ 7. $\binom{-8}{6}$

4. $\binom{-3}{5}$ 8. $\binom{-\frac{3}{2}}{6}$

5. $\binom{-\frac{1}{2}}{4}$

9. Determine the value of $\binom{-1}{n}$ when n is any odd natural number.

10. Determine the value of $\binom{-1}{n}$ when n is any even natural number.

CHAPTER SUMMARY

THE PRINCIPLE OF CHOICE is as follows:

Let $A_1, A_2, \ldots, A_n$ be a sequence of k situations in which choices need to be made. If N_1 is the number of different choices (decisions) to be made with respect to A_1, if N_2 is the number of choices with respect to A_2, ..., and if N_k is the number of choices with respect to A_k, then the total number of possible sequences of choices, one from each of the k situations, is

$$N_1 \cdot N_2 \cdot N_3 \cdot \cdots \cdot N_k.$$

The following results have been deduced.

1. The number of subsets of a set containing n elements is 2^n.
2. The number of arrangements of n distinct objects taken r at a time is given by the formula

$$P(n, r) = n(n - 1) \cdots (n - r + 1)$$

or

$$P(n, r) = \frac{n!}{(n - r)!}.$$

3. The number of arrangements of n objects, not necessarily distinct, with n_i the number of copies of the ith kind such that $n_1 + n_2 + \cdots + n_k = n$ is given by the formula

$$\frac{n!}{n_1! n_2! \cdots n_k!}.$$

4. The number of subsets of r elements which can be formed from a set of n elements is given by the formula

$$C(n, r) = \frac{P(n, r)}{r!}.$$

Alternative notations are

$$C(n, r) = \binom{n}{r} = \frac{n!}{(n - r)! r!}.$$

5. The BINOMIAL THEOREM was inferred from the results of items (3) and (4) and proved by induction. It states that for any natural number n,

$$(x + y)^n = \binom{n}{n} x^n + \binom{n}{n - 1} x^{n-1} y + \cdots + \binom{n}{1} x y^{n-1} + \binom{n}{0} y^n.$$

An alternative notation is achieved for the coefficients by the substitution of $\binom{n}{n - r}$ for $\binom{n}{r}$. This gives us

$$(x + y)^n = \binom{n}{0} x^n + \binom{n}{1} x^{n-1} y + \cdots + \binom{n}{n - 1} x y^{n-1} + \binom{n}{n} y^n.$$

The forms in Σ notation are, respectively,

$$(x + y)^n = \sum_{i=0}^{n} \binom{n}{n-i} x^{n-i} y^i \quad \text{and} \quad (x + y)^n = \sum_{i=0}^{n} \binom{n}{i} x^{n-i} y^i.$$

The chapter concluded with a brief examination of the evaluation of $\binom{n}{r}$ when n is a nonzero rational number and r is a natural number by use of the formula

$$\binom{n}{r} = \frac{n(n - 1)(n - 2) \cdots (n - r + 1)}{r!}.$$

REVIEW EXERCISES

1. How many different arrangements can be made from 12 different letters taken 4 at a time?
2. Find the number of different arrangements of the letters in the word POLYNOMIAL.
3. How many different messages can be sent using 15 different flags if each message consists of 3 flags in a certain order?
4. How many different four-digit numbers can be formed from the digits 1, 2, 3, 4, 5, 6 if (a) the digits cannot be repeated, (b) the digits can be repeated?
5. Give the number of subsets of a set of 11 elements.
6. How many committees of 6 can be formed from 7 Democrats and 8 Republicans if each party is to have equal representation?
7. How many subsets of 7 elements can be formed from a set of 15?
8. What is the probability of throwing a total of 6 with 2 dice?

◄ Each of 50 capsules in a bowl contains a slip of paper with a different number from 1 to 50 written on it. In Exercises 9–12, determine the probability of the indicated result.

9. The number on the slip will be a multiple of 3.
10. The number on the slip will be a multiple of 7.
11. The number on the slip will be a prime number.
12. The number on the slip is not a multiple of 5.
13. Calculate $\binom{12}{4}$.
14. Calculate $P(11, 6)$ and $C(11, 6)$.
15. Find the value of $P(15, 4)$.

16. Compute $\binom{\frac{1}{2}}{6}$ and $\binom{-4}{7}$.

17. Write the expansion of $(r + 25)^5$ and simplify each term.

18. Write the expansion of $(2x - \frac{1}{2}y)^6$ and simplify each term.

19. Give in simplified form the fifth term of $(x - 4y)^{11}$.

20. Give the fourth term of $(1 - x)^{12}$.

21. Give in simplified form the last 3 terms of $(2z - 1)^{20}$.

22. Write the terms of $(x + y)^{15}$ which have 1365 as coefficient.

23. Write the term of $(a - b)^{12}$ which has -924 as coefficient.

24. Using four terms of a binomial expansion, determine an approximate value of $(0.98)^6$.

25. Using four terms of a binomial expansion determine an approximate value of $(1.03)^4$.

GOING FURTHER: READING AND RESEARCH

The Binomial Theorem has been confined in this chapter to the expansion of $(x + y)^n$ for $n \in N$. It is possible, however, to extend the Binomial Theorem to include expansions in which n is not restricted to the set of natural numbers. One can extend Formula (2), Section 9.4, for example, and write

$$(1 + x)^{\frac{1}{2}} = \binom{\frac{1}{2}}{0} + \binom{\frac{1}{2}}{1}x + \binom{\frac{1}{2}}{2}x^2 + \binom{\frac{1}{2}}{3}x^3 + \cdots + \binom{\frac{1}{2}}{n}x^n + \cdots .$$

Can you offer a conjecture as to where such an expansion might end? How many terms would the series contain?

At the end of Chapter 1 it was suggested that an *infinite* series could *converge* to a real number. Thus it is possible to show that meaningful results can be obtained by using formulas such as the one just suggested.

For example, if we let $x = 1$ in the above formula for $(1 + x)^{\frac{1}{2}}$, we get

$$(1 + 1)^{\frac{1}{2}} = \binom{\frac{1}{2}}{0} + \binom{\frac{1}{2}}{1} + \binom{\frac{1}{2}}{2} + \binom{\frac{1}{2}}{3} + \cdots .$$

Compute the first four terms on the right and note also that

$$(1 + 1)^{\frac{1}{2}} = \sqrt{2}.$$

Does this suggest a method for finding approximations to irrational square roots? Compare your answer to $\sqrt{2}$ obtained above with 1.4142 (which is the value of $\sqrt{2}$ correct to four decimal places).

Try using the first four terms of the binomial expansion of $(1 + x)^{\frac{1}{2}}$ to calculate an approximate value for $\sqrt{3}$.

Since $\sqrt{1.02} = (1 + 0.02)^{\frac{1}{2}}$, a binomial expansion can be used to calculate rational approximations to such numbers as $\sqrt{1.02}$, $\sqrt{1.04}$, and $\sqrt[3]{0.98}$. You might like to experiment along these lines and compare your results with those obtained using tables of roots.

As a final illustration consider the rational expression $\dfrac{1}{1 + x}$. Dividing 1 by $1 + x$, we get

$$
\begin{array}{r}
1 - x + x^2 - x^3 + x^4 - \cdots \\[2pt]
\hline
1 + x \,\big)\, 1 \\
\underline{1 + x} \\
- x \\
\underline{- x - x^2} \\
+ x^2 \\
\underline{+ x^2 + x^3} \\
- x^3 \\
\underline{- x^3 - x^4} \\
x^4 \cdots .
\end{array}
$$

It appears that the process could go on forever, that is, that the quotient is an infinite series. We know that

$$
\frac{1}{1 + x} = (1 + x)^{-1},
$$

which suggests a binomial expansion $(1 + x)^n$ with $n = -1$. Write the first seven terms of

$$
(1 + x)^{-1} = \binom{-1}{0} + \binom{-1}{1}x + \binom{-1}{2}x^2 + \cdots .
$$

There are many applications of the Binomial Theorem including an important one in probability theory. For further information on this and other questions concerning the Binomial Theorem we suggest reading in the following books.

FEHR, H. F. *Secondary Mathematics—A Functional Approach For Teachers.* Boston, Mass.: D. C. Heath, 1964.

SMITH, D. E. *A Source Book in Mathematics* (2 volumes). New York: Dover, 1929.

NEWMAN, J. R. (Editor), *The World of Mathematics* (Volume 1). New York: Simon and Schuster, 1962.

CHAPTER TEN
CIRCULAR FUNCTIONS

10.1 THE WRAPPING FUNCTION

The succession of day and night, the changing of the seasons, the rise and fall of the tides, the changes in our blood pressure as our heart beats are natural phenomena whose characteristic quality is that they repeat at regular intervals. The measure of this interval is called the **period** of the motion, and the phenomenon itself is called **periodic.**

Periodic functions, often used to describe such phenomena, are functions that repeat themselves in the sense of the following definition.

> **Definition 10.1** A **periodic function** is a function f such that there exists a positive number p with the property that
>
> $$f(x) = f(x + p)$$
>
> for all x in the domain of f. The smallest positive number p such that
>
> $$f(x + p) = f(x)$$
>
> for all x in the domain of f is called the **period** of f.

The most basic of the period functions are the **circular functions.**

Suppose we have a number line that is determined by fixing one point as the origin, 0, and a second point, normally to the right of the origin, as corresponding to the number 1. The distance between these two points is thereby fixed as the unit distance. The points corresponding to all other real numbers are determined as a consequence.

Now suppose that a circle with a radius of length 1 (called a unit circle since the length of its radius is equal to the unit distance) is placed tangent to a number line as in Figure 10-1a so that a point P on the circle coincides with the origin of the number line. If the

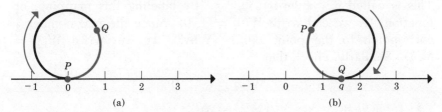

(a)   (b)

Figure 10-1

421

circle is rolled along the line as shown in Figure 10-1b, each point of the circle will coincide with a point on the number line. Since the circumference of the circle is 2π, each point on the circle corresponds to a number between 0 and 2π. If we continue to roll the circle along the number line, each point on the circle will correspond to many points on the number line. For example, P on the circle will correspond to 0, 2π, 4π, . . . on the number line. If Q on the circle corresponds initially to the point q on the number line, then Q will correspond to all points $q + n(2\pi)$ for $n \in N$ on the number line. If the circle is rolled in the opposite direction from 0, a correspondence is established between negative numbers and the points on the circle.

If you choose, you can think of the circle as being fixed and of the real number line as being flexible. If we "wrap" the real number line around the circle as in Figure 10-2, we have defined a function that maps each real number into a unique point of the circle. Such a function is called a **wrapping function.**

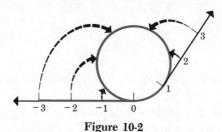

Figure 10-2

We begin our study of circular functions by assuming that a correspondence can be established that associates with each real number a unique point on a unit circle. For convenience, let us consider a unit circle $u^2 + v^2 = 1$ with its center at the origin of the uv-plane and a real number line parallel to the v-axis with its origin at the point (1, 0) as shown in Figure 10-3. (The Cartesian plane and the real number line are to have the same unit of distance.)

We now "wrap" the real number line around the unit circle and thus associate each real number with a unique point on the circle. This is called a **standard mapping.** By labeling this mapping, or function, W, we can write $W(0) = (1, 0)$. Since the real number $\frac{\pi}{2}$ corresponds to the point $B(0, 1)$ (Why?), we can write $W(\frac{\pi}{2}) = (0, 1)$. Similarly, check that

$$W(\pi) = (-1, 0),$$
$$W\left(\frac{3\pi}{2}\right) = (0, -1),$$

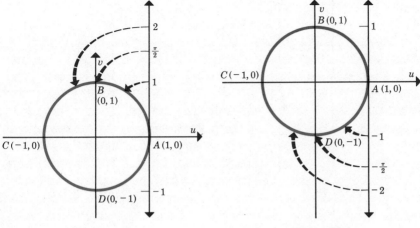

Figure 10-3   Figure 10-4

and

$$W(2\pi) = (1, \ 0).$$

For numbers greater than 2π, the standard mapping gives the following associations:

$$W\left(2\pi + \frac{\pi}{2}\right) = W\left(\frac{5\pi}{2}\right) = (0, \ 1),$$

$$W(2\pi + \pi) = W(3\pi) = (-1, \ 0),$$

$$W\left(2\pi + \frac{3\pi}{2}\right) = W\left(\frac{7\pi}{2}\right) = (0, \ -1),$$

and

$$W(2\pi + 2\pi) = W(4\pi) = (1, \ 0).$$

For negative numbers, as shown in Figure 10-4, we have, for example,

$$W\left(-\frac{\pi}{2}\right) = (0, \ -1),$$

$$W(-\pi) = (-1, \ 0),$$

$$W\left(-\frac{3\pi}{2}\right) = (0, \ 1),$$

and

$$W(-2\pi) = (1, \ 0).$$

If a number x is mapped under W into a point whose coordinates on the unit circle are $(u, \ v)$, we have

$$W(x) = (u, \ v).$$

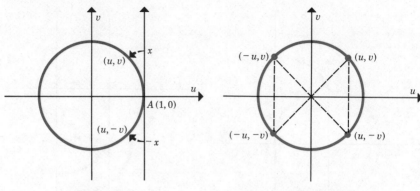

Figure 10-5   Figure 10-6

Then

$$W(-x) = (u, -v)$$

as shown in Figure 10-5.

Furthermore, we have, because of the symmetries of a circle as shown in Figure 10-6, that for all real numbers x,

$$W(x + \pi) = (-u, -v),$$
$$W(x - \pi) = (-u, -v),$$
$$W(\pi - x) = (-u, v),$$

and

$$W(-x - \pi) = (-u, v)$$

as illustrated in Figure 10-7a for $0 < x < \frac{\pi}{2}$ and in Figure 10-7b for $\pi < x < \frac{3\pi}{2}$. (In Figure 10-7b, note that both u and v are negative numbers.)

Example 1   If $W(x) = (\frac{3}{5}, \frac{4}{5})$, find (1) $W(-x)$ and (2) $W(x + \pi)$.

Solution:

(1) $W(-x) = \left(\dfrac{3}{5}, -\dfrac{4}{5}\right).$

(2) $W(x + \pi) = \left(-\dfrac{3}{5}, -\dfrac{4}{5}\right).$

Example 2   If $W(x) = (-\frac{5}{13}, \frac{12}{13})$, find (1) $W(-x)$ and (2) $W(x + \pi)$.

Solution:

(1) $W(-x) = \left(-\dfrac{5}{13}, -\dfrac{12}{13}\right).$

(2) $W(x + \pi) = \left(\dfrac{5}{13}, -\dfrac{12}{13}\right).$

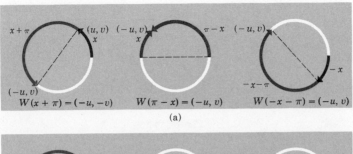

(a)

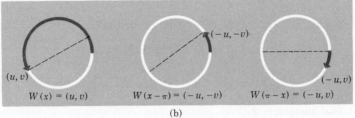

(b)

Figure 10-7

Example 3 If $W(5x) = \left(\dfrac{\sqrt{7}}{4}, \dfrac{3}{4}\right)$, find (1) $W(-5x)$ and (2) $W(-5x + \pi)$.

Solution:

$$(1) \qquad W(-5x) = \left(\frac{\sqrt{7}}{4},\ -\frac{3}{4}\right).$$

$$(2) \qquad W(-5x + \pi) = \left(-\frac{\sqrt{7}}{4}, \frac{3}{4}\right).$$

Example 4 Given that $W(\tfrac{\pi}{6}) = \left(\dfrac{\sqrt{3}}{2}, \dfrac{1}{2}\right)$, find (1) $W(-\tfrac{\pi}{6})$, (2) $W(\tfrac{5\pi}{6})$, and (3) $W(\tfrac{7\pi}{6})$.

Solution:

$$(1) \qquad W\left(-\frac{\pi}{6}\right) = \left(\frac{\sqrt{3}}{2},\ -\frac{1}{2}\right),$$

$$(2) \qquad W\left(\frac{5\pi}{6}\right) = W\left(\pi - \frac{\pi}{6}\right) = \left(-\frac{\sqrt{3}}{2}, \frac{1}{2}\right),$$

and

$$(3) \qquad W\left(\frac{7\pi}{6}\right) = W\left(\pi + \frac{\pi}{6}\right) = \left(-\frac{\sqrt{3}}{2},\ -\frac{1}{2}\right).$$

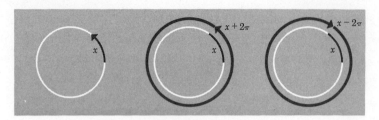

Figure 10-8

Note also, as suggested in Figure 10-8, that

$$W(x) = W(x + 2\pi) = W(x - 2\pi)$$

and

$$W(-x) = W(-x + 2\pi) = W(-x - 2\pi)$$

for all real numbers x.

In general, if P is a point on the unit circle corresponding to a real number x, then P is also the point that corresponds to the real numbers $x \pm 2\pi, x \pm 4\pi, \ldots$. Thus

$$W(x) = W(x \pm n[2\pi]) \qquad \text{for all } n \in N.$$

Since $W(x + 2\pi) = W(x)$ and $W(x + p) \neq W(x)$ for $|p| < 2\pi$, we conclude that W is a periodic function with period 2π.

EXERCISES 10.1A

1. Consider a circle with a radius of length 1 tangent to a real number line so that a point Q on the circle coincides with the origin of the real number line as in Figure 10-9. What point of the number line is tangent to the circle if the circle rolls
   (a) one revolution in the positive direction?
   (b) two revolutions in the positive direction?
   (c) one-half of a revolution in the positive direction?

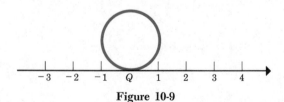

Figure 10-9

(d) one-half of a revolution in the negative direction?

(e) three-fourths of a revolution in the negative direction?

◀Exercises 2-10 refer to Figure 10-10 where we assume that the real number line R has been wrapped around the circle.

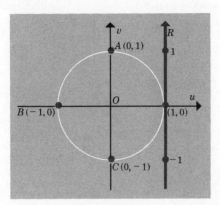

Figure 10-10

2. What is the smallest positive number that corresponds to point A?

3. What is the largest negative number that corresponds to point A?

4. Give three other numbers that correspond to point A.

5. What is the smallest positive number that corresponds to point B?

6. What is the largest negative number that corresponds to point B?

7. Name three other numbers that correspond to point B.

8. What is the smallest positive number that corresponds to point C?

9. What is the largest negative number that corresponds to point C?

10. Give three other numbers that correspond to point C.

◀In Exercises 11–20, find $W(x)$ for each given value of x.

11. 0

12. π

13. 2π

14. $\dfrac{\pi}{2}$

15. $-\pi$

16. $-\dfrac{\pi}{2}$

17. $\dfrac{15\pi}{2}$

18. $-\dfrac{5\pi}{2}$

19. 1000π

20. -200π

◀ In Exercises 21–26, assume that $W(y) = (-\frac{3}{5}, \frac{4}{5})$. Find each of the following.

21. $W(-y)$ **24.** $W(\pi - y)$
22. $W(y + \pi)$ **25.** $W(y + 80\pi)$
23. $W(y - 2\pi)$ **26.** $W(-y - \pi)$

◀ In Exercises 27–31, assume that $W(t) = (\frac{5}{13}, -\frac{12}{13})$. Find each of the following,

27. $W(-t)$ **30.** $W(t - 2\pi)$
28. $W(-t - \pi)$ **31.** $W(t - 40\pi)$
29. $W(t + \pi)$

◀ In Exercises 32–46, assume that the coordinates of point $P = W(x)$ are (a, b) and that $x \geq 0$. Find the coordinates of the points that correspond to each of the following real numbers.

32. $-x$ **37.** $2\pi - x$ **42.** $-x - 2\pi$
33. $x + 2\pi$ **38.** $2\pi + x$ **43.** $-x + \pi$
34. $x + \pi$ **39.** $\pi - x$ **44.** $-x + 2\pi$
35. $x - \pi$ **40.** $\pi + x$ **45.** $-x - 5\pi$
36. $x - 2\pi$ **41.** $-x - \pi$ **46.** $-x + 3\pi$

47. Give five examples of periodic phenomena and specify an approximate period for each.

◀ For Exercises 48–51, consider a unit circle in the uv-plane. For $x \in R$ and $-\frac{\pi}{2} < x < 0$, let $W(x) = (a, b)$, $a, b \in R$. Find each of the following by using the symmetries of a circle.

48. $W(x + \pi)$ **50.** $W(-x - \pi)$
49. $W(x + 2\pi)$ **51.** $W(x - \pi)$

◀ For Exercises 52–55, consider a unit circle in the uv-plane. For $y \in R$ and $\frac{3\pi}{2} < y < 2\pi$, let $W(y) = (r, s)$, $r, s \in R$. Find each of the following by using the symmetries of a circle.

52. $W(-y)$ **54.** $W(y + 6\pi)$
53. $W(y + \pi)$ **55.** $W(-y - \pi)$

You will recall from your previous work in mathematics that the degree measure of a right angle is 90. Hence if P is a point on the unit circle corresponding to the number $\frac{\pi}{4}$, we have $m \angle AOP = 45$ as in Figure 10-11. Here, as elsewhere in this chapter, the measure of an angle will be considered as being given in degrees unless specified otherwise. We know $OP = 1$. Why? Since $m \angle ROP = m \angle OPR$, the

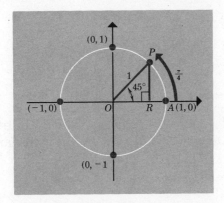

Figure 10-11

triangle ROP is isosceles and $OR = RP$. Hence we have, by the Pythagorean Theorem,

$$(OR)^2 + (RP)^2 = 1.$$

If we let $OR = RP = a$, we have

$$a^2 + a^2 = 1,$$
$$2a^2 = 1,$$
$$a^2 = \frac{1}{2},$$
$$|a| = \sqrt{\frac{1}{2}} = \frac{\sqrt{2}}{2}.$$

Thus

$$OR = \frac{\sqrt{2}}{2} \quad \text{and} \quad RP = \frac{\sqrt{2}}{2}.$$

Hence the coordinates of P are $\left(\dfrac{\sqrt{2}}{2}, \dfrac{\sqrt{2}}{2}\right)$, and

$$W\!\left(\frac{\pi}{4}\right) = \left(\frac{\sqrt{2}}{2}, \frac{\sqrt{2}}{2}\right).$$

In a similar way you can find $W(\frac{\pi}{6})$, $W(\frac{\pi}{3})$, $W(\frac{5\pi}{6})$, $W(-\frac{\pi}{6})$, and numerous other functional values.

Example 5 Find $W(\frac{5\pi}{6})$ by using a right triangle.

Solution: Since $\frac{5\pi}{6} = \frac{5}{12}(2\pi)$, we have

$$m\angle POA = \frac{5}{12}(360) = 150.$$

Thus

$$m\angle POC = 180 - 150 = 30.$$

Triangle ROP of Figure 10-12 is a 30, 60, 90 triangle. Hence $RP = \frac{1}{2}(OP) = \frac{1}{2}$. If we let $OR = a$, by the Pythagorean Theorem,

$$a^2 + \left(\frac{1}{2}\right)^2 = 1^2.$$

Solving for a, we get $|a| = \frac{\sqrt{3}}{2}$. Hence OR, the undirected distance from O to R, is equal to $\frac{\sqrt{3}}{2}$, and so the coordinates of R are $\left(-\frac{\sqrt{3}}{2}, 0\right)$. Thus the coordinates of P are $\left(-\frac{\sqrt{3}}{2}, \frac{1}{2}\right)$. Accordingly,

$$W\left(\frac{5\pi}{6}\right) = \left(-\frac{\sqrt{3}}{2}, \frac{1}{2}\right).$$

Example 6   Find $W\left(-\frac{7\pi}{6}\right)$ using the results of Example 5.

Solution: Recall that

$$W(x) = W(x + 2\pi).$$

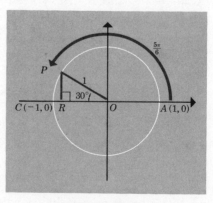

Figure 10-12

Hence

$$W\left(-\frac{7\pi}{6}\right) = W\left(-\frac{7\pi}{6} + 2\pi\right) = W\left(\frac{5\pi}{6}\right) = \left(-\frac{\sqrt{3}}{2}, \frac{1}{2}\right).$$

Example 7   Find $W(-\frac{17\pi}{4})$.

Solution: Since

$$-\frac{17\pi}{4} = -\frac{\pi}{4} - 4\pi,$$

we have

$$W\left(-\frac{17\pi}{4}\right) = W\left(-\frac{\pi}{4}\right).$$

Recall that

$$W\left(\frac{\pi}{4}\right) = \left(\frac{\sqrt{2}}{2}, \frac{\sqrt{2}}{2}\right).$$

Thus

$$W\left(-\frac{\pi}{4}\right) = \left(\frac{\sqrt{2}}{2}, -\frac{\sqrt{2}}{2}\right)$$

as shown in Figure 10-13. (Compare Figures 10-11 and 10-13.)

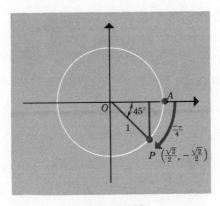

Figure 10-13

EXERCISES 10.1B

◀In Exercises 1–8, find the functional value by using a diagram involving a right triangle.

1. $W\left(\frac{\pi}{6}\right)$ 2. $W\left(\frac{\pi}{3}\right)$

3. $W\left(-\dfrac{\pi}{6}\right)$ 6. $W\left(-\dfrac{2\pi}{3}\right)$

4. $W\left(-\dfrac{\pi}{3}\right)$ 7. $W\left(\dfrac{3\pi}{4}\right)$

5. $W\left(\dfrac{2\pi}{3}\right)$ 8. $W\left(-\dfrac{3\pi}{4}\right)$

9. Using the results of Exercise 1, find

  (a) $W\left(\dfrac{\pi}{6}+\pi\right)$ (b) $W\left(\dfrac{\pi}{6}-\pi\right)$

10. Using the results of Exercise 2, find

  (a) $W\left(\dfrac{\pi}{3}-2\pi\right)$ (b) $W\left(\dfrac{\pi}{3}-\pi\right)$

11. Using the results of Exercise 7, find

  (a) $W\left(\dfrac{3\pi}{4}+\pi\right)$ (b) $W\left(-\dfrac{3\pi}{4}-\pi\right)$

12. Using the results of Exercise 1, find

  (a) $W\left(-\dfrac{\pi}{6}\right)$ (c) $W\left(\dfrac{7\pi}{6}\right)$

  (b) $W\left(-\dfrac{5\pi}{6}\right)$ (d) $W\left(\dfrac{13\pi}{6}\right)$

13. Using the results of Exercise 2, find

  (a) $W\left(-\dfrac{\pi}{3}\right)$ (c) $W\left(-\dfrac{7\pi}{3}\right)$

  (b) $W\left(-\dfrac{2\pi}{3}\right)$ (d) $W\left(-\dfrac{46\pi}{3}\right)$

◀ In Exercises 14–20, use Figure 10-14 to find approximations for each of the functional values.

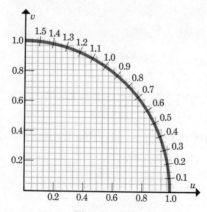

Figure 10-14

Example $W(1.0) \approx (0.54, 0.84)$.

14. $W(0.3)$ 18. $W(-0.4)$
15. $W(0.7)$ 19. $W(-1.1)$
16. $W(1.2)$ 20. $W(-1.2 + \pi)$
17. $W(1.5)$

21. CHALLENGE PROBLEM. In each case find a number x such that $0 \le x \le 2\pi$ and $W(x) = W(y)$.

    (a) $y = 3\pi$ (d) $y = 5$

    (b) $y = \dfrac{5\pi}{2}$ (e) $y = 2\pi - 5$

    (c) $y = -\dfrac{9\pi}{2}$

10.2 THE CIRCULAR FUNCTIONS

The wrapping function, W, as you have seen, associates with each real number an ordered pair of numbers. It is, however, more convenient to work with functions whose range is a set of single numbers rather than a set of ordered pairs. We therefore define two separate functions, the *sine* function and the *cosine* function. These functions, abbreviated sin and cos, are called **circular** functions.

> *Definition 10.2* If x is a real number such that $W(x) = (u, v)$ where $P(u, v)$ is a point on the unit circle $u^2 + v^2 = 1$, then
>
> $$\sin x = v \quad \text{and} \quad \cos x = u$$
>
> where sin x is an abbreviation for sine(x) and cos x is an abbreviation for cosine(x). Thus the domain of both the **sine and cosine functions** is the set of all real numbers.

By Definition 10.2, we have $W(x) = (\cos x, \sin x)$. (See Figure 10-15.) We also have, since $u^2 + v^2 = 1$,

$$(\cos x)^2 + (\sin x)^2 = 1 \quad \text{for all } x \in R$$

which is usually written as

(I) $$\sin^2 x + \cos^2 x = 1.$$

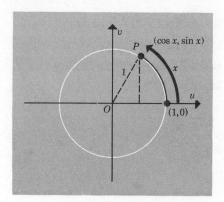

Figure 10-15

Equation (I) which is true for all $x \in R$ is called an **identity** and expresses one of the most fundamental relationships between the circular functions. (Unless otherwise stated, it should be assumed that the domain of the variable or variables in an identity involving circular functions is the entire set of real numbers.)

If $W(x) = (u, v)$, we know that $W(x + 2\pi) = W(x - 2\pi) = (u, v)$. (Why?) Hence

$$\cos x = u = \cos (x + 2\pi) = \cos (x - 2\pi)$$

and

$$\sin x = v = \sin (x + 2\pi) = \sin (x - 2\pi).$$

Similarly, for all $n \in I$,

$$\sin (x + 2n\pi) = \sin x$$

and

$$\cos (x + 2n\pi) = \cos x.$$

Also if

$$W(x) = (u, v),$$

then

$$W(-x) = (u, -v)$$

as indicated in Figure 10-16a for $x > 0$ and in (b) for $x < 0$. Since $\cos x = u$ and $\cos(-x) = u$, we have the identity

(II) $$\cos (-x) = \cos x.$$

Since $\sin x = v$ and $\sin (-x) = -v$, we have

(III) $$\sin (-x) = -\sin x.$$

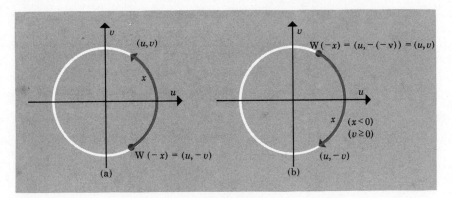

Figure 10-16

Other identities that follow from the symmetries of a circle are

(IV) $$\cos (x \pm \pi) = -\cos x$$

and

(V) $$\sin (x \pm \pi) = -\sin x$$

where, for example, $\cos (x \pm \pi) = -\cos x$ is an abbreviation for

$$\cos (x + \pi) = -\cos x \qquad \text{and} \qquad \cos (x - \pi) = -\cos x.$$

Example 1 If $\sin x = \frac{4}{5}$ and $0 < x < \frac{\pi}{2}$, find (1) $\sin (-x)$, (2) $\sin (x + \pi)$, and (3) $\sin (-x - \pi)$.
Solution:

(1) $$\sin (-x) = -\sin x = -\frac{4}{5}.$$

(2) $$\sin (x + \pi) = -\sin x = -\frac{4}{5}.$$

(3) $$\sin (-x - \pi) = \sin ([-x] - \pi) = -\sin [-x]$$
$$= -\left(-\frac{4}{5}\right) = \frac{4}{5}.$$

Example 2 If $\cos y = -\frac{5}{13}$ and $\frac{\pi}{2} < y < \pi$, find (1) $\cos (-y)$, (2) $\cos (y + 2\pi)$, and (3) $\cos (\pi - y)$.

Solution:

(1) $$\cos{(-y)} = \cos{y} = -\frac{5}{13}.$$

(2) $$\cos{(y + 2\pi)} = \cos{y} = -\frac{5}{13}.$$

(3) $$\cos{(\pi - y)} = \cos{[\pi + (-y)]} = -\cos{(-y)}$$
$$= -\left(-\frac{5}{13}\right) = \frac{5}{13}.$$

In Section 10.1 we found $W(x)$ for values of x such as π, $-\pi$, $\frac{\pi}{2}$, and $\frac{\pi}{3}$. Table 10-1 shows values of $W(x)$ for some of these special values of x between 0 and π. Since $W(x) = (\cos{x}, \sin{x})$, we can readily form Tables 10-2 and 10-3 from Table 10-1.

TABLE 10-1		**TABLE 10-2**		**TABLE 10-3**	
x	$W(x)$	x	$\cos{x}$	x	$\sin{x}$
0	(1, 0)	0	1	0	0
$\frac{\pi}{6}$	$\left(\frac{\sqrt{3}}{2}, \frac{1}{2}\right)$	$\frac{\pi}{6}$	$\frac{\sqrt{3}}{2}$	$\frac{\pi}{6}$	$\frac{1}{2}$
$\frac{\pi}{4}$	$\left(\frac{\sqrt{2}}{2}, \frac{\sqrt{2}}{2}\right)$	$\frac{\pi}{4}$	$\frac{\sqrt{2}}{2}$	$\frac{\pi}{4}$	$\frac{\sqrt{2}}{2}$
$\frac{\pi}{3}$	$\left(\frac{1}{2}, \frac{\sqrt{3}}{2}\right)$	$\frac{\pi}{3}$	$\frac{1}{2}$	$\frac{\pi}{3}$	$\frac{\sqrt{3}}{2}$
$\frac{\pi}{2}$	(0, 1)	$\frac{\pi}{2}$	0	$\frac{\pi}{2}$	1
$\frac{2\pi}{3}$	$\left(-\frac{1}{2}, \frac{\sqrt{3}}{2}\right)$	$\frac{2\pi}{3}$	$-\frac{1}{2}$	$\frac{2\pi}{3}$	$\frac{\sqrt{3}}{2}$
$\frac{3\pi}{4}$	$\left(-\frac{\sqrt{2}}{2}, \frac{\sqrt{2}}{2}\right)$	$\frac{3\pi}{4}$	$-\frac{\sqrt{2}}{2}$	$\frac{3\pi}{4}$	$\frac{\sqrt{2}}{2}$
$\frac{5\pi}{6}$	$\left(-\frac{\sqrt{3}}{2}, \frac{1}{2}\right)$	$\frac{5\pi}{6}$	$-\frac{\sqrt{3}}{2}$	$\frac{5\pi}{6}$	$\frac{1}{2}$
π	(-1, 0)	π	-1	π	0

The u-axis and the v-axis divide the Cartesian plane into **quadrants** as shown in Figure 10-17. The values of the sine function and the cosine function associated with the points (1, 0), (0, 1), (−1, 0), and

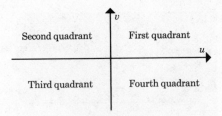

Figure 10-17

$(0, -1)$ are called quadrantal values of the functions and are

$$\cos 0 = 1, \qquad \sin 0 = 0$$

$$\cos \frac{\pi}{2} = 0, \qquad \sin \frac{\pi}{2} = 1$$

$$\cos \pi = -1, \qquad \sin \pi = 0$$

$$\cos \frac{3\pi}{2} = 0, \qquad \sin \frac{3\pi}{2} = -1$$

since $W(0) = (1, 0)$, $W(\frac{\pi}{2}) = (0, 1)$, $W(\pi) = (-1, 0)$, and $W(\frac{3\pi}{2}) = (0, -1)$. The range of values of the sine and cosine functions is shown in Figure 10-18.

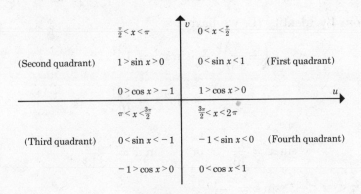

Figure 10-18

Example 3   If $0 < x < \frac{\pi}{2}$ and $\sin x = \frac{4}{5}$, find $\cos x$.

Solution: Since $0 < x < \frac{\pi}{2}$, x is in quadrant I. (See Figure 10-19.) By Identity (I), we have

$$(\cos x)^2 + \left(\frac{4}{5}\right)^2 = 1$$

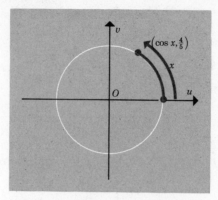

Figure 10-19

and so
$$\cos^2 x = \frac{9}{25}.$$

Hence
$$\cos x = \frac{3}{5} \quad \text{or} \quad \cos x = -\frac{3}{5}.$$

Since $0 < x < \frac{\pi}{2}$, $\cos x = \frac{3}{5}$.

Example 4 If $\pi < x < \frac{3\pi}{2}$ and $\cos x = -\frac{5}{13}$, find $\sin x$.

Solution: By Identity (I), we have
$$\left(-\frac{5}{13}\right)^2 + (\sin x)^2 = 1$$

and so
$$\sin^2 x = \frac{144}{169}.$$

Hence
$$\sin x = \frac{12}{13} \quad \text{or} \quad \sin x = -\frac{12}{13}.$$

Since $\pi < x < \frac{3\pi}{2}$, $\sin x = -\frac{12}{13}$.

EXERCISES 10.2A

◀In Exercises 1–30, find $\sin x$ and $\cos x$ for each given number x.

1. 0   3. π 5. 2π

2. $\frac{\pi}{2}$ 4. $\frac{3\pi}{2}$ 6. 4π

7. $-\pi$

8. -2π

9. $-\dfrac{\pi}{2}$

10. $-\dfrac{3\pi}{2}$

11. $\dfrac{\pi}{3}$

12. $\dfrac{2\pi}{3}$

13. $-\dfrac{\pi}{3}$

14. $-\dfrac{2\pi}{3}$

15. $\dfrac{4\pi}{3}$

16. $-\dfrac{4\pi}{3}$

17. $\dfrac{5\pi}{3}$

18. $-\dfrac{5\pi}{3}$

19. $\dfrac{15\pi}{2}$

20. $-\dfrac{15\pi}{2}$

21. 3π

22. -3π

23. $\dfrac{\pi}{4}$

24. $-\dfrac{\pi}{4}$

25. $\dfrac{3\pi}{4}$

26. $\dfrac{5\pi}{4}$

27. $\dfrac{7\pi}{4}$

28. $-\dfrac{3\pi}{4}$

29. $-\dfrac{7\pi}{4}$

30. 1000π

◀ In Exercises 31–40, find, using Figure 10-14 (page 432), an approximation for each value.

31. $\sin 0.5$

32. $\cos 0.6$

33. $\sin 1.5$

34. $\cos 0.3$

35. $\sin 1.0$

36. $\cos 1.0$

37. $\sin 0.2$

38. $\cos 0.2$

39. $\sin 0.4$

40. $\cos 0.4$

41. If $\sin 0.3 = 0.295$, find $\sin (0.3 + \pi)$.

42. If $\cos 0.5 = 0.878$, find $\cos (-\pi - 0.5)$.

43. If $\sin 0.75 = 0.681$, find $\sin (0.75 - 2\pi)$.

44. If $\cos 0.85 = 0.751$, find $\cos (\pi - 0.85)$.

45. If $\sin 301.5 = y$, find $\sin (2\pi - 301.5)$.

◀ In Exercises 46–49, verify that the following statements are true.

46. $2 \sin \dfrac{\pi}{3} \cos \dfrac{\pi}{3} = \sin \dfrac{2\pi}{3}$

47. $\cos \dfrac{\pi}{3} = 2 \cos^2 \dfrac{\pi}{6} - 1$

48. $\cos \dfrac{\pi}{2} = \cos^2 \dfrac{\pi}{4} - \sin^2 \dfrac{\pi}{4}$

49. $\sin \dfrac{\pi}{3} = 2 \sin \dfrac{\pi}{6} \cos \dfrac{\pi}{6}$

50. If $0 < x < \dfrac{\pi}{2}$ and $\cos x = \dfrac{3}{5}$, find $\sin x$.

51. If $0 < x < \dfrac{\pi}{2}$ and $\sin x = \dfrac{5}{13}$, find $\cos x$.

52. If $\dfrac{\pi}{2} < y < \pi$ and $\sin y = \dfrac{4}{5}$, find $\cos y$.

53. If $\dfrac{\pi}{2} < x < \pi$ and $\cos x = -\dfrac{8}{17}$, find $\sin x$.

54. If $\pi < y < \dfrac{3\pi}{2}$ and $\sin y = -\dfrac{12}{13}$, find $\cos y$.

55. If $\dfrac{3\pi}{2} < x < 2\pi$ and $\cos x = \dfrac{15}{17}$, find $\sin x$.

56. If $0 < y < \pi$ and $\cos y = \dfrac{\sqrt{7}}{4}$, find $\sin y$.

57. If $-\dfrac{\pi}{2} < 3x < 0$ and $\sin 3x = -\dfrac{4}{5}$, find $\cos 3x$.

58. If $6\pi < 2x < \dfrac{13\pi}{2}$ and $\cos 2x = \dfrac{2\sqrt{13}}{13}$, find $\sin 2x$.

59. If $-\dfrac{3\pi}{2} < 2y < -\pi$ and $\sin 2y = \dfrac{\sqrt{5}}{5}$, find $\cos 2y$.

60. If $-\dfrac{111\pi}{2} < t < -55\pi$ and $\cos t = -\dfrac{\sqrt{10}}{10}$, find $\sin t$.

We now define four other circular functions: the *tangent,* the *cotangent,* the *secant,* and the *cosecant* (which are basic to the study of circular functions).

Definition 10.3 If x is a real number such that $W(x) = (u, v)$ where $P(u, v)$ is a point on the unit circle $u^2 + v^2 = 1$, then

$$\text{tangent } x = \frac{v}{u} \qquad \text{whenever } u \neq 0,$$

$$\text{cotangent } x = \frac{u}{v} \qquad \text{whenever } v \neq 0,$$

$$\text{secant } x = \frac{1}{u} \qquad \text{whenever } u \neq 0,$$

$$\text{cosecant } x = \frac{1}{v} \qquad \text{whenever } v \neq 0.$$

The abbreviations for these functions are tan, cot, sec, and csc, respectively.

Clearly, since tan x and sec x are not defined when $u = 0$, certain quadrantal values such as tan $\frac{\pi}{2}$, tan $\frac{3\pi}{2}$, sec $\frac{\pi}{2}$, and sec $\frac{3\pi}{2}$ are not defined. Specifically, tan $(\frac{\pi}{2} \pm n\pi)$ and sec $(\frac{\pi}{2} \pm n\pi)$ for $n \in N$ are not defined. Describe the values of x for which cot x and csc x are not defined.

Example 5   If $0 < x < \frac{\pi}{2}$ and cos $x = \frac{1}{2}$, determine the functional values of x for the other five circular functions.

Solution: By Identity (I),

$$\sin^2 x + \left(\frac{1}{2}\right)^2 = 1.$$

Hence

$$\sin x = \frac{\sqrt{3}}{2}. \quad \left(\text{Sin } x \neq -\frac{\sqrt{3}}{2} \text{ since } 0 < x < \frac{\pi}{2}.\right)$$

We have cos $x = u = \dfrac{1}{2}$ and sin $x = v = \dfrac{\sqrt{3}}{2}$.   Hence

$$\tan x = \frac{v}{u} = \sqrt{3},$$

$$\cot x = \frac{u}{v} = \frac{1}{\sqrt{3}} = \frac{1}{\sqrt{3}} \cdot \frac{\sqrt{3}}{\sqrt{3}} = \frac{\sqrt{3}}{3},$$

$$\sec x = \frac{1}{u} = 2,$$

and

$$\csc x = \frac{1}{v} = \frac{2}{\sqrt{3}} = \frac{2}{\sqrt{3}} \cdot \frac{\sqrt{3}}{\sqrt{3}} = \frac{2\sqrt{3}}{3}.$$

Example 6   If $0 > x > -\dfrac{\pi}{2}$ and sin $x = -\dfrac{\sqrt{2}}{2}$, determine the other five circular functional values of x.

Solution: A diagram such as Figure 10-20 can help considerably.   By Identity (I),

$$\cos^2 x + \left(-\frac{\sqrt{2}}{2}\right)^2 = 1.$$

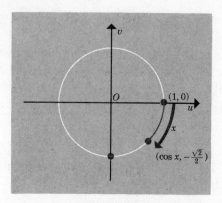

Figure 10-20

Hence

$$\cos x = \frac{\sqrt{2}}{2}. \qquad \left(\text{Cos } x \neq -\frac{\sqrt{2}}{2}. \text{ Why?}\right)$$

We have $u = \dfrac{\sqrt{2}}{2}$ and $v = -\dfrac{\sqrt{2}}{2}$. Thus

$$\tan x = \frac{v}{u} = -1,$$

$$\cot x = \frac{u}{v} = -1,$$

$$\sec x = \frac{1}{u} = \frac{2}{\sqrt{2}} = \sqrt{2},$$

and

$$\csc x = \frac{1}{v} = -\frac{2}{\sqrt{2}} = -\sqrt{2}.$$

Example 7 Find the five other circular functional values of x if $\tan x = \frac{3}{4}$ and $\cos x < 0$.

Solution: If $\cos x < 0$, then x corresponds to a point in the second or third quadrants. If $\tan x > 0$, then x corresponds to a point in the first or third quadrants. Hence, since $\tan x = \frac{3}{4} > 0$ and $\cos x < 0$, x must correspond to a point (u, v) in the third quadrant as shown in Figure 10-21.

Since $\tan x = \dfrac{v}{u}$, $\dfrac{v}{u} = \dfrac{3}{4}$. Since $u \neq 0$, we may divide both sides of $u^2 + v^2 = 1$ by u^2 and get

$$1 + \frac{v^2}{u^2} = \frac{1}{u^2}.$$

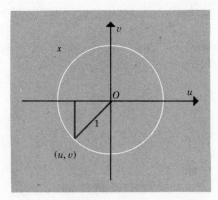

Figure 10-21

Replacing $\dfrac{v^2}{u^2}$ by $\dfrac{9}{16}$, we obtain

$$1 + \frac{9}{16} = \frac{1}{u^2}.$$

Hence

$$u = \frac{4}{5} \qquad \text{or} \qquad u = -\frac{4}{5}.$$

Since $u = \cos x < 0$ is given, $u = -\frac{4}{5}$. Replacing u by $-\frac{4}{5}$ in $\dfrac{v}{u} = \dfrac{3}{4}$, we find $v = -\frac{3}{5}$. Using u and v, we write

$$\sin x = -\frac{3}{5}, \qquad \cos x = -\frac{4}{5},$$

$$\cot x = \frac{4}{3}, \qquad \sec x = -\frac{5}{4},$$

and

$$\csc x = -\frac{5}{3}.$$

Example 8   If $\sin x = 1$, find $\cos x$, $\cot x$, and $\csc x$. What are the possible values of x?

Solution: If $\sin x = 1$, we have by Identity (I), $1^2 + \cos^2 x = 1$ so that $\cos x = 0$. Hence $u = 0$ and $v = 1$. Thus

$$\cot x = \frac{u}{v} = 0$$

and

$$\csc x = \frac{1}{v} = 1.$$

Note that $\tan x$ and $\sec x$ are undefined since $u = 0$.

Since $W(x) = (0, 1)$, it follows that $x = \frac{\pi}{2}$ if $0 < x < 2\pi$. But we may also have $x = \frac{\pi}{2} \pm 2n\pi$ for any $n \in N$. Hence if we seek only the values of x such that $0 \leq x \leq 2\pi$, then $x = \frac{\pi}{2}$; if we wish to determine all possible values of x, we may take $x = \frac{\pi}{2} \pm 2n\pi$ for any $n \in N$. That is,

$$\{x : x \in R, 0 \leq x \leq 2\pi, \text{ and } \sin x = 1\} = \left\{\frac{\pi}{2}\right\};$$

but

$$\{x : x \in R \text{ and } \sin x = 1\} = \left\{\frac{\pi}{2} \pm 2n\pi : n \in N\right\}$$

or, alternatively,

$$\left\{\frac{\pi}{2} + 2n\pi : n \in I\right\}$$

where I is the set of integers $\{\ldots, -2, -1, 0, 1, 2, \ldots\}$.

The definitions of the six trigonometric functions lead immediately to the following identities:

(VI) $\qquad \tan x = \dfrac{\sin x}{\cos x} \qquad (x \neq \frac{\pi}{2} + n\pi \text{ for } n \in I)$,

(VII) $\qquad \cot x = \dfrac{\cos x}{\sin x} \qquad (x \neq n\pi \text{ for } n \in I)$,

(VIII) $\qquad \sec x = \dfrac{1}{\cos x} \qquad (x \neq \frac{\pi}{2} + n\pi \text{ for } n \in I)$,

and

(IX) $\qquad \csc x = \dfrac{1}{\sin x} \qquad (x \neq n\pi \text{ for } n \in I)$.

EXERCISES 10.2B

◀ In Exercises 1–10, find the functional values of the tangent, cotangent, secant, and cosecant functions for each given number.

1. $\dfrac{\pi}{6}$ 4. $\dfrac{2\pi}{3}$

2. $\dfrac{\pi}{4}$ 5. $\dfrac{3\pi}{4}$

3. $\dfrac{\pi}{3}$ 6. $-\dfrac{\pi}{6}$

7. $-\dfrac{2\pi}{3}$ **9.** $\dfrac{7\pi}{6}$

8. $-\dfrac{3\pi}{4}$ **10.** $-\dfrac{21\pi}{4}$

◀ In Exercises 11–30, find all x, $0 \le x < 2\pi$, that satisfy the given equation.

11. $\sin x = 1$ **21.** $\cot x = -1$

12. $\cos x = 0$ **22.** $\csc x = -\dfrac{2\sqrt{3}}{3}$

13. $\cos x = 1$ **23.** $\sin x = -1$

14. $\sin x = \dfrac{1}{2}$ **24.** $\cot x = -\dfrac{\sqrt{3}}{3}$

15. $\tan x = 1$ **25.** $\cos x = \dfrac{\sqrt{2}}{2}$

16. $\csc x = 2$ **26.** $\cot x = -\sqrt{3}$

17. $\sec x = -2$ **27.** $\tan x = 0$

18. $\sin x = -\dfrac{\sqrt{3}}{2}$ **28.** $\sec x = \sqrt{2}$

19. $\cos x = -\dfrac{\sqrt{2}}{2}$ **29.** $\sin x = -\dfrac{1}{2}$

20. $\csc x = \sqrt{2}$ **30.** $\cos x = -1$

◀ In Exercises 31–40, find the designated functional value using the given information.

31. $\cos x$, $\tan x = 1$ and $\sin x < 0$.

32. $\sin x$, $\cos x = -\dfrac{3}{5}$ and $\sin x > 0$

33. $\tan x$, $\sin x = \dfrac{5}{13}$ and $\tan x < 0$

34. $\cot x$, $\cos x = \dfrac{4}{5}$ and $\sin x < 0$

35. $\sec x$, $\sin x = -\dfrac{8}{17}$ and $\cos x < 0$

36. $\sin x$, $\tan x = -\dfrac{3}{4}$ and $\cos x > 0$

37. $\cos x$, $\cot x = \dfrac{5}{12}$ and $\sin x < 0$

38. $\sec x$, $\tan x = \dfrac{8}{15}$ and $\dfrac{3\pi}{2} > x > \pi$

39. $\csc x$, $\cot x = -1$ and $\pi > x > \dfrac{\pi}{2}$

40. $\tan x$, $\sin x = -\dfrac{8}{17}$ and $2\pi > x > \dfrac{3\pi}{2}$

41. Give the domain and range of the sine function.

42. Give the domain and range of the cosine function.

43. Give the domain and range of the tangent function.

44. Give the domain and range of the secant function.

45. Show that $(\tan x)^2 + 1 = (\sec x)^2$ for all $x \in R$.

46. If $f(-x) = f(x)$ for every x in the domain of f, f is called an **even function**. Show that the cosine is an even function.

47. If $f(-x) = -f(x)$ for every x in the domain of f, f is called an **odd function**. Show that the sine is an odd function.

48. Show that $f : f(x) = 3x^3 + x$ is an odd function.

49. Show that $g : g(x) = 2x^4 + 3x^2 - 5$ is an even function.

◀In Exercises 50–60, find the numbers between 0 and 2π in the solution set of each equation.

50. $\sin 2x = \dfrac{1}{2}$

51. $\cos 3x = \dfrac{1}{2}$

52. $2 \sin 3r = 1$

53. $3 \tan s = \sqrt{3}$

54. $\cos x = 2 \sin \dfrac{\pi}{4} \cos \dfrac{\pi}{4}$

55. $\sin x = 2 \sin \dfrac{\pi}{3} \cos \dfrac{\pi}{3}$

56. $\cos y = \cos \dfrac{\pi}{3} \cos \dfrac{\pi}{6} - \sin \dfrac{\pi}{3} \sin \dfrac{\pi}{6}$

57. $\cos t = \cos^2 \dfrac{\pi}{6} - \sin^2 \dfrac{\pi}{6}$

58. $\sin \theta = \sin \dfrac{\pi}{6} \cos \dfrac{\pi}{3} + \cos \dfrac{\pi}{6} \sin \dfrac{\pi}{3}$

59. $\cos \theta = \cos \dfrac{\pi}{2} \cos \dfrac{\pi}{4} + \sin \dfrac{\pi}{2} \sin \dfrac{\pi}{4}$

60. $\tan \theta = \dfrac{2 \tan \frac{\pi}{6}}{1 - \tan^2 \frac{\pi}{6}}$

10.3 GRAPHS OF THE CIRCULAR FUNCTIONS

The graph of the sine function consists of all points $P(x, y)$ such that each ordered pair (x, y) is an element of the set

$$\{(x, y) : y = \sin x\}.$$

As we have done in other chapters, we shall use the expression "graph of $y = \sin x$" as an abbreviation of the longer statement.

We have already found many elements of the set

$$\{(x, y) : y = \sin x\}.$$

Values for $\sin x$ for some numbers x such that $0 \leq x \leq \pi$ are listed in Table 10-4 where decimal approximations have been used for all irrational numbers. We then use the identity

$$\sin (x + \pi) = -\sin x$$

to obtain values of $\sin x$ in the interval $\pi < x < 2\pi$. Figure 10-22 shows these points graphed on the Cartesian plane. Because each

TABLE 10-4

x	$\sin x$
0	0
$\dfrac{\pi}{6}$	0.5
$\dfrac{\pi}{4}$	0.71
$\dfrac{\pi}{3}$	0.87
$\dfrac{\pi}{2}$	1
$\dfrac{2\pi}{3}$	0.87
$\dfrac{3\pi}{4}$	0.71
$\dfrac{5\pi}{6}$	0.5
π	0

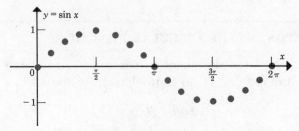

Figure 10-22

point on the real number line has a unique mapping under W, there is a point on the graph of $y = \sin x$ that corresponds to each point on the x-axis. The result is a smooth curve as shown in Figure 10-23. Since $\sin x = \sin (x + n \cdot 2\pi)$ for all integers n, the graph of

$$\{(x, y) : y = \sin x\}$$

for all $x \in R$ will consist of endless repetitions of Figure 10-23 as indicated in Figure 10-24.

The graph of the cosine function is constructed in a similar way. Figure 10-25 shows the graph of $y = \cos x$.

Each segment of the graph of the sine function as in Figure 10-23 is called a **sine wave** and is one **cycle** of the graph.

In general, when a periodic function has a maximum value M and a minimum value m, the **amplitude** of the function is $\dfrac{M - m}{2}$. Thus

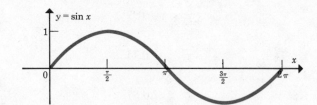

Figure 10-23

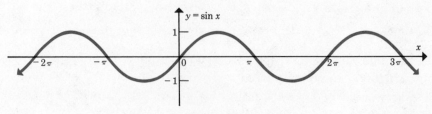

Figure 10-24

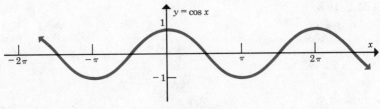

Figure 10-25

the amplitude of the sine wave is 1 since

$$\frac{1 - (-1)}{2} = 1;$$

and the period of the sine function is 2π. (Recall Definition 10.1.)

Example 1   Give the amplitude and period of the curve in Figure 10-26. (Note that the scales on the two axes are different.)

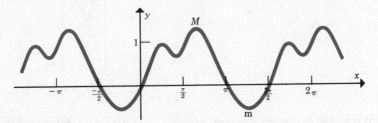

Figure 10-26

Solution: $M = 1.3$ and $m = -0.5$. The amplitude is

$$\frac{1.3 - (-0.5)}{2} = \frac{1.3 + 0.5}{2} = 0.9.$$

The period is $\frac{3\pi}{2}$.

The tangent, cotangent, secant, and cosecant functions are also periodic functions as you can show. Since the values of u and v on the unit circle $u^2 + v^2 = 1$ repeat in a regular fashion, then $\tan x = \dfrac{v}{u}$, for example, will also repeat in a regular fashion. The graph of the function $y = \tan x$ confirms this. To construct this graph we first construct Table 10-5 of some ordered pairs $(x, \tan x)$. Using this table, we can sketch one cycle of the graph. (See Figure 10-27.)

TABLE 10-5

x	$\tan x$
$-\dfrac{\pi}{2}$	—
$-\dfrac{\pi}{3}$	-1.7
$-\dfrac{\pi}{4}$	-1
$-\dfrac{\pi}{6}$	-0.58
0	0
$\dfrac{\pi}{6}$	0.58
$\dfrac{\pi}{4}$	1
$\dfrac{\pi}{3}$	1.7
$\dfrac{\pi}{2}$	—

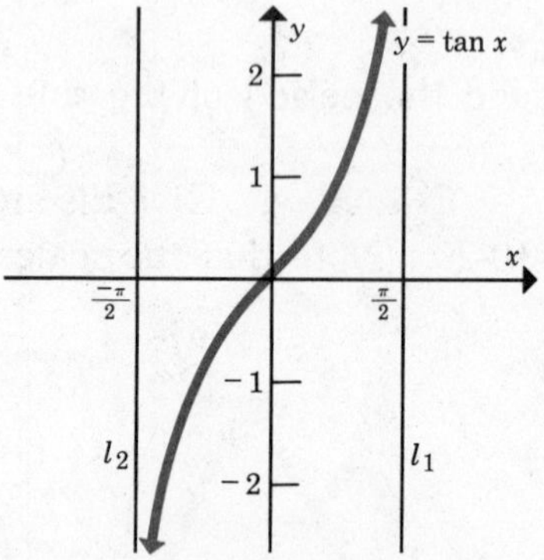

Figure 10-27

We have, by definition, $\tan x = \dfrac{v}{u}$ $(u \neq 0)$. As x approaches $\frac{\pi}{2}$, u approaches 0. Since v remains approximately 1, $\tan x$ becomes increasingly large as x approaches $\frac{\pi}{2}$. We can make the difference $|\frac{\pi}{2} - x|$ as small a positive number as we wish and, as we do so, $\tan x$ becomes increasingly large. A similar observation holds as x approaches $-\frac{\pi}{2}$. Thus, the lines $l_1 : x = \frac{\pi}{2}$ and $l_2 : x = -\frac{\pi}{2}$ are asymptotes of the graph of the equation $y = \tan x$ for $-\frac{\pi}{2} < x < \frac{\pi}{2}$. (Compare the asymptotes of the tangent curve with the asymptotes of hyperbolas discussed in Section 7.5.)

Since the domain of the tangent function does not contain numbers, x, for which $\cos x = 0$, the domain is

$$\left\{ x : x \in R \text{ and } x \neq \frac{\pi}{2} + n\pi, \, n \in I \right\}.$$

Accordingly, there are no points on the graph whose first coordinate is $\frac{\pi}{2} + n\pi$ for any $n \in I$. What is the range of the tangent function?

The graph of $y = \tan x$ for all x in the domain of the function repeats each cycle. (See Figure 10-28.) The period of the graph is π.

The graph of the secant function shown in Figure 10-29 can be obtained in a manner similar to that of the tangent function. Note that the range of the graph is $\{y : y \in R \text{ and } |y| \geq 1\}$. What is the domain?

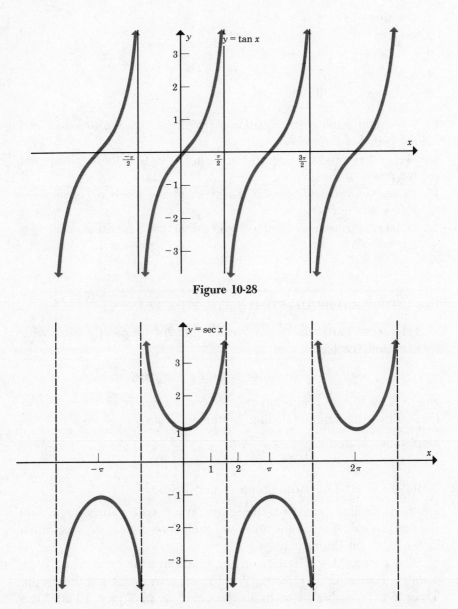

Figure 10-28

Figure 10-29

EXERCISES 10.3

◄ In Exercises 1–10, sketch the graph of each equation. Give the amplitude (in Exercises 1, 2, 7, 8) and the period of each graph.

1. $y = \sin x$ 6. $y = \csc x$
2. $y = \cos x$ 7. $y = \sin (-x)$
3. $y = \tan x$ 8. $y = -\cos x$
4. $y = \cot x$ 9. $y = \tan (-x)$
5. $y = \sec x$ 10. $y = -\sec x$

11. Sketch the graph of the equation $y = 2 \sin x$. Give the amplitude and the period of the graph.
12. Sketch the graph of the equation $y = \sin x + 1$. Give the amplitude and the period of the graph.
13. CHALLENGE PROBLEM. Sketch the graph of the function f defined by $f(x) = \sin x + \cos x$.
14. CHALLENGE PROBLEM. Sketch the graph of the function g defined by $g(x) = 2(\cos x - \sin x)$.

10.4   THE GRAPHS OF COMPOSITE FUNCTIONS

Engineers, particularly electrical engineers, frequently work with equations of the form

$$y = A \sin Bx$$

where A and B are real numbers. As you can see from the examples that follow, the function defined by the equation $y = A \sin Bx$ has an amplitude $|A|$ and a period $\dfrac{2\pi}{|B|}$.

Example 1   Draw the graph of $y = 3 \sin x$.

Solution: To find points on the graph, we choose values for x, find the corresponding value for $\sin x$, and then determine $3 \sin x$ as shown in the table on the next page.

The coordinates of points on the graph are the ordered pairs $(x, 3 \sin x)$. Hence $(0, 0)$, $(\frac{\pi}{6}, 1.5)$, $(\frac{\pi}{4}, 2.13)$, etc., are points on the graph. These points enable us to draw the curve as in Figure 10-30. The graph shows that the amplitude is 3 and the period is 2π.

x	$\sin x$	$3 \sin x$		x	$2x$	$\cos 2x$
0	0	0		0	0	1
$\dfrac{\pi}{6}$	0.5	1.5		$\dfrac{\pi}{6}$	$\dfrac{\pi}{3}$	$\dfrac{1}{2}$
$\dfrac{\pi}{4}$	0.71	2.13		$\dfrac{\pi}{4}$	$\dfrac{\pi}{2}$	0
$\dfrac{\pi}{3}$	0.87	2.61		$\dfrac{\pi}{3}$	$\dfrac{2\pi}{3}$	$-\dfrac{1}{2}$
$\dfrac{\pi}{2}$	1	3		$\dfrac{\pi}{2}$	π	-1
$\dfrac{2\pi}{3}$	0.87	2.61		$\dfrac{2\pi}{3}$	$\dfrac{4\pi}{3}$	$-\dfrac{1}{2}$
$\dfrac{3\pi}{4}$	0.71	2.13		$\dfrac{3\pi}{4}$	$\dfrac{3\pi}{2}$	0
$\dfrac{5\pi}{6}$	0.5	1.5		$\dfrac{5\pi}{6}$	$\dfrac{5\pi}{3}$	$\dfrac{1}{2}$
π	0	0		π	2π	1

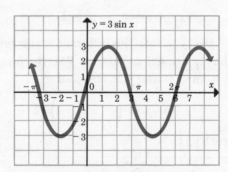

Figure 10-30

Example 2   Draw the graph of $y = \cos 2x$.

Solution: We form the table above by listing assorted values of x, finding $2x$, and then finding $\cos 2x$.

The coordinates of points on the graph are the ordered pairs $(x, \cos 2x)$. Using the points from the table, we draw the graph as in Figure 10-31. The graph shows that the amplitude is 1 and the period is π.

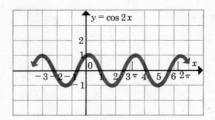

Figure 10-31

Example 3 Draw the graph of $y = 3 \cos 2x$.

Solution: The function will have an amplitude of 3 and a period of π. (Why?) By substituting a few values for x, we find that the points $(0, 3)$, $(\frac{\pi}{6}, \frac{3}{2})$, $(\frac{\pi}{4}, 0)$, and $(\frac{\pi}{3}, -\frac{3}{2})$ are on the graph. Using these points and our knowledge of the amplitude and period, we can sketch the graph as seen in Figure 10-32.

Recall that a translation is a function that associates with each point (x, y) the point $(x + h, y + k)$. The graph of the equation $y = \sin x + 1$ is the image of the graph of the equation $y = \sin x$ under the translation

$$(x, y) \longrightarrow (x, y + 1)$$

as shown in Figure 10-33. (Note the different scales on the two axes.)

The graph of the equation $y = \sin (x - \frac{\pi}{2})$ is the image of the graph of the equation $y = \sin x$ under the translation

$$(x, y) \longrightarrow \left(x - \frac{\pi}{2}, y\right)$$

as shown in Figure 10-34.

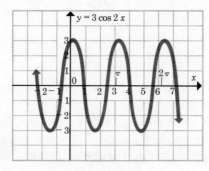

Figure 10-32

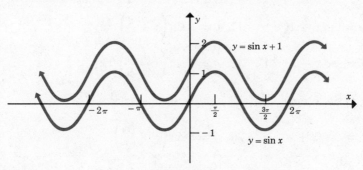

Figure 10-33

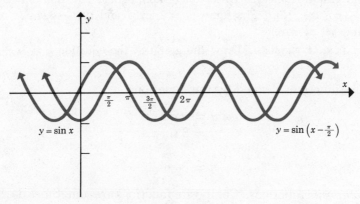

Figure 10-34

EXERCISES 10.4A

◀ In Exercises 1–10, draw the graph of the given equation. Give the period and amplitude of the function defined by the equation.

1. $y = \sin 3x$ 6. $y = \cos(-x)$

2. $y = \cos 2x$ 7. $y = 2 \sin 3x$

3. $y = 2 \sin x$ 8. $y = 3 \cos 2x$

4. $y = 3 \cos x$ 9. $y = 4 \sin \dfrac{1}{2}x$

5. $y = \sin \dfrac{1}{2}x$ 10. $y = 3 \cos \dfrac{1}{3}x$

◀ In Exercises 11–20, draw the graph of each equation.

11. $y = \sin x + 2$ 12. $y = \cos x - 1$

13. $y = \tan x + 1$ **17.** $y = 2 \sin \left(x + \dfrac{\pi}{2} \right)$

14. $y = \sec x - 1$ **18.** $y = 2 \sin \left(x - \dfrac{\pi}{2} \right)$

15. $y = \sin (x - \pi)$ **19.** $y = - \sin (x + \pi)$

16. $y = \cos (x + \pi)$ **20.** $y = 2 \sin \left(x + \dfrac{3\pi}{2} \right)$

21. Draw the graph of the function $f : f(x) = |\sin x|$.

22. CHALLENGE PROBLEM. Draw the graph of the equation $y = 2 \sin 3x - 3 \cos 2x$. (*Hint:* First draw $y = 2 \sin 3x$ and $y = -3 \cos 2x$ on the same set of axes.)

23. CHALLENGE PROBLEM. Draw the graph of the equation $y = \frac{1}{2}x + \sin x$.

The functions defined by the equations

$$y = \cos 2x,$$
$$y = 3 \sin x,$$

and

$$y = 3 \cos 2x$$

are *composite* functions. Composite functions are sometimes described as being "functions of functions." Recall the procedure by which we found points on the graph of $y = \cos 2x$ (Example 2). First, we chose a value for x, then we found $2x$, and finally we found the value of the cosine function that was associated with $2x$. If we express each step as a mapping, we have

$$x \longrightarrow 2x \longrightarrow \cos 2x.$$

If f is a function such that $f(x) = 2x$ and g is a function such that $g(x) = \cos x$, then $g[f(x)] = \cos 2x$. The notation $g[f(x)]$ signifies that first we find $f(x) = 2x$ and then we find $g(2x)$. A single function that defines the same mapping as two separate functions is called the composite of the two separate functions. Thus if the function h is defined by the equation $h(x) = \cos 2x$, then

$$h(x) = g[f(x)].$$

We write $h = g \circ f$.

The function F defined by the equation $y = 3 \sin x$ defines the same mapping as the composite $g \circ f$ of the two functions f and *g* where

$$f : f(x) = \sin x \qquad \text{and} \qquad g : g(x) = 3x.$$

Figure 10-35 shows a diagram of the mappings. Thus $F(x) = g[f(x)]$.
We can also indicate that F is a composite function by writing

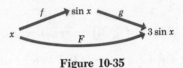

Figure 10-35

$F = g \circ f$.

  Note that $g \circ f \neq f \circ g$. If $f(x) = \sin x$ and $g(x) = 3x$, for $F = g \circ f$
we have

$$F(x) = g[f(x)] = 3 \sin x,$$

whereas for $F = f \circ g$ we have

$$F(x) = f[g(x)] = \sin 3x.$$

Hence we conclude that the operation of composition of functions is
not commutative.

> **Definition 10.4** Given two functions, f and g, the
> function F defined by $F(x) = g[f(x)]$ is called a **composite**
> of f and g and is denoted by $g \circ f$. The domain of F
> is the set of all elements x in the domain of f for which
> $f(x)$ is in the domain of g. The operation of forming
> a composite is called **composition**.

  The function G defined by the equation $y = 3 \sin 2x$ defines the
same mapping as the composite $h \circ F$ of the two functions h and F
where

$$F : F(x) = \sin 2x \qquad \text{and} \qquad h : h(x) = 3x.$$

But F, as we have seen, is a composite function itself. Thus $F = f \circ g$
where

$$f : f(x) = \sin x \qquad \text{and} \qquad g : g(x) = 2x.$$

Hence we can write $G = h \circ F = h \circ (f \circ g)$. Does

$$h \circ (f \circ g) = (h \circ f) \circ g?$$

  Examples 4 and 5 show that composition of functions need not
involve circular functions.

Example 4   Let $f : f(x) = 2x$ and $g : g(x) = x^2 + x + 1$. Let $F = f \circ g$ and $G = g \circ f$. Find (1) $F(3)$ and (2) $G(3)$.

Solution: (1) Since $F = f \circ g$, $F(3) = f[g(3)]$. We have

$$g(3) = 9 + 3 + 1 = 13$$

and hence

$$F(3) = f[g(3)] = f(13) = 26.$$

(2) Since $G = g \circ f$, $G(3) = g[f(3)]$. We have $f(3) = 6$ and hence

$$G(3) = g[f(3)] = 36 + 6 + 1 = 43.$$

Once again $f \circ g \neq g \circ f$.

Example 5   Given $f : f(x) = x^2 + x + 1$, $g : g(x) = x + 2$, and $h : h(x) = -2x$, find (1) $f \circ g$, (2) $g \circ h$, (3) $(f \circ g) \circ h$, and (4) $f \circ (g \circ h)$.
Solution:

1. $f[g(x)] = (x + 2)^2 + (x + 2) + 1 = x^2 + 5x + 7$.
   Let $f \circ g = F$. Then

$$F : F(x) = x^2 + 5x + 7.$$

2. $g[h(x)] = (-2x) + 2 = -2x + 2$. Let $G = g \circ h$. Then

$$G : G(x) = -2x + 2.$$

3. $(f \circ g)[h(x)] = F[h(x)] = (-2x)^2 + 5(-2x) + 7 = 4x^2 - 10x + 7$.
   Let $H = (f \circ g) \circ h$. Then

$$H(x) = 4x^2 - 10x + 7.$$

4. $f \circ [(g \circ h)(x)] = f[G(x)] =$
   $(-2x + 2)^2 + (-2x + 2) + 1 = 4x^2 - 10x + 7$. Hence

$$f \circ (g \circ h) = (f \circ g) \circ h.$$

Note that $f \circ g$ is not a symbol related to the product of two functional values, that is,

$$f \circ g \neq (F : F(x) = f(x) \cdot g(x)).$$

EXERCISES 10.4B

◀ In Exercises 1–10, find the value or a defining equation of each expression if $f : f(x) = x^2 + 1$ and $g : g(x) = x - 2$ for all $x \in R$.

1. $f[g(2)]$ 6. $f[g(x)]$

2. $g[f(2)]$ 7. $g[f(x)]$

3. $g[f(0)]$ 8. $(f \circ g)[g(x)]$

4. $g[g(1)]$ 9. $f[(g \circ g)(x)]$

5. $f[f(0)]$ 10. $\dfrac{f[g(x)] - f[g(1)]}{g(x)}$ for $x \neq 2$

◀In Exercises 11–20, find the value or a defining equation for each expression if $f : f(x) = x + \pi$ and $g : g(x) = \sin x$.

11. $f[g(0)]$ 16. $f[g(-\pi)]$

12. $g[f(0)]$ 17. $f[g(-x)]$

13. $f[g(\pi)]$ 18. $-f[g(x)]$

14. $g[f(\pi)]$ 19. $f[g(2x)]$

15. $-f[g(\pi)]$ 20. $\dfrac{g[f(\pi)] + g[f(x - \pi)]}{g(x)}$ for $g(x) \neq 0$

◀In Exercises 21–24, $f : f(x) = \dfrac{1}{x}$ for all nonzero $x \in R$.

21. Find $f[f(1)]$.   23. Find $f[f(\pi)]$.

22. Find $f[f(-3)]$.   24. Find $f[f(x)]$.

25. The function $h : h(x) = 2 \cos 3x$ can be expressed as the composition $h = k \circ g \circ f$ of three functions f, g, and k. Define f, g, and k.

26. Let $f : f(x) = x + 2$, $g : g(x) = x - 3$, and $h : h(x) = x^2$ for all $x \in R$. Find expressions for
    (a) $(f \circ g)(x)$ (c) $[(f \circ g) \circ h](x)$
    (b) $(f \circ h)(x)$ (d) $[f \circ (g \circ h)](x)$

27. **CHALLENGE PROBLEM.** Let S be the set of all linear functions $f : f(x) = ax + b$, $a, b \in R$. Prove that
    (a) S is closed under composition of functions;
    (b) the operation of composition is associative in S.

28. **CHALLENGE PROBLEM.** Prove that $[(f \circ g) \circ h](x) = [f \circ (g \circ h)](x)$ for any functions f, g, and h and all x for which the composites are defined, that is, show that composition of functions is associative.

10.5 ADDITION FORMULAS

The circular functions we have studied so far, such as the sine function, involve one variable. Functions such as $\cos(x + y)$ or $\sin(x - y)$ that involve two variables have considerable importance. It is useful to have formulas that express the functions of the sum or difference of two variables in terms of functions of the separate variables.

If, under the standard mapping, the real number x is associated with point R on the unit circle and the real number y is associated with point S as in Figure 10-36, then we have the length of arc $AR = x$ and the length of arc $AS = y$. If the length of arc $RP =$ length of arc $AS = y$, then we have the length of arc $(AR + RP) = x + y$. Thus we have, under the standard mapping, $W(x + y) = P$. Conversely, if $W(x + y) = P$, then the length of arc $AS =$ length of arc RP. (Note that length of arc refers to arc measure expressed as a real number.)

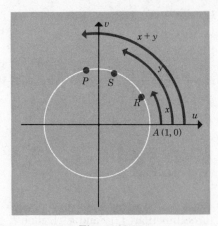

Figure 10-36

We have

$$W(x) = (\cos x, \sin x),$$
$$W(y) = (\cos y, \sin y),$$

and

$$W(x + y) = \big(\cos (x + y), \sin (x + y)\big).$$

Also

$$W(-y) = (\cos y, \, - \sin y)$$

since $\cos (-y) = \cos y$ and $\sin (-y) = -\sin y$ for all $y \in R$.

If $W(-y) = T$ as in Figure 10-37, $RT = AP$ since, in the same circle, arcs having the same length subtend chords having the same length. Recall that the formula

$$PQ = \sqrt{(x_1 - x_2)^2 + (y_1 - y_2)^2}$$

gives the distance PQ between two points $P(x_1, y_1)$ and $Q(x_2, y_2)$. If we now apply this formula for the distance between the two points R and T, we get

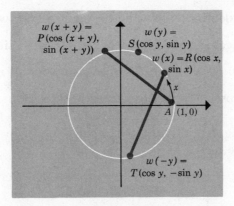

Figure 10-37

$(RT)^2$

$$= (\cos x - \cos y)^2 + (\sin x + \sin y)^2$$
$$= (\cos^2 x - 2 \cos x \cos y + \cos^2 y)$$
$$\qquad\qquad + (\sin^2 x + 2 \sin x \sin y + \sin^2 y)$$
$$= (\cos^2 x + \sin^2 x) + (\cos^2 y + \sin^2 y)$$
$$\qquad\qquad\qquad - 2 \cos x \cos y + 2 \sin x \sin y.$$

Since $\sin^2 x + \cos^2 x = 1$ for all $x \in R$,

$$(RT)^2 = 2 - 2 \cos x \cos y + 2 \sin x \sin y.$$

Similarly, we have

$$(AP)^2 = [1 - \cos (x + y)]^2 + \sin^2 (x + y)$$
$$= 1 - 2 \cos (x + y) + \cos^2 (x + y) + \sin^2 (x + y)$$
$$= 2 - 2 \cos (x + y).$$

Since $AP = RT$,

$$2 - 2 \cos (x + y) = 2 - 2 \cos x \cos y + 2 \sin x \sin y$$

or

(X) $$\cos (x + y) = \cos x \cos y - \sin x \sin y.$$

(Recall that unless otherwise stated we are to assume that R is the domain of the variables in an identity involving circular functions.)

Identity (X) is very important because it serves as a starting point for the development of many other useful identities.

First, if $x = y$, we have

(Xa) $$\cos 2x = \cos^2 x - \sin^2 x.$$

Next, since Identity (X) is valid for all $x, y \in R$, we have

$$\cos(x - y) = \cos[x + (-y)] = \cos x \cos(-y) - \sin x \sin(-y).$$

Since $\cos(-x) = \cos x$ and $\sin(-x) = -\sin x$ for all $x \in R$, we have

(XI) $$\cos(x - y) = \cos x \cos y + \sin x \sin y.$$

Now if, using Identity (I), we replace $\sin^2 x$ by $1 - \cos^2 x$ in Identity (Xa), we have

$$\cos 2x = \cos^2 x - (1 - \cos^2 x) = -1 + 2\cos^2 x$$

or

$$\cos^2 x = \frac{1}{2}(1 + \cos 2x).$$

If we now replace x by $\dfrac{u}{2}$, and hence $2x$ by u, we have

$$\left(\cos \frac{u}{2}\right)^2 = \frac{1}{2}(1 + \cos u).$$

Thus

$$\left|\cos \frac{u}{2}\right| = \sqrt{\frac{1 + \cos u}{2}}.$$

Since the choice of notation for a variable is arbitrary, we can rewrite this identity in the form

(XII) $$\left|\cos \frac{x}{2}\right| = \sqrt{\frac{1 + \cos x}{2}} \qquad \text{for all } x \in R.$$

Thus

$$\cos \frac{x}{2} = \sqrt{\frac{1 + \cos x}{2}} \qquad \text{or} \qquad \cos \frac{x}{2} = -\sqrt{\frac{1 + \cos x}{2}}$$

where the choice of sign is determined by the quadrant in which $\dfrac{x}{2}$ lies.

The examples and exercises that follow indicate a few of the useful applications of these identities.

Example 1 Find $\cos \frac{\pi}{12}$.

Solution: Since $\frac{\pi}{12} = \frac{\pi}{4} - \frac{\pi}{6}$, we have, by Identity (XI),

$$\cos \frac{\pi}{12} = \cos\left(\frac{\pi}{4} - \frac{\pi}{6}\right) = \cos \frac{\pi}{4} \cos \frac{\pi}{6} + \sin \frac{\pi}{4} \sin \frac{\pi}{6}.$$

Then, since $\cos\frac{\pi}{4} = \sin\frac{\pi}{4} = \frac{\sqrt{2}}{2}$, $\cos\frac{\pi}{6} = \frac{\sqrt{3}}{2}$, and $\sin\frac{\pi}{6} = \frac{1}{2}$, we have

$$\cos\frac{\pi}{12} = \frac{\sqrt{2}}{2}\cdot\frac{\sqrt{3}}{2} + \frac{\sqrt{2}}{2}\cdot\frac{1}{2} = \frac{\sqrt{6}}{4} + \frac{\sqrt{2}}{4} = \frac{\sqrt{6}+\sqrt{2}}{4}.$$

Example 2 Find $\cos\frac{\pi}{8}$.

Solution: By Identity (XII), we have

$$\left|\cos\frac{\pi}{8}\right| = \sqrt{\frac{1+\cos\frac{\pi}{4}}{2}}.$$

Then, since $\cos\frac{\pi}{4} = \frac{\sqrt{2}}{2}$,

$$\left|\cos\frac{\pi}{8}\right| = \sqrt{\frac{1+\frac{\sqrt{2}}{2}}{2}} = \frac{1}{2}\sqrt{2+\sqrt{2}}.$$

Since $0 < \frac{\pi}{8} < \frac{\pi}{2}$, $\cos\frac{\pi}{8}$ is positive. Hence

$$\cos\frac{\pi}{8} = \frac{1}{2}\sqrt{2+\sqrt{2}} \approx 0.92.$$

Example 3 Show that $\cos\left(\frac{\pi}{2} - y\right) = \sin y$ for all $y \in R$.

Solution: We use Identity (XI) and let $x = \frac{\pi}{2}$. Then

$$\cos\left(\frac{\pi}{2} - y\right) = \cos\frac{\pi}{2}\cos y + \sin\frac{\pi}{2}\sin y$$

$$= 0\cdot\cos y + 1\cdot\sin y = \sin y.$$

EXERCISES 10.5A

◀ In Exercises 1–5, find the functional value.

1. $\cos\dfrac{5\pi}{12}$ 4. $\cos\dfrac{13\pi}{12}$

2. $\cos\dfrac{7\pi}{12}$ 5. $\cos\dfrac{3\pi}{8}$

3. $\cos\dfrac{11\pi}{12}$

6. Show that $\cos\left(\dfrac{\pi}{2} + x\right) = -\sin x$ for all $x \in R$.

7. Show that $\cos\left(\dfrac{3\pi}{2} - t\right) = -\sin t$ for all $t \in R$.

8. Show that $\cos 2x \cos x + \sin 2x \sin x = \cos x$.

◀ In Exercises 9–14, find the functional value given that $\sin x = \frac{3}{5}$ $(0 < x < \frac{\pi}{2})$ and $\cos y = \frac{5}{13}$ $(0 < y < \frac{\pi}{2})$.

9. $\cos\dfrac{y}{2}$ 12. $\cos(x + y)$

10. $\cos 2y$ 13. $\cos\dfrac{x}{2}$

11. $\cos(x - y)$ 14. $\cos 2x$

15. Verify that $\cos(\pi - s) = -\cos s$ for all $s \in R$ by using Identity (X).

16. If $\cos y = \dfrac{4}{5}$ and $0 < y < \dfrac{\pi}{2}$, find $\cos 2y$.

17. If $\cos r = -\dfrac{3}{5}\left(\dfrac{\pi}{2} < r < \pi\right)$ and $\sin s = -\dfrac{5}{13}\left(-\dfrac{\pi}{2} < s < 0\right)$, find $\cos(r + s)$.

18. If $\cos x = -\dfrac{5}{13}\left(\dfrac{\pi}{2} < x < \pi\right)$ and $\sin y = \dfrac{3}{5}\left(0 < y < \dfrac{\pi}{2}\right)$, find $\cos(x + y)$.

19. If $\cos 3x = -\dfrac{\sqrt{7}}{4}\left(\pi < 3x < \dfrac{3\pi}{2}\right)$, find $\sin 3x$.

20. Show that, for $r, s \in R$,

$$\cos\left(\frac{\pi}{2} - r - s\right) = \sin(r + s).$$

We now develop identities, similar to Identities (X), (XI), and (XII), concerning the sine function. First, by Example 3,

(XIII) $\cos\left(\dfrac{\pi}{2} - y\right) = \sin y$ for all $y \in R$.

Since y is arbitrary, we have for all x and $y \in R$

(1) $\cos\left(\dfrac{\pi}{2} - x\right) = \sin x,$

(2) $\cos\left[\dfrac{\pi}{2} - (x + y)\right] = \sin(x + y),$

(3) $\cos\left[\dfrac{\pi}{2} - (x - y)\right] = \sin(x - y),$

(4) $$\cos\left[\frac{\pi}{2} - \left(\frac{\pi}{2} - x\right)\right] = \sin\left(\frac{\pi}{2} - x\right).$$

Equation (4) is equivalent to

(XIV) $$\sin\left(\frac{\pi}{2} - x\right) = \cos x.$$

Equation (3) leads to a formula for $\sin(x - y)$:

$$\sin(x - y) = \cos\left[\frac{\pi}{2} - (x - y)\right]$$
$$= \cos\left[\left(\frac{\pi}{2} - x\right) + y\right]$$
$$= \cos\left(\frac{\pi}{2} - x\right)\cos y - \sin\left(\frac{\pi}{2} - x\right)\sin y$$
$$= \sin x \cos y - \cos x \sin y.$$

Hence

(XV) $$\sin(x - y) = \sin x \cos y - \cos x \sin y.$$

Similarly, Equation (2) yields

(XVI) $$\sin(x + y) = \sin x \cos y + \cos x \sin y.$$

If $x = y$, we have

(XVIa) $$\sin 2x = 2 \sin x \cos x.$$

Using Identity (I), if we replace $\cos^2 x$ by $1 - \sin^2 x$ in Identity (Xa), we obtain

$$\cos 2x = \cos^2 x - \sin^2 x,$$
$$\cos 2x = (1 - \sin^2 x) - \sin^2 x = 1 - 2\sin^2 x,$$

so that

$$\sin^2 x = \frac{1}{2}(1 - \cos 2x).$$

Again, if we replace x by $\frac{u}{2}$ and hence $2x$ by u, we have

$$\left(\sin\frac{u}{2}\right)^2 = \frac{1}{2}(1 - \cos u).$$

Thus

$$\left|\sin\frac{u}{2}\right| = \sqrt{\frac{1 - \cos u}{2}}.$$

Since the choice of notation for a variable is arbitrary, we can rewrite this identity as

(XVII) $$\left| \sin \frac{x}{2} \right| = \sqrt{\frac{1 - \cos x}{2}}.$$

Whether $\sin \frac{x}{2}$ is positive or negative depends on the quadrant in which $\frac{x}{2}$ lies.

Example 4   Find $\sin \frac{\pi}{12}$.

Solution: Since $\frac{\pi}{12} = \frac{\pi}{3} - \frac{\pi}{4}$, we have, by Identity (XV),

$$\sin \frac{\pi}{12} = \sin \left(\frac{\pi}{3} - \frac{\pi}{4} \right)$$

$$= \sin \frac{\pi}{3} \cos \frac{\pi}{4} - \cos \frac{\pi}{3} \sin \frac{\pi}{4}$$

$$= \frac{\sqrt{3}}{2} \cdot \frac{\sqrt{2}}{2} - \frac{1}{2} \cdot \frac{\sqrt{2}}{2}$$

$$= \frac{\sqrt{2}}{4} (\sqrt{3} - 1) \approx 0.26.$$

Example 5   Find $\sin \frac{3\pi}{8}$.

Solution: If we take $\frac{x}{2} = \frac{3\pi}{8}$, then $x = \frac{3\pi}{4}$. Thus, by Identity (XVII),

$$\left| \sin \frac{3\pi}{8} \right| = \sqrt{\frac{1 - \cos \frac{3\pi}{4}}{2}}$$

$$= \sqrt{\frac{1 - \left(-\frac{\sqrt{2}}{2} \right)}{2}} = \sqrt{\frac{1 + \frac{\sqrt{2}}{2}}{2}}$$

$$= \sqrt{\frac{2 + \sqrt{2}}{4}} = \frac{1}{2} \sqrt{2 + \sqrt{2}}.$$

Since $0 < \frac{3\pi}{8} < \frac{\pi}{2}$, $\sin \frac{3\pi}{8} > 0$. Thus

$$\sin \frac{3\pi}{8} = \frac{1}{2} \sqrt{2 + \sqrt{2}} \approx 0.92.$$

EXERCISES 10.5B

◀In Exercises 1–5, find the functional value.

1. $\sin \dfrac{5\pi}{12}$ 4. $\sin \dfrac{9\pi}{8}$

2. $\sin \dfrac{7\pi}{12}$ 5. $\sin \dfrac{11\pi}{12}$

3. $\sin \dfrac{\pi}{8}$

6. Show that $\sin \left(\dfrac{\pi}{2} + x \right) = \cos x$.

7. Simplify $\sin \left(\dfrac{7\pi}{2} - x \right)$.

◀In Exercises 8–12, find the functional value, given that $\sin x = \frac{3}{5}$ and $0 < x < \frac{\pi}{2}$.

8. $\cos x$ 11. $\sin 4x$

9. $\sin 2x$ 12. $\sin \dfrac{1}{2}x$

10. $\sin 3x$

◀In Exercises 13–17, find the functional value, given that $\sin y = \frac{5}{13}$ and $\frac{\pi}{2} < y < \pi$.

13. $\cos y$ 16. $\sin 4y$

14. $\sin 2y$ 17. $\sin \dfrac{y}{2}$

15. $\sin 3y$

18. Show that

$$\sin (x + y) + \sin (x - y) = 2 \sin x \cos y.$$

19. Verify that $\sin (x \pm \pi) = -\sin x$ using the procedures of Section 10.5.

20. Find x if $\sin x = \sin \frac{\pi}{6} \cos \frac{2\pi}{5} + \cos \frac{\pi}{6} \sin \frac{2\pi}{5}$.

From Identity (VI)

$$\tan x = \frac{\sin x}{\cos x} \qquad \left(x \neq \frac{\pi}{2} + n\pi \text{ for } n \in I \right).$$

Hence, for $x + y \in R$ and $(x + y) \neq \frac{\pi}{2} + n\pi$ for $n \in I$, we have

$$\tan (x + y) = \frac{\sin (x + y)}{\cos (x + y)}.$$

Then, by Identities (XVI) and (X),

$$\tan (x + y) = \frac{\sin x \cos y + \cos x \sin y}{\cos x \cos y - \sin x \sin y}.$$

Dividing numerator and denominator of the right-hand member by $\cos x \cos y$ ($\cos x \cos y \neq 0$), we obtain

(XVIII) $$\tan (x + y) = \frac{\tan x + \tan y}{1 - \tan x \tan y}.$$

Similarly, we obtain for $(x - y) \neq \frac{\pi}{2} + n\pi$ for $n \in I$,

(XIX) $$\tan (x - y) = \frac{\tan x - \tan y}{1 + \tan x \tan y}.$$

Although formulas for $\cot (x + y)$, $\sec (x + y)$, and $\csc (x + y)$ exist, we shall not derive them here.

For easy reference we list the more important identities. (Remember that the domain of the variable or variables is assumed to be R unless stated otherwise.) Previously we demonstrated that

(I) $\sin^2 x + \cos^2 x = 1.$

(II) $\cos (-x) = \cos x.$

(III) $\sin (-x) = -\sin x.$

(IV) $\cos (x \pm \pi) = -\cos x.$

(V) $\sin (x \pm \pi) = -\sin x.$

(VI) $\tan x = \dfrac{\sin x}{\cos x}$ $(x \neq \dfrac{\pi}{2} + n\pi$ for $n \in I).$

(VII) $\cot x = \dfrac{\cos x}{\sin x}$ $(x \neq n\pi$ for $n \in I).$

(VIII) $\sec x = \dfrac{1}{\cos x}$ $(x \neq \dfrac{\pi}{2} + n\pi$ for $n \in I).$

(IX) $\csc x = \dfrac{1}{\sin x}$ $(x \neq n\pi$ for $n \in I).$

In Section 10.5 we have shown that

(XVI) $\sin (x + y) = \sin x \cos y + \cos x \sin y.$

(XV) $\sin (x - y) = \sin x \cos y - \cos x \sin y.$

(X) $\cos (x + y) = \cos x \cos y - \sin x \sin y.$

(XI) $\cos (x - y) = \cos x \cos y + \sin x \sin y.$

(XVIII) $\tan (x + y) = \dfrac{\tan x + \tan y}{1 - \tan x \tan y}$

$$\left(x + y \neq \frac{\pi}{2} + n\pi \text{ for } n \in I\right).$$

(XIX) $\tan (x - y) = \dfrac{\tan x - \tan y}{1 + \tan x \tan y}$

$$\left(x - y \neq \frac{\pi}{2} + n\pi \text{ for } n \in I\right).$$

(XVIa) $\sin 2x = 2 \sin x \cos x.$

(Xa) $\cos 2x = \cos^2 x - \sin^2 x.$

(XIV) $\sin \left(\dfrac{\pi}{2} - x\right) = \cos x.$

(XIII) $\cos \left(\dfrac{\pi}{2} - x\right) = \sin x.$

(XVII) $\left| \sin \dfrac{x}{2} \right| = \sqrt{\dfrac{1 - \cos x}{2}}.$

(XII) $\left| \cos \dfrac{x}{2} \right| = \sqrt{\dfrac{1 + \cos x}{2}}.$

EXERCISES 10.5C

◀ In Exercises 1–6, find the functional value, given that $\sin x = \frac{3}{5}$ $(0 < x < \frac{\pi}{2})$ and $\sin y = \frac{5}{13}$ $(0 < y < \frac{\pi}{2})$.

1. $\tan x$ 4. $\tan (x - y)$

2. $\tan y$ 5. $\tan 2x$

3. $\tan (x + y)$ 6. $\tan 2y$

◀ In Exercises 7–10, find the functional value given that $\tan x = \sqrt{3}$ and $0 < x < \frac{\pi}{2}$.

7. $\sin 2x$ **9.** $\sin \dfrac{x}{2}$

8. $\cos 2x$ **10.** $\cos \dfrac{x}{2}$

◀ In Exercises 11–14, find the functional value, given that $\tan x = -2\sqrt{2}$ and $0 > x > -\frac{\pi}{2}$,

11. $\sin x$ **13.** $\sin 2x$
12. $\cos x$ **14.** $\cos 2x$

15. Simplify $\cos\left(\dfrac{5\pi}{2} - x\right)$.

16. Simplify $\sin\left(\dfrac{11\pi}{2} - x\right)$.

17. Find x if $0 < x < \dfrac{\pi}{2}$ and $\sin 3x \cos 2x + \cos 3x \sin 2x = 1$.

18. Find t if $0 < t < \dfrac{\pi}{2}$ and $\sin 2t \cos t - \cos 2t \sin t = \dfrac{1}{2}$.

19. Find $\sin y$ if $\sin \dfrac{y}{2} = \dfrac{3}{5}$ and $0 < y < \dfrac{\pi}{2}$.

20. Find $\sin n$ if $\cos \dfrac{n}{2} = -\dfrac{4}{5}$ and $0 > n > -\dfrac{\pi}{2}$.

◀ In Exercises 21–25, find the functional value, given that $\sin t = \frac{4}{5}$ and $\frac{\pi}{2} < t < \pi$.

21. $\cos t$ **24.** $\cos \dfrac{t}{2}$

22. $\tan t$ **25.** $\sin 2t$

23. $\sin \dfrac{t}{2}$

26. Find $\cos \dfrac{3\pi}{8}$.

27. Find $\sin \dfrac{23\pi}{12}$.

28. Find $\tan \dfrac{11\pi}{12}$.

29. Show that if $x \in R$ and $x \neq n\pi$ for $n \in I$, then

$$\tan \frac{x}{2} = \frac{1 - \cos x}{\sin x}.$$

30. Show that if $x \in R$ and $x \neq (2n + 1)\pi$ for $n \in I$, then

$$\left| \tan \frac{x}{2} \right| = \sqrt{\frac{1 - \cos x}{1 + \cos x}}.$$

31. Show that if $x \in R$ and $x \neq n\frac{\pi}{2}$ for $n \in I$, then

$$\cot 2x = \frac{\cot^2 x - 1}{2 \cot x}.$$

◀ In Exercises 32–35, show that the given statement is an identity.

32. $\cos(x + y) - \cos(x - y) = -2 \sin x \sin y$

33. $\cos x = \cos\left(\frac{x}{2} + \frac{x}{2}\right) = 2 \cos^2 \frac{x}{2} - 1$

34. $\cos(x + y) + \cos(x - y) = 2 \cos x \cos y$

35. $\cos\left(\frac{\pi}{3} + x\right) - \sin\left(\frac{\pi}{6} - x\right) = 0$

36. CHALLENGE PROBLEM. Find $\sin 4x$ if $\sin x = 0.25$ and $0 < x < \frac{\pi}{2}$.

37. CHALLENGE PROBLEM. Show that if $x + y + z = \pi$, then

$$\sin z = \sin x \cos y + \cos x \sin y.$$

38. CHALLENGE PROBLEM. Show that if $x + y + z = 2\pi$, then

$$\tan x + \tan y + \tan z = \tan x \tan y \tan z.$$

10.6   EQUATIONS AND IDENTITIES

So far we have been concerned with equations such as

$$\sin^2 x + \cos^2 x = 1$$

and

$$\sin(x + y) = \sin x \cos y + \sin y \cos x$$

which are commonly called **identities**. By this we mean that for every replacement of the variable by real numbers in the domain of the functions, we obtain a true statement. Thus

$$\sin^2 \frac{\pi}{2} + \cos^2 \frac{\pi}{2} = 1,$$

$$\sin^2 0.7 + \cos^2 0.7 = 1,$$

and

$$\sin^2 \left(\frac{\pi}{2} + 0.1 \right) + \cos^2 \left(\frac{\pi}{2} + 0.1 \right) = 1$$

are true statements. The solution set of the equation

$$\sin^2 x + \cos^2 x = 1$$

is $\{x : x \in R\}$.

An equation such as

$$\sin x = 1,$$

on the other hand, certainly does not have R as its solution set and thus is not called an identity. Indeed, we know that in the interval $0 \leq x \leq 2\pi$, $\sin x = 1$ only when $x = \frac{\pi}{2}$. Since the circular functions are periodic, the solution set also contains $-\frac{3\pi}{2}, \frac{5\pi}{2}, \frac{9\pi}{2}$ and so on. If the domain of the replacement set is not restricted, the solution set is an infinite set. We write that the solution set of the equation $\sin x = 1$ is

$$\left\{ x : x = \frac{\pi}{2} + 2n\pi \text{ for } n \in I \right\}.$$

A similar distinction exists among algebraic equations. For example, the equation

$$x^2 - 1 = (x + 1)(x - 1)$$

is an algebraic identity because every replacement of x by a real number forms a true statement; in the equation $2x + 3 = 5$, however, the only replacement of x that forms a true statement is 1.

To prove an identity involving circular functions means verifying that an equation becomes a true statement for every replacement of the variables in the domain of the functions involved. The usual procedure for verifying that an equation is an identity is to transform one member of the equation into the identical form of the second member by using the properties of the real numbers, the definitions of the circular functions, or previously proven identities. The statement of the identity does not usually include explicit mention of the domain of the various functions; however, any restrictions must be kept in mind to prevent using improper operations.

Example 1   Prove that $1 + \tan^2 x = \sec^2 x$.

Solution: (The domain of the tangent and the secant function does not contain the numbers $\frac{\pi}{2} + n\pi$ for $n \in I$.) From the unit circle we have

$$u^2 + v^2 = 1.$$

Since $u = 0$ if and only if $x = \frac{\pi}{2} + n\pi$ for $n \in I$ and these values of x are excluded, we have

$$1 + \frac{v^2}{u^2} = \frac{1}{u^2}.$$

Since $\tan x = \dfrac{v}{u}$ and $\sec x = \dfrac{1}{u}$ by definition, we have

(XX) $1 + \tan^2 x = \sec^2 x.$

The proof of the identity

(XXI) $1 + \cot^2 x = \csc^2 x$

is similar.

Example 2   Prove that

$$\left(\frac{1 + \csc x}{\csc x}\right)\left(\frac{\csc x - 1}{\csc x}\right) = \cos^2 x.$$

Solution: For $x \neq n\pi$ where $n \in I$, we have

$$\left(\frac{1 + \csc x}{\csc x}\right)\left(\frac{\csc x - 1}{\csc x}\right) = \left(\frac{1}{\csc x} + 1\right)\left(1 - \frac{1}{\csc x}\right)$$

$$= (\sin x + 1)(1 - \sin x)$$

$$= 1 - \sin^2 x = \cos^2 x.$$

Note that the proof involves the properties of the real numbers and the two identities

$$\sin x = \frac{1}{\csc x} \qquad \text{and} \qquad \sin^2 x + \cos^2 x = 1.$$

Note also that $|\csc x| \geq 1$ for all x in the domain of the function. Hence $\csc x \neq 0$ for all real numbers x in the domain.

Example 3   Prove that $\tan x \sin 2x = 2 \sin^2 x$.

Solution: Although, on occasion, there may be shorter and more elegant procedures, it is generally expedient to transform one member of a conjectured identity into an expression involving only the sine function and the cosine function. Thus for all $x \in R$ except $x = (2n + 1)\frac{\pi}{2}$ for $n \in I$,

$$\tan x \sin 2x = \frac{\sin x}{\cos x}(2 \sin x \cos x) = 2 \sin^2 x.$$

Although the restrictions or "worry points" have been mentioned in each of the foregoing examples, in practice the restrictions are not mentioned unless specifically required.

To solve an equation involving circular functions means, as it always has, to find the solution set. Again we may use the properties of the real numbers, the definitions of the circular functions, and previously proven identities.

Equations involving circular functions, however, are impossible to categorize in any general way and thus systematic techniques for solving all equations cannot be displayed. We shall look at a few of the simpler types of equations and show some methods of handling them. In all but the simplest cases, however, some ingenuity and a bit of perseverance are required.

Example 4   Find the solution set of $\sin 2x = 2 \sin x$.

Solution: Since $\sin 2x = 2 \sin x \cos x$ for all $x \in R$, we have

$$2 \sin x \cos x = 2 \sin x$$

so that

$$2 \sin x \cos x - 2 \sin x = 0.$$

Hence

$$\sin x(\cos x - 1) = 0.$$

Thus

$$\sin x = 0 \qquad \text{or} \qquad \cos x - 1 = 0.$$

Since $\sin x = 0$ only when $x = 0$, π, 2π, etc., and $\cos x = 1$ only when $x = 0$, 2π, etc., the solution set is $\{x : x = n\pi \text{ and } n \in I\}$.

Frequently, you are asked only to find the elements in the solution set for $0 \leq x < 2\pi$.

Example 5   Find the solution set of

$$2 \sin^2 x - 3 \sin x + 1 = 0 \qquad \text{for } 0 \leq x < 2\pi.$$

Solution: The equation is a quadratic equation in sin x and hence we may use any of the three usual methods of solving a quadratic: factoring, completing the square, or use of the quadratic formula. We have

$$2 \sin^2 x - 3 \sin x + 1 = (2 \sin x - 1)(\sin x - 1).$$

Thus

$$2 \sin x - 1 = 0 \qquad \text{or} \qquad \sin x - 1 = 0.$$

If $\sin x = \frac{1}{2}$ and $0 \le x \le 2\pi$, we have $x = \frac{\pi}{6}$ or $\frac{5\pi}{6}$. If $\sin x = 1$ and $0 \le x \le 2\pi$, we have $x = \frac{\pi}{2}$. Hence the solution set is $\{\frac{\pi}{6}, \frac{5\pi}{6}, \frac{\pi}{2}\}$ for $0 \le x < 2\pi$.

Example 6   Solve $\sin x = \sqrt{1 - \cos^2 x}$ for $0 \le x < 2\pi$.

Solution: We have

$$\sin^2 x = 1 - \cos^2 x \qquad \text{for all } x \in R.$$

For all $x \in R$, $\sqrt{1 - \cos^2 x} \ge 0$, but $\sin x \ge 0$ if and only if $0 \le x \le \pi$. Hence the solution set is $\{x : 0 \le x \le \pi\}$.

Sometimes the equation

$$\sin x = \pm \sqrt{1 - \cos^2 x},$$

which is an abbreviation for the longer expression

$$\sin x = \sqrt{1 - \cos^2 x} \quad \text{or} \quad \sin x = -\sqrt{1 - \cos^2 x},$$

is called an identity. We mean by this that

$$\{x : \sin x = \sqrt{1 - \cos^2 x} \quad \text{or} \quad \sin x = -\sqrt{1 - \cos^2 x}\} = R.$$

These examples show the similarity between the techniques you have already used for solving algebraic equations and the techniques used to solve equations involving circular functions. To become really proficient, however, you will need considerable experience.

EXERCISES 10.6

◀ In Exercises 1–10, find the solution set of the given equation.

1. $2 \sin x - 1 = 0$
2. $4 \cos^2 x - 3 = 0$
3. $3 \tan^2 x - 1 = 0$
4. $\sin^2 x - \cos^2 x + 1 = 0$
5. $4 \sin^3 y - \sin y = 0$
6. $2 \sin r \cos r + \sin r = 0$
7. $\cos y + \sin y = 0$
8. $\cos 2t = 0$
9. $2 \sin^2 s - 5 \sin s + 2 = 0$
10. $\cos x = -\sqrt{1 - \sin^2 x}$

◀ In Exercises 11–44, prove the given identity and give the restrictions, if any.

11. $\tan x \cos x = \sin x$

12. $\sin x \tan x + \cos x = \dfrac{1}{\cos x}$

13. $\tan x \sin 2x = 2 \sin^2 x$

14. $\tan x \sin x = \sec x - \cos x$

15. $\dfrac{\sin x}{1 + \cos x} = \dfrac{1 - \cos x}{\sin x}$

16. $\tan^2 x - \cot^2 x = \sec^2 x - \csc^2 x$

17. $(\sin x + \cos x)^2 = 1 + \sin 2x$

18. $1 - 2 \sin^2 x = 2 \cos^2 x - 1$

19. $\sec^2 y - \csc^2 y = (\tan y + \cot y)(\tan y - \cot y)$

20. $\tan x - \tan y = \sin (x - y) \sec x \sec y$

21. $\dfrac{\sin x}{\csc x} + \dfrac{\cos x}{\sec x} = 1$

22. $\dfrac{1}{\sec^2 x} + \dfrac{1}{\csc^2 x} = 1$

23. $\sin^4 x - \cos^4 x = 1 - 2 \cos^2 x$

24. $(\cot y - 1)^2 + (\cot y + 1)^2 = 2 + 2 \cot^2 y$

25. $\cot y + \tan y = (\csc^2 y + \sec^2 y) \sin y \cos y$

26. $\dfrac{1}{1 - \sin x} + \dfrac{1}{1 + \sin x} = 2 \sec^2 x$

27. $\cot^2 y - \cos^2 y = \cot^2 y \cos^2 y$

28. $(1 + \tan^2 x)(1 - \sin^2 x) = 1$

29. $\sec^2 x - \sin^2 x \sec^2 x = 1$

30. $\dfrac{\sin x}{1 + \cos x} + \dfrac{1 + \cos x}{\sin x} = 2 \csc x$

31. $\dfrac{\cot y - \cos y}{\cot y + \cos y} = \dfrac{1 - \sin y}{1 + \sin y}$

32. $\dfrac{\tan^3 x}{1 + \tan^2 x} \cdot \dfrac{1 + \cot^2 x}{\cot x} = \tan^2 x$

33. $\dfrac{\sin x + \tan x}{\cot x + \csc x} = \sin x \tan x$

34. $\dfrac{1}{\tan x + \cot x} = \dfrac{\sin x}{\sec x}$

35. $\dfrac{\sin y}{1 - \cos y} = \dfrac{1 + \cos y}{\sin y}$

36. $\dfrac{\sec x + 1}{\tan x} = \dfrac{\tan x}{\sec x - 1}$

37. $\cot x - \tan x = 2 \cot 2x$

38. $\sec^2 x = 2(1 - \tan x \cot 2x)$

39. $\dfrac{\sin 2x}{1 + \cos 2x} = \tan x$

40. $\dfrac{\sin 2y}{1 - \cos 2y} = \cot y$

41. $\tan x + \cot 2x = \csc 2x$

42. $\dfrac{1 - \cos 2x}{1 + \cos 2x} = \tan^2 x$

43. $\tan x = (1 + \sec x) \tan \dfrac{x}{2}$

44. $\dfrac{1 + \sin 2x}{\cos x + \sin x} = \dfrac{\cos 2x}{\cos x - \sin x}$

45. Solve $\cos x = \sqrt{1 - \sin^2 x}$.

46. Solve $\cos x = -\sqrt{1 - \sin^2 x}$.

47. If $x + y + z = \pi$, prove that $\sin x = \sin (y + z)$.

48. Show that for $x \in R$ and $x \neq n\pi$ for $n \in I$,

$$1 + \cot^2 x = \csc^2 x.$$

49. CHALLENGE PROBLEM. Prove that $\sin^4 3x - \cos^4 3x = 1 - 2 \cos^2 3x$.

50. CHALLENGE PROBLEM. Prove that $\sin^6 x + \cos^6 x = 1 - 3 \sin^2 x \cos^2 x$.

CHAPTER SUMMARY

The WRAPPING FUNCTION W maps each real number into a unique point of the unit circle $u^2 + v^2 = 1$. Thus $W(x) = (u, v)$ for all $x \in R$. The wrapping function is a PERIODIC function. Since for all $x \in R$, $W(x) = W(x + 2\pi)$ and $W(x + p) \neq W(x)$ for $0 < p < 2\pi$, the period of the function is 2π.

The six CIRCULAR functions, SINE, COSINE, TANGENT, COTANGENT, SECANT, and COSECANT, are also periodic functions. By definition

$$\sin x = v,$$

$$\cos x = u,$$

$$\tan x = \frac{v}{u} \qquad (u \neq 0),$$

$$\cot x = \frac{u}{v} \qquad (v \neq 0),$$

$$\sec x = \frac{1}{u} \qquad (u \neq 0),$$

and

$$\csc x = \frac{1}{v} \qquad (v \neq 0),$$

where $W(x) = (u, v)$ on the unit circle $u^2 + v^2 = 1$. The period of the sine, cosine, secant, and cosecant functions is 2π; the period of the tangent and cotangent functions is π.

The graphs of the six circular functions consist of CYCLES that repeat. The AMPLITUDE of a periodic graph is equal to $\frac{M - m}{2}$, where M is the maximum value of the function and m is the minimum value of the function.

A COMPOSITE function is a function of a function. If F is a composite function of f and g, then $F = g \circ f$ or $F = f \circ g$. The operation of composition is not commutative, that is, $f \circ g$ does not always equal $g \circ f$. The function defined by $y = 2 \sin 3x$ may be viewed as a composite function. The composite function defined by $y = A \sin Bx$ has an amplitude of $|A|$ and a period of $\frac{2\pi}{|B|}$.

Equations involving circular functions may be IDENTITIES or open sentences. An identity over the set of real numbers is a sentence whose solution set is the intersection of the domains of the functions in the expression. To prove an identity, we reduce or transform one member of the equation to an expression identical to the other member. To solve an open sentence involving circular functions is to find the solution set of the open sentence.

The more important identities are the following:

(I) $\qquad\qquad\qquad \sin^2 x + \cos^2 x = 1.$

(II) $\qquad\qquad\qquad \cos(-x) = \cos x.$

(III) $\qquad\qquad\qquad \sin(-x) = -\sin x.$

(IV) $\qquad\qquad\qquad \cos(x \pm \pi) = -\cos x.$

(V) $\qquad\qquad\qquad \sin(x \pm \pi) = -\sin x.$

(VI) $\qquad \tan x = \dfrac{\sin x}{\cos x} \qquad \left(x \neq \dfrac{\pi}{2} + n\pi \text{ for } n \in I \right).$

(VII) $\qquad \cot x = \dfrac{\cos x}{\sin x} \qquad \left(x \neq n\pi \text{ for } n \in I \right).$

(VIII) $$\sec x = \frac{1}{\cos x} \qquad \left(x \neq \frac{\pi}{2} + n\pi \text{ for } n \in I\right).$$

(IX) $$\csc x = \frac{1}{\sin x} \qquad (x \neq n\pi \text{ for } n \in I).$$

(X) $\cos(x + y) = \cos x \cos y - \sin x \sin y.$

(XI) $\cos(x - y) = \cos x \cos y + \sin x \sin y.$

(XII) $$\left|\cos \frac{x}{2}\right| = \sqrt{\frac{1 + \cos x}{2}}.$$

(XIII) $$\cos\left(\frac{\pi}{2} - x\right) = \sin x.$$

(XIV) $$\sin\left(\frac{\pi}{2} - x\right) = \cos x.$$

(XV) $\sin(x - y) = \sin x \cos y - \cos x \sin y.$

(XVI) $\sin(x + y) = \sin x \cos y + \cos x \sin y.$

(XVII) $$\left|\sin \frac{x}{2}\right| = \sqrt{\frac{1 - \cos x}{2}}.$$

(XVIII) $$\tan(x + y) = \frac{\tan x + \tan y}{1 - \tan x \tan y}.$$

(XIX) $$\tan(x - y) = \frac{\tan x - \tan y}{1 + \tan x \tan y}.$$

(XX) $1 + \tan^2 x = \sec^2 x.$

(XXI) $1 + \cot^2 x = \csc^2 x.$

REVIEW EXERCISES

1. If W is the wrapping function, name the point corresponding to each of the following:

   (a) $W(\pi)$ (c) $W(4\pi)$

   (b) $W\left(-\frac{\pi}{2}\right)$ (d) $W\left(-\frac{7\pi}{2}\right)$

2. If $W(y) = (-\frac{4}{5}, \frac{3}{5})$, find each of the following:

   (a) $W(-y)$ (c) $W(\pi - y)$

   (b) $W(y - \pi)$ (d) $W(y + 10\pi)$

◀In Exercises 3–10, find the functional value.

3. $\sin \dfrac{\pi}{6}$ 7. $\tan \dfrac{\pi}{4}$

4. $\cos \dfrac{\pi}{3}$ 8. $\tan \dfrac{3\pi}{4}$

5. $\sin \left(-\dfrac{\pi}{4} \right)$ 9. $\sec \dfrac{2\pi}{3}$

6. $\cos \left(-\dfrac{7\pi}{3} \right)$ 10. $\csc \left(-\dfrac{7\pi}{6} \right)$

◀In Exercises 11–16, find the functional value if $\cos y = -\frac{12}{13}$ and $\pi > y > \frac{\pi}{2}$.

11. $\cos(-y)$ 14. $\tan y$
12. $\cos(y - \pi)$ 15. $\sec y$
13. $\sin y$ 16. $\csc y$

17. Draw the graph of $\cos x$.
18. Draw the graph of $\tan x$.
19. Give the amplitude and period of the curve defined by the equation $y = 2 \sin 3x$.
20. Give the amplitude and period of the curve defined by the equation $y = 3 \cos \dfrac{x}{2}$.
21. Give the domain and range of the tangent function.
22. Give the domain and range of the cosecant function.
23. If $-\dfrac{3\pi}{2} < 2r < -\pi$ and $\sin 2r = \dfrac{\sqrt{7}}{4}$, find

    (a) $\cos 2r$ (b) $\tan 2r$
24. If $f : f(x) = x + 1$, $g : g(x) = x^2 + 1$, and $h : h(x) = -2x$, find
    (a) $f \circ g$ (b) $g \circ h$ (c) $f \circ (g \circ h)$
25. If $f : f(x) = x^2 + 1$ and $g : g(x) = x - 3$, find
    (a) $f[g(2)]$ (b) $(f \circ g)[g(0)]$
26. If $\sin t = \frac{3}{5}$ and $0 < t < \frac{\pi}{2}$, find

    (a) $\sin 2t$ (b) $\sin \dfrac{t}{2}$

◀In Exercises 27–32, find the functional value, given that $\sin x = \frac{4}{5} \left(0 < x < \frac{\pi}{2} \right)$ and $\sin y = -\frac{3}{5} \left(0 > y > -\frac{\pi}{2} \right)$.

27. $\sin(x + y)$ 30. $\cos 2y$
28. $\cos(x - y)$ 31. $\tan(x + y)$
29. $\sin 2x$ 32. $\tan 2y$

◀In Exercises 33–40, prove the given identity. Give the restrictions on x, if any.

33. $\tan x \sin x \cos x + \cot x \sin x \cos x = 1$

34. $2 \cos x = \dfrac{\sin 2x}{\sin x}$

35. $\dfrac{1 - \sin 2x}{\cos 2x} = \dfrac{1 - \tan x}{1 + \tan x}$

36. $\tan x + \cot x = \dfrac{2}{\sin 2x}$

37. $\cot 2x = \dfrac{\cot x - \tan x}{2}$

38. $\dfrac{1 + \sin 2x}{1 - \sin 2x} = \left(\dfrac{1 + \tan x}{1 - \tan x}\right)^2$

39. $\dfrac{1 - \sin x}{\cos x} = \dfrac{\cos x}{1 + \sin x}$

40. $\sin 3x = 3 \sin x - 4 \sin^3 x$

◀In Exercises 41–45, find the elements between 0 and 2π in the solution set of the equation.

41. $2 \sin^2 x = 1$

42. $\sin 2x - \sin x = 0$

43. $\cos 2y + \sin y = 1$

44. $4 \cos^2 y - 4 \sin y = 1$

45. $\cot^2 2t = 3$

GOING FURTHER: READING AND RESEARCH

One of the reasons for the importance of the cosine and the sine functions is the fact that many functions F, over a restricted domain, may have their values $F(x)$ approximated as closely as we desire by a series (see Chapter 1) of the form

$$a_0 + a_1 \cos x + a_2 \cos 2x + \cdots + b_1 \sin x + b_2 \sin 2x + b_3 \sin 3x + \cdots,$$

where $a_0, a_1, \ldots, b_1, b_2, \ldots$ are real numbers. Such a series is called a **Fourier series** in honor of Joseph Fourier (1768–1830) who first developed them.

The following example may give some indication of the power of the idea. Consider the function F such that

$$F(x) = \begin{cases} \dfrac{\pi}{4} x & \text{when } 0 \le x \le \dfrac{\pi}{2} \\[2mm] \dfrac{\pi}{4}(\pi - x) & \text{when } \dfrac{\pi}{2} < x \le \pi. \end{cases}$$

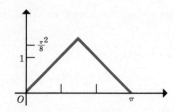

The figure shows a graph of F. By "Fourier analysis" it can be shown that this function can have its values approximated by the series

$$\sin x - \frac{\sin 3x}{3^2} + \frac{\sin 5x}{5^2} - \frac{\sin 7x}{7^2} + \cdots .$$

Fourier series are discussed in more detail in the following books:

KOLMOGOROV, A. and LAVRENT'ev (Editors), *Mathematics, Its Content, Methods, and Meaning.* Cambridge, Mass.: M.I.T. Press, 1963.

NEWMAN, J. R. (Editor), *The World of Mathematics* (4 volumes). New York: Simon and Schuster, 1956.

BELL, E. T., *Mathematics, Queen and Servant of Science.* New York: McGraw-Hill, 1951.

Goldstone Tracking Station

CHAPTER ELEVEN
EXPONENTIAL AND LOGARITHMIC FUNCTIONS

11.1 INTEGRAL EXPONENTS

The language of mathematics is precise. It is also compact. Many ingenious symbols have been devised to save time and space in the presentation of mathematical concepts, theorems, and formulas. One well-known "labor saver" is the exponent. In this chapter we shall review and prove some of the important properties of natural number exponents and subsequently examine the way in which the exponent concept can be extended to include exponents from larger sets of numbers.

You have probably had some experience in developing by plausible, although informal, arguments the following basic properties.

If x and y are any real numbers, then for any natural numbers m and n,

(P–1) $x^m x^n = x^{m+n}$,

(P–2)   for $x \neq 0$, $\dfrac{x^m}{x^n} = x^{m-n}$ if $m > n$, and $\dfrac{x^m}{x^n} = \dfrac{1}{x^{n-m}}$ if $n > m$,

(P–3) $(x^n)^m = x^{nm}$,

(P–4) $(xy)^n = x^n y^n$,

(P–5) $\left(\dfrac{x}{y}\right)^n = \dfrac{x^n}{y^n}$ if $y \neq 0$.

It is possible to prove these properties in a formal manner by means of mathematical induction. We shall give a proof of P–4 based on the following definition.

|| **Definition 11.1**   For any real number a we define $a^1 = a$ and $a^{k+1} = a^k a$ for all natural numbers k. ||

This is an example of a **recursive** definition such as those you have encountered in Chapter 1. From this definition we see that

$$a^2 = a^1 a, \qquad a^3 = a^2 a = (aa)a, \qquad \text{and so on.}$$

We now wish to prove that

$$(xy)^n = x^n y^n$$

485

for every x and $y \in R$ and every $n \in N$. As in previous induction proofs, we first let S be the set of natural numbers for which the property holds:

$$S = \{n : n \in N \text{ and } (xy)^n = x^n y^n \text{ for all } x, y \in R\}.$$

Then to show that $S = N$ we must, as you recall, prove (1) that $1 \in S$ and (2) that S is inductive, that is, that $k \in S$ implies $k + 1 \in S$.

First, we know that $1 \in S$ since

$$(xy)^1 = xy = x^1 y^1$$

by definition. Assume now that $k \in S$, that is, that

(1) $$(xy)^k = x^k y^k$$

for every x and $y \in R$. We want to show that then $k + 1 \in S$.

Now $k \in S$ if and only if (1) holds for every x and $y \in R$ and $k + 1 \in S$ if and only if

(2) $$(xy)^{k+1} = x^{k+1} y^{k+1}$$

holds for every x and $y \in R$. Thus we wish to show that (1) implies (2).

From (1), by multiplying by xy, we get

(3) $$(xy)^k (xy) = (x^k y^k)(xy).$$

Using the Associative and Commutative Properties on the right-hand side of (3), we can write (3) as

(4) $$(xy)^k (xy) = (x^k x)(y^k y).$$

Applying Definition 11.1 to both sides of (4), we get

$$(xy)^{k+1} = x^{k+1} y^{k+1},$$

which is (2). Hence $k \in S$ does imply that $k + 1 \in S$, S is inductive, and our proof is complete.

Property P–5 can be proved by an argument completely analogous to that used in the proof of P–4. You are asked to do this in the Exercises. The proofs of P–1 and P–3 by induction are more complicated since these properties involve *two* natural numbers, m and n. In the Exercises you are asked to demonstrate parts of the proofs of P–1 and P–3 using a fixed value for m. Later we shall show how a simplified version of P–2 follows from P–1.

Our next job is to extend the domain of exponents from the natural numbers to the integers. In the process we shall be guided by the desirability of defining expressions involving negative and zero exponents so as to have Properties P–1 through P–5 hold.

First, consider the integer 0. How shall we define x^0 for a real number x? By P–1 we know that $x^m x^n = x^{m+n}$ when m and n are natural numbers. If we want P–1 to hold when $m = 0$, we would have to agree that

$$x^0 x^n = x^{0+n} = x^n.$$

What does this suggest? We know that if $ab = b$ for some real number $b \neq 0$, then a is the multiplicative identity 1. Thus it seems reasonable to make the following definition.

> **Definition 11.2** $x^0 = 1$ for every nonzero real number x.

Let us now consider the case where n is a negative integer. How, for example, shall we define x^{-5}? If P–1 is to hold, then we must have

$$x^5 x^{-5} = x^{5+(-5)} = x^0 = 1.$$

In other words, x^{-5} must be the multiplicative inverse of x^5 provided that $x^5 \neq 0$, that is, provided that $x \neq 0$, and, likewise, x^5 must be the multiplicative inverse of x^{-5}. Furthermore, we know that the multiplicative inverse of any nonzero real number a is $\frac{1}{a}$. In view of the above discussion it also seems reasonable to make the following definition.

> **Definition 11.3** For any integer n and for any nonzero real number x
>
> $$x^{-n} = \frac{1}{x^n}.$$

Examples

$$x^{-4} = \frac{1}{x^4} \quad \text{and} \quad x^3 = \frac{1}{x^{-3}}.$$

Using Definition 11.3, it can be shown that Properties P–1 through P–5 also hold when m and n are any integers. We shall present the argument for P–1. To show that this property holds for all integers

m and n under the assumption that it holds for all natural numbers m and n, we need to examine several cases.

To begin with, suppose m and n are both negative integers and $x \neq 0$. We can then write $m = -s$ and $n = -t$ where s and t are positive integers, that is, natural numbers. (For example, if $m = -2$ and $n = -3$, then $s = 2$ and $t = 3$.) It follows that

$$x^m x^n = x^{-s} x^{-t}$$

$$= \frac{1}{x^s} \cdot \frac{1}{x^t} \qquad \text{by Definition 11.3}$$

$$= \frac{1}{x^s x^t}$$

$$= \frac{1}{x^{s+t}} \qquad \text{by P–1 for natural numbers}$$

$$= x^{-(s+t)} \qquad \text{by Definition 11.3}$$

$$= x^{-s-t} = x^{(-s)+(-t)}$$

$$= x^{m+n}.$$

Hence P–1 holds when m and n are both negative integers.

Next assume that m is a positive integer, that n is a negative integer, and again that $x \neq 0$. As before, write $n = -s$. Then

$$x^m x^n = x^m x^{-s} = \frac{x^m}{x^s}.$$

Since s is a natural number, we can use P–2:

$$\text{If } m > s, \ \frac{x^m}{x^s} = x^{m-s} = x^{m+(-s)} = x^{m+n}.$$

$$\text{If } s > m, \ \frac{x^m}{x^s} = \frac{1}{x^{s-m}} = \frac{1}{x^{-(m-s)}} = x^{m-s} = x^{m+n}.$$

A similar proof, of course, holds if n is a positive integer and m is a negative integer.

Finally, if $s = m$, that is, if $n = -m$, then

$$\frac{x^m}{x^s} = 1 = x^0 \qquad \text{from Definition 11.2}$$

and, since $m + n = 0$, we have

$$x^0 = x^{m+n}.$$

Using Definition 11.2, we can readily see that, for $x \neq 0$,

$$x^m x^n = x^{m+n} \qquad \text{if } m = 0 \text{ or } n = 0.$$

For example, if $n = 0$, we have

$$x^m x^n = x^m x^0 = x^{m+0} = x^m = x^{m+n}.$$

From the foregoing discussion we can conclude that:

If m and n are any integers and x is a nonzero real number, then

$$x^m x^n = x^{m+n}.$$

This is essentially Property P–1 with the added restriction that $x \neq 0$. Can you give a reason for this restriction?

Now we can use Definition 11.3 to rewrite P–2 as

$$\frac{x^m}{x^n} = x^m \cdot \frac{1}{x^n} = x^m \cdot x^{-n} \qquad \text{if } x \neq 0$$

and then use the extended version of P–1 to get

$$\frac{x^m}{x^n} = x^{m+(-n)} = x^{m-n} = \frac{1}{x^{-(m-n)}} = \frac{1}{x^{n-m}}$$

for $x \neq 0$ and any integers m and n. (Note that, because of Definition 11.3, we do not need to distinguish between the case $m > n$ and the case $m < n$.)

In the Exercises you are asked to supply proofs for the extension of P–3 through P–5 to integral exponents. The arguments are similar to those used for P–1.

We conclude this section with a few illustrations of the ways in which the properties of exponents may be used to simplify expressions.

Example 1

$$(x^{-2}y^2)^3 = \left(\frac{y^2}{x^2}\right)^3 = \frac{y^6}{x^6} = \left(\frac{y}{x}\right)^6.$$

Example 2

$$\frac{(5x^2y)^2}{5x^{-2}} = \frac{5^2 x^4 y^2}{\dfrac{5}{x^2}} = \frac{5^2 x^4 y^2}{1} \cdot \frac{x^2}{5} = 5x^6 y^2.$$

Example 3

$$(10^{-3})^2 \cdot (10^{-2})^{-2} = \left(\frac{1}{10^3}\right)^2 \cdot 10^{(-2)(-2)} = \frac{1}{10^6} \cdot 10^4 = \frac{1}{10^2} = \frac{1}{100}.$$

As an alternative method for Example 3, we have

$$(10^{-3})^2 \cdot (10^{-2})^{-2} = 10^{(-3)(2)} \cdot 10^{(-2)(-2)}$$

$$= 10^{-6} \cdot 10^4 = 10^{-2} = \frac{1}{100}.$$

EXERCISES 11.1

◀ In Exercises 1–25, use the properties of exponents to simplify the given expressions. In each case give answers without negative exponents.

1. $x^3 x^2 x$

2. $y^5 y^{-4} y$

3. $(x^{-3})^{-2}$

4. $(x^{-3} y^2)^{-2}$

5. $\left(\dfrac{a}{b}\right)^{-2}$

6. $(a + b)^3 (a + b)^{-2}$

7. $(x + y)^0 - 1$

8. $\dfrac{(a - b)^2}{(a - b)^{-2}}$

9. $\dfrac{(4x^3 y^2)^2}{4x^{-3}}$

10. $\left(\dfrac{3a^3 b^2}{a}\right)^{-2}$

11. $\left(\dfrac{2x^2 y^3}{x^3}\right)^{-2}$

12. $\left(\dfrac{2r^3 s}{3r^2 s^3}\right)^{-3}$

13. $(xy)^{-3}(x^2 y)^2$

14. $\left(\dfrac{1}{x}\right)^{-2} \left(\dfrac{1}{y}\right)^2 (xy)^{-1}$

15. $(a^2 b)^{-2}(a^{-2} b)^{-1}$

16. $\left(\dfrac{3}{r}\right)^{-2} \left(\dfrac{3}{s}\right)^2$

17. $\left(\dfrac{a^2 b}{a^{-1}}\right)\left(\dfrac{b^2 a}{b}\right)^{-2}$

18. $\left(\dfrac{x^2 y^{-1}}{y^2}\right)^{-3} (xy)^2$

19. $\left(\dfrac{x^3 y}{z}\right)^0 \left(\dfrac{x^{-1} y^3}{z}\right)^{-1}$

20. $\left(\dfrac{x}{y}\right)^{-2} \left(\dfrac{y}{x}\right)^3 \left(\dfrac{x}{y}\right)^{-1}$

21. $\left(\dfrac{rst^{-1}}{r^3}\right)^{-3} \left(\dfrac{r^4}{st}\right)^{-2}$

22. $\left(\dfrac{a^{-2}}{b^{-3}}\right)\left(\dfrac{2b}{a}\right)^3 \left(\dfrac{1}{a^4}\right)^{-1}$

23. $\left(\dfrac{x + y}{x^2}\right)^0 \left(\dfrac{1}{x + y}\right)^{-1} \left(\dfrac{1}{x + y}\right)$

24. $\left(\dfrac{a^4 b^2 c}{b^3}\right)^{-2} \left(\dfrac{1}{b}\right)\left(\dfrac{c}{a}\right)^3$

25. $\left(\dfrac{x^2 y^{-2}}{z^{-2}}\right)^2 \left(\dfrac{1}{xy}\right)^{-3}$

◀In Exercises 26–45, evaluate each of the expressions, giving all answers without exponents.

26. $2^{-3} \cdot 2$

27. $10^2 \cdot 10^{-4} \cdot 10$

28. $(-3)^0 \cdot 4^2$

29. $\left(\frac{3}{4}\right)^{-2}\left(\frac{1}{3}\right)^{-1}$

30. $\dfrac{7^{-1}}{7^{-3}}$

31. $3^{-2} \cdot 27$

32. $(8^{-2})^{-1}$

33. $\left(\frac{2}{3}\right)^{-1}\left(\frac{3}{2}\right)^{2}$

34. $\left(\frac{5}{6}\right)^{-2}\left(\frac{1}{3}\right)^{-1}$

35. $(10^{-4})^2(10^{-2})^{-3}$

36. $(5^{-2})(6)^{-1}(30)^2$

37. $(4^{-2})^3\left(\frac{1}{2}\right)^{-4}$

38. $\left(\frac{1}{3}\right)^{-3}(6)^3\left(\frac{1}{2}\right)^{-2}$

39. $\left(\frac{1}{4^{-2}}\right)^{2}\left(\frac{2}{3}\right)^{-4}$

40. $\left(\frac{1}{10}\right)^{-2}\left(\frac{1}{100}\right)^{-1}\left(\frac{1}{10}\right)^{3}$

41. $\left(\frac{3}{2}\right)^{2}\left(\frac{2}{3}\right)^{-2}\left(\frac{4}{9}\right)$

42. $\left(\frac{1}{4}\right)^{2}\left(\frac{1}{16}\right)^{-2}\left(\frac{2}{3}\right)^{-1}$

43. $\left(\frac{1}{9}\right)^{-2}\left(\frac{16}{27}\right)^{-1}\left(\frac{4}{3}\right)$

44. $\left(\frac{3}{5}\right)^{2}\left(\frac{1}{3}\right)^{-2}\left(\frac{5}{9}\right)^{-2}$

45. $\left(\frac{4^{-2}}{5^{-1}}\right)^{3}\left(\frac{1}{2^{-3}}\right)^{2}$

◀In Exercises 46–50, you may assume that Properties P–1 through P–5 hold for natural numbers m and n.

46. Prove that if m and n are both negative integers and x is any nonzero real number, then
$$(x^n)^m = x^{nm}.$$

47. Prove the conclusion of Exercise 46 for n a positive integer and m a negative integer.

48. Establish the conclusion of Exercise 46 for the case where n is a negative integer and m is a positive integer.

49. Prove that if n is a negative integer and x and y are any nonzero real numbers, then
$$(xy)^n = x^n y^n.$$

50. Prove that if n is a negative integer and x and y are any nonzero real numbers, then
$$\left(\frac{x}{y}\right)^n = \frac{x^n}{y^n}.$$

51. CHALLENGE PROBLEM. Using Definition 11.1, prove by induction that for any real number x and any $n \in N$

$$x^2 x^n = x^{2+n}.$$

52. CHALLENGE PROBLEM. Using Definition 11.1, prove by induction that for any real number x and any $n \in N$

$$(x^2)^n = x^{2n}.$$

53. CHALLENGE PROBLEM. Using Definition 11.1, prove Property P–5 by induction.

11.2 RATIONAL EXPONENTS

We have reviewed the properties of exponents as applied to exponents that are integers. As we extend the domain of exponents to include all rational numbers and, ultimately, all real numbers, we want these properties to be preserved.

For a beginning, let us consider all rational numbers of the form $\frac{1}{n}$ with n a natural number. For example, how should we define $8^{\frac{1}{3}}$? If Property P–3, that is, $(x^m)^n = x^{mn}$, is to hold, then

$$(8^{\frac{1}{3}})^3 = 8^{(\frac{1}{3})(3)} = 8^1 = 8.$$

This suggests that if $x = 8^{\frac{1}{3}}$, then $x^3 = 8$.

We now ask for what real number is the equation $x^3 = 8$ true? We can see that the solution is 2 since if x is a real number and $x^3 = 8$, then $x = 2 = \sqrt[3]{8}$. Thus it seems reasonable to define $x^{\frac{1}{n}}$ as follows.

If n is any natural number such that $n > 1$, then the following definition holds.

$\|$ *Definition* **11.4** $x^{\frac{1}{n}} = \sqrt[n]{x}$ for all real numbers x for which $\sqrt[n]{x}$ is a real number. $\|$

(Recall that, for n odd, $\sqrt[n]{x}$ is the unique real number y such that $y^n = x$; and for n even and x nonnegative, $\sqrt[n]{x}$ is the unique nonnegative real number y such that $y^n = x$. Thus, for example, $\sqrt[5]{32} = 2$ since $2^5 = 32$, $\sqrt[5]{-32} = -2$ since $(-2)^5 = -32$, $\sqrt[4]{81} = 3$ since $3 > 0$ and $3^4 = 81$, and $\sqrt[4]{0} = \sqrt[5]{0} = 0$. By Definition 11.4, we do not have $\sqrt[4]{81} = -3$ since, although $(-3)^4 = 81$, $-3 < 0$. If n is even and x is negative, then $\sqrt[n]{x}$ is not a real number. For example, $\sqrt[4]{-2}$ is not a real number. Finally, recall that we write $\sqrt[2]{x}$ as $\sqrt{x}$.)

By this definition $x^{\frac{1}{2}} = \sqrt{x}$ when $\sqrt{x} \in R$ and $x \geq 0$; also $x^{\frac{1}{3}} = \sqrt[3]{x}$ for all $x \in R$. From this it follows that $25^{\frac{1}{2}} = 5$ and $(-27)^{\frac{1}{3}} = -3$.

Note that in Definition 11.4 we defined $x^{\frac{1}{n}}$ to be equal to $\sqrt[n]{x}$ for all values of x for which $\sqrt[n]{x}$ is a *real* number. Why is this restriction placed on x? Why do we not define $(-4)^{\frac{1}{2}}$ as $\sqrt{-4}$ even though $\sqrt{-4}$ is *not* a real number? Our reason is that we wish Properties P–1 through P–5 to hold and if we allow x to be negative in $x^{\frac{1}{2}}$, complications result. For example, you know that

$$\sqrt{-4}\sqrt{-4} = \sqrt{4}i \cdot \sqrt{4}i = 4 \cdot i^2 = -4.$$

However, if P–4 is to hold for rational number exponents, it should then follow that

$$(-4)^{\frac{1}{2}} \cdot (-4)^{\frac{1}{2}} = [(-4) \cdot (-4)]^{\frac{1}{2}} = 16^{\frac{1}{2}} = 4,$$

and here we have a contradiction.

Let us now consider exponents that are rational numbers $\dfrac{m}{n}$ with $m \in I$ and $n \in N$. How shall we define $x^{\frac{m}{n}}$? Since $\dfrac{m}{n} = \dfrac{1}{n} \cdot m$, it follows that if P–3 is to hold, we must have

$$x^{\frac{m}{n}} = (x^{\frac{1}{n}})^m.$$

This leads us to the following definition.

> *Definition 11.5*
>
> 1. For any real number $x \geq 0$ and for any rational number $\frac{m}{n}$ with $m \in I$ and $n \in N$, we define
>
> $$x^{\frac{m}{n}} = (x^{\frac{1}{n}})^m$$
>
> provided that $x \neq 0$ when $m < 0$.
> 2. For any real number $x < 0$ and for any rational number $\dfrac{m}{n}, m \in I$ and $n \in N$ and with m and n not both even, we define
>
> $$x^{\frac{m}{n}} = (x^{\frac{1}{n}})^m$$
>
> providing that $x^{\frac{1}{n}}$ is defined.

Example 1

$$9^{\frac{3}{2}} = (9^{\frac{1}{2}})^3 = 3^3 = 27.$$

Example 2

$$8^{-\frac{2}{3}} = 8^{\frac{-2}{3}} = (8^{\frac{1}{3}})^{-2} = 2^{-2} = \frac{1}{4}.$$

Example 3

$$(-64)^{\frac{2}{3}} = [(-64)^{\frac{1}{3}}]^2 = (-4)^2 = 16.$$

Why did we specify in part 2 of Definition 11.5 that when $x < 0$, not both m and n be even? Consider the expression

$$(-8)^{\frac{2}{6}}.$$

By Definition 11.5, if no restriction were made on m and n, we could have

$$(-8)^{\frac{2}{6}} = [(-8)^{\frac{1}{6}}]^2.$$

Since 6 is an even number, we know that $(-8)^{\frac{1}{6}} = \sqrt[6]{-8}$ is not a real number. Hence we would conclude that the expression $(-8)^{\frac{2}{6}}$ is not defined.

On the other hand, since $\frac{2}{6} = \frac{1}{3}$, we have

$$(-8)^{\frac{2}{6}} = (-8)^{\frac{1}{3}} = -2.$$

By the restriction that not both m and n be even when $x < 0$, we avoid such contradictions.

The adaptation of Properties P–1 through P–5 to rational exponents leads to certain difficulties. We have already seen that when we allow x to be negative in $x^{\frac{1}{2}}$, then problems can arise. These problems are avoided by insisting, in Definitions 11.4 and 11.5, that $\sqrt[n]{x}$ be a real number. If we apply the same type of restriction, Properties P–1, P–2, P–4, and P–5 will also hold for rational exponents as well as for integral exponents. As we shall see, however, P–3 will lead to contradictions unless an additional condition is placed on x.

The following, then, is a restatement of the Properties P–1 through P–5 as applied to rational number exponents.

If x and y are any nonzero real numbers and if r and s are any rational numbers, then in all cases where the exponential expressions represent real numbers

(P–1) $$x^r x^s = x^{r+s}.$$

(P–2) $$\frac{x^r}{x^s} = x^{r-s}.$$

(P–3) $$(x^r)^s = x^{rs} \qquad \text{if } x > 0.$$

(P–4) $$(xy)^r = x^r y^r.$$

(P–5) $$\left(\frac{x}{y}\right)^r = \frac{x^r}{y^r}.$$

All of these statements can be proved from Definitions 11.4 and 11.5 under the assumption that P–1 through P–5 hold for integral exponents. We shall now give a proof for P–1.

To prove P–1 we let $r = \dfrac{m}{n}$ and $s = \dfrac{u}{v}$, where $m, u \in I, n, v \in N$, $n \geq 2$, and $v \geq 2$. Then we need to show that

$$x^{\frac{m}{n}} x^{\frac{u}{v}} = x^{\frac{m}{n} + \frac{u}{v}}.$$

Now

$$x^{\frac{m}{n}} = x^{\frac{mv}{nv}}$$

and

$$x^{\frac{u}{v}} = x^{\frac{un}{nv}}.$$

Thus we have

$$
\begin{aligned}
x^{\frac{m}{n}} x^{\frac{u}{v}} &= x^{\frac{mv}{nv}} x^{\frac{un}{nv}} \\
&= (\sqrt[nv]{x})^{mv} (\sqrt[nv]{x})^{un} && \text{by Definition 11.5} \\
&= (\sqrt[nv]{x})^{mv+un} && \text{by P–1 for integral exponents} \\
&= x^{\left(\frac{mv+un}{nv}\right)} && \text{by Definition 11.5} \\
&= x^{\left(\frac{m}{n} + \frac{u}{v}\right)},
\end{aligned}
$$

and the proof is complete.

The above argument, however, raises two rather sticky questions. From the hypotheses implied in the statement of P–1 for rational number exponents, we assume that $x^{\frac{m}{n}}$ and $x^{\frac{u}{v}}$ are defined, that is, that $\sqrt[n]{x}$ and $\sqrt[v]{x}$ are real numbers.

Question 1   Can we infer from this, however, that $\sqrt[nv]{x}$ is also a real number, that is, that $x^{\frac{1}{nv}}$ is defined? If not, then the statements

$$x^{\frac{mv}{nv}} = (\sqrt[nv]{x})^{mv},$$
$$x^{\frac{un}{nv}} = (\sqrt[nv]{x})^{un},$$

and

$$(\sqrt[nv]{x})^{mv+un} = x^{\frac{mv+un}{nv}}$$

are not valid. Fortunately, we can show that if $\sqrt[n]{x}$ and $\sqrt[v]{x} \in R$, then $\sqrt[nv]{x} \in R$ as follows.

If $x < 0$, then the existence of $\sqrt[n]{x}$ and $\sqrt[v]{x}$ must imply that n and v are both odd. But if n and v are odd, so is nv. For $x \geq 0$ there is no problem since for $x \geq 0$, $\sqrt[n]{x}$ exists for all natural numbers $n > 1$.

Question 2   Does Definition 11.5 really apply to $x^{\frac{mv}{nv}} x^{\frac{un}{nv}}$? Can we conclude from Definition 11.5 that

$$x^{\frac{mv}{nv}} = (x^{\frac{1}{nv}})^{mv} = (\sqrt[nv]{x})^{mv}$$

and

$$x^{\frac{un}{nv}} = (x^{\frac{1}{nv}})^{un} = (\sqrt[nv]{x})^{mv}?$$

Certainly! The only restriction on the rational exponent $\frac{m}{n}$ in Definition 11.5 is that m and n not be both even when $x < 0$. But we have already remarked in dealing with Question 1 that if $x < 0$, then n and v are both odd and hence nv is also odd.

Properties P–2 through P–5 for rational number exponents can all be proved in an analogous manner using the corresponding properties for integral exponents. Some of these proofs are asked for in the Exercises.

We now wish to show the reason for the restriction $x > 0$ in respect to P–3. Consider the expression

$$[(-9)^2]^{\frac{1}{2}}.$$

If we evaluate this expression in a reasonable way, we would note that

$$(-9)^2 = 81$$

and then write

$$[(-9)^2]^{\frac{1}{2}} = (81)^{\frac{1}{2}} = \sqrt{81} = 9.$$

On the other hand, since $(-9)^2$ is a real number, we would be able to use P–3, if no restrictions existed, to get

$$(-9^2)^{\frac{1}{2}} = (-9)^{2 \cdot \frac{1}{2}} = (-9)^1 = -9.$$

How do we avoid such a contradiction? One way might be to state P–3 as

$$(x^r)^s = x^{rs} \qquad \text{or} \qquad |x^{rs}|.$$

Then, in the example we were just considering, we would have

$$[(-9)^2]^{\frac{1}{2}} = 81^{\frac{1}{2}} = 9$$

on the one hand and

$$[(-9)^2]^{\frac{1}{2}} = |(-9)^{2 \cdot \frac{1}{2}}| = |(-9)^1| = 9$$

on the other, and no contradiction would result.

By considering, however,

$$[(-3)^3]^{\frac{1}{3}} = (-27)^{\frac{1}{3}} = -3$$

and

$$[(-3)^3]^{\frac{1}{3}} = (-3)^{3 \cdot \frac{1}{3}} = (-3)^1 = -3$$

we see that sometimes the absolute value is correct and sometimes not. A complete analysis of when the absolute value is correct is rather tedious to make and we omit it in favor of the simple restriction that $x > 0$. (Of course, in any numerical situation such as the examples just considered, direct computation will reveal any need for absolute values.)

The question remains of how to define x^{-r} if r is a rational number and $x \neq 0$.

Consider the example $8^{-\frac{2}{3}}$. Since the rational number $-\frac{2}{3} = \frac{-2}{3}$, we may apply Definition 11.5 as follows:

$$8^{-\frac{2}{3}} = 8^{\frac{-2}{3}} = (8^{\frac{1}{3}})^{-2} = (\sqrt[3]{8})^{-2} = 2^{-2} = \frac{1}{2^2} = \frac{1}{4}.$$

Now consider $\dfrac{1}{8^{\frac{2}{3}}}$. Again using Definition 11.5 we get

$$\frac{1}{8^{\frac{2}{3}}} = \frac{1}{(8^{\frac{1}{3}})^2} = \frac{1}{2^2} = \frac{1}{4}.$$

From this example it seems that the definition for x^{-r} for rational numbers r is most appropriately stated as follows:

Definition 11.6 For any rational number r and for any nonzero real number x,

$$x^{-r} = \frac{1}{x^r}$$

in all cases where x^r is a real number. (Compare with Definition 11.3.)

Properties P–1 through P–5 can often be applied in several ways in a problem. For example,

$$\left(\frac{16}{25}\right)^{-\frac{1}{2}} = \frac{1}{\left(\dfrac{16}{25}\right)^{\frac{1}{2}}} = \frac{1}{\dfrac{16^{\frac{1}{2}}}{25^{\frac{1}{2}}}} = \frac{1}{\dfrac{4}{5}} = \frac{5}{4}$$

or

$$\left(\frac{16}{25}\right)^{-\frac{1}{2}} = \left[\left(\frac{16}{25}\right)^{\frac{1}{2}}\right]^{-1} = \left[\frac{16^{\frac{1}{2}}}{25^{\frac{1}{2}}}\right]^{-1} = \left[\frac{4}{5}\right]^{-1} = \frac{1}{\frac{4}{5}} = \frac{5}{4}.$$

Rational numbers can, of course, be written as decimals. Thus, for example,

$$81^{1.25} = 81^{\frac{5}{4}} = [81^{\frac{1}{4}}]^5 = (\sqrt[4]{81})^5 = 3^5 = 243.$$

EXERCISES 11.2

◄In Exercises 1–25, evaluate each of the expressions, giving all answers without exponents.

1. $8^{-\frac{2}{3}}$

2. $25^{-\frac{1}{2}}$

3. $81^{\frac{3}{4}}$

4. $81^{-\frac{1}{4}}$

5. $\left(\frac{4}{9}\right)^{-\frac{1}{2}}$

6. $\left(\frac{1}{125}\right)^{-\frac{1}{3}}$

7. $\left(\frac{1}{125}\right)^{\frac{2}{3}}$

8. $64^{-\frac{2}{3}}$

9. $64^{-\frac{1}{6}}$

10. $32^{\frac{3}{5}}$

11. $729^{-\frac{2}{3}}$

12. $(8^{\frac{2}{3}})^{-\frac{1}{2}}$

13. $(343)^{-\frac{2}{3}}$

14. $\left(\frac{1}{216}\right)^{-\frac{1}{3}}$

15. $(1000)^{-\frac{2}{3}}$

16. $(10^{\frac{1}{3}})^6$

17. $(243)^{\frac{2}{5}}$

18. $\left(\frac{1}{243}\right)^{-\frac{1}{5}}$

19. $(128)^{\frac{2}{7}}$

20. $(128)^{-\frac{3}{7}}$

21. $36^{2.5}$

22. $\left(\frac{1}{16}\right)^{-1.25}$

23. $\left(\frac{4}{9}\right)^{-3.5}$

24. $625^{0.75}$

25. $32^{1.4}$

◄In Exercises 26–31, write the given expression in simplest form.

26. $\dfrac{2x^{-1} + x}{x^{-2}}$

27. $\dfrac{3a^2 + a^{-2}}{a^{-1}}$

28. $\dfrac{y^{-1} - y^{-2}}{y^{-3}}$

29. $\dfrac{b^{-2}}{b^{-1} + b^{-2}}$

30. $\dfrac{x^{-1} + y^{-1}}{(xy)^{-1}}$

31. $\dfrac{\left(\frac{1}{y}\right)^{-1} + y^{-2}}{y^2 + y^{-1}}$

32. Show that if $n \in N$ and $n > 1$, then

$$\sqrt[n]{xy} = \sqrt[n]{x}\sqrt[n]{y}$$

for all positive real numbers x and y.

33. Show that if n is an odd positive integer and $n > 1$, then

$$\sqrt[n]{xy} = \sqrt[n]{x}\sqrt[n]{y}$$

for all real numbers x and y.

34. Show that if $n \in N$ and $n > 1$, then

$$\sqrt[n]{\frac{x}{y}} = \frac{\sqrt[n]{x}}{\sqrt[n]{y}}$$

for all positive real numbers x and y.

35. Show that if n is an odd positive integer and $n > 1$, then

$$\sqrt[n]{\frac{x}{y}} = \frac{\sqrt[n]{x}}{\sqrt[n]{y}}$$

for all real numbers x and y with $y \neq 0$.

36. Use the results of Exercise 35 to evaluate $\sqrt[3]{\frac{125}{27}}$.

◀ In Exercises 37–40, assume that properties P–1 through P–5 hold for all integral exponents.

37. Prove P–2 for rational exponents.
38. Prove P–3 for rational exponents.
39. Prove P–4 for rational exponents.
40. Prove P–5 for rational exponents.

11.3 REAL NUMBER EXPONENTS AND EXPONENTIAL FUNCTIONS

We have defined $a^{\frac{m}{n}}$ for any positive real number a and any rational number $\frac{m}{n}$ with $m \in I$ and $n \in N$. Is it possible to define a^r where r is any real number in such a way as to preserve the properties of exponents and give meaning to $5^{\sqrt{2}}$, $2^{\sqrt{3}}$, 3^{π}, etc.?

To do this we shall have to use an approach somewhat different from that used for defining integral and rational exponents, that is, integral and rational powers of numbers.

Suppose we begin by considering the expression 2^x. First, let us make a clear distinction between this expression and another one, x^2, somewhat similar in appearance. For x^2 the base is the variable x and the exponent is the number 2. In 2^x the roles are reversed. Here the exponent is the variable and the base is 2.

You are quite familiar with the graph of the equation

$$y = x^2.$$

It is illustrated, in part, in Figure 11-1.

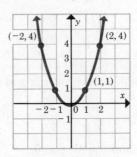

Figure 11-1

Let us now consider the equation

$$y = 2^x.$$

From our previous study we know that the function f defined by

$$f : f(x) = 2^x$$

is defined for every rational number x.

Partial graphs of $y = 2^x$ are shown in Figure 11-2a and b.

Here are a few examples to illustrate the way in which the table on the left of the graph was constructed.

x	y
-3	0.125
-2.5	0.18
-2	0.25
-1.5	0.35
-1	0.5
-0.5	0.71
0	1
0.5	1.41
1	2
1.5	2.83
2	4
2.5	5.66
3	8

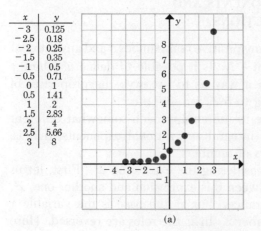

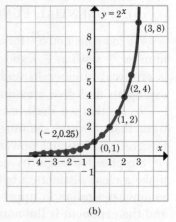

(a)   (b)

Figure 11-2

For $x = -3$,

$$2^x = 2^{-3} = \frac{1}{2^3} = \frac{1}{8} = 0.125.$$

For $x = -1.5$,

$$2^x = 2^{-1.5} = 2^{-\frac{3}{2}} = (2^{\frac{1}{2}})^{-3} = \frac{1}{(2^{\frac{1}{2}})^3} = \frac{1}{2\sqrt{2}} \cdot \frac{\sqrt{2}}{\sqrt{2}}$$

$$= \frac{\sqrt{2}}{4} \approx \frac{1.414}{4} \approx 0.35.$$

For $x = 1.5$,

$$2^x = 2^{\frac{3}{2}} = (2^3)^{\frac{1}{2}} = \sqrt{8} \approx 2.83.$$

As stated before, the points on the graph correspond to ordered pairs $(x, 2^x)$ for rational values of x.

We now try to give meaning to values of

$$f : f(x) = 2^x$$

when x is replaced by an irrational number. To focus on the problem let us consider in particular the case where $x = \sqrt{2}$. Our big question is, "What do we mean by $2^{\sqrt{2}}$?"

Let us attack the question on two fronts: first, from the standpoint of the graph. We can plot a few points corresponding to some rational values of x as in Figure 11-2a. As more points are plotted, these points tend to approximate a curve as in Figure 11-2b. Suppose we now assume that points have been plotted for all rational numbers x and that these points have been connected by a smooth curve that fills in all the gaps. Does it seem reasonable to let $f(\sqrt{2}) = 2^{\sqrt{2}}$ be the real number that corresponds to the y-coordinate of the "filled in" graph directly above the point corresponding to $\sqrt{2}$ on the x-axis? The curve drawn in Figure 11-3 suggests the idea.

In attacking the question on the second front, let us look at a numerical approximation of the number $2^{\sqrt{2}}$. We can infer either by means of the graph or by a few evaluations that $f(x) = 2^x$ increases as x increases. On this basis, then, we may assume that if, for example, x lies between 1 and 2, that is, $1 < x < 2$, it follows that $f(x) = 2^x$ must lie between 2^1 and 2^2, that is, $2 < 2^x < 4$.

In general, we may assume that for any rational numbers a and b such that $a < b$, if x is a real number and $a < x < b$, then $2^a <$

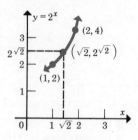

Figure 11-3

$2^x < 2^b$. With this assumption in mind we can proceed to approximate $2^{\sqrt{2}}$ as follows.

We know that $1.4 < \sqrt{2} < 1.5$. Thus $2^{1.4} < 2^{\sqrt{2}} < 2^{1.5}$. Since 1.4 and 1.5 are rational numbers, the numbers $2^{1.4}$ and $2^{1.5}$ are defined and can be approximated as follows:

$$2^{1.5} = 2^{\frac{3}{2}} = (2^3)^{\frac{1}{2}} = \sqrt{8} \approx 2.828.$$
$$2^{1.4} = 2^{\frac{14}{10}} = 2^{\frac{7}{5}} = 2^{7 \cdot \frac{1}{5}} = (2^7)^{\frac{1}{5}} = (128)^{\frac{1}{5}}.$$

To find an approximation for $(128)^{\frac{1}{5}}$ we could use the method of Chapter 6 to solve the equation

$$x^5 = 128 \qquad \text{or} \qquad x^5 - 128 = 0.$$

It turns out that an approximate solution is 2.639 and hence the following inequality holds:

$$2.639 < 2^{1.4} < 2^{\sqrt{2}} < 2^{1.5} < 2.829.$$

If we continue the refinement one step further, we obtain

$$2.657 < 2^{1.41} < 2^{\sqrt{2}} < 2^{1.42} < 2.676$$

since it is known that $1.41 < \sqrt{2} < 1.42$.

Eventually, by ascertaining that $1.4142 < \sqrt{2} < 1.4143$, we arrive at the inequality

$$2.665 < 2^{1.4142} < 2^{\sqrt{2}} < 2^{1.4143} < 2.666.$$

This tells us that

$$2^{\sqrt{2}} \approx 2.665$$

to three decimal places.

By two intuitively reasonable processes, then, we have been led to the conjecture that

$$f : f(x) = 2^x$$

associates with every real number x a unique real number 2^x. Although the proof lies beyond the scope of this book, we shall henceforth proceed on the assumption that this conjecture is valid.

Having considered the function f defined by $f(x) = 2^x$, which we call an **exponential function,** we can in a similar way give meaning to a^x for any positive real number a and any real number x. Furthermore, we could also show that Properties P–1 through P–5 continue to hold if the domain of the exponents is extended to include all real numbers and if the base numbers are positive.

Thus we may restate the hypotheses for these properties and write: If a, b, x, $y \in R$ and a, $b > 0$, then

(P–1) $$a^x a^y = a^{x+y}.$$

(P–2) $$\frac{a^x}{a^y} = a^{x-y}.$$

(P–3) $$(a^x)^y = a^{xy}.$$

(P–4) $$(ab)^x = a^x b^x.$$

(P–5) $$\left(\frac{a}{b}\right)^x = \frac{a^x}{b^x}.$$

The proofs that these properties hold for irrational exponents involve a considerable amount of detail as well as a precise definition of a^x for x an irrational number. We shall omit the proofs for these properties, but, as with the existence of $f : f(x) = 2^x$, we shall in what follows be operating on the assumption that P–1 through P–5 do, in fact, hold for all real number exponents.

We shall now consider the more general exponential function

$$g : g(x) = b^x$$

where the base b is any positive real number other than 1. ($h : h(x) = 1^x$ is certainly of little importance since $1^x = 1$ for any real number x.)

For a base b, where $0 < b < 1$, the graph takes on a different appearance from the one for $y = 2^x$, although there is a definite relationship between, for example, the graph of $y = (\frac{1}{2})^x$ and the graph $y = 2^x$. You can explore this relationship as an Exercise. Meanwhile we shall concentrate our attention on equations of the form

$$y = b^x \qquad \text{for } b > 1.$$

To study $g : g(x) = b^x$ for $b > 1$ and $b \neq 2$ it is not necessary to begin all over again. We can, instead, show that $g : g(x) = b^x$ is closely related to $f : f(x) = 2^x$ and that the values of g can be ascertained from the values of f.

Let us return now to the graph of $y = 2^x$, which is reproduced in Figure 11-4. The graph approaches closer and closer to the x-axis as x takes on larger and larger (in absolute value) negative values. On the other hand, as x takes on larger positive values the graph rises rapidly. We say that the value of the function increases without limit as x increases.

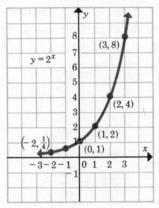

Figure 11-4

From these remarks and the graph, we may conjecture that the range of $f : f(x) = 2^x$ is the set of *all* positive real numbers. What does this tell us? It enables us, theoretically at least, to do the following. Given any positive real number whatsoever, it is always possible to find on the graph of $y = 2^x$ a point whose y-coordinate is the given number. Thus if r is any positive real number, a real number s exists such that $2^s = r$.

For example, given that $r = 32$, what is s? Since $2^5 = 32$, we may conclude that s in this case is 5. If $r = 128$, what is s? If $r = \frac{1}{2}$ and $2^s = r$, what is s?

We now need to consider how to convert a number in the form b^x, $b \neq 2$ to a number in the form 2^y.

Suppose, then, that we take the function g defined by $g(x) = b^x$ where $b \in R$ and $b > 1$. From the previous discussion we know that $b = 2^s$ for some real number s. From the equation $b = 2^s$ and Property P–3 of exponents which says that $(a^s)^r = a^{sr}$ if $a > 0$, we can now assert that

$$b^x = (2^s)^x = 2^{sx}.$$

For example, if $b = 4$, then $4 = 2^2$ and $4^x = 2^{2x}$; and if $b = 8$, then $8 = 2^3$ and $8^x = 2^{3x}$. Thus

$$4^5 = 2^{10} \qquad \text{and} \qquad 8^5 = 2^{15}.$$

The problem of converting 3^x to an expression of the form 2^y is a little more complicated than these simple examples of conversion when $b = 4$ or $b = 8$ might indicate. Suppose, for example, that $x = 3^{0.5}$. We seek a number y such that

$$3^{0.5} = 2^y.$$

If we had a table showing the values of 2^x for every real number x, the correct answer could, of course, be read directly from the table. Such a table, although very helpful, would assume rather large, in fact infinite, proportions. For our purposes we can, however, use the following greatly abbreviated table which gives approximate values of 2^x for certain values of x between 0 and 1.

To find a real number s such that $2^s = 3$, we proceed as follows. Since $3 = 2 \cdot \frac{3}{2} = 2(1.5)$, we can write

$$2^s = 3$$

as

(1) $$2^s = 2(1.5).$$

TABLE 11-1

x	2^x	x	2^x
0.01	1.007	0.45	1.366
0.02	1.014	0.50	1.414
0.03	1.021	0.55	1.464
0.04	1.028	0.60	1.516
0.05	1.035	0.65	1.569
0.10	1.072	0.70	1.624
0.15	1.110	0.75	1.682
0.20	1.149	0.80	1.741
0.25	1.189	0.85	1.802
0.30	1.231	0.90	1.866
0.35	1.274	0.95	1.932
0.40	1.319	1.00	2.000

If we now find a real number r such that $2^r = 1.5$, then, by P-1,

(2) $$2^s = 2 \cdot 2^r = 2^{1+r}.$$

From the table we see that $2^x \approx 1.515$ if $x = 0.60$. Thus, when $r = 0.6$, $2^r \approx 1.5$. From (2) we see that $s = 1 + r$, that is, if $r = 0.6$, then $s \approx 1.6$. Hence

$$2^{1.6} \approx 3.$$

From this it follows that

$$3^{0.5} \approx (2^{1.6})^{0.5}$$
$$\approx 2^{0.8}. \quad \text{(Property P–3)}$$

This answer, as you can see, is a rough approximation. For a more accurate result we can use a process known as **linear interpolation** which we consider in Section 11.4.

EXERCISES 11.3

◀In Exercises 1–10, determine the value of s which satisfies the given equation. Use the table when needed.

1. $2^s = 64$ 2. $2^s = \dfrac{1}{2}$

3. $2^s = 256$ 4. $2^s = \dfrac{1}{16}$

5. $2^s = \dfrac{1}{128}$ 6. $2^s = \dfrac{1}{1024}$

7. $2^s \approx 2.482$ (*Hint:* Let $2^s = 2^{1+r} = 2 \cdot 2^r$.)
8. $2^s \approx 2.638$ (*Hint:* Let $2^s = 2^{1+r} = 2 \cdot 2^r$.)
9. $2^s \approx 4.440$ (*Hint:* Let $2^s = 2^{2+r} = 2^2 \cdot 2r$.)
10. $2^s \approx 11.312$ (*Hint:* Let $2^s = 2^{3+r} = 2^3 \cdot 2^r$.)

11. Using the results of the discussion immediately preceding these Exercises, determine x such that $2^x \approx 9$.
12. As in Exercise 11, determine x such that $2^x \approx 3^{1.2}$.
13. Draw the graph of $f : f(x) = 1^x$ for $x \in R$.

14. Make a sketch of the graph of $f : f(x) = \left(\dfrac{1}{2}\right)^x$.

$$\left(\text{Hint: } \left(\frac{1}{2}\right)^x = \frac{1}{2^x} = 2^{-x}.\right)$$

15. Make a sketch of the graph of $f : f(x) = 3^x$.

In Exercises 16–20, copy and complete the equations. Write all answers with positive exponents. (Assume x, y, $z > 0$.)

16. $x^{\sqrt{2}} \cdot x^{\frac{\sqrt{2}}{2}} = x^{?}$

17. $(y^{\frac{\sqrt{3}}{2}})^{\sqrt{3}} = y^{?}$

18. $\dfrac{z^{\sqrt{2}}}{z^{\frac{1}{\sqrt{2}}}} = z^{?}$

19. $x^{\pi} \cdot y^{\pi} = (xy)^{?}$

20. $\dfrac{x^{-\sqrt{5}}}{x^{-\frac{1}{\sqrt{5}}}} = \dfrac{1}{x^{?}}$

11.4 LINEAR INTERPOLATION

In Section 11.3, we found an approximate solution to the equation

$$2^r = 1.5$$

by noting that $2^{0.60} \approx 1.516$. It was suggested that greater accuracy could be obtained by linear interpolation—a process you have already used in Chapter 6 when estimating roots of polynomial equations. We shall review the process in general terms and then apply it to some of the examples previously considered.

Suppose that we have a function $f : f(x)$ which is defined for all real numbers and that the following table gives functional values for certain values of x.

TABLE 11-2

x	$f(x)$
1	2.5
2	5.5
3	10.5
4	17.5
5	26.5

Suppose we now wish to estimate a value of $f(x)$ for some number x such as 3.6 which is not included in the table. How can we determine an approximate value for $f(3.6)$?

Figure 11-5 shows the points A, B, C, D, and E corresponding to the data of Table 11-2. (Note that the scale on the $f(x)$ axis differs from the scale on the x-axis.) Since we have not been given functional values for any values of x other than those given in Table 11-2, we cannot be certain what the rest of the graph looks like. It could conceivably be of the shape indicated by either one of the dashed lines as well as many others. We will assume, however, that the solid line

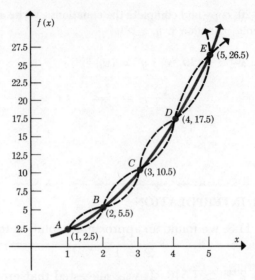

Figure 11-5

in Figure 11-5 does indicate the shape of the actual graph between $x = 1$ and $x = 5$.

To estimate a value of $f(x)$ for $3 < x < 4$ we shall assume that a reasonable approximation to its graph between $(3, 10.5)$ and $(4, 17.5)$ can be made by constructing a line segment between $(3, 10.5)$ and $(4, 17.5)$. We could then use this line segment to determine an approximation to $f(3.6)$. (See Figure 11-6.)

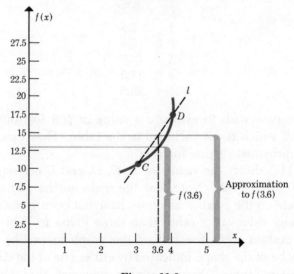

Figure 11-6

As we did in Chapter 6, we can find an equation of the line l containing $(3, 10.5)$ and $(4, 17.5)$ by first calculating the slope of l. We have

$$\text{slope of } l = \frac{17.5 - 10.5}{4 - 3} = 7.$$

Then, selecting the point $(3, 10.5)$, we have, for an equation of l,

(1) $$y - 10.5 = 7(x - 3).$$

The coordinates of any point on this line must satisfy (1). Hence we find $f(3.6)$ by substituting 3.6 for x and solving for y. This gives us

$$y - 10.5 = 7(3.6 - 3)$$

so that

$$y = 7(0.6) + 10.5 = 14.7.$$

Thus our estimate is that

$$f(3.6) = 14.7.$$

To test the accuracy of this estimate (and all methods of approximation should be carefully examined) we suppose that the "unknown" function we really had in mind was

$$f : f(x) = x^2 + 1.5.$$

(You can check that this function does yield the values in Table 11-2.) If we substitute 3.6 for x in this equation, we get

$$f(3.6) = (3.6)^2 + 1.5 = 14.46.$$

Thus our approximation, $f(3.6) \approx 14.7$, is really quite close. Certainly, when we use a straight line as a "stand in" for a curve that is not a straight line we must expect a margin of error. This is a fact always to remember when using linear interpolation on nonlinear functions.

You may greatly simplify the procedure just used by means of the scheme suggested by the following array which shows the table entries corresponding to $x = 3$ and $x = 4$. The outer bracket on the left shows

	x	$f(x)$	
	3	10.5	
0.6			d
1	3.6	?	7
	4	17.5	

the difference between the values of x, that is, $4 - 3 = 1$, and on the right the difference between the functional values,

$$f(4) - f(3) = 17.5 - 10.5 = 7.$$

The difference between 3 and 3.6, that is 0.6, is shown on the left inner bracket. To find d, the difference between $f(3)$ and $f(3.6)$, we use the proportion

$$\frac{0.6}{1} = \frac{d}{7}$$

from which we get

$$d = 7(0.6) = 4.2.$$

Since

$$f(3.6) = f(3) + d,$$

we have

$$f(3.6) = 10.5 + 4.2 = 14.7.$$

By examining the way in which the proportion is constructed, you can undoubtedly see the connection between the linear equation and the shortcut process.

It should be noted that the method of linear interpolation may not provide a good approximation if the points determining the line are not "close together." Consider, for example, the function $f : f(x) = x^2$. (See Figure 11-7.)

If we try to find an approximation for $f(0)$ by using the line which joins $(1, f(1))$ and $(-1, f(-1))$, the results could be disastrous—as shown in Figure 11-7.

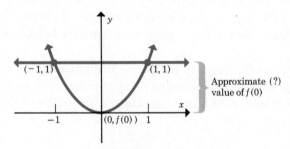

Figure 11-7

Let us now consider an interpolation example based on Table 11-1 which is reproduced here for convenience.

Suppose that we wish to find an approximate value of 2^x when

TABLE 11-1

x	2^x	x	2^x
0.01	1.007	0.45	1.366
0.02	1.014	0.50	1.414
0.03	1.021	0.55	1.464
0.04	1.028	0.60	1.516
0.05	1.035	0.65	1.569
0.10	1.072	0.70	1.624
0.15	1.110	0.75	1.682
0.20	1.149	0.80	1.741
0.25	1.189	0.85	1.802
0.30	1.231	0.90	1.866
0.35	1.274	0.95	1.932
0.40	1.319	1.00	2.000

$x = 0.795$. Since 0.795 lies between 0.75 and 0.80, we use the points (0.75, 1.68) and (0.80, 1.74) as illustrated in Figure 11-8.

As before, we connect the two points with a straight line. The slope of the line is

$$\frac{1.74 - 1.68}{0.80 - 0.75} = \frac{0.06}{0.05} = \frac{6}{5}.$$

An equation of the line is therefore

$$y - 1.68 = \frac{6}{5}(x - 0.75).$$

To find y when $x = 0.795$, we get, by substitution,

$$y - 1.68 = \frac{6}{5}(0.795 - 0.75) = 0.054.$$

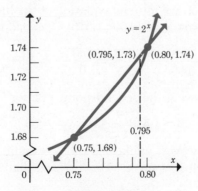

Figure 11-8

Hence

$$y = 1.68 + 0.054 = 1.734 \approx 1.73,$$

and thus

$$2^{0.795} \approx 1.73.$$

The tabular procedure is shown below.

$$
\begin{array}{ccc}
 & x & 2^x \\
\end{array}
$$

$$
0.05 \left[\begin{array}{c} 0.045 \left[\begin{array}{cc} 0.75 & 1.68 \\ 0.795 & \boxed{?} \end{array} \right] d \\ \\ 0.80 \quad 1.74 \end{array} \right] 0.06
$$

From the array we see that

$$\frac{0.045}{0.05} = \frac{d}{0.06}.$$

Hence

$$d = (0.06)\,\frac{0.045}{0.05} = 0.054$$

and so

$$2^{0.795} \approx 1.68 + 0.05 = 1.73.$$

Linear interpolation can be applied in two directions. For example, suppose that we wish to find the value of x for which $2^x = 1.5$ (the example discussed on page 506).

From Table 11-1 we note that

$$2^{0.55} = 1.464 \quad \text{and} \quad 2^{0.60} = 1.516.$$

With this information we form an equation of the line passing through the points $(0.55, 1.464)$ and $(0.60, 1.516)$. The slope is

$$\frac{1.516 - 1.464}{0.60 - 0.55} = \frac{0.052}{0.05} = 1.04.$$

Thus we have, choosing the second point just for variety,

$$y - 1.516 = 1.04(x - 0.60).$$

In this instance, however, unlike the other examples, we wish to find x when 1.5 is substituted for y. Thus we see that

$$1.5 - 1.516 = 1.04x - 0.624,$$
$$1.04x = 0.608,$$

and, finally, that to two decimal places,

$$x \approx 0.58.$$

(Recall that our "rough" approximation of Section 11.3 gave us $x \approx 0.60$.)

The tabular procedure also reveals the "two-way" aspect of the process.

$$
\begin{array}{cc}
x & 2^x
\end{array}
$$

$$
0.05 \left[\begin{array}{c} 0.55 \quad 1.464 \\ d \left[\begin{array}{cc} \boxed{?} & 1.5 \\ 0.60 & 1.516 \end{array} \right] 0.016 \end{array} \right] 0.052
$$

Notice that in this instance we seek a value of x corresponding to a given value of 2^x. The proportion is

$$\frac{d}{0.05} = \frac{0.016}{0.052}.$$

Hence

$$d = (0.05)\frac{0.016}{0.052} \approx 0.0154;$$

$$x \approx 0.60 - 0.0154 = 0.5846,$$

or, to two decimal places, $x \approx 0.58$.

Now let us consider a final example involving an exponential function for a base other than 2. We wish to find $g(x) = 6^x$ for $x = 0.3$. The first job is to find s such that $2^s = 6$. This can be carried out as follows:

$$6 = 2^2 \cdot (1.5)$$

and

$$2^r \approx 1.5$$

if $r = 0.58$ as before. Hence

$$6 \approx 2^2 \cdot 2^{0.58} = 2^{2.58}.$$

It follows that

$$6^{0.3} \approx (2^{2.58})^{0.3} \approx 2^{0.77}.$$

Using linear interpolation, we find that $2^{0.77} \approx 1.740$ and hence

$$6^{0.3} \approx 1.740.$$

EXERCISES 11.4

1. Verify by linear interpolation that $2^{0.78} \approx 1.716$.

◀ In Exercises 2–10, by linear interpolation in Table 11-1, find the approximate value of 2^x, rounded to three decimal places, for each of the given values of x.

2. $x = 0.23$	**7.** $x = 0.51$
3. $x = 0.42$	**8.** $x = 0.032$
4. $x = 0.66$	**9.** $x = 0.08$
5. $x = 0.93$	**10.** $x = 0.015$
6. $x = 0.72$	

◀ In Exercises 11–20, using linear interpolation in Table 11-1, determine approximately the value of x, rounded to three decimal places, for which 2^x has the indicated value.

11. $2^x = 1.25$	**16.** $2^x = 1.03$
12. $2^x = 1.60$	**17.** $2^x = 1.10$
13. $2^x = 1.34$	**18.** $2^x = 1.75$
14. $2^x = 1.20$	**19.** $2^x = 1.375$
15. $2^x = 1.90$	**20.** $2^x = 1.125$

◀ In Exercises 21–30, determine approximately the value of s, rounded to two decimal places, for which 2^s has the indicated value. (*Hint:* You may find some results in Exercises 11–20 helpful.)

21. $2^s = 5$	**26.** $2^s = 9$
22. $2^s = 7$	**27.** $2^s = 11$
23. $2^s = 10$	**28.** $2^s = 20$
24. $2^s = 12$	**29.** $2^s = 18$
25. $2^s = 14$	**30.** $2^s = 40$

◀ In Exercises 31–40, determine the approximate value, rounded to two decimal places, of the given exponential function $g : g(x) = b^x$ for the indicated values of b and x.

31. $6^{0.2}$	**36.** $6^{0.08}$
32. $3^{0.4}$	**37.** $5^{0.4}$
33. $5^{0.2}$	**38.** $7^{0.07}$
34. $7^{0.2}$	**39.** $20^{0.1}$
35. $10^{0.2}$	**40.** $18^{0.2}$

11.5 INVERSE FUNCTIONS

If we have a function f with domain the set of real numbers defined by

$$f(x) = 3x + 5,$$

we can determine $f(x)$ for each value of x by substitution. For example, when $x = 7$, $f(x) = 26$; when $x = -4$, $f(x) = -7$; and so on.

One is frequently confronted with the reverse question: For a given value of the function, what is the corresponding value of x? For an equation such as

$$f(x) = 3x + 5$$

the question has a ready answer. As an example, for what value of x does $f(x) = 11$? By inspection we can infer that the answer is 2 since

$$11 = 3(2) + 5.$$

On the other hand, it might be more efficient to proceed first as follows. Let $y = f(x)$, that is,

$$y = 3x + 5.$$

Then

$$y - 5 = 3x$$

and hence

$$x = \frac{1}{3}(y - 5)$$

for each $y \in R$.

This new equation tells us explicitly what the value of x is when y is any given number. Thus if $y = 6$, then $x = \frac{1}{3}$; if $y = 14$, then $x = 3$; and so on. It can also be seen that this equation defines another function

$$g : g(y) = \frac{1}{3}(y - 5)$$

for all $y \in R$ or, since the symbol used to denote the variable is arbitrary, we may write

$$g : g(x) = \frac{1}{3}(x - 5)$$

for all $x \in R$.

As another example, suppose that we are given the function

$$f : f(x) = \frac{1}{2}x - 7$$

with domain the real numbers and want an equation that would tell us explicitly the value of x corresponding to a given functional value $f(x)$. To find such an equation we could let

$$f(x) = \frac{1}{2}x - 7 = y.$$

Then

$$y + 7 = \frac{1}{2}x,$$

and so

$$x = 2y + 14 \qquad \text{for each } y \in R.$$

This last equation again defines another function, in this case

$$g : g(y) = 2y + 14 \qquad \text{for all } y \in R$$

or, alternatively,

$$g : g(x) = 2x + 14 \qquad \text{for all } x \in R.$$

Let us now look at the two functions $f : f(x) = \frac{1}{2}x - 7$ and $g : g(x) = 2x + 14$ jointly. Let $x = 8$. Then

$$f(8) = \frac{1}{2}(8) - 7 = -3$$

and

$$g(-3) = 2(-3) + 14 = 8.$$

Furthermore,

$$f(22) = \frac{1}{2}(22) - 7 = 4$$

and

$$g(4) = 2(4) + 14 = 22.$$

Thus we can say that the function f maps 8 into -3 and the function g maps -3 back into 8. Similarly, f maps 22 into 4 and g maps 4 back into 22.

Let us look again at the first example where

$$f : f(x) = 3x + 5 \qquad \text{for all } x \in R$$

and

$$g : g(x) = \frac{1}{3}(x - 5) \qquad \text{for all } x \in R.$$

Now using the number 10, we see that

$$f(10) = 3(10) + 5 = 35$$

and

$$g(35) = \frac{1}{3}(35 - 5) = \frac{1}{3}(30) = 10.$$

In this instance f maps 10 into 35 and g maps 35 back into 10.

Informally we could say that the mapping g in both examples is the reverse of the mapping f. In symbols

$$10 \xrightarrow{f} 35 \xrightarrow{g} 10$$

and, in general,

$$x \xrightarrow{f} y \xrightarrow{g} x.$$

We call g the **inverse** of f. The concept of an inverse mapping, or function, is of considerable importance in mathematics. Before continuing the discussion of inverse functions, however, let us review briefly what is meant by composition of functions—a topic that you studied in Chapter 10. (We will assume, unless otherwise stated, that all of the functions considered here have as domain the set R of all real numbers.)

Suppose that we have a function

$$f : f(x) = x^2 + 5$$

and another function

$$g : g(x) = 3x - 1.$$

Then

$$f(6) = 6^2 + 5 = 41$$

and

$$g(41) = 3(41) - 1 = 122.$$

We can write

$$g[f(6)] = 122$$

or

$$[g \circ f](6) = 122,$$

and define the composite function $g \circ f = h$ as

$$h : h(x) = [g \circ f](x) = g[f(x)].$$

As stated in Chapter 10, the domain of h is the set of all numbers x in the domain of f for which $f(x)$ is in the domain of g, and so, in this case, the domain of h is R.

To determine a function $h = g \circ f$ if functions f and g are given, we proceed as in the following example.

Let $f(x) = 2x + 1$ and $g(x) = x^2$. Then

$$h(x) = g[f(x)] = g(2x + 1) = (2x + 1)^2$$

and so

$$h : h(x) = (2x + 1)^2.$$

The domain of h is R because for all $x \in R$, $f(x) \in R$ and hence $f(x) \in$ domain, R, of g.

EXERCISES 11.5A

◀In Exercises 1–10, define the composite function $h = g \circ f$ for the given functions f and g. Then evaluate $h(x) = g[f(x)]$ for each of the given values of x.

1. $f : f(x) = 3x - 8$, $g : g(x) = 2x + 10$
   (a) $x = 1$, (b) $x = 0$, (c) $x = -1$
2. $f : f(x) = x^2$, $g : g(x) = x + 4$
   (a) $x = -3$, (b) $x = -2$, (c) $x = 0$
3. $f : f(x) = x^2 + 3$, $g : g(x) = x - 7$
   (a) $x = 0$, (b) $x = 12$, (c) $x = -5$
4. $f : f(x) = x^2 - x$, $g : g(x) = x + 5$
   (a) $x = 5$, (b) $x = -7$, (c) $x = 10$
5. $f : f(x) = x^3 + x$, $g : g(x) = x^2$
   (a) $x = 1$, (b) $x = 4$, (c) $x = -1$
6. $f : f(x) = \dfrac{1}{x}$, $g : g(x) = x^2 + 1$

   (a) $x = 2$, (b) $x = -2$, (c) $x = \dfrac{1}{2}$

7. $f : f(x) = \dfrac{1}{x^2}$, $g : g(x) = x^{-2}$

   (a) $x = 5$, (b) $x = -3$, (c) $x = \dfrac{1}{3}$

8. $f : f(x) = x^{\frac{1}{2}}$, $g : g(x) = x^{-3}$
   (a) $x = 1$, (b) $x = 9$, (c) $x = 25$
9. $f : f(x) = x^{\frac{1}{3}}$, $g : g(x) = x^{\frac{3}{2}}$
   (a) $x = 27$, (b) $x = 64$, (c) $x = 5$
10. $f : f(x) = x^3$, $g : g(x) = -x^{\frac{1}{3}}$

    (a) $x = 5$, (b) $x = \dfrac{1}{2}$, (c) $x = -\dfrac{2}{3}$

Let us return now to the example where

$$f : f(x) = \frac{1}{2}x - 7$$

and

$$g : g(x) = 2x + 14.$$

Consider $[g \circ f](x)$ and let $x = 20$. Then

$$[g \circ f](20) = g[f(20)] = g\left[\frac{1}{2}(20) - 7\right] = g(3)$$

$$= 2(3) + 14 = 20.$$

Let $x = 4$. We then have

$$[g \circ f](4) = g[f(4)] = g\left[\frac{1}{2}(4) - 7\right] = g(-5)$$

$$= 2(-5) + 14 = 4.$$

In both cases the composite function maps the number into itself.

Now consider the situation in reverse, that is, consider the composite function

$$f \circ g.$$

For this case let us start with an element in the domain of g, say 8. This gives us

$$[f \circ g](8) = f[g(8)] = f[2(8) + 14] = f(30)$$

$$= \frac{1}{2}(30) - 7 = 8.$$

Thus

$$[f \circ g](8) = f[g(8)] = 8.$$

On the basis of these examples we may infer that for the functions

$$f : f(x) = \frac{1}{2}x - 7 \qquad \text{and} \qquad g : g(x) = 2x + 14,$$

the composite function $g \circ f$ and the composite function $f \circ g$ both map any element x in the domains of $g \circ f$ and $f \circ g$ into itself. Thus it seems reasonable to state that

$$g \circ f = f \circ g = I,$$

where I is the identity mapping, that is,

$$I : I(x) = x.$$

These ideas, then, suggest the following definition.

> **Definition 11.7**   If f and g are functions such that
>
> 1. $$[g \circ f](x) = g[f(x)] = x$$
>
> for every element x in the domain of f and such that
>
> 2. $$[f \circ g](x) = f[g(x)] = x$$
>
> for every element x in the domain of g, then f and g
> are said to be **inverses** of each other.

Using conditions 1 and 2, it can be shown that the domain of g is the range of f and the domain of f is the range of g.

A "picture" of a mapping f and its inverse g would look something like Figure 11-9.

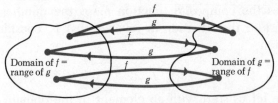

Figure 11-9

From Definition 11.7 and the relationship

$$f \circ g = g \circ f = I$$

it would seem appropriate to designate the inverse function g of f as f^{-1}. Thus $f \circ f^{-1} = f^{-1} \circ f = I$.

EXERCISES 11-5B

◄In Exercises 1–6, a function f and its inverse f^{-1} are given. Compute the functional values $[f \circ f^{-1}](x)$ and $[f^{-1} \circ f](x)$ for the indicated number x. Use the form shown in the example.

Example. $f : f(x) = 3x + 6,\ f^{-1} : f^{-1}(x) = \dfrac{1}{3}x - 2,\ x = 7.$

Solution:

$$[f \circ f^{-1}](7) = f[f^{-1}(7)] = f\left[\frac{1}{3}(7) - 2\right] = f\left(\frac{1}{3}\right) = 3\left(\frac{1}{3}\right) + 6 = 7;$$

$$[f^{-1} \circ f](7) = f^{-1}[f(7)] = f^{-1}[3(7) + 6] = f^{-1}(27) = \frac{1}{3}(27) - 2 = 7$$

1. $f : f(x) = 2x - 8$, $f^{-1} : f^{-1}(x) = \frac{1}{2}x + 4$, $x = 11$

2. $f : f(x) = \frac{1}{3}x - 12$, $f^{-1} : f^{-1}(x) = 3x + 36$, $x = 15$

3. $g : g(x) = \frac{2x - 7}{3}$, $g^{-1} : g^{-1}(x) = \frac{3x + 7}{2}$, $x = \frac{1}{3}$

4. $h : h(x) = \frac{5x + \frac{1}{2}}{2}$, $h^{-1} : h^{-1}(x) = \frac{2x - \frac{1}{2}}{5}$, $x = 20$

5. $g : g(x) = x^3 + 1$, $g^{-1} : g^{-1}(x) = \sqrt[3]{x - 1}$, $x = 9$

6. $f : f(x) = \frac{1}{2}x^3$, $f^{-1} : f^{-1}(x) = \sqrt[3]{2x}$, $x = 1$

◄ In Exercises 7–10, two functions are given. Determine whether or not the functions are inverses of each other.

7. $f : f(x) = 7x - \frac{1}{2}$, $g : g(x) = \frac{1}{7}x - 2$

8. $h : h(x) = x^3 + 3$, $g : g(x) = \sqrt[3]{x} - 3$

9. $g : g(x) = \frac{1}{3}x - \frac{1}{2}$, $f : f(x) = 3x + \frac{3}{2}$

10. $f : f(x) = \frac{2x - 7}{3}$, $g : g(x) = \frac{3}{2}x + \frac{7}{2}$

11. **CHALLENGE PROBLEM.** Given the function $f : f(x) = x^2$, does the function f have an inverse? Give reasons for your answer.

12. **CHALLENGE PROBLEM.** Do all functions of the form $f : f(x) = ax^2 + bx + c$, with a, b, and c real numbers, have inverses? Explain.

WHAT FUNCTIONS HAVE INVERSES?

We must now ask a basic question: "Under what conditions does a function f have an inverse?" Let us look at the function f with domain R the set of real numbers and with $f : f(x) = x^2$. For this function we know that $f(2) = 4$ and $f(-2) = 4$ where $2 \in R$ and $-2 \in R$. Now suppose we consider the number 4 which is an element in the range of f. If we wanted a function g which would "return"

4 to both 2 and -2, we would certainly be asking for the impossible. We cannot have both $g(4) = 2$ and $g(4) = -2$.

The difficulty in finding an inverse in this case stems from the fact that $f : f(x) = x^2$ is a *many-to-one* (in this case two-to-one) mapping. That is, for any positive real number a there are two real numbers, $\sqrt{a}$ and $-\sqrt{a}$, such that

$$f(\sqrt{a}) = f(-\sqrt{a}) = a.$$

Thus, for example,

$$f(2) = f(-2) = 4$$

and

$$f\left(\frac{1}{2}\right) = f\left(-\frac{1}{2}\right) = \frac{1}{4}.$$

We may reasonably conjecture, then, that the existence of an inverse function for a function f requires that f be a one-to-one mapping, that is, that f have the property that for all a and b in the domain of f

$$f(a) = f(b)$$

implies $a = b$.

The question immediately arises, "How can we tell whether or not a given function is a one-to-one mapping?" One of the easiest ways of answering this question is to examine the graphs of a varied assortment of functions. Let us first consider the graph of the polynomial function f defined by the equation

$$f(x) = 2x^3 - 3x^2 - 12x + 13.$$

The graph is illustrated in Figure 11-10. (In Figure 11-10, the scales on the x- and y-axes are different in order to accommodate more points.)

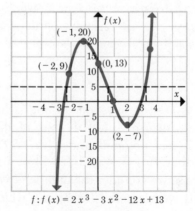

$$f : f(x) = 2x^3 - 3x^2 - 12x + 13$$

Figure 11-10

Recall that, by definition, a function associates with each element x in the domain a *unique* value $f(x)$ in the range. The function

$$f : f(x) = 2x^3 - 3x^2 - 12x + 13$$

with domain the set of real numbers satisfies this condition. However, the dashed horizontal line intersecting the graph of f tells us that there are three distinct values of x for which $f(x) = 5$. We know then that f is not one-to-one since we have more than one value of x for which $f(x) = 5$.

Consider now the graphs of the functions

$$\begin{array}{ll} j : j(x) = x^3, & g : g(x) = -x^3, \\ h : h(x) = x + 1, & k : k(x) = 2, \end{array}$$

each with domain the set R of real numbers as illustrated in Figure 11-11.

It should be apparent from Figure 11-11 that the graphs of j, g, and h are graphs of functions which are one-to-one mappings. On the other hand, from the graph of $k : k(x) = 2$, we see that every real number x is associated with the value 2 of $k(x)$. Hence $k : k(x) = 2$ is most assuredly *not* a one-to-one mapping.

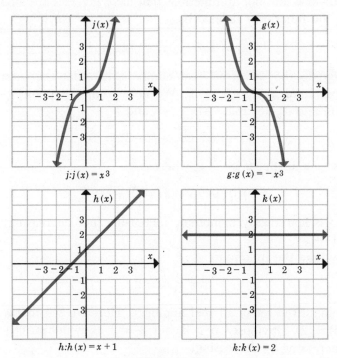

Figure 11-11

From this discussion and these illustrations, what conclusions may be drawn regarding one-to-one functions?

The three graphs j, g, and h which describe one-to-one functions have a common characteristic. The functions in each case are examples of what we call strictly **monotonic** functions. The functions j and h are called strictly monotonic **increasing**, whereas g is strictly monotonic **decreasing**.

This means, roughly speaking, that for j and h the functional values increase with each increase in x, and that for g the functional values decrease with each increase in x. In slightly more technical language we say that a function f is strictly monotonic increasing on an interval $a < x < b$ if for all real numbers x_1 and x_2 in this interval such that $x_2 > x_1$, it follows that $f(x_2) > f(x_1)$.

Now see if you can define a strictly monotonic decreasing function.

It is possible to define functions that are not strictly monotonic but that are one-to-one and therefore have an inverse. However, we know that a strictly monotonic (decreasing or increasing) function must be one-to-one. For this reason we say that the property of being strictly monotonic is a *sufficient*, but not a *necessary*, condition for a function to have an inverse. That is, any strictly monotonic function has an inverse, but not all functions that have inverses are monotonic.

For the function f defined by

$$f(x) = 2x^3 - 3x^2 - 12x + 13$$

with domain the set of all real numbers, consider the interval on the x-axis from, say, -3 to 3. We note from the points $(-2, 9)$, $(-1, 20)$, and $(0, 13)$ that $-1 > -2$ and

$$f(-1) = 20 > f(-2) = 9.$$

So far so good! But $0 > -1$, whereas

$$f(0) = 13 < f(-1) = 20.$$

Hence f is not strictly monotonic increasing (nor decreasing) on the interval $-3 < x < 3$.

Now take the function

$$k : k(x) = 2$$

with domain the set of all real numbers. Here we see that $k(3) = 2$ and $k(5) = 2$. Obviously, $5 > 3$, but $2 = 2$. k is not strictly monotonic increasing or decreasing!

The inverse of

$$h : h(x) = x + 1$$

is

$$h^{-1} : h^{-1}(x) = x - 1$$

since

$$[h^{-1} \circ h](x) = h^{-1}[h(x)] = h^{-1}(x + 1) = (x + 1) - 1 = x$$

and also

$$[h \circ h^{-1}](x) = h[h^{-1}(x)] = h(x - 1) = (x - 1) + 1 = x.$$

What is the inverse of

$$j : j(x) = x^3?$$

It is

$$j^{-1} : j^{-1}(x) = \sqrt[3]{x}.$$

To show that j^{-1} is the inverse of j, we note that

$$[j^{-1} \circ j](x) = j^{-1}[j(x)] = j^{-1}(x^3) = \sqrt[3]{x^3} = x$$

and also

$$[j \circ j^{-1}](x) = j[j^{-1}(x)] = j(\sqrt[3]{x}) = (\sqrt[3]{x})^3 = x.$$

What about the inverse of

$$g : g(x) = -x^3?$$

Check that it is

$$g^{-1} : g^{-1}(x) = -\sqrt[3]{x}.$$

The problem of determining the inverse function for a given function f can be difficult if f, for example, is defined by an equation such as

$$f(x) = x^3 + x + 5.$$

However, for some functions such as the ones we have been considering, the process is fairly simple. Here are two other examples.

Example 1 Given $f : f(x) = 2x + 5$ with domain R, find f^{-1}.

Solution: We write

$$y = 2x + 5,$$
$$y - 5 = 2x,$$
$$\frac{y - 5}{2} = x,$$

and conclude that f^{-1} is defined by

$$f^{-1} : f^{-1}(x) = \frac{x-5}{2} \quad \text{with domain } R.$$

Check:

$$[f^{-1} \circ f](x) = f^{-1}(2x + 5) = \frac{(2x+5)-5}{2} = x$$

and

$$[f \circ f^{-1}](x) = f\left(\frac{x-5}{2}\right) = 2\left(\frac{x-5}{2}\right) + 5 = x.$$

Example 2   Given $g : g(x) = \frac{1}{4}x^3 - \frac{1}{3}$ with domain R, find g^{-1}.
Solution: We write

$$y = \frac{1}{4}x^3 - \frac{1}{3}$$

$$y + \frac{1}{3} = \frac{1}{4}x^3$$

$$4\left(y + \frac{1}{3}\right) = x^3$$

$$\sqrt[3]{4\left(y + \frac{1}{3}\right)} = x$$

and conclude that g^{-1} is defined by

$$g^{-1} : g^{-1}(x) = \sqrt[3]{4\left(x + \frac{1}{3}\right)} \quad \text{with domain } R.$$

Check:

$$[g^{-1} \circ g](x) = g^{-1}\left(\frac{1}{4}x^3 - \frac{1}{3}\right) = \sqrt[3]{4\left[\left(\frac{1}{4}x^3 - \frac{1}{3}\right) + \frac{1}{3}\right]} = \sqrt[3]{x^3} = x$$

and

$$[g \circ g^{-1}](x) = g\left(\sqrt[3]{4\left(x + \frac{1}{3}\right)}\right) = \frac{1}{4}\left[\sqrt[3]{4\left(x + \frac{1}{3}\right)}\right]^3 - \frac{1}{3}$$

$$= \frac{1}{4}\left[4\left(x + \frac{1}{3}\right)\right] - \frac{1}{3} = x.$$

EXERCISES 11.5C

◀In Exercises 1–17, give the inverse function f^{-1} for each of the given functions with domain the set of all real numbers. (Example: $f : f(x) = 2x; f^{-1} : f^{-1}(x) = \frac{1}{2}x$.)

1. $f : f(x) = 3x$

2. $f : f(x) = 7x + 2$

3. $f : f(x) = 3x - 12$

4. $f : f(x) = \frac{1}{3}x + 4$

5. $f : f(x) = \frac{3}{4}x - \frac{1}{3}$

6. $f : f(x) = 2x^3$

7. $f : f(x) = -2x^3$

8. $f : f(x) = x^5$

9. $f : f(x) = 2x^3 - 1$

10. $f : f(x) = \frac{1}{2}x^3 + \frac{1}{3}$

11. $g : g(x) = 2x^3 + 3$

12. $h : h(x) = \frac{x^3 + 1}{3}$

13. $r : r(x) = \frac{1}{4}x^3 - \frac{1}{2}$

14. $h : h(x) = 2x^5 - 7$

15. $s : s(x) = \frac{x^5 + 2}{3}$

16. $f : f(x) = \frac{1}{4}x^3 - 1$

17. $f : f(x) = \frac{x^5}{5} + \frac{1}{2}$

18. Given the function $f : f(x) = x^2 + 1$, specify a domain for which f^{-1} exists. Define f^{-1} and indicate the domain of f^{-1}.

19. Proceed as in Exercise 18 for the function $f : f(x) = \frac{1}{x}$.

20. Proceed as in Exercise 18 for the function $f : f(x) = \frac{1}{x - 5}$.

11.6 GRAPHS OF INVERSE FUNCTIONS

Much can be learned about the relationship between a function f and its inverse f^{-1} through the study of their respective graphs. Consider our example of

$$f : f(x) = \frac{1}{2}x - 7$$

and

$$f^{-1} : f(x) = 2x + 14.$$

The graphs of these two functions appear side by side in Figure 11-12. Study these graphs carefully. See if you can devise a method for drawing one graph when given the other by "tracing" through transparent paper.

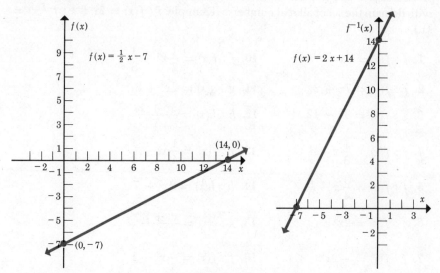

Figure 11-12

Figure 11-13 shows the two functions

$$f : f(x) = x^2 \qquad \text{(domain, nonnegative real numbers)}$$

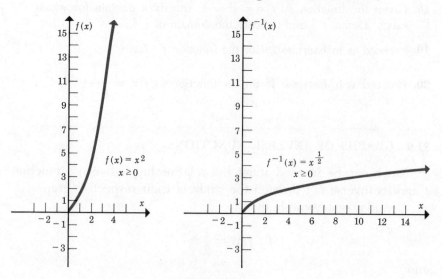

Figure 11-13

and

$$f^{-1} : f^{-1}(x) = x^{\frac{1}{2}} \qquad \text{(domain, nonnegative real numbers)}.$$

Does your method of tracing work for this pair of graphs?

EXERCISES 11.6

◀In Exercises 1–10, define the function f^{-1} for each given function f and draw graphs of f and f^{-1}. Unless otherwise indicated, assume that the domain of f is R.

1. $f : f(x) = 3x + 5$ 6. $f : f(x) = 3x^2, \ x \geq 0$

2. $f : f(x) = \dfrac{1}{2}x - 4$ 7. $f : f(x) = x^3$

3. $f : f(x) = x$ 8. $f : f(x) = x^2 + 3$

4. $f : f(x) = \dfrac{5 - x}{3}$ 9. $f : f(x) = \dfrac{2 + x}{x}, \ x > 0$

5. $f : f(x) = \dfrac{1}{x}, \ x > 0$ 10. $f : f(x) = x^{\frac{1}{2}} + 1, \ x \geq 0$

11.7 LOGARITHMIC FUNCTIONS

Let us now return to the exponential function

$$f : f(x) = 2^x$$

which we discussed in Section 11.3. Since whenever $x_1 > x_2$, $f(x_1) > f(x_2)$, f is strictly monotonic increasing. Furthermore, we have made plausible the fact that for every positive real number a there exists a unique real number r such that $a = 2^r$.

This, in effect, is saying that a function f^{-1} exists whose domain is the set of all positive real numbers, whose range is the set of all real numbers, and which is the inverse of f. How can we describe this particular function f^{-1}?

Let us first consider some examples. Since f is defined as

$$f : f(x) = 2^x,$$

then f^{-1} must be such that for each $x \in R$, $f^{-1}(2^x) = x$.
For example, if $x = 3$, then

$$f(3) = 2^3 = 8 \qquad \text{and} \qquad f^{-1}(8) = 3.$$

For the case where $x = 5$, since $f(5) = 2^5 = 32$, we see that $f^{-1}(32)$ must be equal to 5.

When $x = \frac{1}{2}$,

$$f\left(\frac{1}{2}\right) = 2^{\frac{1}{2}} = \sqrt{2}.$$

Therefore the function f^{-1} must be such that

$$f^{-1}(\sqrt{2}) = \frac{1}{2}.$$

When $x = -3$,

$$f(-3) = 2^{-3} = \frac{1}{2^3} = \frac{1}{8}.$$

Hence for f^{-1} we must have

$$f^{-1}\left(\frac{1}{8}\right) = -3.$$

Now that we have seen a few of the tasks that f^{-1} is expected to do, how can we construct such a "hard-working" function?

Let us have another look at the graph of

$$f : f(x) = 2^x$$

as reproduced in Figure 11-14.

If we now start from points on the y-axis and project horizontal lines to the curve, we can find, approximately, the values of f^{-1}. Thus,

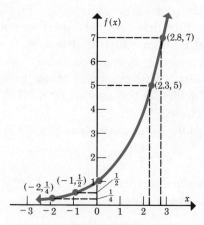

Figure 11-14

for example, $f^{-1}(5) \approx 2.3$, $f^{-1}(7) \approx 2.8$, and so on. We also see that

$$f^{-1}(1) = 0, \qquad f^{-1}\left(\frac{1}{2}\right) = -1, \qquad \text{and} \qquad f^{-1}\left(\frac{1}{4}\right) = -2.$$

The graph of f^{-1} in Figure 11-15 shows the general appearance of the function f^{-1}.

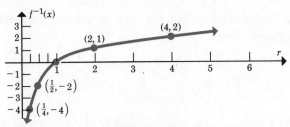

Figure 11-15

In your more advanced courses in mathematics you may discover that it is possible to describe f^{-1} by means of a certain kind of infinite series—something like a polynomial with an unlimited number of terms. However, for our purposes we shall merely designate f^{-1} by a name. The name generally used in mathematics to denote the inverse function of an exponential function is **logarithmic function**, usually abbreviated as **log function**. Thus we say that if f is described as

$$f : f(x) = 2^x$$

for any real number x, then we define f^{-1} as

$$f^{-1} : f^{-1}(x) = \log_2(x)$$

for any positive real number x. We read "$\log_2(x)$" as "the logarithm of x to the base 2."

The subscript 2 on the word log indicates that $\log_2$ is the inverse of the exponential function with a *base* 2. Generalizing, we have the following definition.

Definition 11.8 If f is a function such that

$$f : f(x) = b^x, \, b \in R, \, b > 0, \text{ and } b \neq 1,$$

then

$$f^{-1} : f^{-1}(x) = \log_b(x).$$

Because the inverse principle works two ways, we could also say that if l is defined as

$$l : l(x) = \log_b (x),$$

then l^{-1} may be defined as

$$l^{-1} : l^{-1}(x) = b^x.$$

We now use the fact that $f[f^{-1}(x)] = x$ from Definition 11.8 to reason as follows. Since

$$l[l^{-1}(x)] = x,$$

then

$$\log_b (b^x) = x.$$

Letting $b = 2$ and $x = 3$, we can say that $\log_2 (2^3) = 3$. In other words, since $2^3 = 8$, we know that $\log_2 (8) = 3$. Note that we have written $\log_2 (8)$ with parentheses to emphasize the fact that $\log_2 (x)$ describes a function. Henceforth for brevity, however, we shall omit parentheses and write simply $\log_2 x$, $\log_3 x$, and so on.

To familiarize ourselves with the log functions we should work with a number of examples. Here are a few.

Example 1   Find $\log_2 32$.

Solution: The job is essentially that of finding $l(32)$ where $l^{-1}(x) = 2^x$. Since we know that $32 = 2^5$, we can write

$$\log_2 32 = \log_2 2^5 = 5.$$

Example 2   Find $\log_3 9$.

Solution: Since the base in this case is 3, we know that $l^{-1}(x) = 3^x$. Recognizing that $9 = 3^2$, we write

$$\log_3 9 = \log_3 3^2 = 2.$$

Example 3   Find $\log_8 2$.

Solution: Here we are dealing with a base of 8, so our inverse function can be described as

$$l^{-1} : l^{-1}(x) = 8^x.$$

Question: For what value of x does $8^x = 2$? Experience with exponents tells us that $8^{\frac{1}{3}} = 2$ and we thus have

$$\log_8 2 = \log_8 8^{\frac{1}{3}} = \frac{1}{3}.$$

Example 4 Find $\log_{\frac{1}{2}} 8$.
Solution: Since $(\frac{1}{2})^{-3} = 8$, $\log_{\frac{1}{2}} 8 = -3$.

Example 5 Given that $\log_b 125 = 3$, what number does the base b represent?
Solution: Since $\log_b 125 = 3$, then

$$l^{-1}(3) = b^3 = 125.$$

The problem thus reduces to the question of what number has a cube equal to 125. Since the answer to this question is 5, we know that

$$\log_5 125 = 3.$$

Example 6 Find $\log_{10} 3$.
Solution: Since the base in this case is 10, we know that $l^{-1}(x) = 10^x$. We want $10^x = 3$. Is there any rational number x such that $10^x = 3$? Suppose that $10^{\frac{p}{q}} = 3$ where $\frac{p}{q}$ is a fraction in lowest terms with $q \in N$ and $p \in I$. Then we would have

$$(10^{\frac{p}{q}})^q = 3^q$$

so that

$$10^p = 3^q.$$

Assume first that p is negative and let $p = -s$, where s is a positive integer. Then $10^{-s} = \dfrac{1}{10^s}$ *and* $\dfrac{1}{10^s} < 1$. Since q is a positive integer, we have $3^q > 1$ and hence we cannot have $\dfrac{1}{10^s} = 3^q$.

If, on the other hand, p is a positive integer, then $10^p \geq 10$. Since q is a positive integer, it is easy to see that the decimal numeral for 3^q must have a unit's digit that is either 3, 9, 7, or 1 ($3^1 = 3$, $3^2 = 9$, $3^3 = 27$, $3^4 = 81$, $3^5 = 243$, $3^6 = 729$, etc.) and is certainly never 0. But the unit's digit of 10^n is 0 for all $n \in N$.

We conclude, then, that there is no rational number x such that

$\log_{10} 3 = x$. Thus $\log_{10} 3$ is an irrational number. It can be shown, however, that

$$\log_{10} 3 \approx 0.48, \quad \text{that is,} \quad 10^{0.48} \approx 3.$$

EXERCISES 11.7

◀In Exercises 1–30, find the value of the given logarithm.

1. $\log_3 27$

2. $\log_2 64$

3. $\log_{\frac{1}{3}} 81$

4. $\log_5 125$

5. $\log_9 3$

6. $\log_{16} 2$

7. $\log_6 36$

8. $\log_7 49$

9. $\log_2 128$

10. $\log_3 243$

11. $\log_4 256$

12. $\log_2 1$

13. $\log_5 1$

14. $\log_7 1$

15. $\log_{49} 7$

16. $\log_{32} 2$

17. $\log_{10} 100$

18. $\log_{10} 1000$

19. $\log_{10} 1{,}000{,}000$

20. $\log_{10} \dfrac{1}{10}$

21. $\log_{\frac{1}{2}} \dfrac{1}{8}$

22. $\log_3 \dfrac{1}{27}$

23. $\log_5 \dfrac{1}{25}$

24. $\log_8 4$

25. $\log_{27} 9$

26. $\log_{10} 0.001$

27. $\log_{10} 0.00001$

28. $\log_{25} 125$

29. $\log_{216} 36$

30. $\log_{10} \dfrac{1}{1{,}000{,}000}$

◀In Exercises 31–50, find the value of x which will make the given sentence true if this value is a rational number. Otherwise represent the answer as the value of a logarithmic function.

Example: $27^x = 9$ *Answer:* $\dfrac{2}{3}$

Example: $11^x = 7$ *Answer:* $\log_{11} 7$

Example: $\log_x 36 = 2$ *Answer:* 6

31. $2^x = 128$

32. $3^x = 729$

33. $4^x = 256$

34. $9^x = 27$

35. $4^x = 8$

36. $6^x = \dfrac{1}{36}$

37. $81^x = 3$

38. $13^x = 5$

39. $343^x = 7$

40. $2^x = \dfrac{1}{64}$

41. $7^{\log_7 11} = x$

42. $10^x = 0.1$

43. $100^x = 10$

44. $10^x = 0.001$

45. $17^x = 16$ **48.** $\log_x 1000 = 3$
46. $\log_x 49 = 2$ **49.** $12^{\log_x 5} = 5$
47. $\log_x 125 = 3$ **50.** $\log_x 0.01 = -2$

51. CHALLENGE PROBLEM. Prove that $\log_{10} 2$ is an irrational number.

11.8 PROPERTIES AND APPLICATIONS OF LOGARITHMS

By definition, we know that if

$$b^l = r,$$

then

$$\log_b r = l.$$

As an example, since

$$3^4 = 81,$$

then

$$\log_3 81 = 4.$$

Thus

$$3^{(\log_3 81)} = 81$$

and, in general,

$$b^{(\log_b r)} = r.$$

We see that values of logarithmic functions (or, for short, logarithms) are related closely to real number exponents. On this basis it seems reasonable to assume that the properties of exponents, P–1 through P–5, might be used to develop similar properties for logarithms.

Before proceeding with the discussion of these properties, however, we must realize that we have not proved P–1 through P–5 for real number exponents. Thus at this time we must regard the following conclusions concerning logarithms as based on assumptions or hypotheses about real number exponents rather than on proven fact. In more advanced courses, rigorous proofs are given.

Assuming, then, that P–1 through P–5 hold for real number exponents, we assert that the following properties hold for all positive real numbers x and y and any real number r. (b is a fixed real number > 0 and $\neq 1$.)

(L–1) $\log_b xy = \log_b x + \log_b y.$

(L–2) $\log_b x^r = r \log_b x.$

(L–3) $\log_b \left(\dfrac{x}{y}\right) = \log_b x - \log_b y$ if $y \neq 0.$

Using P–1 and the theorem (which you will be asked to prove in the Exercises) that if $b^r = b^s$ with $b > 0$, $b \neq 1$, and r and s real numbers, then $r = s$, we will prove L–1.

We first recall that

$$(1) \qquad b^{\log_b r} = r \qquad \text{for any positive real number } r.$$

Thus

$$b^{\log_b x} = x \qquad \text{and} \qquad b^{\log_b y} = y,$$

and so

$$b^{\log_b x} \cdot b^{\log_b y} = xy.$$

By (1), however,

$$xy = b^{\log_b xy}$$

and, by P–1,

$$b^{\log_b x} \cdot b^{\log_b y} = b^{\log_b x + \log_b y}.$$

Hence

$$b^{\log_b x + \log_b y} = b^{\log_b xy}$$

and finally, using our theorem about $b^r = b^s$,

$$\log_b xy = \log_b x + \log_b y.$$

The proofs for L–2 and L–3 are similar and are left for the Exercises. Properties L–1, L–2, and L–3 can, as we shall see, be utilized in performing many computations.

Since most computations carried out in practice are related to the decimal system of notation, that is, notation with a base 10, we shall, for the most part, use a base 10 in working with logarithms. A value of the function $\log_b x$ with $b = 10$ is usually called a **common** logarithm. In what follows we shall assume for convenience that in writing $\log x$ we shall mean $\log_{10} x$.

We begin the demonstration with a few simple computations that could be easily performed in other ways but which illustrate general methods. Later you will have an opportunity to deal with more challenging situations.

Let us suppose as a start that we wish to find the product

$$37 \cdot 52$$

using logarithms to help in the computation. (We might imagine for the moment that we had completely forgotten the multiplication tables!)

By L–1 we know that

$$\log (37 \cdot 52) = \log 37 + \log 52.$$

All well and good, but what are the values of log 37 and log 52, respectively?

Table I at the back of the book gives values of logarithms for a certain set of real numbers. We shall discuss this table and how to use it shortly. Meanwhile let us take for granted that

$$\log 37 \approx 1.5682$$

and

$$\log 52 \approx 1.7160$$

Thus

$$\log 37 + \log 52 \approx 3.2842$$

By L–1 we know that

$$\log 37 + \log 52 = \log (37 \cdot 52)$$

and hence

$$\log (37 \cdot 52) \approx 3.2842.$$

It remains, then, to find the number x such that $\log x = 3.2842$. It can also be seen from Table I (plus a little interpolation) that if $\log x = 3.2842$, then $x \approx 1924$. (You may want to check this result by multiplication.)

In this illustration we used the symbol $\approx$ in giving values of logarithms from the table. We did this to emphasize the fact that the entries in Table I are four-place approximations to the actual logarithms, except for log 1 which is actually equal to 0. Henceforth, however, we shall normally use the equal sign (=) for convenience. You must keep in mind, however, that in most calculations with logarithms we will be dealing with approximations.

For a second illustration we shall use Property L–2. Suppose we wish to find an approximate value for the cube root of 7:

$$\sqrt[3]{7} = 7^{\frac{1}{3}}.$$

By L–3 we know that

$$\log 7^{\frac{1}{3}} = \frac{1}{3} \log 7.$$

Again referring to Table I, we see that

$$\log 7 = 0.8451.$$

Thus

$$\frac{1}{3} \log 7 = 0.2817,$$

and so

$$\log \sqrt[3]{7} = 0.2817.$$

Using the table again, we are able to ascertain that if $\log x = 0.2817$, then $x = 1.913$. Actually, the cube root of 7 correct to six decimal places is known to be 1.912931. So 1.913 is a quite respectable approximation.

Using a table of logarithms and Property L–2, we can now find approximations for exponential functions involving somewhat complicated exponents. To find an approximate value of 5^π, for example, we have

$$\log 5^\pi = \pi \log 5.$$

Since $\log 5 = .6990$ and $\pi \approx 3.14$, we find that

$$\log 5^\pi \approx (3.14)(0.6990) \approx 2.1949.$$

From the table we see that if

$$\log x = 2.1959,$$

then

$$x = 156.$$

Thus

$$5^\pi \approx 156.$$

Having investigated some of the general ideas, let us now investigate the problem of how to use Table I. Suppose, for example, we want to find $\log 7.84$. We look for 78 in the column on the extreme left and scan the top row for 4. The entry to the right of 78 and directly under 4 is 8943. This tells us that the logarithm of $7.84 = 0.8943$. Note that in this table, as in most logarithm tables, the decimal points are omitted. Thus in using Table I we must supply our own. We need, then, to assume that there is a decimal point between the two digits in each entry of the left-hand column and also to assume a decimal point at the left of each four-digit numeral in the other columns. Our example illustrates this. We found that the entry 8943 was associated with 784, from which we inferred that

$$\log 7.84 = 0.8943.$$

Table I, whose entries represent four-place decimals, actually supplies us with logarithms of numbers from 1 to 9.99. For logarithms of numbers between 0 and 1 or numbers greater than 9.99 we shall need to apply Properties L–1 and L–3 as you will see.

Practice reading the table. Check to see that $\log 4.62 = 0.6646$, $\log 3.70 = 0.5682$, $\log 8.37 = 0.9227$, and $\log 1.02 = 0.0086$.

As we indicated, common logarithms of positive real numbers outside the interval from 1.00 to 9.99 can be found by using Property L–1. From what follows you will also see the reason for using base 10 for logarithms.

In our first example we used $\log 37$. Note that $37 = 3.7 \cdot 10$. Hence, by L–1,

$$\log (3.7 \cdot 10) = \log 3.7 + \log 10.$$

But $\log 10 = 1$. (Why?) So

$$\log 37 = (\log 3.7) + 1 = 0.5682 + 1 = 1.5682,$$

as indicated earlier.

To find $\log 257$, for example, we use the fact that $257 = 2.57 \cdot 100$. So, noting that $\log 100 = 2$ (explain this), we have

$$\log 257 = (\log 2.57) + \log 100 = \log 2.57 + 2$$

or $0.4099 + 2 = 2.4099$.

EXERCISES 11.8A

◀In Exercises 1–29, determine the logarithm of the given number using Table I.

1. 591	**11.** 3650	**21.** 2,420,000
2. 352	**12.** 4210	**22.** $6.31 \cdot 10^4$
3. 24,100	**13.** 6570	**23.** 4120
4. 15,600	**14.** 21,700	**24.** 5130
5. 46.7	**15.** 78,200	**25.** 217,000
6. 35.8	**16.** 4.36	**26.** 186,000
7. 21.3	**17.** 7.92	**27.** 647,000,000
8. 92.8	**18.** 187,000	**28.** 213,000,000
9. $1.15 \cdot 10^4$	**19.** 315,000	**29.** 36,000,000,000
10. $2.14 \cdot 10^5$	**20.** 1,260,000	

30. Prove that if $b^r = b^s$ with $b > 1$, then $r = s$. (*Hint:* Use the property that $f : f(x) = b^x$ is strictly monotonic increasing for $b > 1$.)

To find the logarithm of numbers between 0 and 1, we can apply Property L–3. Two examples should clarify the essentials. First find log 0.00381. We begin by noting that

$$0.00381 = \frac{3.81}{1000}.$$

But by L–3, we know that

$$\log 0.00381 = \log 3.81 - \log 1000.$$

Since log 1000 = 3, this gives us

$$\begin{aligned}
\log 0.00381 &= \log 3.81 - 3 \\
&= 0.5809 - 3 \\
&= -2.4191.
\end{aligned}$$

The fact that the logarithm of a number less than 1 is negative should come as no surprise. (See the graph of $y = \log_2 r$ in Figure 11-15.)

As a second example, note that

$$\begin{aligned}
\log 0.487 &= \log \frac{4.87}{10} \\
&= \log 4.87 - \log 10 \\
&= 0.6875 - 1 \\
&= -0.3125.
\end{aligned}$$

We now come to the second phase of table usage, namely that of determining x when $\log x$ is given.

In one of our examples we found that $\log \sqrt[3]{7} = 0.2817$. From the table we see that

$$\log 1.91 = 0.2810 \quad \text{and} \quad \log 1.92 = 0.2833.$$

To find x such that $\log x = 0.2817$, we go back once more to linear interpolation, that is, we estimate a value for x on the assumption (faulty though it is) that the function we are dealing with has as its graph a straight line.

We have two points

$$(1.91, 0.2810) \quad \text{and} \quad (1.92, 0.2833).$$

A linear equation based on these points is

$$y - 0.2810 = 0.23(x - 1.91).$$

From this equation we see, after a bit of computation, that when $y = 0.2817$, we have $x = 1.913$. Using proportions, we get

$$
\begin{array}{cc}
x & \log x
\end{array}
$$

$$
0.01 \left[\; d \left[\begin{array}{cc} 1.91 & 0.2810 \\ \boxed{?} & 0.2817 \end{array} \right] 0.007 \right] 0.0023
$$

$$
\begin{array}{cc}
1.92 & 0.2833
\end{array}
$$

$$
\frac{d}{0.01} = \frac{0.007}{0.0023},
$$

so that $d = 0.003$.

Thus we find, as before, that when $\log x = 0.2817$, then

$$
x = 1.91 + 0.003 = 1.913.
$$

This tells us that

$$
\log 1.913 \approx 0.2817.
$$

Hence $\sqrt[3]{7} \approx 1.913$.

If we seek a value of x for which $\log x = 3.2842$, as in our first example involving $\log (37 \cdot 52)$, we reason as follows:

$$
3.2842 = 0.2842 + 3.
$$

Assume now that $x = y \cdot z$ for two positive real numbers y and z. Then

$$
\log x = \log (y \cdot z) = \log y + \log z.
$$

If $\log x = 0.2842 + 3$, we may let $\log y = 0.2842$ and let $\log z = 3$. From this we see, after interpolation, that

$$
y \approx 1.924
$$

and

$$
z = 1000.
$$

Therefore

$$
x = y \cdot z = 1924.
$$

Complicated computations can be performed using all three properties at once. For example, to find

$$
\frac{\sqrt{3.14} \cdot \sqrt{5.17}}{\sqrt[3]{12.3}}
$$

we note that

$$\log\left(\frac{\sqrt{3.14} \cdot \sqrt{5.17}}{\sqrt[3]{12.3}}\right) = \frac{1}{2}\log 3.14 + \frac{1}{2}\log 5.17 - \frac{1}{3}\log 12.3.$$

Laws of exponents may also be used to advantage. To compute

$$\sqrt[3]{\frac{27.1}{0.32}},$$

for example, we write

$$\sqrt[3]{\frac{27.1}{0.32}} = \left(\frac{27.1}{0.32}\right)^{\frac{1}{3}}$$

and

$$\log\left(\frac{27.1}{0.32}\right)^{\frac{1}{3}} = \frac{1}{3}(\log 27.1 - \log 0.32).$$

To finish the computation for this example we can proceed as follows. From Table I we note that $\log 2.71 = 0.4330$. Hence

$$\log 27.1 = \log (10 \cdot 2.71) = \log 10 + \log 2.71.$$

Thus

$$\log 27.1 = 1 + 0.4330 = 1.4330.$$

Similarly,

$$\log 0.32 = \log\left(3.2 \cdot \frac{1}{10}\right),$$

that is,

$$\log 3.2 + \log\frac{1}{10} = 0.5051 - 1, \qquad \left(\text{since } \log\frac{1}{10} = -1. \text{ Why?}\right)$$

$$= -0.4949.$$

We now calculate

$$\frac{1}{3}(0.4330 - (-0.4949)) = \frac{1}{3}(0.4330 + 0.4949)$$

$$= \frac{1}{3}(0.9279) = 0.3093.$$

For the final answer we seek a number x such that $\log x = 0.3093$. With a bit of interpolation we find that $x = 2.038$. Hence $\sqrt[3]{\dfrac{27.1}{0.32}}$ $= 2.038.$

Computation with logarithms often presents unforeseen difficulties. For example, if we seek a number x such that $\log x = -0.4162$, how do we use the table? One device is to note that

$$-0.4162 = 0.5838 - 1.$$

Hence if $\log x = 0.5838 - 1$, we may write $x = \dfrac{y}{z}$ and use L–3 as follows. Since

$$\log \frac{y}{z} = \log y - \log z,$$

we may assume that $\log y = 0.5838$ and $\log z = 1$. From this we get $y = 3.835$ and $z = 10$, which tells us that

$$x = \frac{y}{z} = \frac{3.835}{10} = 0.3835.$$

This example suggests a possible device for dealing with negative logarithms, that is, logarithms of numbers between 0 and 1. However, in practice, it is usually advisable to avoid involvement with negative logarithms whenever feasible.

For example, to find $\sqrt[3]{0.05}$, we could write

$$\sqrt[3]{0.05} = \sqrt[3]{\frac{50}{1000}} = \frac{\sqrt[3]{50}}{10},$$

then compute $\sqrt[3]{50}$ by logarithms and divide the result by 10. Similarly, in the previous example, $\sqrt[3]{\dfrac{27.1}{0.32}}$ could have been written as $\sqrt[3]{\dfrac{271}{3.2}}$.

To find $\sqrt{\dfrac{1}{0.006}}$, we could write

$$\sqrt{\frac{1}{0.006}} = \sqrt{\frac{10,000}{60}} = \frac{100}{\sqrt{60}}.$$

From this we get

$$\log \frac{100}{\sqrt{60}} = \log 100 - \frac{1}{2} \log 60$$

$$= 2 - \frac{1}{2}(1.7782) = 2 - 0.8891 = 1.1109.$$

Here, as in a previous example, we let $x = yz$. Then

$$\log x = \log yz = \log y + \log z = 1 + 0.1109.$$

Thus

$$\log y = 1, \quad \log z = 0.1109, \quad y = 10, \quad z = 1.291,$$

and finally,

$$x = 12.91.$$

Another kind of difficulty occurs in a problem such as that of finding

$$N = \frac{256}{917}.$$

We have $\log N = \log 256 - \log 917$ and so we write

$$\log 256 = 2.4082$$
$$(-) \log 917 = 2.9624$$
$$\overline{\log N = \boxed{?}}$$

But now we have a problem since $2.4082 < 2.9624$. We overcome this difficulty by writing

$$\log 256 = 3.4082 - 1$$
$$(-) \log 917 = 2.9624$$
$$\overline{\log N = 0.4458 - 1}$$

Property L–2 can be used to solve exponential equations. For example, to find the solution of

$$5^x = 12,$$

we write

$$\log 5^x = \log 12$$
$$x \log 5 = \log 12$$
$$x = \frac{\log 12}{\log 5} = \frac{1.0792}{0.6990} \approx 1.54.$$

The final quotient can be computed by ordinary division or (naturally!) by logarithms.

EXERCISES 11.8B

◀In Exercises 1–24, perform the computations using Table I of common logarithms. Where necessary use linear interpolation to achieve four-place accuracy.

1. 7^4

2. 12^3

3. $\sqrt{12}$

4. $10^{\frac{1}{3}}$

5. $15^{\frac{2}{3}}$

6. $6^{\frac{1}{4}}$

7. 35^3

8. $\sqrt{29}$

9. $\sqrt{65}$

10. $\sqrt[3]{20}$

11. $\sqrt[5]{5} \cdot \sqrt[3]{4}$

12. $\dfrac{19.1}{25.2}$

13. $\dfrac{6.97}{17.3}$

14. 10^{π} (Use $\pi \approx 3.14$.)

15. $\sqrt{(0.015)(2.17)}$

16. $\sqrt{\dfrac{3.86}{0.154}}$

17. $\sqrt{11} \cdot \sqrt[3]{5}$

18. $\left(\dfrac{3}{5}\right)^6$

19. π^5 (Use $\pi \approx 3.14$.)

20. $\sqrt{\pi}$ (Use $\pi \approx 3.14$.)

21. $\sqrt[3]{\dfrac{(6.8)(0.785)}{(3.86)^2}}$

22. $\sqrt{\dfrac{(452)(16.4)}{1.86}}$

23. $\sqrt[3]{\dfrac{(0.06)(1750)}{12.6}}$

24. $\sqrt{(35.7)(286)(0.71)}$

25. Prove L-2. [*Hint:* Let $\log_b x^r = N$, then $b^N = x^r$, and $(b^N)^{\frac{1}{r}} = (x^r)^{\frac{1}{r}}$.]

26. Prove L-3. (*Hint:* Let $\dfrac{x}{y} = x \cdot y^{-1}$ and use L-2.)

27. Solve the equation $10^x = 7$.

28. Solve the equation $5^x = 20$.

29. Solve the equation $7^x = 100$.

30. A formula for the area of a triangle in terms of the three sides is

$$\text{area} = \left[\frac{p}{2}\left(\frac{p}{2} - a\right)\left(\frac{p}{2} - b\right)\left(\frac{p}{2} - c\right)\right]^{\frac{1}{2}}$$

where a, b, and c are the lengths of the sides and p is the perimeter. Use logarithms to compute the area of a triangle the lengths of whose sides are 52 ft., 71 ft., and 43 ft., respectively.

31. The volume of a sphere is given by the formula

$$V = \frac{4}{3}\pi r^3.$$

Use logarithms to calculate the volume of a sphere with a radius r of length 72.3 in. (Use $\pi \approx 3.14$.)

32. In the Thrifty Citizens Savings Bank, 5% interest is compounded quarterly. The amount which accumulates over a period of years is given by the formula

$$A = P\left[1 + \frac{1}{4}(0.05)\right]^{4y},$$

where P is the principal deposited and y is the number of years on deposit. If $5000 is deposited initially, what will the amount be at the end of ten years?

33. Under the conditions of Exercise 32, what will $10,000 amount to at the end of seven years?

34. Population growth is often measured by means of an exponential function. The population of Hoodville after y years is given by the formula

$$P \approx P_0 \cdot (2.72)^{\frac{y}{10}},$$

where P_0 is the population to start with and y is the period of growth in years. If Hoodville had a population of 24,000 in 1960, what is the expected population for 1974?

35. CHALLENGE PROBLEM. The graph of $j : j(x) = x^3$ is given in Figure 11-11, page 523. Using this graph, some tracing paper, and your knowledge of inverse functions, draw the graph of j^{-1}.

36. CHALLENGE PROBLEM. Solve the equation

$$3^{x+1} = \sqrt[3]{2^{3x-2}}.$$

37. CHALLENGE PROBLEM. Figure 11-16 illustrates the graph of the function $f : f(x) = e^x$, where $e \approx 2.72$ is an irrational number that plays an impor-

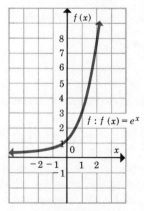

Figure 11-16

tant role in many branches of higher mathematics. Use this graph and your knowledge of inverse functions to construct the graph of $f^{-1} : f^{-1}(x) = \log_e x$. (Functional values of $\log_e x$ for $x \in R$ and $x > 0$ are called **natural logarithms**.)

CHAPTER SUMMARY

The following properties and definitions relating to EXPONENTS, EXPONENTIAL FUNCTIONS, INVERSE FUNCTIONS, and LOGARITHMIC FUNCTIONS have been examined in this chapter.

If x and y are any rational numbers and if r and s are any rational numbers, then in all cases where the exponential expressions are real numbers,

P–1 $$x^r x^s = x^{r+s}.$$

P–2 $$\frac{x^r}{x^s} = x^{r-s}.$$

P–3 $$(x^r)^s = x^{rs} \text{ if } x > 0.$$

P–4 $$(xy)^r = x^r y^r.$$

P–5 $$\left(\frac{x}{y}\right)^r = \frac{x^r}{y^r}.$$

The same properties apply to real number exponents if the further restriction is made that the base numbers are positive.

> *Definition* 11.1 $x^1 = x$ and if $k \in N$, $x^k \cdot x = x^{k+1}$.
>
> *Definition* 11.2 $x^0 = 1$ for $x \neq 0$.
>
> *Definition* 11.3 and 11.6 For any rational number r and for any nonzero real number x, $x^{-r} = \dfrac{1}{x^r}$ in all cases where x^r is a real number.
>
> *Definition* 11.4 $x^{\frac{1}{n}} = \sqrt[n]{x}$ where $n \in N$ and x is any real number for which $\sqrt[n]{x}$ is a real number.
>
> *Definition* 11.5
> 1. For any real number $x \geq 0$ and for any rational number $\frac{m}{n}$ with $m \in I$ and $n \in N$, we define
> $$x^{\frac{m}{n}} = (x^{\frac{1}{n}})^m$$
> provided that $x \neq 0$ when $m < 0$.

2. For any real number $x < 0$ and for any rational number $\frac{m}{n}$, $m \in I$ and $n \in N$ and with m and n not both even, we define

$$x^{\frac{m}{n}} = (x^{\frac{1}{n}})^m$$

providing that $x^{\frac{1}{n}}$ is defined.

Definition 11.7   If f and g are functions related in such a way that

$$[g \circ f](x) = g[f(x)] = x$$

for every element x in the domain of f and

$$[f \circ g](x) = f[g(x)] = x$$

for every element x in the domain of g, then f and g are said to be the inverses of each other.

The following conditions are equivalent:
1. f is a one-to-one mapping.
2. f has an inverse.
If f is strictly monotonic increasing or strictly montonic decreasing, then (1) and (2) both hold.

Definition 11.8   If f is a function such that $f(x) = b^x$ $(b > 0, b \neq 1)$, then the inverse function f^{-1} is

$$f^{-1} : f^{-1}(x) = \log_b x,$$

and

(L–1) $\log_b (xy) = \log_b x + \log_b y;\ (x > 0,\ y > 0)$
(L–2) $\log_b x^r = r \log_b x;\ (x > 0)$

(L–3) $\log_b \dfrac{x}{y} = \log_b x - \log_b y;\ (x > 0, y > 0)$

Properties L–1, L–2, and L–3 can be used to facilitate arithmetic computations.

The chapter also included a discussion of LINEAR INTERPOLATION and the graphing of inverse functions.

REVIEW EXERCISES

◀ In Exercises 1–5, simplify the given expressions, writing all answers without negative exponents.

1. $(x^{-2}y^3)^{-2}$ 4. $\left(\dfrac{s^{-1}}{2t^3}\right)^{-2}$

2. $(a^2b^{-3})^{-4}$ 5. $\left(\dfrac{x^3y^{-2}}{x^0 + y^0}\right)^{-3}$

3. $(2x^{-1})^{-2}$

◀ In Exercises 6–15, evaluate each of the given expressions.

6. $27^{\frac{2}{3}}$ 11. $\log_3 81$

7. $81^{-\frac{1}{4}}$ 12. $\log_5 \dfrac{1}{125}$

8. $100^{-\frac{3}{2}}$ 13. $\log_2 128$

9. $32^{-\frac{3}{5}}$ 14. $\log_{10} 1{,}000{,}000$

10. $729^{-\frac{1}{3}}$ 15. $\log_{10} 0.001$

◀ In Exercises 16–18, consider the functions $f : f(x) = x^2 - x$, $g : g(x) = x + 7$, and $h : h(x) = x^2 + 1$, each with domain R, and find the given functional value.

16. $[f \circ g](2)$ and $[g \circ f](2)$ 18. $[g \circ h](2)$ and $[h \circ g](2)$

17. $[f \circ h](2)$ and $[h \circ f](2)$

◀ In Exercises 19–25, give the inverse function f^{-1} for each of the given functions with domain the set of all real numbers.

19. $f : f(x) = 4x$ 23. $h : h(x) = \dfrac{3 - x}{2}$

20. $f : f(x) = 8x + 3$ 24. $g : g(x) = \dfrac{1}{2}x^3 - 1$

21. $f : f(x) = 6x - 12$ 25. $h : h(x) = \dfrac{8x^3 + 5}{2}$

22. $g : g(x) = \dfrac{1}{3}x + 8$

26. Construct a graph of $f : f(x) = 3^x$.
27. Construct a graph of $f^{-1} : f^{-1}(r) = \log_3 r$.

◄In Exercises 28–35, change the given descriptions of logarithmic functions to descriptions of exponentials and the given descriptions of exponential functions to descriptions of logarithmic functions.

28. $y = 7^x$ **32.** $z = x^x$

29. $a = b^x$ **33.** $y = s^{3t}$

30. $s = \log_b t$ **34.** $17 = \pi^r$

31. $l = \log_3 r$ **35.** $D = \log_{10} w$

◄In Exercises 36–40, change the given equations to equations involving common logarithms.

36. $x = \dfrac{\sqrt[3]{y}}{\sqrt{z}}$ *Answer:* $\log x = \dfrac{1}{3}\log y - \dfrac{1}{2}\log z$

37. $a = b^5\left(\dfrac{\sqrt{c}}{\sqrt[3]{d}}\right)$ **39.** $z = \dfrac{\sqrt{uv}}{\sqrt[3]{wx}}$

38. $r = \dfrac{\sqrt[3]{s} \cdot \sqrt{t}}{w^4}$ **40.** $\sqrt[3]{x} = 7\sqrt[4]{y}$

◄In Exercises 41–45, give in each case the value of the variable x which will make the given sentence true.

41. $\log_x 128 = 7$ **44.** $8^x = 50$

42. $216^x = 6$ **45.** $12^x = 13$

43. $\log_{10} x = 5$

◄In Exercises 46–49, perform the computations using logarithms and, where appropriate, linear interpolation.

46. 13^5 **48.** $\sqrt{\dfrac{3}{4}}$

47. $6^{\frac{1}{3}}$ **49.** $\sqrt{(1.21)(42.7)}$

50. Using the compound interest formula $A = P\left[1 + \dfrac{1}{4}(0.05)\right]^{4y}$, find the sum of money which $7000 will amount to in 15 years.

51. CHALLENGE PROBLEM. Using the formula of Exercise 50, calculate how many years it takes for a deposit to double itself.

GOING FURTHER: READING AND RESEARCH

Many phenomena in nature are related to exponential functions. Among these are growth curves and radioactive decay. For example, the number of bacteria in a culture is known to increase at a rate proportional to the number present. To determine the number x present at any given time we can use the formula

$$x = ne^{kt}.$$

Here $e \approx 2.72$ (see Challenge Problem 37 in Exercises 11.8B), t is the time in minutes, n is the original number of bacteria, and k is a constant determined by the specific culture under consideration.

As an example, suppose it takes 10 minutes for the original number of bacteria to double. When will the number of bacteria be 5 times what it was originally? Using your knowledge of exponential and logarithmic functions, see if you can follow the steps in the calculation. We begin with

$$x = ne^{kt}.$$

After 10 minutes, $x = 2n$, and so

$$2n = ne^{10k},$$
$$2 = e^{10k},$$
$$\log_e 2 = \log_e e^{10k},$$
$$\log_e 2 = 10k \qquad \text{(Why?)}$$

(1)
$$\frac{1}{10} \log_e 2 = k.$$

Now, when the number of bacteria is 5 times what it was originally, we have

$$5n = ne^{kt},$$
$$5 = e^{kt},$$
$$\log_e 5 = \log_e e^{kt},$$
$$\log_e 5 = kt,$$

and

$$t = \frac{\log_e 5}{k}.$$

From (1) we then get

$$t = \frac{\log_e 5}{\frac{1}{10} \log_e 2} = 10 \frac{\log_e 5}{\log_e 2}.$$

From a table of natural logarithms we find that $\log_e 5 \approx 1.61$ and $\log_e 2 \approx 0.69$. Hence

$$t = 10 \cdot \frac{1.61}{0.69} \approx 23.$$

Thus the time required is approximately 23 minutes.

The rate of decomposition of a radioactive substance is proportional to the mass present. To determine how much of a given sample of n grams will remain undecomposed at the end of t minutes if half of the material decomposes in 10 minutes, we can use the formula

$$x = ne^{-kt}.$$

How much material will be left at the end of 20 minutes if we start with 100 grams?

See if you can solve this problem using a table of natural logarithms and a development somewhat similar to that of the first example. For further investigation of this and related types of problems, we suggest you do some reading from books in the following list.

SCHOOL MATHEMATICS STUDY GROUP, *Radioactive Decay*, Pasadena: A. E. Vroman, 1969.

WOLF, G., *Isotopes in Biology*. New York: Academic Press.

HERBERG, T. and J. D. BRISTOL, *Elementary Mathematical Analysis* (Chapter 6). Boston, Mass.: D. C. Heath, 1967.

RICHMOND, D. E., *Mathematical Models of Growth and Decay*. Twenty-eighth Yearbook of the National Council of Teachers of Mathematics, Washington, D.C. National Council of Teachers of Mathematics, 1963.

CHAPTER TWELVE
TRIGONOMETRIC
FUNCTIONS

12.1 ANGLES AND ANGLE MEASURE

Trigonometry, the study of triangles, probably originated when man began to establish land boundaries. Later, trigonometry became increasingly important as an aid to navigation and astronomy. Today, however, its major significance lies in the area of mathematical analysis where the application depends on the functional aspects of trigonometry.

We begin our study of trigonometric functions by expanding the concept of an angle with which you are familiar from your study of geometry. An angle is still defined as the union of two rays with a common endpoint called the **vertex**. It is useful, however, to consider one ray, called the **initial** ray, as fixed and the other ray, called the **terminal** ray, as able to rotate. As the terminal ray moves, it generates angles. In Figure 12-1, the curved arrow indicates that the ray $\overrightarrow{OP}$ moved from a position coinciding with $\overrightarrow{OQ}$ in a counterclockwise direction to form angle QOP (denoted as $\angle QOP$).

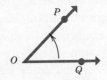

Figure 12-1

If an angle is generated by moving the terminal ray in a counterclockwise direction, the measure of the angle formed is considered as positive. If an angle is generated by moving the terminal ray in a clockwise direction, the measure of the angle formed is considered as negative. We call such angles **directed** or **signed** angles. An angle as considered in this chapter will always be a directed angle.

Other rotations can be associated with $\angle QOP$ as indicated in Figure 12-2.

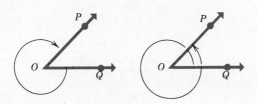

Figure 12-2

Angles having the same initial and terminal sides and the same vertex are called **coterminal** angles. Given an initial ray $\overrightarrow{OQ}$, the terminal ray $\overrightarrow{OP}$ together with $\overrightarrow{OQ}$ determines an infinite set of coterminal angles, each having the same vertex, the same initial ray, and the same terminal ray.

An angle QOP is in **standard position** if its vertex is the origin of the Cartesian plane and its initial ray $\overrightarrow{OQ}$ coincides with the non-negative ray of the x-axis as in Figure 12-3.

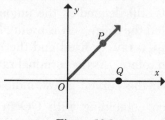

Figure 12-3

We shall restrict our discussion to angles in the standard position. This, however, does not restrict the application of our conclusions since every angle in the plane is congruent to an angle in standard position. Each angle in standard position intercepts a unique arc on the unit circle. Thus in Figure 12-4, $\angle QOP$ intercepts $\overset{\frown}{AB}$, $\angle QOP'$ intercepts $\overset{\frown}{AC}$, and $\angle QOP''$ intercepts $\overset{\frown}{AD}$. We use the measure of the arc as the measure of the angle.

The unit that is used to measure an arc is quite arbitrary. To assign any unit we use the procedure of Chapter 10. A number line is wrapped around the circle and thus associates a unique real number with each point on the circle. This number is the measure of the arc

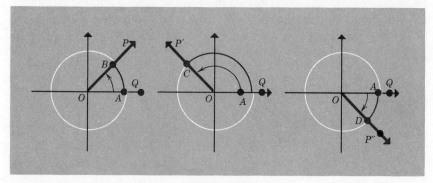

Figure 12-4

length from the point (1, 0). The angle QOP that intercepts arc $\overset{\frown}{AU}$, where U is the point associated with the number 1, becomes thereby the unit angle that determines the measure of other angles. (See Figure 12-5.)

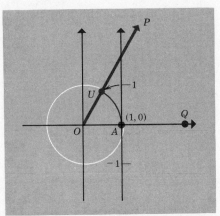

Figure 12-5

If the units on the number line are the same units as those on the x- and y-axes, then 2π units are contained in the circle. In this instance the unit of angle measure is the *radian*. Note that the length of the arc $\overset{\frown}{AU}$, intercepted by $\angle QOP$, is the same as the length of $\overline{OA}$.

Definition 12.1 If an angle has its vertex at the center of a circle and intercepts an arc whose length is equal to the radius of the circle, then the measure of the angle is one **radian**.

In Figure 12-5, the measure in radians of $\angle QOP$ is 1. We write $M\angle QOP = 1$. Thus $M\angle A$ refers to the radian measure of $\angle A$. In Figure 12-6, the measure in radians of $\angle QOP$ is -2 since the length of the arc of the unit circle swept out by $\angle QOP$ in a clockwise direction is 2. We write $M\angle QOP = -2$.

If the units on the number line are chosen so that exactly 360 units are contained in the circle, then the unit of angle measure is the **degree**. In Figure 12-7, the measure in degrees of $\angle QOP$ is 70. We write $m\angle QOP = 70$. Thus $m\angle QOP$ refers to the degree measure of $\angle QOP$. (Note that M indicates radian measure; m, degree measure.)

Figure 12-8 shows five other angles and their degree measure.

The radian and degree units of measure are related to each other

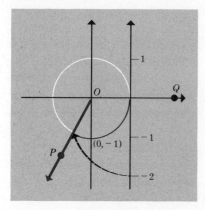

Figure 12-6

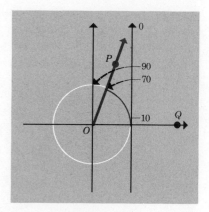

Figure 12-7

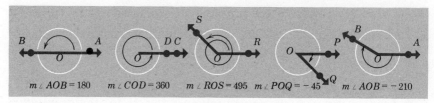

$m \angle AOB = 180$ $m \angle COD = 360$ $m \angle ROS = 495$ $m \angle POQ = -45$ $m \angle AOB = -210$

Figure 12-8

in the sense that, for example, centimeters and inches are related to each other. We write 2.54 cm. = 1 in. to express the fact that a measurement of 2.54 cm. represents the same length as a measurement of 1 in. Since there are 2π units of radian measure contained in one circle and 360 units of degree measure, 2π radians correspond to 360 degrees. To express this relationship we write

$$2\pi \text{ radians} = 360 \text{ degrees}$$

to indicate that an angular measure in radians of 2π is equivalent to an angular measure in degrees of 360. In the same sense, we write

(1) $$1 \text{ radian} = \frac{180}{\pi} \text{ degrees}$$

and

(2) $$1 \text{ degree} = \frac{\pi}{180} \text{ radians.}$$

Formulas (1) and (2) are useful in making conversions from degree measure to radian measure or vice versa. If more exact measure is not needed, the approximations

$$1 \text{ radian} \approx 57.3 \text{ degrees}$$

and

$$1 \text{ degree} \approx 0.0175 \text{ radian}$$

may be used.

Example 1   If $M \angle AOC = 2$, find $m \angle AOC$.

Solution: 2 radians $= 2(\frac{180}{\pi})$ degrees. Thus $m \angle AOC = \frac{360}{\pi}$. Also

$$m \angle AOC \approx 2(57.3) = 114.6.$$

The degree is subdivided into 60 equal parts called minutes and the minute is subdivided into 60 equal parts called seconds. Thus

$$1 \text{ degree} = 60 \text{ minutes}$$

and

$$1 \text{ minute} = 60 \text{ seconds}.$$

Although, strictly speaking, the measure of an angle should be expressed as a number, it is customary to write, as in Example 1, $m \angle AOC = 114°36'$ rather than $m \angle AOC = 114.6$.

Example 2   Find $M \angle AOB$ if $m \angle AOB = -45$.

Solution: 45 degrees $= 45(\frac{\pi}{180})$ radians. Thus

$$M \angle AOB = \frac{-45\pi}{180} = -\frac{\pi}{4}.$$

Since $\frac{\pi}{4} \approx 0.785$,

$$M \angle AOB \approx -0.785.$$

The radian measure of angles is very useful because a simple relation exists between the length of an arc of a circle and the radian measure of the central angle subtended by the arc (that is, the angle whose vertex is at the center of the circle and whose sides intercept the arc). If $\angle POQ$ intercepts $\overset{\frown}{AB}$ on the unit circle and $\overset{\frown}{A'B'}$ on a concentric circle of radius r, as in Figure 12-9, then

(3)
$$\frac{\text{length of } \overset{\frown}{AB}}{2\pi} = \frac{\text{length of } \overset{\frown}{A'B'}}{2\pi r}.$$

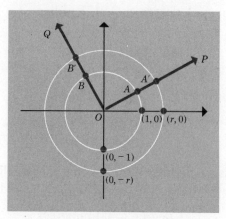

Figure 12-9

If the measure in radians of $\angle POQ$ is t, then the length of $\overset{\frown}{AB}$ is t. If we replace the length of $\overset{\frown}{AB}$ by t in Equation (3) and simplify, we have

$$\frac{t}{1} = \frac{\text{length of } \overset{\frown}{A'B'}}{r}$$

so that

$$\text{length of } \overset{\frown}{A'B'} = tr.$$

Letting $s = \text{length of } \overset{\frown}{A'B'}$, we have

(4)
$$s = tr.$$

Example 3   Find the length of an arc on a circle of radius 3 in. if the arc subtends a central angle of 2.5 radians.

Solution: By (4) we have

$$s = (2.5)(3) = 7.5.$$

The length of the arc in inches is, therefore, 7.5.

Example 4   Find the measure of a central angle that subtends an arc of 10 ft. on a circle whose radius is 2 ft.

Solution: By (4), we have $10 = t(2)$. Hence $t = 5$ and the radian measure of the angle is 5.

_____

It is customary to use degrees in surveying and in similar applications that involve triangles. We shall study some of these applications in Section 12.3. The radian unit of measure, however, is more useful than the degree unit in engineering and advanced mathematics. This is particularly true in the development of calculus.

Example 5 Find the length of an arc on a circle of radius 6 cm. if the arc subtends a central angle of 120°.

Solution: We have $120° = \frac{\pi}{180}(120) = \frac{2\pi}{3}$ radians. Then, by (4), we have

$$s = 6 \cdot \frac{2\pi}{3} = 4\pi.$$

The length of the arc is 4π cm.

_____

EXERCISES 12.1

◀ In Exercises 1–10, express each radian measure as an equivalent degree measure.

1. $\frac{\pi}{2}$ 5. 3   8. 0.50

2. π 6. -2 9. 0.15

3. $\frac{\pi}{3}$ 7. 2π 10. 0.01

4. 2

◀ In Exercises 11–20, express each degree measure as an equivalent radian measure.

11. 30°   15. $-20°$ 18. $-70°$

12. 60°   16. 360°   19. 125°10′

13. 15°30′   17. 720°   20. $-70°40′$

14. 10°

21. A wheel turns through 20 revolutions. Through how many radians does the wheel turn?

22. A wheel turns through 20 radians. Through how many revolutions does the wheel turn?

23. A space vehicle orbits the earth once every 110 minutes. Through how many radians does the space vehicle orbit in 24 hours?

24. Through how many radians does the minute hand of a clock revolve in 40 minutes?

25. If a boy is running around a circular track at the rate of four revolutions per half-hour, what is the boy's rate in radians per minute?

◀ In Exercises 26–35, find the length of the arc $\widehat{AB}$ subtended by a central angle AOB on a circle with radius of length r.

26. $M\angle AOB = \pi, r = 5$ **31.** $m\angle AOB = 60, r = 3$

27. $M\angle AOB = 2, r = 10$ **32.** $m\angle AOB = 120, r = 1.5$

28. $M\angle AOB = 1, r = 4.68$ **33.** $m\angle AOB = 45, r = 2.7$

29. $M\angle AOB = -2\pi, r = 1$ **34.** $m\angle AOB = 72, r = 5$

30. $M\angle AOB = -10, r = 2$ **35.** $m\angle AOB = 208, r = 4.9$

◀ In Exercises 36–40, find the measure in radians of a central angle AOB which intercepts $\widehat{AB}$ on a circle with radius of length r.

36. Length of $\widehat{AB} = 3, r = 5$ **39.** Length of $\widehat{AB} = 6\pi, r = 2$

37. Length of $\widehat{AB} = 10, r = 2$ **40.** Length of $\widehat{AB} = -2\pi, r = 6$

38. Length of $\widehat{AB} = 3\pi, r = \dfrac{1}{2}$

◀ In Exercises 41–48, find the length of the radius of the circle on which the given central angle intercepts the given arc.

41. $M\angle AOB = 1, \widehat{AB} = 1$ **45.** $m\angle AOB = 80, \widehat{AB} = 4.5$

42. $M\angle AOB = 1.5, \widehat{AB} = 4.5$ **46.** $m\angle AOB = 220, \widehat{AB} = 36$

43. $M\angle AOB = 4.3, \widehat{AB} = 12.9$ **47.** $M\angle COD = 3.2, \widehat{CD} = 512$

44. $m\angle AOB = 120, \widehat{AB} = 6$ **48.** $m\angle COD = 162, \widehat{CD} = 0.0054$

49. CHALLENGE PROBLEM. The unit of circular measure used by artillery engineers in the army is the **mil**. Find a definition of the mil. Give conversion formulas comparing mil measure with degree measure. Write approximately 100 words giving your opinion why mil measure is preferred by artillery engineers.

50. CHALLENGE PROBLEM. A unit of circular measure that was originally introduced in France as part of the metric system of measure is the **grad**. 1 grad = 90 degrees. Give conversion formulas comparing grad measure and radian measure. Write approximately 100 words giving the history of grad measure after doing research either in your school library or some other source.

12.2 TRIGONOMETRIC FUNCTIONS

On the unit circle, an angle whose measure is 1 radian intercepts an arc whose length is 1 unit. Thus in Figure 12-10, $\angle AOB$ where $M\angle AOB = 1$ intercepts $\overset{\frown}{AB}$ whose length is 1. If B is the point (u, v), then $\sin 1 = v$ and $\cos 1 = u$.

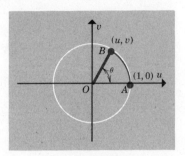

Figure 12-10

Since for all central angles of the unit circle the number that is the radian measure of the angle is equal to the number that is the measure of the length of the arc intercepted by the angle, we have functions that associate each angle θ with the numbers $\sin x$ and $\cos x$. (The Greek letters θ (theta) and ϕ (phi) are often used to designate angles in the same way that x and y are used to designate numbers.)

> **Definition 12.2** If the measure of angle θ in radians is x, we define $\sin \theta = \sin x$ and $\cos \theta = \cos x$. The sine function of θ and the cosine function of θ are called **trigonometric functions**.

As a consequence of Definition 12.2, the expression $\sin \frac{\pi}{6} = \frac{1}{2}$ can have two meanings: (1) that the sine of the real number $\frac{\pi}{6}$ is $\frac{1}{2}$ and (2) that the sine of the angle whose radian measure is $\frac{\pi}{6}$ is $\frac{1}{2}$. It is seldom necessary to discriminate between the two meanings; when it is, the context usually makes clear which meaning is being used.

If $\sin \frac{\pi}{6} = \frac{1}{2}$ refers to angle measure, we can write equivalently $\sin 30° = \frac{1}{2}$ because $30° = \frac{\pi}{6}$ radians. Note that the expression "$\sin 30°$" is an abbreviation for the longer statement "$\sin \theta$ when the measure of θ in degrees is 30." When we write sin 30°, we are using the symbol 30° as a name for the angle.

The definitions of the other trigonometric functions that we shall use follow from the agreement that

$$\tan \theta = \frac{\sin \theta}{\cos \theta} \qquad \text{when } \cos \theta \neq 0,$$

$$\sec \theta = \frac{1}{\cos \theta} \qquad \text{when } \cos \theta \neq 0,$$

$$\csc \theta = \frac{1}{\sin \theta} \qquad \text{when } \sin \theta \neq 0,$$

and

$$\cot \theta = \frac{1}{\tan \theta} \qquad \text{when } \tan \theta \neq 0.$$

Each of the statements we have made about circular functions can be rephrased as a statement about trigonometric functions. For example, the identity

$$\sin^2 x + \cos^2 x = 1$$

can be expressed as

$$\sin^2 \theta + \cos^2 \theta = 1.$$

Similarly, we have seen that the solution set of the equation

$$\sin 2x = 2 \sin x$$

is

$$\{x : x = n\pi \text{ and } n \in I\}.$$

Thus it follows that the solution set of the equation $\sin 2\theta = 2 \sin \theta$ is

$$\{\theta : \theta = 180°n \text{ and } n \in I\}.$$

Note that we have written $\{\theta : \theta = 180°n \text{ and } n \in I\}$. If θ is the name of an angle, which it is, we should, in place of $\theta = 180°$, write $m \angle \theta = 180$, which means the degree measure of θ is 180, 180 being a number. When we use minutes and, possibly, seconds, this notation leads to a dilemma. If we write, for example, $m \angle \theta = 66°42'$, we err one way; if we write $\theta = 66°42'$, we err another way. Since any student of mathematics is likely to see both notations in textbooks and in the classroom, he should be familiar with both. From time to time, we shall use both notations as well as the statement "The measure of θ is 66°42'."

Angle θ									
Degrees	Radians	sin θ	csc θ	tan θ	cot θ	sec θ	cos θ		
23° 00′	.4014	.3907	2.559	.4245	2.356	1.086	.9205	1.1694	67° 00′
10	043	934	542	279	337	088	194	665	50
20	072	961	525	314	318	089	182	636	40
30	.4102	.3987	2.508	.4348	2.300	1.090	.9171	1.1606	30
40	131	.4014	491	383	282	092	159	577	20
50	160	041	475	417	264	093	147	548	10
24° 00′	.4189	.4067	2.459	.4452	2.246	1.095	.9135	1.1519	66° 00′
		cos θ	sec θ	cot θ	tan θ	csc θ	sin θ	Radians	Degrees
									Angle θ

Figure 12-11

Crude tables of trigonometric functions have been in existence for a long time. Following the invention and development of calculus, mathematicians were able to refine these tables to a remarkable degree. Table II at the back of the book lists values of $\sin \theta$, $\csc \theta$, $\tan \theta$, $\cot \theta$, $\sec \theta$, and $\cos \theta$ by 10 minute intervals. Except for a few special values such as $\sin 0°$, $\sin 30°$, $\sin 90°$, $\cos 0°$, $\cos 60°$, $\cos 90°$, $\tan 0°$, $\tan 45°$, the values are irrational numbers and each entry in the table gives an approximation accurate to four decimal places. (A part of a page from Table II is shown in Figure 12-11.) Column headings for angles whose measure in degrees is less than 45 are at the top of the page and we read *down* the columns. Thus we see, for example, that $\sin 23°30' = .3987$ to four significant figures. Rather, however, than write "$\sin 23°30' \approx .3987$ or the longer statement "$\sin 23°30' = .3987$ to four significant figures," we simply write $\sin 23°30' = .3987$ as is the common practice.

Column headings for angles whose measure in degrees is greater than 45 are at the bottom of the page and we read *up* the columns. Thus $\sin 66°30' = .9171$. Note that $\sin 23°30' = .3987 = \cos 66°30'$ and $\sin 66°30' = .9171 = \cos 23°30'$. The fact that $\sin (90° - \theta) = \cos \theta$, $\tan (90° - \theta) = \cot \theta$, and $\sec (90° - \theta) = \csc \theta$ has made it possible to use one entry with two functions and hence abbreviate the tables.

To find a value such as $\sin 66°33'$ we again use linear interpolation as in Chapter 11.

Example 1   Find $\sin 66°33'$.

Solution: The arrangement displayed simplifies the construction of the interpolation proportion.

$$10'\left[3'\left[\begin{array}{ll}\sin 66°30' & .9171 \\ \sin 66°33' & \\ 66°40' & .9182\end{array}\right]d\right].0011$$

$$\frac{3}{10} = \frac{d}{.0011}$$

d = .0003 to four significant figures

Hence sin 66°33′ = .9171 + .0003 = .9174.

Note again that we write sin 66°33′ = .9174 rather than sin 66°33′ ≈ .9174 or sin 66°33′ = .9174 to four significant digits.

To find values of the trigonometric functions of angles greater than 90 or less than 0, we make use of such identities as

$$\sin(\theta \pm 180) = -\sin\theta, \quad \sin(180 - \theta) = \sin\theta$$

and

$$\cos(\theta \pm 180) = -\cos\theta = \cos(180 - \theta).$$

Example 2   Find sin 126°50′.

Solution: We have

$$\sin(180° - 126°50') = \sin 53°10'.$$

From Table II, we have sin 53°10′ = .8004. Hence sin 126°50′ = .8004.

Example 3   Find cos 214°20′.

Solution: We have

$$\cos(214°20' - 180°) = -\cos 34°20'.$$

From Table II, we have cos 34°20′ = .8258. Hence cos 214°20′ = −.8258.

Example 4   Find sin (−316°).

Solution: We use the identity sin (θ ± 360°) = sin θ. Thus

$$\sin(-316°) = \sin(-316° + 360°) = \sin 44°.$$

Since sin 44° = .6947, we have sin (−316°) = .6947.

For every angle θ, there is an angle ϕ, 0 ≤ ϕ ≤ 90, such that the absolute value of any trigonometric function of θ is equal to the value of the same trigonometric function of ϕ. Thus, for example,

$$|\sin 126°50'| = \sin 53°10' \qquad \text{because } 180° - 126°50' = 53°10',$$
$$|\cos 214°20'| = \cos 34°20' \qquad \text{because } 214°20 - 180° = 34°20',$$

and

$$|\sin (-316°)| = \sin 44° \qquad \text{because } 360° - 316° = 44°.$$

The angle ϕ is called the **reference** angle. To find the measure of ϕ, we can use the formula

$$|m\angle\theta + 180n| = m\angle\phi \qquad \text{for some } n \in I.$$

Example 5   Find the measure of the reference angle ϕ if
(1) $m\angle\theta = 126$,
(2) $m\angle\theta = 234$, and
(3) $m\angle\theta = 306$.

Solution:
1. $|126 - 180| = |-54| = 54$.
   Hence $m\angle\phi = 54$.
2. $|234 - 180| = 54$.
   Hence $m\angle\phi = 54$.
3. $|306 - 360| = |-54| = 54$.
   Hence $m\angle\phi = 54$.

Figure 12-12 shows the relationship among the four angles of Example 5.

Once the measure of the reference angle, ϕ, is known, tables are used to find the functional values. The functional values of the angle θ, however, are positive or negative according to the quadrant in which the terminal ray of the angle θ lies. Thus, in Example 5, $|\sin 234°| = \sin 54°$, but $\sin 234° = -\sin 54°$ because the terminal ray of an angle whose degree measure is 234 lies in the third quadrant.

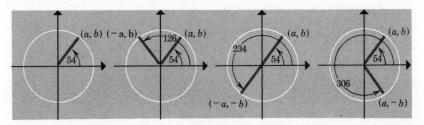

Figure 12-12

On occasion we refer to "an angle of the third quadrant" or "a third quadrant angle," which are abbreviations for the longer statement "an angle whose terminal ray lies in the third quadrant when the angle is in standard position." Similar meanings are attached to the phrases "first quadrant angle," "second quadrant angle," and "fourth quadrant angle."

Example 6 Find, using tables, tan 124°40′.

Solution: Since 180° − 124°40′ = 55°20′, we have

$$|\tan 124°40'| = \tan 55°20'.$$

From Table II, we find that tan 55°20′ = 1.446 and thus we know that

$$|\tan 124°40'| = 1.446.$$

For angles greater than 90° and less than 180°, that is, second quadrant angles, the tangent is negative. Hence tan 124°40′ = −1.446.

EXERCISES 12.2

◀ In Exercises 1–15, find the reference angle associated with each angle.

1. 63°	**6.** 141°19′	**11.** −217°51′
2. 208°	**7.** 283°46′	**12.** 163°16′
3. −54°	**8.** −128°43′	**13.** 352°8′
4. 128°	**9.** 761°15′	**14.** 432°48′
5. 315°	**10.** −63°36′	**15.** −78°43′

◀ In Exercises 16–35, find, using tables, the functional value.

16. sin 28°	**26.** tan 46°18′
17. tan 42°	**27.** cot 43°32′
18. cos 63°20′	**28.** tan 82°43′
19. −sin 18°50′	**29.** sin 30°5′
20. cos 124°40′	**30.** cos 78°22′
21. tan 68°30′	**31.** sin 146°21′
22. sin 330°40′	**32.** cos 211°52′
23. cos (−42°30′)	**33.** sin (−308°43′)
24. sin 128°43′	**34.** cos 411°17′
25. cos (−128°43′)	**35.** tan 136°45′

◀ In Exercises 36–55, find, using tables, the smallest positive angle θ that satisfies the given equation.

36. $\sin \theta = .9613$	**46.** $\sin \theta = .6186$
37. $\cos \theta = .5831$	**47.** $\cos \theta = .7477$
38. $\tan \theta = .7445$	**48.** $\tan \theta = .7498$
39. $\sin \theta = -.6293$	**49.** $\sin \theta = -.3800$
40. $\sin \theta = .5204$	**50.** $\sin \theta = .4201$
41. $\cos \theta = .8765$	**51.** $\cos \theta = .5202$
42. $\sin \theta = -.5206$	**52.** $\sin \theta = -.4203$
43. $\cos \theta = -.8767$	**53.** $\cos \theta = .5644$
44. $\tan \theta = -.7278$	**54.** $\cot \theta = 1.565$
45. $\cos \theta = .4209$	**55.** $\cot \theta = .4306$

12.3 SOLUTION OF RIGHT TRIANGLES

The application of trigonometric functions to right triangles depends on the properties of similar triangles. Suppose that the ray $\overrightarrow{OP}$ contains the point $A(a, b)$ as in Figure 12-13. The ray $\overrightarrow{OP}$ also intersects the unit circle at some point $A'(u, v)$. A line segment from A perpendicular to $\overrightarrow{OQ}$ meets $\overrightarrow{OQ}$ at point $C(a, O)$ and a line segment from A' perpendicular to $\overrightarrow{OQ}$ meets $\overrightarrow{OQ}$ at point C' (u, O) forming two right triangles $\triangle AOC$ and $\triangle A'OC'$ each having θ as one of its angles. Accordingly, $\triangle AOC$ is similar to $\triangle A'OC'$ so that

$$(1) \qquad \frac{A'C'}{OA'} = \frac{AC}{OA} \quad \text{and} \quad \frac{OC'}{OA'} = \frac{OC}{OA}.$$

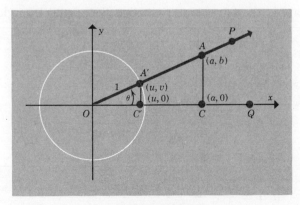

Figure 12-13

Equations (1) may be written as

(2) $$\frac{v}{1} = \frac{b}{OA} \quad \text{and} \quad \frac{u}{1} = \frac{a}{OA}.$$

Since $v = \sin \theta$ and $u = \cos \theta$, by replacement in Equations (2), we have the **trigonometric ratios**

(3) $$\sin \theta = \frac{b}{OA} \quad \text{and} \quad \cos \theta = \frac{a}{OA}.$$

These trigonometric ratios can be applied to all right triangles since each right triangle is congruent to a right triangle having the vertex of one of its acute angles at the origin and the vertex of the right angle on the positive ray, $\overrightarrow{OQ}$, of the x-axis.

It is customary to describe a right triangle with the notation shown in Figure 12-14. The choice of letters is arbitrary, of course; other letters, such as D, E, and F, might be used to designate the vertices of the angles. We assume, however, that if the A, B, C notation is used, then $m \angle ACB = 90$.

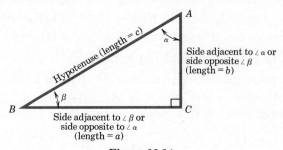

Figure 12-14

If we apply the trigonometric ratios (3) to right triangle ABC, we have

$$\sin \beta = \frac{\text{length of side opposite } \angle \beta}{\text{length of hypotenuse}} = \frac{b}{c}$$

and

$$\cos \beta = \frac{\text{length of side adjacent to } \angle \beta}{\text{length of hypotenuse}} = \frac{a}{c}.$$

Similarly, we have

$$\sin \alpha = \frac{\text{length of side opposite } \angle \alpha}{\text{length of hypotenuse}} = \frac{a}{c}$$

and

$$\cos \alpha = \frac{\text{length of side adjacent to } \angle \alpha}{\text{length of hypotenuse}} = \frac{b}{c}.$$

In a similar manner, it can be shown that each of the four other trigonometric functions also leads to a trigonometric ratio. For $\angle BAC$ in $\triangle ABC$, they are

$$\tan \alpha = \frac{\text{length of side opposite } \angle \alpha}{\text{length of side adjacent to } \angle \alpha} = \frac{a}{b},$$

$$\cot \alpha = \frac{\text{length of side adjacent to } \angle \alpha}{\text{length of side opposite } \angle \alpha} = \frac{b}{a},$$

$$\sec \alpha = \frac{\text{length of hypotenuse}}{\text{length of side adjacent to } \angle \alpha} = \frac{c}{b},$$

and

$$\csc \alpha = \frac{\text{length of hypotenuse}}{\text{length of side opposite } \angle \alpha} = \frac{c}{a}.$$

The opposite-adjacent-hypotenuse terminology facilitates the application of trigonometric functions to right triangles.

Example 1 The ray $\overrightarrow{OP}$ contains the point $P(3, 7)$. Find the measure of the angle that $\overrightarrow{OP}$ makes with $\overrightarrow{OX}$, the positive ray of the x-axis.

Solution: A perpendicular from the point $P(3, 7)$ meets $\overrightarrow{OX}$ at point $D(3, 0)$ as shown in Figure 12-15. In the right triangle POD with

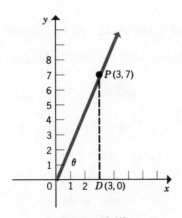

Figure 12-15

$\angle POD = \theta$, we have

$$\tan \theta = \frac{PD}{OD} = \frac{7}{3} = 2.333$$

to four significant digits. From Table II, we find that the measure of $\angle POD$ is $66°50'$.

Example 2 In the right triangle ABC, $m\angle ABC = 34$ and the length of the hypotenuse is 78.2 ft. Find the measure of the other angles and, to the nearest foot, of the other two sides. (Finding the measure of the three angles and the lengths of the three sides of a triangle is called *solving the triangle*.)

Solution: Using a labeled diagram as in Figure 12-16, we have

$$(1) \qquad \sin 34° = \frac{b}{c} \qquad \text{and} \qquad (2) \qquad \cos 34° = \frac{a}{c}.$$

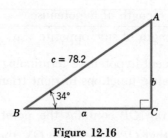

Figure 12-16

From Table II, we find $\sin 34° = .5592$. Hence, replacing $\sin 34°$ by .5592 and c by 78.2 in (1), we have

$$.5592 = \frac{b}{78.2}$$

so that

$$b = (78.2)(.5592) = 44 \quad \text{to two significant figures.}$$

From Table II, we find $\cos 34° = .8290$. Hence (2) becomes

$$.8290 = \frac{a}{78.2}$$

so that

$$a = (.8290)(78.2) = 65 \quad \text{to two significant figures.}$$

Finally, $m\angle BAC = 90 - m\angle ABC = 90 - 34 = 56$. Thus we have

$$m\angle BAC = 56.$$
$$AC = 44 \text{ ft.}$$
$$BC = 65 \text{ ft.}$$

Even though $(78.2)(.5592) = 43.72944$, note that we did not write $AC = 43.72944$ ft., which would be ridiculous! Since the angle measure was expressed to the nearest degree, the measure is "rounded off" to two significant figures. In general, angle measure to the nearest degree corresponds roughly to two significant figures. Angle measure to the nearest 10 minutes corresponds roughly to three significant figures, and angle measure to the nearest minute corresponds roughly to four significant figures.

In the applications of trigonometric ratios, it is customary to round off answers to the same number of significant figures as given in the data. Hence, in the example above, we may write

$$b = (78.2)(.5592) = 44$$

with the understanding that the answer of 44 is correct to two significant figures and not that $(78.2)(.5592)$ is actually equal to 44.

Example 3 Find the length of the hypotenuse and the measures of the acute angles of a right triangle whose legs have lengths of 41.2 in. and 32.6 in.

Solution: Using a labeled diagram as in Figure 12-17, we have

$$\tan \angle BAC = \frac{a}{b} = \frac{41.2}{32.6} = 1.264.$$

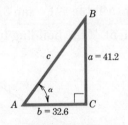

Figure 12-17

From Table II, using interpolation, we find $m \angle BAC = 51°39'$. Then

$$m \angle ABC = 90° - 51°39' = 89°60' - 51°39' = 38°21'.$$

Since $\sin \alpha = \dfrac{41.2}{c}$ and $\sin \alpha = \sin 51°39' = .7842$ from Table II and interpolation, we have

$$.7842 = \frac{41.2}{c}.$$

Hence

$$c = \frac{41.2}{.7842} = 52.6 \quad \text{to three significant figures.}$$

Thus, to three significant figures, we have

$$\text{length of hypotenuse} = 52.6 \text{ in.}$$
$$m\angle BAC = 51°40'.$$
$$m\angle ABC = 38°20'.$$

Note that the final "rounding off" to the proper number of significant figures has not been done until the final step. This is recommended procedure in dealing with computations involving numbers obtained from measurements.

The ability to solve right triangles can be applied to problems of **triangulation,** the indirect measurement of distances.

Example 4   At a point 225 ft. from the base of a building, the measure of the angle between the horizontal and the line to the top of the building (the **angle of elevation**) is 31°20′. Find the height of the building.

Solution: Figure 12-18 represents the situation of the problem. Thus $\tan 31°20' = \frac{h}{225}$.   From   Table   II, $\tan 31°20' = .6088$.   Hence $(.6088)(225) = h$. The height of the building to three significant digits is 137 ft.

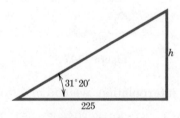

31°20′

225

Figure 12-18

Example 5   Radar station A on the seacoast observes a ship 28° east of south. At the same time radar station B 100 miles directly south of radar station A locates the same ship as 62° east of north. Assuming that the coast line is a straight line between A and B, find how far the ship is from the coast.

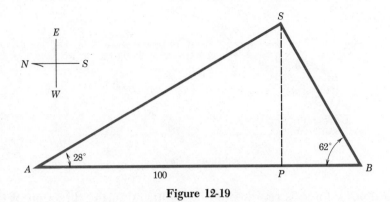

Figure 12-19

Solution: Note that $28° + 62° = 90°$. Hence $\triangle ASB$ is a right triangle as shown in Figure 12-19 and, in $\triangle ASB$, $\cos 28° = \dfrac{AS}{100}$. Hence

$$AS = (.8829)(100) = 88.29.$$

The line SP represents the distance of the ship from the shore. $\triangle ASP$ is a right triangle and, in $\triangle ASP$,

$$\sin 28° = \frac{SP}{AS} = \frac{SP}{88.29}.$$

Hence

$$SP = (.4695)(88.29) = 42. \qquad \text{(Why two significant digits?)}$$

The ship is 42 miles from shore.

The essential step in these examples is the discovery and construction of a right triangle that can be solved. A scale drawing diagram can be used as a check. In Figure 12-19, for example, which is a scale drawing, actual measurement will show that $AB = 3.5$ in. and $SP = 1.5$ in. Then if we let x be the distance in miles that the ship is from shore, we have

$$\frac{3.5}{1.5} = \frac{100}{x}.$$

Solving for x, we get $x \approx 43$—a reasonable check.

Example 6   As a wheelbarrow is pushed, its handle makes an angle of $34°$ with the ground. If the force, in the direction of the handle, with which the wheelbarrow is pushed is 100 lb., what is the force propelling the wheelbarrow forward?

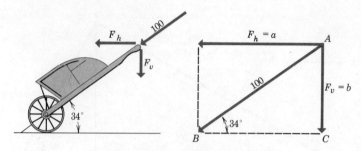

Figure 12-20

Solution: A force is represented by an arrow in the direction of the force. The length of the arrow is proportional to the magnitude of the force. The force, F, of 100 lb. is the sum of two components, F_h and F_v, one horizontal and one vertical as shown in Figure 12-20. By solving the right triangle ABC, we find $F_h = a$ and $F_v = b$. In $\triangle ACB$, we have

$$\cos B = \frac{a}{AB} \quad \text{and} \quad \sin B = \frac{b}{AB}.$$

Since $AB = 100$ and $m \angle B = 34$, we have, by using Table II,

$$a = 100(.8290) = 83 \quad \text{to two significant figures}$$

and

$$b = 100(.5592) = 56 \quad \text{to two significant figures.}$$

Hence $F_h = a = 83$ lb. and $F_v = b = 56$ lb. Since F_h is the force propelling the wheelbarrow forward, the answer to the problem is 83 lb.

EXERCISES 12.3

1. Give the value for each trigonometric ratio as it applies to $\triangle ABC$.

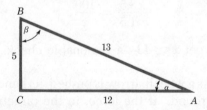

(a) $\sin \beta$ (d) $\cos \beta$
(b) $\cos \alpha$ (e) $\tan \beta$
(c) $\sin \alpha$ (f) $\tan \alpha$

2. Give the value of each trigonometric ratio as it applies to △*DEF*.

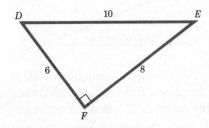

(a) sin ∠*DEF*   (d) cos ∠*FDE*
(b) cos ∠*EDF*   (e) tan ∠*DEF*
(c) tan ∠*EDF*   (f) sec ∠*EDF*

3. The ray $\overrightarrow{OP}$ from the origin of the Cartesian plane contains the point (8, 6). Find the measure of the angle whose sides are $\overrightarrow{OP}$ and $\overrightarrow{OX}$, the positive ray of the *x*-axis.

4. The ray $\overrightarrow{OP}$ in the Cartesian plane contains the point (−4, 5). Find the measure of the angle whose sides are $\overrightarrow{OP}$ and $\overrightarrow{OX}$, the positive ray of the *x*-axis.

5. The ray $\overrightarrow{OP}$ contains the point (8, 6) and the ray $\overrightarrow{OQ}$ contains the point (8, −10). Find the measure of the angle *POQ*.

6. Let *P* be a point on a circle whose radius is *r* and whose center is the origin of the Cartesian plane. Show that *P* is the point ($r \cos \theta$, $r \sin \theta$) where θ is the angle whose sides are $\overrightarrow{OP}$ and the positive ray of the *x*-axis.

◀ In Exercises 7–14, the given information applies to a right triangle *ABC* where *a* = *BC*, *b* = *AC*, *c* = *AB*, and *m*∠*ACB* = 90. In each exercise, solve the triangle.

7. *m*∠*BAC* = 67°, *c* = 42   **11.** *c* = 103, *a* = 87.0
8. *m*∠*BAC* = 48°10′, *a* = 362   **12.** *m*∠*BAC* = 67°40′, *c* = 432
9. *m*∠*ABC* = 16°30′, *c* = 571   **13.** *m*∠*CBA* = 37°20′, *a* = 123
10. *a* = 46.0, *b* = 32.0   **14.** *a* = 821, *b* = 436

15. A wire 72 ft. long is stretched from level ground to the top of a pole 54 ft. high. Find the measure of the angle between the pole and the wire.

16. A kite string 120 yd. long makes with the ground an angle whose measure is 42°. How high is the kite? (Assume that the kite string is not curved.)

17. From the top of a rock that rises vertically 320 ft. out of the water, the angle between the horizontal and the line of sight of a boat (the **angle of depression**) has a measure of 24°. Find the distance of the boat from the base of the rock.

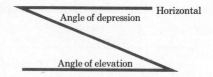

18. An artillery officer on the top of a mountain 2800 ft. above a level plane sees a target on the plane. If the measure of the angle of depression (see Exercise 17) of the target is 32°, how far is the target from the officer?

19. At a point 185 ft. from the base of a tree, the measure of the angle of elevation (see Exercise 17) of the top is 55°40′. How tall is the tree?

20. Find the angles of intersection of the diagonals of a rectangle 6.2 ft. wide and 13.6 ft. long.

21. Find the radius of a regular octagon each side of which is 8.2 in.

22. A ladder 18 ft. long leans against a wall. If the foot of the ladder is $5\frac{1}{2}$ ft. from the wall, what angle does the ladder make with the ground?

23. In right triangle ACB ($m \angle C = 90$), let h be the length of the altitude from C to $\overline{AB}$. If $AB = c$, prove that

$$h = \frac{c}{\cot A + \cot B}.$$

24. A student wanting to apply his knowledge of trigonometry does so in the following manner. Using a steel tape measure, he marks off a point 75 ft. from a side of his school building. At that point he estimates the measure of the angle of elevation of the top of the building to be 40°. He finds, using tables, that tan 40° = .8391. Since (.8391)(75) = 62.9325, the student reports the building as being 62.9325 ft. high. Comment on the reliability and accuracy of the report.

25. A force F of 72 lb. is pulling on an object at O at an angle of 32° with the horizontal. Find F_h, the horizontal component of F and F_v, the vertical component of F.

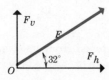

26. A jet plane is headed due east at 520 miles per hour. A wind of 55 miles per hour is blowing from due north. Find the actual speed, v, of the plane and the direction of its motion.

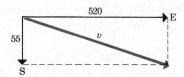

27. A barrel weighing 320 lb. is resting on a smooth inclined plane whose angle of elevation is 24°. What is the minimum force $(-F_d)$ pushing in the direction of the plane that will prevent the barrel from moving? (Ignore the force of friction.)

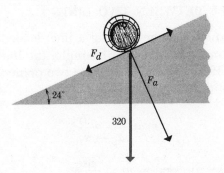

◀In Exercises 28–32, the computations are such that the use of logarithms may be advisable.

28. From the top of a tower 138.7 ft. high, a point is sighted on the ground. If the measure of the angle of depression (see Exercise 17) is 9°26′, how far is the point from the foot of the tower?

29. An airplane is directly over a check station. At a second check station 15.35 miles from the first, an observer notes that the angle of elevation of the airplane is 12°36′. Find the height in feet of the airplane above the check station. Assume that both check stations are at sea level.

30. Points *A* and *B* are directly opposite each other on the banks of a river. At point *C* 325.6 yd. along the river from *A*, the measure of ∠*ACB* is 53°14′. How wide is the river at point *A*? (Assume that there are no bends in the river and that the banks are parallel.)

31. From the top of a lighthouse, the measures of the angles of depression (see Exercise 17) of two boats are 9°38′ and 6°13′. If the observation is made from a point 48 ft. 6 in. above the water, how far apart are the boats? Assume that the boats are in line with the foot of the lighthouse.

32. CHALLENGE PROBLEM. A surveyor at point *A* at sea level measures the angle of elevation of the top of a mountain. The measure is 8°56′. The surveyor moves directly toward the mountain a distance of 7132 ft. and makes a second observation at sea level. Now the measure of the angle of elevation is 11°18′. How high is the mountain?

33. **CHALLENGE PROBLEM.** Two radar stations are 136.5 miles apart on a line that points 18°12′ west of north. If the bearing of a ship is 33°42′ east of south from one station and 56°19′ east of north from the second station, how far is the ship from a coast guard station which is 48.50 miles directly north of the more southerly radar station?

12.4 THE ANGLE BETWEEN TWO LINES

You have learned that the slope of a line defined by an equation of the form $y = mx + b$ is m. For example, the line $l : y = 2x + 3$ has a slope of 2. If (x_1, y_1) and (x_2, y_2) are two points on a nonvertical line $l : y = mx + b$, then

$$m = \frac{y_2 - y_1}{x_2 - x_1}.$$

Sometimes this is expressed as $m = \dfrac{\text{rise}}{\text{run}}$.

Each line l_1, not parallel to the x-axis, intersects the x-axis at some point $P(x_1, 0)$. The **angle of inclination** of line l_1 is defined to be the positive angle θ, $0 < m\angle\theta < 180$, whose vertex is P, whose initial ray is the positive part of the x-axis to the right of P, and whose terminal ray is on l_1 as shown in Figure 12-21.

If a line segment is drawn perpendicular to the x-axis from point $R(x_2, y_2)$, $R \neq P$, on l_1 to $S(x_2, 0)$ on the x-axis, a right triangle RSP

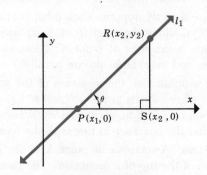

Figure 12-21

is formed. We have, in Figure 12.2,

$$\tan \theta = \frac{RS}{PS} = \frac{y_2}{x_2 - x_1}$$

and, also,

$$m = \frac{y_2}{x_2 - x_1}.$$

Thus if $0 < m\angle\theta < 90$, we have

$$\tan \theta = m.$$

On the other hand, if $90 < m\angle\theta < 180$ as in Figure 12-22, then

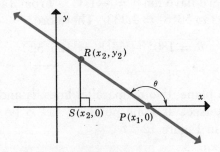

Figure 12-22

$x_2 - x_1 < 0$, $y_2 > 0$ and hence $\dfrac{y_2}{x_2 - x_1} = m$ is a negative number.

Thus

$$\tan \theta = -\tan(180 - \theta) = -\frac{y_2}{|x_2 - x_1|}$$

$$= -\frac{y_2}{-(x_2 - x_1)} = \frac{y_2}{x_2 - x_1} = m.$$

Accordingly, for $90 < m\angle\theta < 180$, $\tan \theta = m$. Thus we conclude that for all θ such that $0 < m\angle\theta < 180$ we have

(1) $\tan \theta = m.$

If we now adopt the convention that, for a line parallel to the x-axis, the angle of inclination has measure 0, we see that, since $\tan 0 = 0$, Equation (1) holds for all θ such that $0 \le m\angle\theta < 180$. That is, the slope of a line not parallel to the y-axis equals the tangent of the angle of inclination of the line. If a line is parallel to the y-axis, then it is perpendicular to the x-axis so that $m\angle\theta = 90$. The slope of such a line, by agreement, is not defined.

Example 1 Find the measure of the angle of inclination θ of a line, l, that contains the two points $(4, 3)$ and $(1, 7)$.

Solution: The slope, m, of l is

$$m = \frac{y_2 - y_1}{x_2 - x_1} = \frac{7 - 3}{1 - 4} = \frac{4}{-3} \approx -1.333.$$

Since $\tan\theta = m$, we have $\tan\theta = -1.333$. From Table II and interpolation, we find $\tan 53°8' = 1.333$. Therefore

$$\theta = 180° - 53°8' = 126°52'.$$

In a Cartesian plane, two nonparallel lines, l_1 and l_2, not parallel to either axis, intersect at some point P. Let ϕ be the angle from l_1 to l_2 as shown in Figure 12-23.

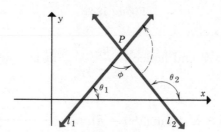

Figure 12-23

If the angle of inclination of l_1 is θ_1 and the angle of inclination of l_2 is θ_2, then we have

$$\theta_1 + \phi + (180 - \theta_2) = 180$$

so that

$$\phi = \theta_2 - \theta_1.$$

Recall from Chapter 10 [Identity (XIX)] that

$$\tan(x - y) = \frac{\tan x - \tan y}{1 + \tan x \tan y}.$$

Replacing x and y by θ_2 and θ_1, we have

$$(2) \qquad \tan \phi = \tan (\theta_2 - \theta_1) = \frac{\tan \theta_2 - \tan \theta_1}{1 + \tan \theta_2 \tan \theta_1}$$

or, equivalently,

$$(3) \qquad \tan \phi = \tan (\theta_2 - \theta_1) = \frac{m_2 - m_1}{1 + m_2 m_1},$$

where m_2 is the slope of l_2 and m_1 is the slope of l_1. Note that ϕ was defined as the angle *from* l_1 *to* l_2. If ϕ were the angle *from* l_2 *to* l_1, then we would obtain the formula

$$\tan \phi = \frac{m_1 - m_2}{1 + m_1 m_2}.$$

Here, of course, we must have $1 + m_2 m_1 \neq 0$. Later we will discuss what happens when $1 + m_2 m_1 = 0$.

Example 2   Find the measure of the angle between the two lines

$$l_1 : 2x - 3y = 6$$

and

$$l_2 : 4x + 5y = 8.$$

Solution: The slope of $l_1 : y = \frac{2}{3}x - 2$ is $\frac{2}{3}$ and the slope of $l_2 : y = -\frac{4}{5}x + \frac{8}{5}$ is $-\frac{4}{5}$. If ϕ is the angle of intersection, then

$$\tan \phi = \frac{m_1 - m_2}{1 + m_1 m_2}.$$

Thus

$$\tan \phi = \frac{\frac{2}{3} + \frac{4}{5}}{1 - \frac{8}{15}} = \frac{\frac{22}{15}}{\frac{7}{15}} = \frac{22}{7} \approx 3.143.$$

From Table II (using interpolation), we have $\phi = 72°21'$. Note that if we had

$$\tan \phi = \frac{m_2 - m_1}{1 + m_2 m_1},$$

then we would have $\tan \phi = -\frac{22}{7}$ and $\phi = 180° - 72°21' = 107°39'$. Lines l_1 and l_2 form four angles when they intersect: two have a measure of $72°21'$ and two have a measure of $107°39'$ as shown in Figure 12-24.

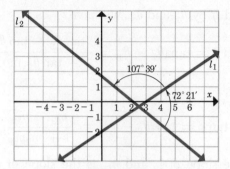

Figure 12-24

Let us now consider the case when $1 + m_1 m_2 = 0$ or, equivalently, when $m_1 m_2 = -1$. The fact that $\tan \theta$ is not defined when $\theta = 90°$ suggests that $m_1 m_2 = -1$ when l_1 and l_2 are perpendicular.

Figure 12-25 shows two lines l_1 and l_2 which are at right angles to each other intersecting at the point P. Suppose that the slopes of l_1 and l_2 are m_1 and m_2, respectively.

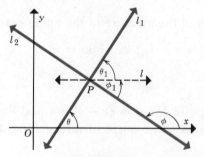

Figure 12-25

We construct the line l through P parallel to the x-axis so that $m \angle \theta = m \angle \theta_1$ in Figure 12-25. Thus $m_1 = \tan \theta_1$ and $m_2 = \tan \phi$.

Since l_1 and l_2 intersect at right angles, it follows that $\theta_1 + \phi_1 = 90$, $\theta_1 = 90 - \phi_1$, and

$$m_1 = \tan (90 - \phi_1) = \cot \phi_1.$$

Also, since $\phi + \phi_1 = 180$, we have

$$m_1 = \cot (180 - \phi).$$

However, since $\cot (180 - \phi) = -\cot \phi$,

$$m_1 = -\cot \phi = -\frac{1}{\tan \phi} = -\frac{1}{m_2}.$$

Thus $m_1 m_2 = -1$.

We have shown that if two lines, neither of which is parallel to the x- or y-axis, are perpendicular, then the product of the slopes of the two lines is -1. The converse, namely, that if the product of the slopes of two lines is equal to -1, then the lines are perpendicular, is also true. (The proof is left as an exercise.) If one of the two lines is parallel to the y-axis, then the slope of the line is undefined. Hence it is not necessary to place a restriction on the statement of the converse.

Example 3 Find an equation of the line l that contains $(4, 8)$ and is perpendicular to the line $l_1 : 2x + 3y = 9$.

Solution: The slope of line is $l_1 : y = -\frac{2}{3}x + 3$ is $-\frac{2}{3}$. If m is the slope of l, then we have $(-\frac{2}{3})(m) = -1$ so that $m = \frac{3}{2}$. Using the form $y - h = m(x - k)$, we have

$$y - 8 = \frac{3}{2}(x - 4)$$

or, equivalently,

$$3x - 2y = -4.$$

EXERCISES 12.4

◀ In Exercises 1–10, find the angle of inclination of the line which passes through each pair of points.

1. $(2, 3)$, $(6, 7)$ **6.** $(\sqrt{3}, 1)$, $(3\sqrt{3}, 4)$

2. $(-1, 3)$, $(4, 5)$ **7.** $(4\sqrt{2}, -1)$, $(-3\sqrt{2}, 5)$

3. $(-2, -3)$, $(4, -6)$ **8.** $(\pi, 4)$, $(3\pi, 12)$

4. $(2, 0)$, $(2, 7)$ **9.** $(5, \sqrt{7})$, $(8, -\sqrt{7})$

5. $(3, 7)$, $(5, 7)$ **10.** $(2, \sqrt{2})$, $(4, \sqrt{8})$

11. A triangle has its vertices at the points $(2, 3)$, $(8, 4)$, and $(4, 9)$. Find the measures of the angles of the triangle.

12. A triangle has its vertices at the points $(-2, -3)$, $(4, 5)$, and $(-1, 6)$. Find the measures of the angles of the triangle.

◀ In Exercises 13–18, find the measure of the smallest positive angle between the two lines defined by the given equations.

13. $y = 2x$, $y = \frac{1}{2}x + 1$ **14.** $y = \frac{1}{2}x$, $y = 3x + 4$

15. $2x + y = 1, 3x - y = 1$ **17.** $x = 3, x - \sqrt{2}y = 3$

16. $y = 3, x = y$ **18.** $4x - 3y = 1, 2x + y = 4$

19. Find an equation of the line l_1 which passes through the point $(4, 2)$ and is perpendicular to the line $l_2 : 2x + 3y = 6$.

20. Find an equation of the line l_1 which passes through the point $(3\frac{1}{2}, -1\frac{1}{4})$ and is perpendicular to the line $l_2 : 4x - 5y = 8$.

21. Show, using a suitable diagram, that if one line of two intersecting lines is parallel to the x-axis, then ϕ, the angle of intersection, is congruent to θ, the angle of inclination of the nonparallel line.

22. CHALLENGE PROBLEM. Prove that if the product of the slopes of two lines is equal to -1, then the lines are perpendicular.

23. CHALLENGE PROBLEM. Find the distance from the origin to the line $l_1 : 3x - 4y = -25$.

24. CHALLENGE PROBLEM. The angle between lines l_1 and l_2 measures $45°$. If an equation defining l_1 is $4x - y = 6$ and line l_2 contains the point $(4, 5)$, find an equation defining l_2.

12.5 LAW OF COSINES

Recall from your study of geometry that all the measures of a triangle are determined when we have (1) the measure of two sides and the included angle, (2) the measure of three sides, or (3) the measure of one side and two adjacent angles. We now develop ways of solving such triangles.

For convenience we shall represent the lengths of $\overline{BC}$, $\overline{AC}$, and $\overline{AB}$ by a, b, and c, respectively, and represent $\angle BAC$, $\angle ABC$, and $\angle ACB$ by α, β, and γ, respectively, as in Figure 12-26.

Let $\triangle ABC$ be placed so that one of the angles is in standard position as in Figure 12-27. If we place the vertex of $\angle ACB$ at the

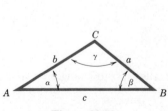

Figure 12-26

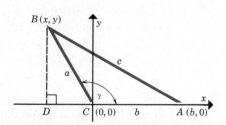

Figure 12-27

origin and place $\overline{AC}$ along the positive x-axis, then the coordinates of A, B, and C are $(b, 0)$, (x, y), and $(0, 0)$, respectively. In the right triangle BCD where D is the foot of the perpendicular from B to $\overleftrightarrow{AC}$, we see that $BD = y$, $DC = |x|$. Hence $a^2 = x^2 + y^2$. By using the formula for the distance between two points, we have

$$(BA)^2 = (b - x)^2 + y^2$$

so that, since $BA = c$,-

$$c^2 = b^2 - 2bx + x^2 + y^2.$$

Also in right triangle BDC, $m \angle BCD = 180 - m \angle \gamma$. Thus we have $x^2 + y^2 = a^2$ and $a \cos \gamma = -(DC) = x$ since

$$\frac{DC}{a} = \cos (180 - \gamma)$$

and

$$\cos (180 - \gamma) = -\cos \gamma.$$

(Note that $\cos (180 - \gamma)$ is an abbreviation of the longer expression $\cos (180 - m \angle \gamma)$. The shorter form is commonly used and is similar to the use of the expression "$\sin 30°$" which is a shortened form for "sine of the angle whose measure in degrees is 30.")

Replacing $x^2 + y^2$ by a^2 and x by $a \cos \gamma$, we have

$$c^2 = a^2 + b^2 - 2ab \cos \gamma.$$

If we place $\angle BAC$ at the origin in standard position, name the angle at A, α, and proceed as above, we obtain

$$a^2 = b^2 + c^2 - 2bc \cos \alpha.$$

If we place $\angle CBA$ at the origin in standard position, name the angle at B, β, and proceed as above, we obtain

$$b^2 = a^2 + c^2 - 2ac \cos \beta.$$

Since the symbols attached to the angles of triangle ABC are arbitrary, one statement, called the *Law of Cosines*, serves all three possibilities.

The Law of Cosines: $c^2 = a^2 + b^2 - 2ab \cos \gamma.$

Note that when $m \angle ACB = 90$, we have $c^2 = a^2 + b^2$ since $\cos 90° = 0$. Because of this, the Law of Cosines is sometimes called the generalized Pythagorean Theorem.

Example 1 In $\triangle ABC$, find c if $a = 10.0$, $b = 13.0$, and $m \angle ACB = 72°30'$.

Solution: Note that we are given the measure of two sides and the measure of the included angle and hence there is a unique solution. We have

$$c^2 = a^2 + b^2 - 2ab \cos \gamma.$$

Hence

$$c^2 = 100 + 169 - 2(10)(13)(.3007) = 190.8.$$

Thus $c = \sqrt{190.8} \approx 13.8$. Figure 12-28 shows the triangle.

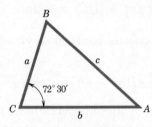

Figure 12-28

Example 2 In $\triangle ABC$, $a = 15$, $b = 9.0$, and $c = 8.0$. Find $m \angle ABC$.

Solution: Note that we are given the measures of three sides and hence there is a unique solution. We use the form

$$b^2 = c^2 + a^2 - 2ac \cos \beta.$$

We have

$$81 = 64 + 225 - 2(15)(8) \cos \beta.$$

Hence

$$\cos \beta = \frac{208}{240} = .8667$$

and so, to the nearest degree,

$$m \angle ABC = 30°.$$

If the lengths of three sides of a triangle ABC are a, b, and c, the area of the triangle can be found by using *Heron's Formula*.

Heron's Formula. The area, K, of a triangle is given by

$$K = \sqrt{s(s - a)(s - b)(s - c)}$$

where $s = \dfrac{a + b + c}{2}$ is the **semiperimeter**. This formula can be derived using the Law of Cosines and a formula for $\tan \dfrac{\theta}{2}$. (See Exercise 25 of Exercises 12.5.)

Example 3 Find the area of $\triangle ABC$ if $a = 15$ in., $b = 9.0$ in., and $c = 8.0$ in.

Solution:

$$s = \frac{15 + 9 + 8}{2} = 16,$$

$$K = \sqrt{16(1)(7)(8)} = \sqrt{896} \approx 30.$$

The area of $\triangle ABC$ is 30 sq. in. to the nearest square inch.

EXERCISES 12.5

◄In Exercises 1–10, the data refer to a $\triangle ABC$.

1. Find c, given that $a = 5.0$, $b = 7.0$, $m\angle\gamma = 45$.
2. Find b, given that $c = 3.0$, $a = 7.0$, $m\angle\beta = 60$.
3. Find a, given that $c = 6.0$, $b = 8.0$, $m\angle\gamma = 120$.
4. Find $m\angle\alpha$, given that $a = 6.0$, $b = 8.0$, $c = 11$.
5. Find $m\angle\beta$, given that $b = 1.2$, $a = 2.5$, $c = 2.2$.
6. Find $m\angle\gamma$, given that $b = 72$, $a = 38$, $c = 39$.
7. Find a, given that $c = 13$, $b = 15$, $m\angle\alpha = 132$.
8. Find b, given that $c = 11$, $a = 21$, $m\angle\beta = 54$.
9. Find c, given that $a = 14$, $b = 18$, $m\angle\gamma = 32$.
10. Find $m\angle\beta$, given that $a = 21.3$, $b = 12.6$, $c = 18.9$.

◄In Exercises 11–14, find the area of $\triangle ABC$ using Heron's Formula.

11. $a = 3$, $b = 4$, $c = 5$
12. $a = 5$, $b = 7$, $c = 8$
13. $b = 14$, $c = 9.0$, $a = 16$
14. $c = 21$, $b = 32$, $a = 41$

15. Find the length of the shorter diagonal of a parallelogram if the measures of two sides are 8.0 in. and 12 in. and the measure of the included angle is 73°.
16. Find the area of a parallelogram if the measures of two sides and the shorter diagonal are 3.6 cm., 7.0 cm., and 5.2 cm., respectively.

17. An airplane leaves an airport and flies in a straight line at the rate of 550 miles per hour. A second airplane leaves at the same time and flies in a straight line at the rate of 320 miles per hour. If the airplanes are 1800 miles apart in 3 hours, what is the angle between the paths of their flights?

18. Two ships leave a harbor at the same time. One heads northeast at a speed of 15 miles per hour, the other heads south at a speed of 18 miles per hour. After 24 hours, how far apart are they?

19. An isoceles trapezoid with base angles measuring 75° has equal sides each of whose lengths is 8 in. If the length of the base is 20 in., find the length of the diagonal.

20. A right triangle is inscribed in a circle. If the lengths of the two legs are 1.6 in. and 2.3 in., find the measures of the angles of the triangle.

21. The lengths of two sides of a parallelogram are 23 in. and 41 in. and the length of the longer diagonal is 56 in. Find the measure of the angle between the diagonal and the longer side.

22. Two forces of 2.0 lb. and 3.0 lb. act on an object and produce a single force of 4.5 lb. Find the measure of each angle between the direction of the force and the direction of the single force.

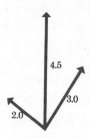

23. Show that in $\triangle ABC$,

$$1 + \cos \alpha = \frac{(b + c + a)(b + c - a)}{2bc}.$$

24. Show that in $\triangle ABC$,

$$1 - \cos \alpha = \frac{(a - b + c)(a + b - c)}{2bc}.$$

25. CHALLENGE PROBLEM. Show that K, the area of $\triangle ABC$, is given by

$$K = \sqrt{s(s - a)(s - b)(s - c)}$$

where $s = \dfrac{a + b + c}{2}$. (*Hint:* Use the results of Exercises 23 and 24.)

12.6 LAW OF SINES

Let $\angle BCA$ of $\triangle ABC$ be placed in standard position as in Figure 12-29 so that the coordinates of C are $(0, 0)$ and $\overline{CA}$ lies along the

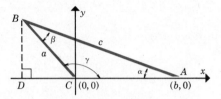

Figure 12-29

positive ray of the x-axis. Also, the coordinates of A are $(b, 0)$. Using right triangle BCD, we see that the coordinates of B are $(a \cos \gamma, a \sin \gamma)$ and that $BD = a \sin (180 - \gamma) = a \sin \gamma$. If $\overline{BD}$ is the altitude from B to $\overline{AC}$, then K, the area of triangle ABC, is given by the formula

$$K = \frac{1}{2}(BD)(AC).$$

Since $BD = a \sin (180 - \gamma) = a \sin \gamma$ and $AC = b$, we have

$$K = \frac{1}{2} ab \sin \gamma.$$

If we place $\angle ABC$ in standard position and proceed as above, we obtain

$$K = \frac{1}{2} ac \sin \beta.$$

If we place $\angle BAC$ in standard position and proceed as above, we get

$$K = \frac{1}{2} bc \sin \alpha.$$

Thus we have

$$K = \frac{1}{2}ab \sin \gamma = \frac{1}{2}bc \sin \alpha = \frac{1}{2}ac \sin \beta.$$

Note that when $m \angle ACB = 90$, then $K = \frac{1}{2}ab$. (Why?)

Example 1   Find the area of $\triangle ABC$ if $m\angle BAC = 108$, $AC = 13$ in., and $AB = 9.0$ in.

Solution: Note that we are given the measures of two sides and the included angle. We have

$$K = \frac{1}{2}bc \, \sin \alpha = \frac{1}{2}(13)(9)(\sin 108).$$

For $108°$, the reference angle is $72°$ and $\sin 72° = .9511$. Thus

$$K = \frac{1}{2}(13)(9)(.9511) = 56.$$

The area in square inches is 56 to two significant figures.

If we divide each member of the equation

$$\frac{1}{2}bc \, \sin \alpha = \frac{1}{2}ac \, \sin \beta = \frac{1}{2}ab \, \sin \gamma$$

by $\frac{1}{2}abc$, we obtain the *Law of Sines*.

The Law of Sines: $\dfrac{\sin \alpha}{a} = \dfrac{\sin \beta}{b} = \dfrac{\sin \gamma}{c}.$

Note that when $m\angle ACB = 90$, we have $\sin \gamma = 1$ so that $\sin \alpha = \dfrac{a}{c}$ and $\sin \beta = \dfrac{b}{c}$ and that these are the statements that were made in Section 12.2 about right triangles.

Example 2   Find b in $\triangle ABC$ if $c = 17$, $m\angle CAB = 42$, and $m\angle CBA = 37$.

Solution: Note that we are given the measure of one side and two adjacent angles. If $m\angle \alpha = 42$ and $m\angle \beta = 37$, then

$$m\angle \gamma = 180 - (42 + 37) = 101.$$

By substituting in the formula

$$\frac{\sin \gamma}{c} = \frac{\sin \beta}{b}$$

and using the fact that

$$\sin 101° = \sin (180 - 101)° = \sin 79°,$$

we have

$$\frac{.9816}{17} = \frac{.6018}{b}$$

so that, to two significant figures,

$$b = \frac{17(.6018)}{.9816} = 10.$$

Example 3   If, in $\triangle ABC$, $m\angle\beta = 42$, $b = 9.0$, and $c = 12$, find $m\angle ACB$.

Solution: By the Law of Sines,

$$\frac{\sin \beta}{b} = \frac{\sin \gamma}{c}.$$

Hence

$$\sin \gamma = \frac{c \sin \beta}{b} = \frac{12(.6691)}{9} = .8921.$$

Thus $m\angle\gamma = 63$ or $m\angle\gamma = 180 - 63 = 117$. Figure 12-30 shows the two solutions.

Note that in Example 3 we are given the measures of two sides and the measure of one angle which is adjacent to one side but opposite the other. This is often called the *ambiguous case* since there may be no triangle, one triangle, or two triangles that fit the data. The number of possibilities depends on the relationships between $m\angle ABC$ and the lengths of $\overline{AB}$ and $\overline{AC}$.

If $90 \le m\angle\beta < 180$, as in Figure 12-31, there is (1) one solution if and only if $b > c$ and there is (2) no solution if and only if $b \le c$.

If $0 < m\angle\beta < 90$, as in Figure 12-32, then (1) there is one solution

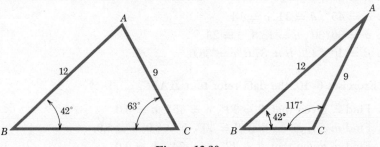

Figure 12-30

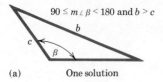

(a) One solution

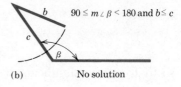

(b) No solution

Figure 12-31

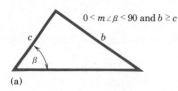

(a)

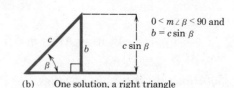

(b) One solution, a right triangle

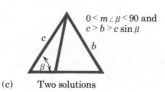

(c) Two solutions

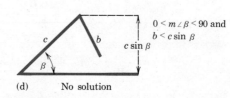

(d) No solution

Figure 12-32

if $b \geq c$, (2) there is one solution, a right triangle, if $b = c \sin \beta$, (3) there are two solutions if and only if $c > b > c \sin \beta$, and (4) there are no solutions if and only if $b < c \sin \beta$.

EXERCISES 12.6

◀ In Exercises 1–5, the data apply to a $\triangle ABC$. In each exercise determine the number of triangles that fit the data.

1. $\beta = 30°$, $b = 3$, $c = 6$
2. $\beta = 45°$, $b = 6$, $c = 10$
3. $\alpha = 45°$, $a = 11$, $c = 13$
4. $\alpha = 36°30'$, $a = 13.8$, $c = 23$
5. $\beta = 108°42'$, $b = 37.6$, $c = 38.6$

◀ In Exercises 6–10, the data refer to a $\triangle ABC$.

6. Find c, given that $\beta = 30°$, $\alpha = 45°$, $a = 5.0$.
7. Find $m\angle\gamma$, given that $\beta = 30°$, $b = 8.0$, $c = 5.0$.
8. Find a, given that $\beta = 42°$, $\gamma = 74°$, $b = 3.0$.

9. Find $\angle \alpha$, given that $a = 36$, $\beta = 102°$, $c = 48$.

10. Find b, given that $\beta = 120°$, $a = 14$, $c = 12$.

◀ In Exericses 11–15, the use of logarithms is advised. The data refer to a $\triangle ABC$.

11. Find a, given that $\beta = 40°37'$, $\alpha = 32°14'$, $b = 6.763$.

12. Find γ, given that $c = 493.2$, $b = 389.1$, $\beta = 36°18'$.

13. Find b, given that $a = 248.3$, $c = 397.6$, $\beta = 36°42'$.

14. Find a, given that $\alpha = 21°38'$, $\beta = 69°13'$, $c = 737.4$.

15. Find c, given that $a = 516.3$, $b = 226.9$, $\gamma = 43°41'$.

16. Find the area of each of the triangles of Exercise 3.

17. Find the area of the triangle of Exercise 9.

18. Find the area of the triangle of Exercise 12.

19. A 15-ft. ladder, as it leans against a building, makes an angle of 52° with the ground. If a 25-ft. ladder is used to reach the same point on the building, what angle must the longer ladder make with the ground?

20. The diagonals of a parallelogram are 32 in. and 46 in. long and form an angle whose measure is $52°14'$. Find the lengths of the sides of the parallelogram.

21. The straight bank of a river runs due south. At one point, a tree on the opposite side of the river is sighted as being $42°30'$ east of south. At a second point, 375 ft. south of the first point, the same tree is sighted as being $72°40'$ east of north. How wide is the river?

22. The longer diagonal of a parallelogram makes an angle of $23°10'$ with one side and of $38°40'$ with a second side. If the length of the diagonal is 427 cm., what is the area of the parallelogram?

23. From the top of an observation tower, 174.0 ft. above sea level, two boats are seen due north of the observer. If the measures of the angles of depression of the boats are $15°16'$ and $9°32'$, respectively, how far apart are the boats?

24. Show that

$$K = \frac{1}{2}a^2\left(\frac{\sin \beta \, \sin \gamma}{\sin \alpha}\right)$$

where K is the area of $\triangle ABC$.

25. **CHALLENGE PROBLEM.** Show that

$$K = \frac{1}{2}c^2\left(\frac{\sin \alpha \, \sin \beta}{\sin (\alpha + \beta)}\right)$$

where K is the area of $\triangle ABC$.

12.7 THE INVERSE TRIGONOMETRIC FUNCTIONS

In Chapter 11, where inverse functions were discussed, you learned that two functions f and g are inverses of each other if and only if $f[g(x)] = x$ for all x in the domain of g and $g[f(x)] = x$ for all x in the domain of f or, using the language of composition of functions, if and only if

$$f \circ g = g \circ f = I$$

where I is the identity function, that is, $I(x) = x$ for all x in the domain of f and g. If g is the inverse of f, we write $g = f^{-1}$. Each element in the range of g must be in the domain of f, and each element in the range of f must be in the domain of g.

Do the trigonometric functions have inverses? For example, is there an inverse sine function, which we denote by $\sin^{-1}$, such that

$$\sin^{-1}(\sin x) = x$$

for all $x \in R$?

Let us look at a familiar situation. We know that $\sin \frac{\pi}{6} = \frac{1}{2}$. In other words, the sine function associates $\frac{\pi}{6}$ with $\frac{1}{2}$. Accordingly, the inverse function, if it exists, must associate the number $\frac{1}{2}$ with precisely the number $\frac{\pi}{6}$. Then we could write

$$\sin^{-1}\left(\sin \frac{\pi}{6}\right) = \frac{\pi}{6}.$$

But, as you know, there are many numbers whose sine is $\frac{1}{2}$. For example,

$$\sin \frac{5\pi}{6} = \frac{1}{2} \quad \text{and} \quad \sin\left(-\frac{7\pi}{6}\right) = \frac{1}{2}.$$

Thus, on the one hand, our desire to have

$$\sin^{-1}(\sin x) = x$$

leads us to say that

$$\sin^{-1}\left(\sin \frac{\pi}{6}\right) = \frac{\pi}{6}, \qquad \sin^{-1}\left(\sin \frac{5\pi}{6}\right) = \frac{5\pi}{6},$$

$$\sin^{-1}\left(\sin \frac{13\pi}{6}\right) = \frac{13\pi}{6}, \qquad \sin^{-1}\left[\sin\left(-\frac{7\pi}{6}\right)\right] = -\frac{7\pi}{6},$$

and so on. But, on the other hand, the fact that

$$\sin \frac{\pi}{6} = \sin \frac{5\pi}{6} = \sin \frac{13\pi}{6} = \sin\left(-\frac{7\pi}{6}\right) = \frac{1}{2}$$

would give us

$$\sin^{-1}\frac{1}{2} = \frac{\pi}{6}, \ \sin^{-1}\frac{1}{2} = \frac{5\pi}{6}, \ \sin^{-1}\frac{13\pi}{6} = \frac{1}{2}, \ \sin^{-1}\left(-\frac{7\pi}{6}\right) = \frac{1}{2},$$

and so on, which is quite impossible if $\sin^{-1}$ is to be a function. We must, then, make a decision as to whether $\sin^{-1}\frac{1}{2} = \frac{\pi}{6}$, $\sin^{-1}\frac{1}{2} = \frac{5\pi}{6}$, or some other number. To make such a decision, let us first note that the solution set of the equation

$$\sin y = \frac{1}{2}$$

is the union of the two infinite sets

$$\left\{y : y = \frac{\pi}{6} + 2n\pi, \ n \in I\right\}$$

and

$$\left\{y : y = \frac{5\pi}{6} + 2n\pi, \ n \in I\right\}.$$

Out of this infinite solution set we pick the unique number $y = \frac{\pi}{6}$ which has the property that $-\frac{\pi}{2} \leq y \leq \frac{\pi}{2}$ and call this the **principal value** corresponding to the equation $\sin y = \frac{1}{2}$.

We now say that

$$\sin^{-1}\frac{1}{2} = y$$

where y is the principal value corresponding to the equation $\sin y = \frac{1}{2}$. That is, we say that

$$\sin^{-1}\frac{1}{2} = \frac{\pi}{6}.$$

We generalize this decision in regard to functional values for $\sin^{-1}$ in the following definition.

Definition 12.3 For $x \in R$ and $-1 \leq x \leq 1$,

$$\sin^{-1} x = y$$

if and only if

(1) $\sin y = x$ and   (2) $-\frac{\pi}{2} \leq y \leq \frac{\pi}{2}$.

Since $\sin \frac{\pi}{6} = \frac{1}{2}$ and $-\frac{\pi}{2} \le \frac{\pi}{6} \le \frac{\pi}{2}$, we see that $\sin^{-1} \frac{1}{2} = \frac{\pi}{6}$ is in accord with Definition 12.3. Later we shall discuss the question whether $\sin^{-1}$ is the inverse of the sine function. In the meantime, the following examples illustrate the use of Definition 12.3.

Example 1 Find $\sin^{-1} \dfrac{\sqrt{3}}{2}$.

Solution: If $y = \sin^{-1} \dfrac{\sqrt{3}}{2}$, then

(1) $\sin y = \dfrac{\sqrt{3}}{2}$ and

(2) $-\dfrac{\pi}{2} \le y \le \dfrac{\pi}{2}$.

The solution set for (1) is

$$S = \left\{ y : y = \frac{\pi}{3} + 2n\pi, \, n \in I \right\} \cup \left\{ y : y = \frac{2\pi}{3} + 2n\pi, \, n \in I \right\}.$$

The unique element of S satisfying (2) is $\frac{\pi}{3}$. Hence

$$\sin^{-1} \frac{\sqrt{3}}{2} = \frac{\pi}{3}.$$

Sometimes it is helpful to read a problem such as this as: "Find the angle y whose sine is $\dfrac{\sqrt{3}}{2}$ and such that $-\dfrac{\pi}{2} \le y \le \dfrac{\pi}{2}$." Note that the restriction that $-\dfrac{\pi}{2} \le y \le \dfrac{\pi}{2}$ means that the angle will be the one with the smallest measure in absolute value in the infinite solution set of the equation $\sin y = \dfrac{\sqrt{3}}{2}$.

Example 2 Find $\sin^{-1} \left(-\dfrac{\sqrt{2}}{2} \right)$.

Solution: If $y = \sin^{-1} \left(-\dfrac{\sqrt{2}}{2} \right)$, then

(1) $\sin y = -\dfrac{\sqrt{2}}{2}$ and

(2) $-\frac{\pi}{2} \le y \le \frac{\pi}{2}$.

The solution set for (1) is

$$S = \left\{ y : y = -\frac{\pi}{4} + 2n\pi, \ n \in I \right\} \cup \left\{ y : y = \frac{5\pi}{4} + 2n\pi, \ n \in I \right\}.$$

The unique element of S satisfying (2) is $-\frac{\pi}{4}$. Hence

$$\sin^{-1}\left(-\frac{\sqrt{2}}{2}\right) = -\frac{\pi}{4}.$$

An alternative notation for $\sin^{-1}$ is **arcsin** (read "arc sine"). Arcsine is the older notation since it was first introduced about 1730, whereas the $\sin^{-1}$ notation was introduced about 1813. Both notations have remained in common use. At the end of this chapter you might conjecture why this is so by speculating on the advantages of each. We use the arcsine notation in the next two examples and will use it as an alternative to $\sin^{-1}$ in the rest of the chapter.

Example 3 Find $\cos\left(\arcsin \frac{3}{4}\right)$.

Solution: We can read the problem as "Find the cosine of the angle whose sin is $\frac{3}{4}$." Thus if we let $\theta = \arcsin \frac{3}{4}$, then $\sin\theta = \frac{3}{4}$ and $0 \le \theta \le \frac{\pi}{2}$. We have therefore

$$\cos\theta = \sqrt{1 - \sin^2\theta} = \sqrt{1 - \left(\frac{3}{4}\right)^2} = \sqrt{\frac{7}{16}} = \frac{\sqrt{7}}{4}.$$

Figure 12-33 shows a diagram of the solution.

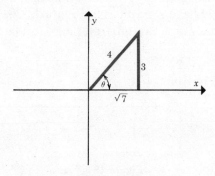

Figure 12-33

Example 4   Find $\tan [\arcsin (-\frac{3}{5}) - \arcsin \frac{5}{13}]$.

Solution: Let $\theta = \arcsin (-\frac{3}{5})$ and $\phi = \arcsin \frac{5}{13}$. Then

$$\sin \theta = -\frac{3}{5} \qquad \text{and} \qquad -\frac{\pi}{2} < \theta < 0.$$

Also

$$\sin \phi = \frac{5}{13} \qquad \text{and} \qquad 0 < \phi < \frac{\pi}{2}.$$

From Figure 12-34a, we obtain $\tan \theta = -\frac{3}{4}$ and from Figure 12-34b we obtain $\tan \phi = \frac{5}{12}$. Hence, using Identity (XIX) of Chapter 10, we have

$$\tan (\theta - \phi) = \frac{\tan \theta - \tan \theta}{1 + \tan \theta \tan \phi} = \frac{-\frac{3}{4} - \frac{5}{12}}{1 + (-\frac{3}{4})(\frac{5}{12})} = -\frac{56}{33}.$$

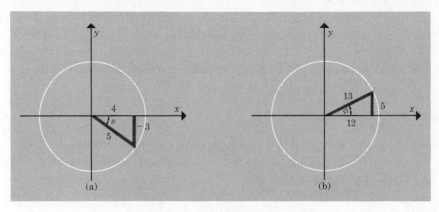

Figure 12-34

The graph of $y = \sin x$ is shown in Figure 12-35. The coordinates of the points on the graph are ordered pairs, $(x, \sin x)$. If we now graph $x = \sin y$, we have a graph as in Figure 12-36. The coordinates of the points on the graph are ordered pairs, $(\sin y, y)$. As the graph

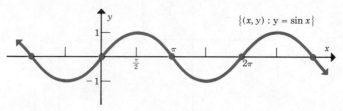

Figure 12-35

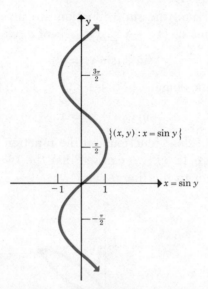

Figure 12-36

shows, each value of $\sin y$ is associated with many values of y. Is the graph in Figure 12-36 the graph of a function? The graph defined by the equation $y = \sin^{-1} x$ as shown in Figure 12-37 is a segment of the graph of the equation $x = \sin y$. Is the graph in Figure 12-37 the graph of a function?

Let s be the function whose domain is the set

$$D = \left\{ x : -\frac{\pi}{2} \leq x \leq \frac{\pi}{2} \right\}$$

and whose range is the set $\{y : y = \sin x$ for $x \in D\}$. The graph of the function s as shown in Figure 12-38 is one segment of the graph of the sine function.

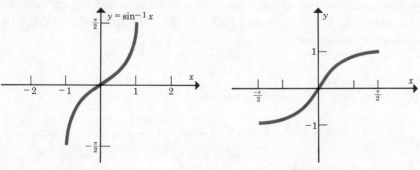

Figure 12-37   **Figure 12-38**

The function s and the $\sin^{-1}$ function are inverses. That is, for every x in the domain, $\{x : -\frac{\pi}{2} \leq x \leq \frac{\pi}{2}\}$, of s, we have

$$\sin^{-1}(\sin x) = x$$

and for every x in the domain, $\{x : -1 \leq x \leq 1\}$, of the $\sin^{-1}$ function, we have

$$\sin(\sin^{-1} x) = x.$$

If we graph the $\sin^{-1}$ function and the function s on the same set of axes as in Figure 12-39, we can see that the two graphs are symmetric with respect to the line $y = x$.

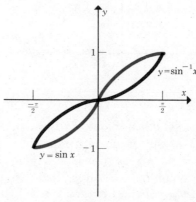

Figure 12-39

We define the inverse cosine function, $\cos^{-1}$, in a similar way to the way we define $\sin^{-1}$. Thus, for example, the equation $\cos y = \frac{1}{2}$ has the infinite solution set

$$\left\{ y : y = \frac{\pi}{3} + 2n\pi, \ n \in I \right\} \cup \left\{ y : y = \frac{5\pi}{3} + 2n\pi, \ n \in I \right\}$$

and we choose the unique number y from this set which has the property that $0 \leq y \leq \pi$ and call this the principal value. Thus, by definition, we have

$$\cos^{-1}\frac{1}{2} = \frac{\pi}{3}.$$

In general, we define $\cos^{-1}$ as follows:

Definition 12.4   For $x \in R$ and $-1 \leq x \leq 1$,

$$\cos^{-1} x = y$$

if and only if

(1) $\cos y = x$ and   (2) $0 \leq y \leq \pi$.

An alternative notation for $\cos^{-1} x$ is arccos x and the function is frequently called the arccosine function. The domain of the inverse cosine function, $\cos^{-1}$, is the set $\{x : -1 \le x \le 1\}$ and the range is the set $\{y : 0 \le y \le \pi\}$. The graph of the function defined by the equation $y = \cos^{-1} x$ as shown in Figure 12-40 is one segment of the graph of the equation $x = \cos y$.

Let c be the function whose domain is the set

$$E = \{x : 0 \le x \le \pi\}$$

and whose range is the set

$$\{y : y = \cos x \text{ for } x \in E\}.$$

The function c and the function $\cos^{-1}$ are inverses. That is, for every x in the domain, $\{x : 0 \le x \le \pi\}$, of c, we have

$$\cos^{-1} (\cos x) = x$$

and for every x in the domain, $\{x : -1 \le x \le 1\}$, of the arccosine function, we have

$$\cos (\cos^{-1} x) = x.$$

In Figure 12-41, the graph of the arccosine function is shown as a solid line and the graph of the function c is shown as a dashed line.

Example 5 Find $\cos^{-1}\left(-\dfrac{\sqrt{3}}{2}\right)$.

Solution: If $y = \cos^{-1}\left(-\dfrac{\sqrt{3}}{2}\right)$, then

$$(1) \qquad \cos y = -\dfrac{\sqrt{3}}{2}$$

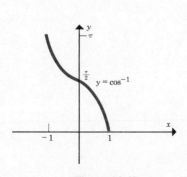

Figure 12-40

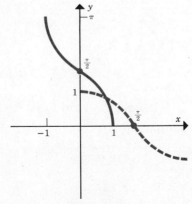

Figure 12-41

and

(2) $0 \leq y \leq \pi$.

The solution set for (1) is

$$S = \left\{ y : y = \frac{5\pi}{6} + 2n\pi,\ n \in I \right\} \cup \left\{ y : y = \frac{7\pi}{6} + 2n\pi,\ n \in I \right\}.$$

The unique element of S satisfying (2) is $\frac{5\pi}{6}$. Hence

$$\cos^{-1}\left(-\frac{\sqrt{3}}{2} \right) = \frac{5\pi}{6}.$$

We define $\tan^{-1} x = \arctan x$ in a similar fashion to the definition of $\sin^{-1} x$ and $\cos^{-1} x$.

Definition 12.5 For $x \in R$,

$$\tan^{-1} x = y$$

if and only if

(1) $\tan y = x$ and   (2) $-\dfrac{\pi}{2} < y < \dfrac{\pi}{2}$.

The domain of the arctangent function is the set $\{ x : x \in R \}$ and the range is the set

$$\left\{ y : \tan y = x \text{ and } -\frac{\pi}{2} < y < \frac{\pi}{2} \right\}.$$

The $\tan^{-1}$ function is the inverse of the tangent function if we restrict the domain of the tangent function to the interval $-\frac{\pi}{2} < x < \frac{\pi}{2}$. That is, for $-\frac{\pi}{2} < x < \frac{\pi}{2}$,

$$\tan^{-1}(\tan x) = x$$

and for $x \in R$,

$$\tan(\tan^{-1} x) = x.$$

Example 6 Find $\tan^{-1}(-\sqrt{3})$.

Solution: If $y = \tan^{-1}(-\sqrt{3})$, then

(1) $\tan y = -\sqrt{3}$

and

(2) $-\dfrac{\pi}{2} < y < \dfrac{\pi}{2}$.

The solution set for (1) is

$$S = \left\{ -\frac{\pi}{3} + 2n\pi,\ n \in I \right\} \cup \left\{ \frac{2\pi}{3} + 2n\pi,\ n \in I \right\}.$$

The unique element of S satisfying (2) is $-\frac{\pi}{3}$. Hence

$$\tan^{-1}(-\sqrt{3}) = -\frac{\pi}{3}.$$

Three other functions, $\sec^{-1}$, $\csc^{-1}$, and $\cot^{-1}$, may be defined in a similar way. We shall not be concerned, however, with these functions.

Example 7 Find $\cos[\arctan(-1)]$.

Solution: Let $\theta = \arctan(-1)$. Then $\tan\theta = -1$ and $-\frac{\pi}{2} < \theta \le 0$. Hence

$$\theta = -\frac{\pi}{4} \quad \text{and} \quad \cos\left(-\frac{\pi}{4}\right) = \frac{\sqrt{2}}{2}.$$

Thus

$$\cos[\arctan(-1)] = \frac{\sqrt{2}}{2}.$$

Example 8 Show that $\arctan\frac{1}{2} + \arctan\frac{1}{3} = \frac{\pi}{4}$.

Solution: Let $\theta = \arctan\frac{1}{2}$ and $\phi = \arctan\frac{1}{3}$. Then $\tan\theta = \frac{1}{2}$ and $0 < \theta < \frac{\pi}{2}$ and $\tan\phi = \frac{1}{3}$ and $0 < \phi < \frac{\pi}{2}$. We have, by Identity (XVIII) of Chapter 10,

$$\tan(\theta + \phi) = \frac{\tan\theta + \tan\phi}{1 - \tan\theta \tan\phi}$$

$$= \frac{\frac{1}{2} + \frac{1}{3}}{1 - \frac{1}{6}} = 1.$$

Since $\tan(\theta + \phi) = 1$ and $\theta + \phi < \pi$, it follows that $\theta + \phi = \frac{\pi}{4}$. Hence

$$\arctan\frac{1}{2} + \arctan\frac{1}{3} = \frac{\pi}{4}.$$

Example 9 Find the solution set of the equation

$$2\tan^{-1} 2x = 3.$$

Solution: Let $\tan\theta = 2x$. Then $2\theta = 3$, or $\theta = \frac{3}{2}$. Hence $\tan\frac{3}{2} = 2x$.

From Table II (using the column of radian measure to find 1.5), we find that tan $1.5 = 14.10$. Hence $2x = 14.10$ and $x = 7.05$.

EXERCISES 12.7

◀In Exercises 1–16, find the value of each inverse function. Use tables if necessary.

1. $\sin^{-1} \dfrac{1}{2}$

2. $\cos^{-1} \dfrac{1}{2}$

3. $\arcsin \dfrac{\sqrt{3}}{2}$

4. $\arccos \left(-\dfrac{\sqrt{3}}{2} \right)$

5. $\tan^{-1} 1$

6. $\tan^{-1} \dfrac{\sqrt{2}}{2}$

7. $\arctan (-1)$

8. $\arctan \sqrt{3}$

9. $\sin^{-1} \left(-\dfrac{\sqrt{3}}{2} \right)$

10. $\cos^{-1} \left(-\dfrac{\sqrt{2}}{2} \right)$

11. $\arcsin \dfrac{\sqrt{2}}{2}$

12. $\arccos 0$

13. $\sin^{-1} (-0.6018)$

14. $\cos^{-1} 0.9572$

15. $\tan^{-1} 0.0670$

16. $\arctan (-11.06)$

◀In Exercises 17–31, find the value of each expression.

17. $\cos \left(\arcsin \dfrac{1}{2} \right)$

18. $\sin \left(\cos^{-1} \dfrac{3}{5} \right)$

19. $\sec \left[\arcsin \left(-\dfrac{\sqrt{3}}{2} \right) \right]$

20. $\tan \left[\arccos \left(-\dfrac{5}{13} \right) \right]$

21. $\cos \left[\tan^{-1} (-2) \right]$

22. $\csc \left[\arctan (-1) \right]$

23. $\sin \left[\arcsin \left(-\dfrac{4}{5} \right) \right]$

24. $\sin (\arccos 0.9572)$

25. $\sin (\arctan 1.033)$

26. $\sec \left[\arccos (-0.2896) \right]$

27. $\sin \left(2 \arctan \dfrac{\sqrt{3}}{3} \right)$

28. $\cos \left[2 \arcsin \left(-\dfrac{3}{5} \right) \right]$

29. $\sin \left(\arccos \dfrac{1}{2} - \arccos \dfrac{\sqrt{3}}{2} \right)$

30. $\cos \left[\arctan (-10) + \arctan 10 \right]$

31. $\tan \left(\sin^{-1} \dfrac{4}{5} + \cos^{-1} \dfrac{4}{5} \right)$

32. Solve $c^2 = a^2 + b^2 - 2ab \cos \gamma$ for γ.

33. Show that, for all real numbers x,

$$\arcsin x + \arccos x = \frac{\pi}{2}.$$

34. Show that

$$\arctan \frac{7}{4} + \arctan \frac{4}{7} = \frac{\pi}{2}.$$

35. Find the solution set of the equation

$$3 \tan^{-1} 3x = 4.$$

36. Find the solution set of the equation

$$2 \sin^{-1} 2x = 3.$$

37. Show that

$$\arctan 2 - \arctan 1 = \arctan \frac{1}{3}.$$

38. Show that, for all real numbers x,

$$\sin (\arctan x) = \frac{x}{\sqrt{1 + x^2}}.$$

39. CHALLENGE PROBLEM. Evaluate $\sin (\frac{1}{2} \arccos x)$.
40. CHALLENGE PROBLEM. Evaluate $\cos (\frac{1}{2} \arcsin \frac{3}{5})$.

CHAPTER SUMMARY

An ANGLE is formed by the union of two rays having a common endpoint called the VERTEX of the angle. An angle θ in STANDARD POSITION has its vertex at the origin of the Cartesian plane and its INITIAL RAY coinciding with the positive ray of the x-axis. The TERMINAL RAY forms a POSITIVE angle when it moves in a counterclockwise direction and a NEGATIVE angle when it moves in a clockwise direction.

The measure of an angle corresponds to the length of the arc intercepted by the angle on the unit circle. The unit of measure is arbitrary. An angle of one DEGREE intercepts an arc which is $\frac{1}{360}$ of the unit circle. An angle of one RADIAN intercepts an arc one unit long which, accordingly, is $\frac{1}{2\pi}$ of the unit circle.

A TRIGONOMETRIC function associates an angle with a real number. The trigonometric sine function is defined by $\sin \theta = \sin x$ where x is the number which is the radian measure of θ. The other five trigonometric functions are similarly defined.

The ANGLE OF INCLINATION of a line is the angle the line forms with the positive direction of the x-axis. The tangent of the angle of inclination is equal to the slope of the line.

An ANGLE OF INTERSECTION, ϕ, of two nonperpendicular intersecting lines can be found by using the identity $\tan \phi = \dfrac{m_1 - m_2}{1 + m_1 m_2}$ where m_1 and m_2 are the slopes of the two lines.

Two lines, neither one parallel to one of the axes, are perpendicular if and only if $m_1 m_2 = -1$.

The LAW OF SINES,

$$\frac{\sin \alpha}{a} = \frac{\sin \beta}{b} = \frac{\sin \gamma}{c},$$

and the LAW OF COSINES,

$$c^2 = a^2 + b^2 - 2ab \cos \gamma,$$

apply to all triangles and are useful in solving triangles, that is, in finding the lengths of the three sides and the measures of the three angles.

To find the area of a triangle, either the formula

$$K = \frac{1}{2} ab \sin \gamma$$

or Heron's Formula,

$$K = \sqrt{s(s - a)(s - b)(s - c)},$$

in which $s = \frac{1}{2}(a + b + c)$, may be used.

The number y (or the angle corresponding to this value of y) in the interval $-\frac{\pi}{2} \le y \le \frac{\pi}{2}$, which is a solution to the equation $\sin y = x$ for some x in the interval $-1 \le x \le 1$, is called the PRINCIPAL VALUE and is denoted by arcsin x.

The equation $y =$ arcsin x describes the ARCSINE FUNCTION whose domain is the set $\{x : -1 \le x \le 1\}$ and whose range is the set

$$\left\{ y : \sin y = x \text{ and } -\frac{\pi}{2} \le y \le \frac{\pi}{2} \right\}.$$

If the domain of the sine function, defined by $y = \sin x$, is restricted to the interval $-\frac{\pi}{2} \le x \le \frac{\pi}{2}$, then the sine function is the inverse of the arcsine function, and vice versa. For $-\frac{\pi}{2} \le x \le \frac{\pi}{2}$, $\sin^{-1}(\sin x) = x$ and for $-1 \le x \le 1$, $\sin(\sin^{-1} x) = x$ where $\sin^{-1} x =$ arcsin x.

The equation $y = \arccos x$ describes the ARCCOSINE FUNCTION whose domain is the set $\{x : -1 \le x \le 1\}$ and whose range is the set

$$\{y : \cos y = x \text{ and } 0 \le y \le \pi\}.$$

If the domain of the cosine function, defined by $y = \cos x$, is restricted to the interval $0 \le x \le \pi$, then the cosine function is the inverse of the arccosine function and vice versa. For $0 \le x \le \pi$, $\cos^{-1}(\cos x) = x$ and for $-1 \le x \le 1$, $\cos(\cos^{-1} x) = x$ where $\cos^{-1} x = \arccos x$.

The equation $y = \arctan x$ describes the ARCTANGENT FUNCTION whose domain is the set $\{x : x \in R\}$ and whose range is the set

$$\left\{y : \tan y = x \text{ and } -\frac{\pi}{2} < y < \frac{\pi}{2}\right\}.$$

If the domain of the tangent function, defined by $y = \tan x$, is restricted to the interval $-\frac{\pi}{2} < x < \frac{\pi}{2}$, then the tangent function is the inverse of the arctangent function and vice versa. For $-\frac{\pi}{2} < x < \frac{\pi}{2}$, $\tan^{-1}(\tan x) = x$, and for $x \in R$, $\tan(\tan^{-1} x) = x$ where $\tan^{-1} x = \arctan x$.

The arcsine, arccosine, and the arctangent functions are called INVERSE trigonometric functions.

REVIEW EXERCISES

1. Express each of the following degree measures as equivalent radian measures:

   (a) 60   (b) 120   (c) -30

2. Express each of the following radian measures as equivalent degree measures:

   (a) $\frac{\pi}{6}$ (b) $-\pi$ (c) 14π

3. Define the trigonometric cosine function.

4. Express each of the six trigonometric functions as ratios involving the sides of a right triangle.

5. In triangle ABC, find AB given that $m\angle ACB = 90°$, $m\angle ABC = 42°30'$, and $b = 14.2$.

6. In triangle DEF, find the area given that $m\angle DEF = 90$, $DE = 18.4$, and $DF = 26.3$.

7. Derive the Law of Cosines.

8. In triangle ABC, find c given that $m \angle ACB = 63°13'$, $a = 42.14$, and $b = 63.91$.

9. Find the area of the triangle the lengths of whose sides are 32, 24, and 18.

10. Find the measure of the largest angle of the triangle in Exercise 9.

11. State the Law of Sines.

12. Find the area of the triangle in Exercise 8.

13. In triangle ABC, find b given that $c = 321.9$, $m \angle CAB = 42°23'$, and $m \angle CBA = 64°18'$.

14. In triangle ABC, find $m \angle ABC$ given that $m \angle CAB = 28°36'$, $b = 9.146$, and $a = 5.437$. (There are two possibilities.)

15. Evaluate

   (a) $\cos^{-1}\left(-\dfrac{1}{2}\right)$ (b) $\sin\left(\arctan\dfrac{5}{12}\right)$

16. Find the measure of an angle of intersection of the two lines

$$4x - 2y = 9 \quad \text{and} \quad y = \frac{5}{3}x + 3.$$

17. Find an equation of the line that contains the point $\left(3\dfrac{1}{2}, -2\dfrac{1}{4}\right)$ and is perpendicular to the line with an equation $3x - 2y = 5$.

18. Graph the function defined by the equation $y = \arccos x$.

19. Find

$$\sin\left[\arctan\frac{3}{4} + \arccos\left(-\frac{3}{5}\right)\right].$$

20. Show that

$$\tan^{-1}\frac{2}{3} + \tan^{-1}\frac{1}{5} = \frac{\pi}{4}.$$

GOING FURTHER: READING AND RESEARCH

In Chapter 7 we saw that a translation is a function that maps each ordered pair (x, y) in the domain into $(x + h, y + k)$ in the range. Thus the ellipse

$$\frac{x^2}{9} + \frac{y^2}{4} = 1$$

is mapped into

$$\frac{(x - 3)^2}{9} + \frac{(y - 4)^2}{4} = 1$$

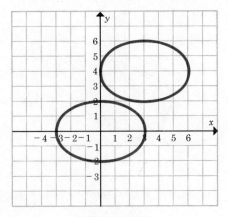

Figure 12-42

under the translation $T: (x, y) \longrightarrow (x + 3, y + 4)$ as shown in Figure 12-42.

A rotation is another kind of a function. Under a rotation each point (x, y) in the domain is mapped into the point $(x \cos \theta - y \sin \theta, x \sin \theta + y \cos \theta)$.

A certain rotation will map the point $(1, 0)$ into the point $(0, 1)$ as in Figure 12-43. What is the measure of θ? Test your guess.

A certain rotation will map the line $l_1 : 4x + 3y = 8$ into the line with an equation $5x' = 8$, relative to the x', y' axes as in Figure 12-44. l_1 is the graph of the equation $5x' = 8$. Can you determine what the measure of θ is?

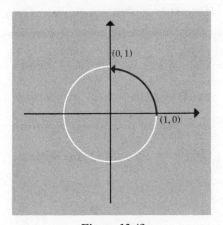

Figure 12-43

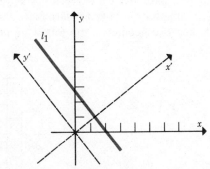

Figure 12-44

What about the rotation that maps one ellipse in Figure 12-45 into the other?

Given two rotations, R and S, is it possible to have a composite function $R \circ S$? If so, is $R \circ S = S \circ R$?

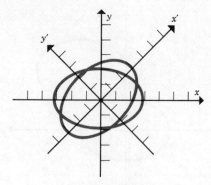

Figure 12-45

The curves defined by the equations

$$19x^2 + 6xy + 11y^2 + 20x - 60y + 80 = 0$$

and

$$2x^2 + y^2 = 2$$

do not seem related at first glance. However, by the function $R \circ T$ where T is a translation that carries $(0, 0)$ into $(-1, 3)$ and R is a rotation through an angle equal to one-half of arctan $\frac{3}{4}$, the first curve is mapped into the second. Figure 12-46 shows the first curve.

By the proper choice of a rotation and translation, each conic section defined by an equation of the form

$$Ax^2 + Bxy + Cy^2 + Dx + Ey + F = 0$$

can be related to a circle, ellipse, or hyperbola whose center is at the origin or to a parabola whose vertex is at the origin.

If you would like to learn more about the use of rotations and translations in analytic geometry, the following books may be helpful.

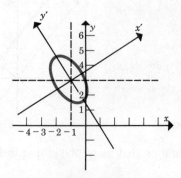

Figure 12-46

WILSON, W. A, and J. I. TRACY, *Analytic Geometry* (3rd ed.). Boston: D. C. Heath, 1949.

School Mathematics Study Group, *Analytic Geometry*, Part 2. Pasadena: A. E. Vroman, Inc., 1968.

MURDOCH, D.C., *Analytic Geometry*, New York: John Wiley and Sons, 1966.

WEXLER, C., *Analytic Geometry—A Vector Approach*, Reading, Mass: Addison-Wesley.

C. Capa—Magnum

CHAPTER THIRTEEN
VECTORS

13.1 DIRECTED LINE SEGMENTS

Quantities such as distance or the measure of an angle that are described by a single real number are called **scalar** quantities. There are many quantities, however, whose description requires more than one real number.

In Chapter 4, for example, when we were discussing complex numbers, we used two real numbers, a and b, to define a complex number. Also in Chapter 7 we used two real numbers to define a translation when discussing conic sections. In the physical sciences, quantities such as force and velocity are frequently represented by arrows because a single number cannot describe both magnitude and direction. Translations, forces, and velocities are called **vector** quantities. Two or more numbers are needed to describe these quantities and, as we will show, they are often represented by directed line segments—sometimes called "arrows."

In a plane, two points A and B determine a unique line segment $\overline{AB}$. If we name one point A as the **initial** point and the other point B as the **terminal** point, we define a **directed line segment,** or simply **directed segment,** which is represented as an arrow in Figure 13-1 and denoted by $\overrightarrow{AB}$. If B is the initial point and A is the terminal point, we write $\overrightarrow{BA}$. Note that $\overrightarrow{AB} \neq \overrightarrow{BA}$ since they have different directions.

A directed segment $\overrightarrow{AB}$ as shown in Figure 13-1 indicates a displacement from A to B. If the coordinates of A are $(2, 2)$ and the coordinates of B are $(5, 4)$, then the displacement has a horizontal or **x-component** along a line parallel to the x-axis and a vertical or **y-component** along a line parallel to the y-axis. For $\overrightarrow{AB}$ in Figure 13-1, the x-component is $5 - 2 = 3$ and the y-component is $4 - 2 = 2$.

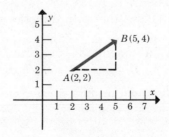

Figure 13-1

615

In general, if the coordinates of A are (x_1, y_1) and the coordinates of B are (x_2, y_2), the x-component of the directed segment $\overrightarrow{AB}$ is $x_2 - x_1$ and the y-component is $y_2 - y_1$.

The length of the directed segment $\overrightarrow{AB}$, called the **magnitude** of $\overrightarrow{AB}$, is defined to be equal to the length, AB, of the segment $\overline{AB}$. (See Figure 13-2.) We can find AB by the use of the distance formula.

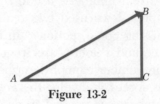

Figure 13-2

If $\overline{AB}$ is not parallel to the x-axis, then the line $\overleftrightarrow{AB}$ containing the directed segment $\overrightarrow{AB}$ meets the x-axis at some point R as shown in Figure 13-3. The direction of $\overrightarrow{AB}$ is defined to be the nonnegative

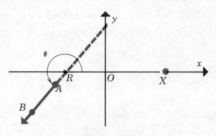

Figure 13-3

measure of the angle θ formed by the ray $\overrightarrow{RB}$ and the ray $\overrightarrow{RX}$ where $\overrightarrow{OX}$ is the positive ray of the x-axis. Since $0 \le \theta < 2\pi$, θ is uniquely defined if $\cos \theta$ and $\sin \theta$ are both specified; $\cos \theta$ and $\sin \theta$ are called the **direction cosines** of $\overrightarrow{AB}$. (Note that we could specify the direction in terms of cosines alone by using the two angles θ and $90 - \theta$ since $\cos (90 - \theta) = \sin \theta$. Hence the term direction *cosines*.)

Example 1 Find the magnitude and direction cosines of $\overrightarrow{AB}$ if the coordinates of A and B are $(3, 2)$ and $(6, 6)$, respectively.

Solution: Using the distance formula, we have

$$AB = \sqrt{(6 - 3)^2 + (6 - 2)^2} = \sqrt{9 + 16} = 5.$$

Thus the magnitude of $\overrightarrow{AB}$ is 5. We denote the magnitude of $\overrightarrow{AB}$ by

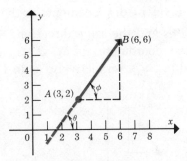

Figure 13-4

$|\overrightarrow{AB}|$. Thus $|\overrightarrow{AB}| = 5$. Since $m\angle\theta = m\angle\phi$ as shown in Figure 13-4, we have

$$\cos\theta = \cos\phi = \frac{x_2 - x_1}{AB} = \frac{3}{5}$$

and

$$\sin\theta = \sin\phi = \frac{y_2 - y_1}{AB} = \frac{4}{5}.$$

If the line $\overleftrightarrow{AB}$ containing $\overrightarrow{AB}$ is parallel to the x-axis, then $\cos\theta = 1$ and $\sin\theta = 0$ when $\overrightarrow{AB}$ and $\overrightarrow{OX}$ have the same direction. If $\overrightarrow{AB}$ and $\overrightarrow{OX}$ have opposite directions, then $\cos\theta = -1$ and $\sin\theta = 0$.

If the length of a directed segment $\overrightarrow{AB}$ is zero, then A and B are the same point. Conversely, if A and B are the same point, then the length of $\overrightarrow{AB}$ is zero.

If A, B, C, and D are the points whose coordinates are $(3, 2)$, $(5, 5)$, $(-5, -1)$, and $(-3, 2)$, respectively, as in Figure 13-5, then $\overrightarrow{AB}$

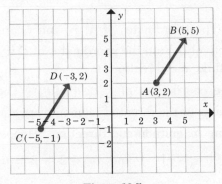

Figure 13-5

and $\overrightarrow{CD}$ indicate **equivalent** displacements in the sense that $\overrightarrow{AB}$ and $\overrightarrow{CD}$ have equal lengths and the same direction.

In general, if the coordinates of A, B, C, and D are $(x_1, \ y_1)$, $(x_2, \ y_2)$, $(x_3, \ y_3)$, and $(x_4, \ y_4)$, respectively, as in Figure 13-6, then

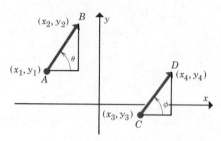

Figure 13-6

$$AB = \sqrt{(x_2 - x_1)^2 + (y_2 - y_1)^2}$$

and

$$CD = \sqrt{(x_4 - x_3)^2 + (y_4 - y_3)^2}.$$

Also

$$\cos \theta = \frac{x_2 - x_1}{AB}, \qquad \sin \theta = \frac{y_2 - y_1}{AB},$$

$$\cos \phi = \frac{x_4 - x_3}{CD}, \qquad \text{and} \qquad \sin \phi = \frac{y_4 - y_3}{CD}.$$

Thus $\overrightarrow{AB}$ has the same magnitude and direction as $\overrightarrow{CD}$ (and hence $\overrightarrow{AB}$ and $\overrightarrow{CD}$ indicate equivalent displacements) if and only if

$$x_2 - x_1 = x_4 - x_3$$

and

$$y_2 - y_1 = y_4 - y_3.$$

Example 2   The coordinates of A, B, and C are $(3, 2)$, $(1, \ -2)$, and $(1, 5)$, respectively. Find the coordinates of D if $\overrightarrow{AB}$ and $\overrightarrow{CD}$ have the same magnitude and direction.

Solution: Let the coordinates of D be $(x_1, \ y_1)$. The horizontal and vertical components of the directed segments are equal since $\overrightarrow{AB}$ and $\overrightarrow{CD}$ have the same magnitude and direction. Hence

$$1 - 3 = x_1 - 1 \qquad \text{and} \qquad -2 - 2 = y_1 - 5.$$

Thus $x_1 = -1$ and $y_1 = 1$. D is the point $(-1, 1)$. Hence $\overrightarrow{CD}$ will be the directed segment from C to D as shown in Figure 13-7.

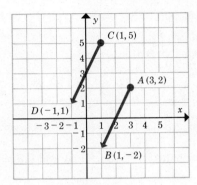

Figure 13-7

EXERCISES 13.1

◀In Exercises 1–10, determine if $\overrightarrow{AB}$ and $\overrightarrow{CD}$ have the same magnitude and direction.

1. $A(0, 0)$, $B(3, 4)$, $C(3, 4)$, $D(6, 8)$
2. $A(0, 0)$, $B(3, 4)$, $C(0, 0)$, $D(-3, -4)$
3. $A(-4, -3)$, $B(0, 0)$, $C(0, 0)$, $D(4, 3)$
4. $A(-3, 2)$, $B(0, 0)$, $C(0, 0)$, $D(3, -2)$
5. $A(0, 0)$, $B(-3, 2)$, $C(0, -2)$, $D(-3, 0)$
6. $A(0, 0)$, $B(5, 6)$, $C(0, 6)$, $D(5, 12)$
7. $A(-3, 0)$, $B(2, -3)$, $C(5, 0)$, $D(0, 3)$
8. $A(3, 4)$, $B(11.5, 15)$, $C(6.5, 9)$, $D(15, 21)$
9. $A(-5, -6)$, $B(-15, 1)$, $C(25, 7)$, $D(35, 14)$
10. $A(-11, 3.9)$, $B(6.9, 4.8)$, $C(8.2, 3.1)$, $D(16.2, 4)$

◀In Exercises 11–20, find the magnitude and direction cosines of each directed segment $\overrightarrow{AB}$.

11. $A(0, 0)$, $B(3, 4)$
12. $A(0, 0)$, $B(-5, 12)$
13. $A(0, 0)$, $B(6, -8)$
14. $A(0, 0)$, $B(5, -12)$
15. $A(-3, -4)$, $B(-3, -4)$

16. $A(3, 4)$, $B(10, 11)$
17. $A(-2, 5)$, $B(8, 9)$
18. $A(4, -8)$, $B(12, 6)$
19. $A(-3, -4)$, $B(8, -10)$
20. $A(3, 4)$, $B(15, 20)$

◀In Exercises 21–25, find P such that $\overrightarrow{AP}$ has the same magnitude and direction as $\overrightarrow{RS}$.

21. $A(0, 0)$, $R(3, 4)$, $S(8, 9)$
22. $A(0, 0)$, $R(-6, 8)$, $S(5, -3)$

23. $A(-5, 4)$, $R(-6, 8)$, $S(5, -3)$

24. $A(3, 8)$, $R(5, -1)$, $S(6, 2)$

25. $A(6, 7)$, $R(0, 0)$, $S(-2.5, 9.8)$

26. Given four points A, B, C, D, no three of which are collinear, list all the directed segments that the four points determine.

27. A directed segment $\overrightarrow{AB}$ has the direction cosines, $\sin\theta = \dfrac{3}{5}$ and $\cos\theta = -\dfrac{4}{5}$. If the coordinates of A are $(6, 8)$ and $AB = 10$, find the coordinates of B.

28. A directed segment $\overrightarrow{CD}$ has the direction cosines, $\cos\theta = -\dfrac{5}{13}$ and $\sin\theta = -\dfrac{12}{13}$. If the coordinates of C are $(-10, -24)$ and $CD = 26$, find the coordinates of D.

29. The coordinates of the four points C, D, E, and F are $(8, -6)$, $(1, 1)$, $(-5, -4)$, and $(3, 4)$, respectively. Show that the line containing $\overrightarrow{CD}$ is perpendicular to the line containing $\overrightarrow{EF}$.

30. The coordinates of the four points A, B, C, and D are $(2, 2)$, $(-2, 7)$, $(4, 4)$, and $(9, 8)$, respectively. Show that the line containing $\overrightarrow{AB}$ is perpendicular to the line containing $\overrightarrow{CD}$.

31. Find two directed segments $\overrightarrow{OP}$ and $\overrightarrow{OP'}$ (O is the origin of the Cartesian plane) that have the same magnitude and direction as $\overrightarrow{AB}$ and $\overrightarrow{CD}$ in Exercise 30. Show that $x_1x_2 + y_1y_2 = 0$ where the coordinates of P and P' are (x_1, y_1) and (x_2, y_2), respectively.

32. **CHALLENGE PROBLEM.** Find the length of the segment $\overline{AB}$ in a space of three dimensions if the coordinates of A and B are $(3, 4, 1)$ and $(6, 9, 4)$. Designate in some way the direction of $\overrightarrow{AB}$.

13.2 VECTORS

As mentioned in the introduction to Section 13.1, we use directed segments to represent vectors. We write "vector $\overrightarrow{AB}$" which is an abbreviation for the longer expression "the vector represented by the directed segment $\overrightarrow{AB}$."

Two vectors $\overrightarrow{AB}$ and $\overrightarrow{CD}$ as pictured in Figure 13-8 that have the same direction and the same magnitude are called **equivalent**. The magnitude and direction of a vector $\overrightarrow{AB}$ are defined as being equal

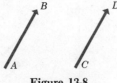

Figure 13-8

to the length and direction, respectively, of the directed line segment $\overrightarrow{AB}$. We denote the magnitude of vector $\overrightarrow{AB}$ as $|\overrightarrow{AB}|$.

A vector whose initial point and terminal point are the same point, such as $\overrightarrow{AA}$, $\overrightarrow{BB}$, and $\overrightarrow{CC}$, is called a **zero** vector.

The set of all vectors equivalent to a given vector as illustrated in Figure 13-9 is called an **equivalence class** of vectors.

If we have a vector $\overrightarrow{AB}$, $A \neq B$, and a point C, then there exists a unique vector $\overrightarrow{CD}$ whose magnitude and direction are the same as $\overrightarrow{AB}$. If C is contained in $\overleftrightarrow{AB}$ as in Figure 13-10a, then $AB = CD$ and the direction from C to D is the same as the direction from A to B. If C is not contained in $\overleftrightarrow{AB}$, then D is the unique point such that $ABDC$ is a parallelogram as shown in Figure 13-10b.

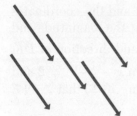

Figure 13-9

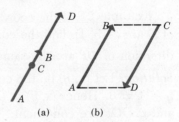

(a)   (b)

Figure 13-10

We use boldface letters such as **u, v,** or **w** as variables whose domain is a set of vectors. Thus statements such as "Let $\mathbf{u} = \overrightarrow{CD}$" or "Let $\mathbf{v} = \overrightarrow{OP}$" are meaningful since any point may be chosen as the initial point of a vector with a fixed magnitude and direction. A zero vector (that is, a vector of magnitude 0) is denoted by **0.**

In each equivalence class of vectors there is a unique vector that is represented by the directed segment whose initial point is O, the origin of the Cartesian plane, and whose terminal point is P. The vector $\overrightarrow{OP}$ is called a **position vector.** Thus in Figure 13-11 each vector pictured is equivalent to the position vector $\overrightarrow{OP}$.

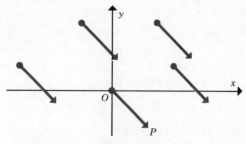

Figure 13-11

Since two position vectors $\overrightarrow{OP}$ and $\overrightarrow{OP'}$ have the same direction and magnitude if and only if $P = P'$, naming the coordinates of point P defines a unique vector $\overrightarrow{OP}$. In turn, of course, a position vector defines an equivalence class of vectors.

Example 1 Each of the four points $P_1(4, 5), P_2(-6, 3), P_3(-3, -5)$, and $P_4(4, -3)$ determine a position vector. Draw the vectors $\overrightarrow{OP_1}$, $\overrightarrow{OP_2}$, $\overrightarrow{OP_3}$, and $\overrightarrow{OP_4}$ determined by the points P_1, P_2, P_3, and P_4.
Solution: See Figure 13-12.

Example 2 If the coordinates of P are $(2, 3)$ and the coordinates of D are $(-5, 1)$, find the coordinates of E so that the magnitude and direction of $\overrightarrow{OP}$ are the same as the magnitude and direction of $\overrightarrow{DE}$.
Solution: Let (x, y) be the coordinates of E. Then $x - (-5) = 2$ and $y - 1 = 3$. Hence $(x, y) = (-3, 4)$. Note in Figure 13-13 that $\triangle DFE$ and $\triangle OQP$ are congruent.

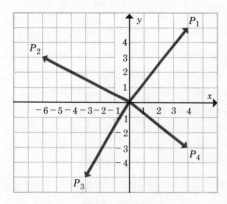

Figure 13-12

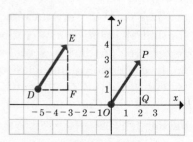

Figure 13-13

The point $P(a, b)$ (a and b not both 0) defines a unique directed line segment $\overrightarrow{OP}$ and hence a unique vector $\overrightarrow{OP}$. The vector defined by this point P is designated by $[a, b]$. If $a = 0$ and $b = 0$, then we have the zero vector $[0, 0]$ which is represented by a single point at the origin. Since the magnitude $|\overrightarrow{AB}|$ of vector $\overrightarrow{AB}$ is equal to the length AB of line segment $\overline{AB}$, the magnitude of vector $[a, b]$ is

$$\|[a, b]\| = \sqrt{a^2 + b^2}.$$

The direction of vector $[a, b]$ is defined by the direction cosines

$$\frac{a}{\sqrt{a^2 + b^2}} \quad \text{and} \quad \frac{b}{\sqrt{a^2 + b^2}}.$$

Whenever we name one position vector, $[a, b]$, we thereby determine the set of all vectors which have the same magnitude and direction. For example, Figure 13-14 shows a few of the vectors contained in the set of vectors which have the same magnitude and direction as the position vector $[2, 3]$. We shall use the notation $[a, b]$ hereafter to denote the set of vectors equivalent to the position vector $\overrightarrow{OP}$ where $P = (a, b)$.

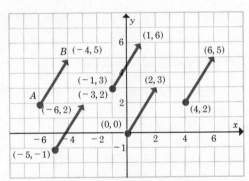

Figure 13-14

If the coordinates of points A and B are $(-6, 2)$ and $(-4, 5)$, respectively, as in Figure 13-14, then the vector $\overrightarrow{AB}$ also defines the same set as $[2, 3]$. In this sense we write

$$\overrightarrow{AB} = [2, 3],$$

meaning that the set of vectors equivalent to $\overrightarrow{AB}$ and the set of vectors equivalent to $[2, 3]$ are the same set. We shall also write $\overrightarrow{AB} = \overrightarrow{CD}$

which denotes not only that $\overrightarrow{AB}$ has the same magnitude and direction as $\overrightarrow{CD}$ but also that any vector in the set of vectors equivalent to $\overrightarrow{AB}$ has the same magnitude and direction as any vector in the set of vectors equivalent to $\overrightarrow{CD}$.

EXERCISES 13.2

◀Exercises 1–5 refer to Figure 13-15. (Read the coordinates of the points from the figure.)

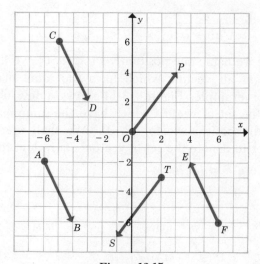

Figure 13-15

1. Find x and y such that $[x, y] = \overrightarrow{OP}$.
2. Name two equivalent vectors.
3. Find the coordinates of R such that $\overrightarrow{OR} = \overrightarrow{CD}$.
4. Find the coordinates of K so that $\overrightarrow{EK} = \overrightarrow{CD}$.
5. Name the equivalence class of vectors that contains $\overrightarrow{AB}$.

◀In Exercises 6–13, the coordinates of a point R are given. If the coordinates of P are $(-3, 5)$, find the coordinates of S so that $\overrightarrow{RS} = \overrightarrow{OP}$.

6. $(-8, 5)$	10. $(0, -13.6)$
7. $(3, -5)$	11. $(-3, -5)$
8. $(11.5, 6.3)$	12. $(\pi, 2\pi)$
9. $(-8, 0)$	13. $(\sqrt{2}, \sqrt{3})$

◀In Exercises 14–21, the coordinates of two points R and S are given in that order. Find x and y so that $[x, y] = \overrightarrow{RS}$.

14. $(1, 2), (5, -7)$ **18.** $(\sqrt{3}, \sqrt{5}), (-5\sqrt{3}, 2\sqrt{5})$

15. $(8, 9), (-3, 4)$ **19.** $(\sqrt{2}, \sqrt{8}), \left(\dfrac{\sqrt{2}}{2}, 2\right)$

16. $(6, 0), (0, -8)$ **20.** $(\sqrt[3]{3}, \sqrt[3]{4}), (\sqrt[3]{81}, \sqrt[3]{32})$

17. $(0, 0), (-\sqrt{2}, \pi)$ **21.** $(\pi, 3\pi), (4\pi, 2\pi)$

22. Prove that $P = P'$ if $\overrightarrow{AB} = \overrightarrow{CP}$ and $\overrightarrow{AB} = \overrightarrow{CP'}$ for fixed points A, B, and C.

23. Prove that $\overrightarrow{AB} = \overrightarrow{DC}$ if $ABCD$ is a parallelogram.

24. Prove that $ABCD$ is a parallelogram if $\overrightarrow{AB} = \overrightarrow{DC}$ and if no three of the points A, B, C, and D are collinear.

25. A translation on the Cartesian plane of the origin to the point $(3, 2)$ is followed by a translation from $(3, 2)$ to $(-1, 6)$. A third translation carries the origin from $(-1, 6)$ to $(-5, 1)$. Represent each translation by a vector. Find a vector which shows the same translation as the sum of the three translations.

26. An airplane is headed southwest and flying at the rate of 500 miles per hour. A wind is blowing from the north at the rate of 50 miles per hour. Represent, by use of vectors, the path of the plane and of the wind. Find a vector that will represent the plane's true course.

27. A boy is rowing directly east across a bay at the rate of 4 miles per hour. The tide in the bay is flowing in a northerly direction at the rate of 6 miles per hour. Find a vector that will represent the boy's true course.

28. Given the parallelogram $ABCD$, list all possible nonzero vectors determined by the four points A, B, C, and D. Specify any vectors that are equivalent.

29. CHALLENGE PROBLEM. In three-space the points A and B have coordinates $(2, -1, 3)$ and $(4, 2, -1)$, respectively. Find the coordinates of P so that $\overrightarrow{OP} = \overrightarrow{AB}$.

30. CHALLENGE PROBLEM. In three-space the points A, B, and C have coordinates $(-3, 4, 5)$, $(2, 1, 4)$, and $(0, 0, 3)$ respectively. Find the point D so that $\overrightarrow{AB} = \overrightarrow{CD}$.

13.3 ADDITION OF VECTORS

We have seen in Chapter 4 that the sum of two complex numbers can be found geometrically by constructing a parallelogram. We define the sum of two vectors so that there will be a similar geometric interpretation.

Definition **13.1**   Let **a** and **b** be any two vectors. Let **a** = $\overrightarrow{AD}$. Choose point B so that **b** = $\overrightarrow{DB}$. Then the sum **a** + **b** = $\overrightarrow{AB}$ as shown in Figure 13-16.

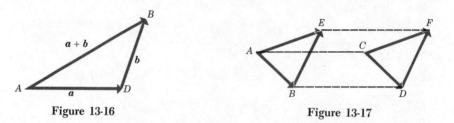

Figure 13-16   Figure 13-17

Figure 13-17 suggests a method for demonstrating that when $\overrightarrow{AB} = \overrightarrow{CD}$ and $\overrightarrow{BE} = \overrightarrow{DF}$, then $\overrightarrow{AB} + \overrightarrow{BE} = \overrightarrow{CD} + \overrightarrow{DF}$.

We call the fact that for all vectors **u**, **v**, **a**, and **b**, if **u** = **v** and **a** = **b**, then **u** + **a** = **v** + **b**, the **substitution property for addition**.

Example 1   Illustrate by a figure the sum $\overrightarrow{OA} + \overrightarrow{OB}$.

Solution: By Definition 13.1, if we choose point E so that $\overrightarrow{AE} = \overrightarrow{OB}$, then $\overrightarrow{OA} + \overrightarrow{OB} = \overrightarrow{OE}$. If $AEBO$ is a parallelogram as shown in Figure 13-18, then $\overrightarrow{AE} = \overrightarrow{OB}$ and hence the sum $\overrightarrow{OA} + \overrightarrow{OB}$ is the vector $\overrightarrow{OE}$ represented by the longer diagonal of the parallelogram.

Example 2   Illustrate by a figure the sum $\overrightarrow{OB} + \overrightarrow{OA}$.

Solution: By Definition 13.1, if we choose point E so that $\overrightarrow{BE} = \overrightarrow{OA}$, then $\overrightarrow{OB} + \overrightarrow{OA} = \overrightarrow{OE}$. Again $\overrightarrow{OA} = \overrightarrow{BE}$ if $AEBO$ is a parallelogram as in Figure 13-19.

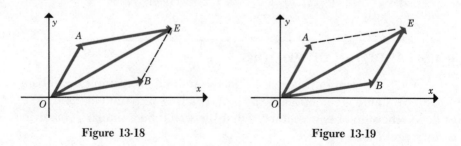

Figure 13-18   Figure 13-19

Examples 1 and 2 indicate that vector addition is commutative and this is indeed true. For all vectors **a** and **b**,

$$\mathbf{a} + \mathbf{b} = \mathbf{b} + \mathbf{a}.$$

By Definition 13.1, we have $\overrightarrow{AD} + \overrightarrow{DD} = \overrightarrow{AD}$. Thus $\overrightarrow{DD}$ serves as the additive identity. However, $\overrightarrow{DD} = \mathbf{0}$. The zero vector, **0**, is the additive identity. For every vector **a**,

$$\mathbf{a} + \mathbf{0} = \mathbf{a}.$$

Example 3   Illustrate by a figure the sum $(\mathbf{a} + \mathbf{b}) + \mathbf{c}$.
Solution: Let $\mathbf{a} + \mathbf{b} = \overrightarrow{AE}$ as in Figure 13-20. Then $\overrightarrow{AE} + \mathbf{c} = \overrightarrow{AF}$. Hence $(\mathbf{a} + \mathbf{b}) + \mathbf{c} = \overrightarrow{AF}$.

Example 4   Illustrate by a figure the sum $\mathbf{a} + (\mathbf{b} + \mathbf{c})$.
Solution: Let $\mathbf{b} + \mathbf{c} = \overrightarrow{BF}$ as in Figure 13-21. Then $\mathbf{a} + \overrightarrow{BF} = \overrightarrow{AF}$. Hence $\mathbf{a} + (\mathbf{b} + \mathbf{c}) = \overrightarrow{AF}$.

As Examples 3 and 4 suggest, vector addition is associative. For all vectors **a**, **b**, and **c**,

$$(\mathbf{a} + \mathbf{b}) + \mathbf{c} = \mathbf{a} + (\mathbf{b} + \mathbf{c}).$$

By Definition 13.1,

$$\overrightarrow{AB} + \overrightarrow{BA} = \overrightarrow{AA} = \mathbf{0} \qquad \text{and} \qquad \overrightarrow{BA} + \overrightarrow{AB} = \overrightarrow{BB} = \mathbf{0}.$$

Thus $\overrightarrow{BA}$ is the additive inverse of $\overrightarrow{AB}$ and $\overrightarrow{AB}$ is the additive inverse of $\overrightarrow{BA}$. Accordingly, we write $\overrightarrow{BA} = -\overrightarrow{AB}$ and $\overrightarrow{AB} = -\overrightarrow{BA}$. Thus for all vectors **a**, there exists a vector $-\mathbf{a}$ such that

$$\mathbf{a} + (-\mathbf{a}) = \mathbf{0}.$$

We call $-\mathbf{a}$ the **additive inverse** of **a**. We are now able to define subtraction of two vectors **a** and **b**.

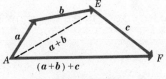

Figure 13-20

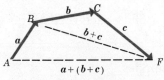

Figure 13-21

> *Definition* **13.2** For all vectors **a** and **b**, **a** − **b** = **a** + (−**b**).

If the sum **a** + **b** is represented by the vector $\overrightarrow{AE}$ as in Figure 13-22, then the difference **a** − **b** = **a** + (−**b**) is represented by $\overrightarrow{AF}$ where $\overrightarrow{DF} = -\overrightarrow{DE}$.

If a parallelogram *OBEA* is formed using the position vectors $\overrightarrow{OB}$ and $\overrightarrow{OA}$ as in Figure 13-23, then $\overrightarrow{OB} + \overrightarrow{OA} = \overrightarrow{OE}$ and $\overrightarrow{OA} - \overrightarrow{OB} = \overrightarrow{BA}$.

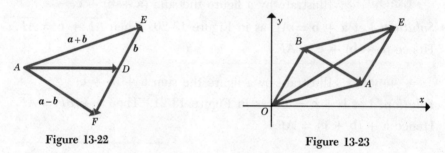

Figure 13-22   Figure 13-23

Example 5 Show that **a** − **b** ≠ **b** − **a** if **a** ≠ **b**.

Solution: Figure 13-24a shows **a** − **b**; Figure 13-24b shows **b** − **a**. Since the directions of the two vectors are different, **a** − **b** ≠ **b** − **a**. Note, however, that **a** − **b** = −(**b** − **a**).

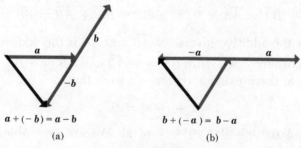

$a + (-b) = a - b$ $b + (-a) = b - a$
(a)   (b)

Figure 13-24

Since each vector **u** possesses an additive inverse −**u**, we can show that

$$\mathbf{u} + \mathbf{v} = \mathbf{u} + \mathbf{w}$$

implies

$$\mathbf{v} = \mathbf{w}.$$

For, if $\mathbf{u} + \mathbf{v} = \mathbf{u} + \mathbf{w}$, then

$$(\mathbf{u} + \mathbf{v}) + (-\mathbf{u}) = (\mathbf{u} + \mathbf{w}) + (-\mathbf{u}),$$
$$[\mathbf{u} + (-\mathbf{u})] + \mathbf{v} = [\mathbf{u} + (-\mathbf{u})] + \mathbf{w},$$
$$\mathbf{0} + \mathbf{v} = \mathbf{0} + \mathbf{w},$$

and

$$\mathbf{v} = \mathbf{w}.$$

This property is called the **cancellation property for vector addition.**

As you have seen, the properties of vector addition are identical with the properties of addition of real numbers. For ready reference we list the properties of vector addition and label them Properties I.

Properties I

For all vectors $\mathbf{u}$, $\mathbf{v}$, and $\mathbf{w}$,

$\mathbf{u} + \mathbf{v}$ is a vector.	Closure
$\mathbf{u} + \mathbf{v} = \mathbf{v} + \mathbf{u}$.	Commutativity
$(\mathbf{u} + \mathbf{v}) + \mathbf{w} = \mathbf{u} + (\mathbf{v} + \mathbf{w})$.	Associativity

There exists a vector $\mathbf{0}$ such that for all vectors $\mathbf{u}$,

$$\mathbf{u} + \mathbf{0} = \mathbf{u}. \qquad\qquad\qquad \text{Additive identity}$$

For any vector $\mathbf{u}$ there exists a vector $-\mathbf{u}$ such that

$$\mathbf{u} + (-\mathbf{u}) = \mathbf{0}. \qquad\qquad\qquad \text{Additive inverse}$$

For all vectors $\mathbf{u}$, $\mathbf{v}$, $\mathbf{w}$, $\mathbf{a}$, and $\mathbf{b}$, if $\mathbf{u} = \mathbf{v}$ and $\mathbf{a} = \mathbf{b}$, then

$$\mathbf{u} + \mathbf{a} = \mathbf{v} + \mathbf{b}; \qquad\qquad \text{Substitution property}$$
if $\mathbf{u} + \mathbf{v} = \mathbf{u} + \mathbf{w}$, then
$$\mathbf{v} = \mathbf{w}. \qquad\qquad\qquad \text{Cancellation property}$$

Since each equivalence class $[x, y]$ has a unique representation $\overrightarrow{OP}$ where the coordinates of P are (x, y), the addition of vectors can be interpreted as the addition of ordered pairs. Let $\mathbf{a} = [x_1, y_1]$ and $\mathbf{b} = [x_2, y_2]$ as in Figure 13-25. Since $\mathbf{a} = \overrightarrow{OP_1}$ and $\mathbf{b} = \overrightarrow{OP_2}$, the coordinates of P_1 are (x_1, y_1) and of P_2 are (x_2, y_2). Let the coordinates of P be (x, y). We have

$$\mathbf{a} + \mathbf{b} = \overrightarrow{OP}.$$

By Definition 13.1, $\overrightarrow{OP_1} = \overrightarrow{P_2P}$, from which it follows that

$$x - x_2 = x_1 - 0$$

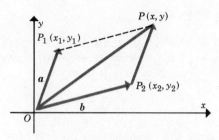

Figure 13-25

and

$$y - y_2 = y_1 - 0.$$

Hence $x = x_1 + x_2$ and $y = y_1 + y_2$. Thus

$$\mathbf{a} + \mathbf{b} = [x_1 + x_2, y_1 + y_2].$$

Do you recall the definition for the sum of two complex numbers? Note the striking similarity between the addition of vectors and the addition of complex numbers.

Example 6 Let $\mathbf{a} = [2, 3]$ and $\mathbf{b} = [-5, 6]$. Find $\mathbf{a} + \mathbf{b}$.

Solution: $\mathbf{a} + \mathbf{b} = [2, 3] + [-5, 6] = [2 + (-5), 3 + 6] = [-3, 9]$.

Vectors can be used to prove theorems of geometry.

Example 7 In the quadrilateral $ABCD$, the diagonals AC and BD intersect at E. If $AE = EC$ and $BE = ED$, prove that $ABCD$ is a parallelogram.

Solution: To prove $ABCD$ in Figure 13-26 is a parallelogram, it is sufficient to prove $\overrightarrow{AB} = \overrightarrow{DC}$ since the statement $\overrightarrow{AB} = \overrightarrow{DC}$ implies both that $AB = DC$ and that $\overleftrightarrow{AB} \parallel \overleftrightarrow{DC}$. We have $\overrightarrow{EB} = \overrightarrow{DE}$ and $\overrightarrow{AE} = \overrightarrow{EC}$. Thus, by the substitution property, we have

$$\overrightarrow{AE} + \overrightarrow{EB} = \overrightarrow{DE} + \overrightarrow{EC}.$$

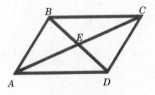

Figure 13-26

However,
$$\vec{AE} + \vec{EB} = \vec{AB}$$

and
$$\vec{DE} + \vec{EC} = \vec{DC}.$$

Hence
$$\vec{AB} = \vec{DC}.$$

EXERCISES 13.3

◀Exercises 1–5 refer to Figure 13-27.

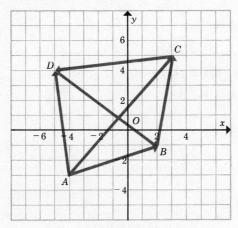

Figure 13-27

1. Name the vector equal to $\vec{AD} + \vec{DB}$.
2. Name the vector equal to $\vec{AB} + \vec{BC}$.
3. Name the vector equal to $\vec{AD} + \vec{DB} + \vec{BC}$.
4. Name the vector equal to $\vec{AD} - \vec{AB}$.
5. Name the vector $\vec{AD} + \vec{DB} + \vec{BC} + \vec{CD}$.

◀**6–10.** In Exercises 6–10, given that A is the point whose coordinates are $(-4, -3)$, B is $(2, -1)$, C is $(3, 5)$, and D is $(-5, 4)$, verify your answers to Exercises 1–5.

◀In Exercises 11–15, express **a** in terms of **u** and **v**.

11.

12.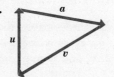

13.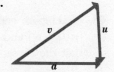

14. 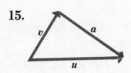   15.

16. Show that the addition of vectors is associative when vectors are defined as ordered pairs.

◄In Exercises 17–21, the four points A, B, C, and D are distinct and no three of them are collinear. Using only two points name an equivalent vector.

17. $\overrightarrow{AB} + \overrightarrow{BC}$

18. $\overrightarrow{AB} - \overrightarrow{AD}$

19. $(\overrightarrow{AD} + \overrightarrow{DB}) - (\overrightarrow{AC} + \overrightarrow{CD})$

20. $(\overrightarrow{AC} - \overrightarrow{BC}) + (\overrightarrow{BD} + \overrightarrow{DA})$

21. $\overrightarrow{AB} + \overrightarrow{BC} + \overrightarrow{CD} + \overrightarrow{DA}$

22. Solve the equation

$$[x, y] + [-6, 7] = [5, -3]$$

for x and y.

23. Solve the equation

$$[-5, 4] + [r, s] = [2, -3]$$

for r and s.

24. Solve the equation

$$[u, v] + \left[3\frac{1}{2}, -2\frac{1}{4}\right] = \left[7\frac{1}{8}, -1\frac{1}{8}\right]$$

for u and v.

25. In the parallelogram $ABCD$, the midpoint of the diagonal $\overline{AC}$ is E. Prove, using vectors, that $BE = ED$.

26. **CHALLENGE PROBLEM.** Show, using vector methods, that the midpoints of the sides of a quadrilateral are vertices of a parallelogram.

27. **CHALLENGE PROBLEM.** Using Figure 13-17, prove that

$$\overrightarrow{AB} + \overrightarrow{BE} = \overrightarrow{CD} + \overrightarrow{DF}.$$

28. **CHALLENGE PROBLEM.** In three-space, the sum of two vectors $\mathbf{u} = [x_1, y_1, z_1]$ and $\mathbf{v} = [x_2, y_2, z_2]$ is $\mathbf{w} = [x_1 + x_2, y_1 + y_2, z_1 + z_2]$. Show that the three vectors $\mathbf{u}$, $\mathbf{v}$, and $\mathbf{w}$ are contained in one plane.

13.4 SCALAR MULTIPLICATION OF VECTORS

According to Definition 13.1, if **a** is a vector, then the sum **a** + **a** is the vector $\overrightarrow{PP_1}$ as in Figure 13-28. If **a** = $\overrightarrow{PD}$, then $PP_1 = 2PD$. Also, the sum **a** + **a** + **a** is the vector $\overrightarrow{PP_2}$ as in Figure 13-29. Since

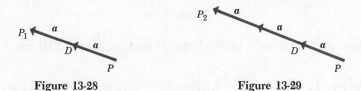

Figure 13-28   Figure 13-29

a = $\overrightarrow{PD}$, then $PP_2 = 3PD$. Our example suggests that we define multiplication of a vector by a number in such a way that

$$1\mathbf{a} = \mathbf{a}, \qquad\qquad (-1)\mathbf{a} = -\mathbf{a},$$
$$2\mathbf{a} = \mathbf{a} + \mathbf{a}, \qquad\qquad (-2)\mathbf{a} = -(\mathbf{a} + \mathbf{a}),$$
$$3\mathbf{a} = \mathbf{a} + \mathbf{a} + \mathbf{a}, \quad \text{and} \quad (-3)(\mathbf{a}) = -(\mathbf{a} + \mathbf{a} + \mathbf{a}).$$

This we now do.

> *Definition* **13.3** If k is a real number and **a** is a vector, then $k\mathbf{a}$ is a vector such that
>
> (1) $k\mathbf{a} = \mathbf{0}$ if $k = 0$ or if $\mathbf{a} = \mathbf{0}$;
> (2) $k\mathbf{a}$ has the same direction as **a** if $k > 0$;
> (3) $k\mathbf{a}$ has the opposite direction as **a** if $k < 0$;
> (4) $|k\mathbf{a}| = |k| \cdot |\mathbf{a}|$.
>
> We call k a **scalar** and $k\mathbf{a}$ a **scalar multiple** of **a**.

Thus, for example,

$$|3\mathbf{a}| = |3||\mathbf{a}| = 3|\mathbf{a}|$$

and

$$|(-3)\mathbf{a}| = |-3||\mathbf{a}| = 3|\mathbf{a}|.$$

Figure 13-30 shows the vector **v** and vectors equivalent to $3\mathbf{v}$, $-1\mathbf{v}$, $\frac{1}{2}\mathbf{v}$, and $-2\mathbf{v}$.

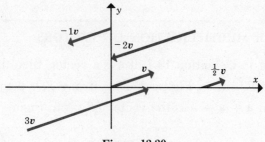

Figure 13-30

Example 1 Show, using directed line segments, that $r(\mathbf{u} + \mathbf{v}) = r\mathbf{u} + r\mathbf{v}$.

Solution: In Figure 13-31, the two triangles are similar since $\dfrac{|\mathbf{u}|}{|r\mathbf{u}|} = \dfrac{|\mathbf{v}|}{|r\mathbf{v}|}$ and $m\angle\theta = m\angle\phi$. Hence

$$\frac{|\mathbf{u}|}{|\mathbf{u} + \mathbf{v}|} = \frac{|r\mathbf{u}|}{|r\mathbf{u} + r\mathbf{v}|}$$

and so

$$|r| \cdot |\mathbf{u} + \mathbf{v}| = |r\mathbf{u} + r\mathbf{v}|.$$

The two vectors $\mathbf{u} + \mathbf{v}$ and $r\mathbf{u} + r\mathbf{v}$ have the same direction. (Why?) Hence $r(\mathbf{u} + \mathbf{v}) = r\mathbf{u} + r\mathbf{v}$.

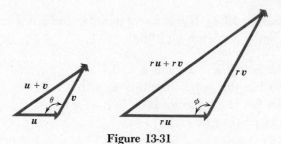

Figure 13-31

As you have seen, any vector has a unique representation as an ordered pair $[x, y]$. Hence, we now define scalar multiplication for vectors as ordered pairs.

Definition 13.4 If $\mathbf{a} = [x, y]$ and k is a real number, then $k\mathbf{a} = [kx, ky]$. We note that

1. $k\mathbf{a} = \mathbf{0}$ if $k = 0$ or $\mathbf{a} = \mathbf{0}$ (that is, if $x = y = 0$);

2. $k\mathbf{a} = k[x, y]$ has the same direction as a if $k > 0$; and

3. $k\mathbf{a} = k[x, y]$ has the opposite direction from $\mathbf{a}$ if $k < 0$.

But also $|k\mathbf{a}| = \sqrt{k^2x^2 + k^2y^2} = |k|\sqrt{x^2 + y^2} = |k||\mathbf{a}|$. Hence

4. $|k\mathbf{a}| = |k||\mathbf{a}|$.

Thus the scalar multiplication introduced in Definition 13.4 is the same as the scalar multiplication introduced in Definition 13.3.

The properties of scalar multiplication which we now list for easy reference and designate as Properties II follow readily from Definition 13.4 and the properties of real numbers. Example 1 presents a proof of one of these properties. Other proofs are asked for in the exercises.

Properties II

For all vectors, **u** and **v,** and for all real numbers, r and s,

$r\mathbf{u}$ is a vector	Closure
$\begin{aligned}0\mathbf{u} &= \mathbf{0}.\\ r\mathbf{0} &= \mathbf{0}.\end{aligned}\Big\}$	Zero properties
$1\mathbf{u} = \mathbf{u}.$	Multiplicative identity
$r(s\mathbf{u}) = (rs)\mathbf{u}.$	Associativity
$\begin{aligned}r(\mathbf{u} + \mathbf{v}) &= r\mathbf{u} + r\mathbf{v}.\\ (r + s)\mathbf{u} &= r\mathbf{u} + s\mathbf{u}.\end{aligned}\Big\}$	Distributivity

Example 2 Prove that $r(\mathbf{u} + \mathbf{v}) = r\mathbf{u} + r\mathbf{v}$ using ordered pair notation.

Solution: Let $\mathbf{u} = [x_1, y_1]$ and $\mathbf{y} = [x_2, y_2]$. Then

$$\begin{aligned}
r\mathbf{u} + r\mathbf{v} &= r[x_1, y_1] + r[x_2, y_2]\\
&= [rx_1, ry_1] + [rx_2, ry_2]\\
&= [rx_1 + rx_2, ry_1 + ry_2]\\
&= [r(x_1 + x_2), r(y_1 + y_2)]\\
&= r[x_1 + x_2, y_1 + y_2]\\
&= r([x_1, y_1] + [x_2, y_2])\\
&= r(\mathbf{u} + \mathbf{v}).
\end{aligned}$$

Example 3   Let $\mathbf{u} = [-2, 3]$, $\mathbf{v} = [5, -2]$, $r = 2$, and $s = -5$. Show that

$$r[s(\mathbf{u} + \mathbf{v})] = r(s\mathbf{u}) + r(s\mathbf{v}).$$

Solution: We have $s(\mathbf{u} + \mathbf{v}) = s\mathbf{u} + s\mathbf{v}$. Hence

$$\begin{aligned} s(\mathbf{u} + \mathbf{v}) &= -5([-2, 3] + [5, -2]) \\ &= (-5)[-2, 3] + (-5)[5, -2] \\ &= [10, -15] + [-25, 10]. \end{aligned}$$

Then

$$\begin{aligned} r[s(\mathbf{u} + \mathbf{v})] &= 2([10, -15] + [-25, 10]) \\ &= 2[10, -15] + 2[-25, 10] \\ &= [20, -30] + [-50, 20]. \end{aligned}$$

Also $r(s\mathbf{u}) = (rs)\mathbf{u}$ and $r(s\mathbf{v}) = (rs)\mathbf{v}$. Hence

$$\begin{aligned} r(s\mathbf{u}) + r(s\mathbf{v}) &= 2\{(-5)[-2, 3]\} + 2\{(-5)[5, -2]\} \\ &= (2)(-5)[-2, 3] + 2(-5)[5, -2] \\ &= (-10)[-2, 3] + (-10)[5, -2] \\ &= [20, -30] + [-50, 20]. \end{aligned}$$

EXERCISES 13.4

◀Exercises 1–5 refer to Figure 13-32.

1. Find $|\mathbf{a}|$.   **3.** Find k_2.   **5.** Find $|k_3\mathbf{a}|$.

2. Find k_1.   **4.** Find k_3.

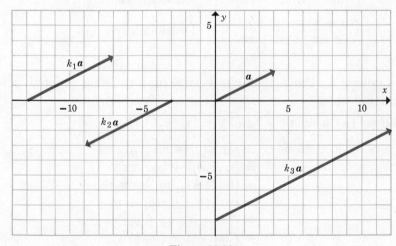

Figure 13-32

◀ In Exercises 6–11, let $\mathbf{a} = [-2, 3]$.

   6. Find $6\mathbf{a}$.   **9.** Find $\sqrt{2}\mathbf{a}$.

   7. Find $-3\mathbf{a}$.   **10.** Find $-5\mathbf{a} + 6\mathbf{a}$.

   8. Find $-\dfrac{1}{2}\mathbf{a}$.   **11.** Find $-6\mathbf{a} - \dfrac{1}{2}\mathbf{a} + 4\dfrac{1}{2}\mathbf{a}$.

◀ In Exercises 12–15, solve for k.

   12. $[6, -1] = k[0.6, -0.1]$ **14.** $[\sqrt{2}, 4] = k[2, \sqrt{32}\,]$

   13. $k\left[\dfrac{3}{4}, \dfrac{9}{10}\right] = \left[\dfrac{1}{2}, \dfrac{3}{5}\right]$ **15.** $k[\pi, 4\pi] = [1, 4]$

◀ In Exercises 16–23, let $\mathbf{u} = [-2, \frac{1}{2}]$ and $\mathbf{v} = [\frac{5}{2}, 7]$. Find the position vector that is equivalent to the indicated vector.

   16. $3\mathbf{u} - 4\mathbf{v}$ **20.** $\pi(\mathbf{u} - \mathbf{v})$

   17. $-1\mathbf{u} + 3\mathbf{v}$ **21.** $0(\mathbf{u} - \mathbf{v})$

   18. $-7(\mathbf{u} - 2\mathbf{v})$ **22.** $\sqrt{2}(\sqrt{3}\mathbf{u} - \sqrt{12}\mathbf{v})$

   19. $\dfrac{1}{2}(4\mathbf{u} - 5\mathbf{v})$ **23.** $\sqrt{5}(\sqrt{15}\mathbf{u} - \sqrt{60}\mathbf{v})$

   24. Prove that $r(s\mathbf{v}) = (rs)\mathbf{v}$ for all vectors $\mathbf{v}$ and real numbers r and s. (See Example 2.)

   25. Prove that $(r + s)\mathbf{u} = r\mathbf{u} + s\mathbf{u}$ for all vectors $\mathbf{u}$ and real numbers r and s.

   26. Find numbers x and y such that

$$[2, 4] = x[4, 2] + y[5, 1].$$

   27. Show that $|\mathbf{u} + \mathbf{v}| = |\mathbf{u}| + |\mathbf{v}|$ if and only if $\mathbf{u}$ and $\mathbf{v}$ have the same direction.

   28. Find numbers x and y such that

$$[13, 9] = x[-1, 3] + y\left[1\frac{1}{2}, 3\frac{1}{2}\right].$$

   29. Solve the following vector equation for r and s:

$$r\left[-1\frac{1}{2}, 2\right] + s\left[2\frac{1}{2}, 3\right] = \left[10\frac{1}{2}, 5\right].$$

◀ Exercises 30–34 refer to Figure 13-33. *ABCD* is a parallelogram. *P, Q, R,* and *S* are the midpoints of the sides. In each exercise, find a vector of the form $r\overrightarrow{OQ} + s\overrightarrow{OP}$ equivalent to the given vector.

   30. $\overrightarrow{OB}$ **33.** $\overrightarrow{CA}$

   31. $\overrightarrow{OC}$ **34.** $\overrightarrow{DB}$

   32. $\overrightarrow{OA}$

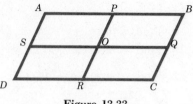

Figure 13-33

35. **CHALLENGE PROBLEM.** Prove, using vectors, that the medians of a triangle meet at a point that divides each median into segments whose lengths have the ratio 2 : 1.

36. **CHALLENGE PROBLEM.** In three-space, the coordinates of points A and B are $(3, 4, 5)$ and $(-2, 4, 8)$. Find the coordinates of point P such that $\overrightarrow{OP} = 3\overrightarrow{AB}$.

13.5 PARALLEL AND PERPENDICULAR VECTORS

If the coordinates of the four points A, B, C, and D are $(-5, 1)$, $(-1, 3)$, $(3, -2)$, and $(5, -1)$, respectively, as in Figure 13-34, then the lines $\overleftrightarrow{AB}$ and $\overleftrightarrow{CD}$, containing the directed segments $\overrightarrow{AB}$ and $\overrightarrow{CD}$, are parallel. Hence, we call $\overrightarrow{AB}$ and $\overrightarrow{CD}$ parallel vectors. Note that $\overrightarrow{AB} = \overrightarrow{OP}$ where P is the point $(4, 2)$ and $\overrightarrow{CD} = \overrightarrow{OP'}$ where P' is the point $(2, 1)$. Thus $\overrightarrow{OP} = 2\overrightarrow{OP'}$. We use this relationship to define parallel vectors.

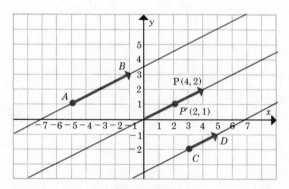

Figure 13-34

Definition 13.5 Let $\mathbf{a} = \overrightarrow{OP} \neq \mathbf{0}$ and $\mathbf{b} = \overrightarrow{OP'} \neq \mathbf{0}$. Then $\mathbf{a}$ is **parallel** to $\mathbf{b}$ if and only if $\overrightarrow{OP} = k\overrightarrow{OP'}$ for some scalar k. By agreement, $\mathbf{0}$ is considered as parallel to every vector.

Example 1 Show that $\mathbf{a} = [5, -10]$ is parallel to $\mathbf{b} = [-1, 2]$.
Solution: We seek a number k such that $\mathbf{a} = k\mathbf{b}$. If $\mathbf{a} = k\mathbf{b}$, the magnitudes of the two equivalent vectors $\mathbf{a}$ and $k\mathbf{b}$ are equal. Hence $|k| = \dfrac{|\mathbf{a}|}{|\mathbf{b}|}$. Thus

$$|k| = \frac{\sqrt{125}}{\sqrt{5}} = \sqrt{25} = 5.$$

We have $\mathbf{a} = -5\mathbf{b}$ (or $[5, -10] = -5[-1, 2]$) and hence $\mathbf{a}$ and $\mathbf{b}$ are parallel.

Example 2 If the coordinates of P and P' are $(3, 4)$ and $(9, 12)$, respectively, show that $\overrightarrow{OP}$ and $\overrightarrow{OP'}$ are parallel. Find r such that $r\overrightarrow{OP} = \overrightarrow{OP'}$.
Solution: If $r\overrightarrow{OP} = \overrightarrow{OP'}$ then $|r| = \dfrac{|OP'|}{|OP|}$. Using the distance formula, we have

$$|r| = \frac{\sqrt{225}}{\sqrt{25}} = 3.$$

We also have $3[3, 4] = [9, 12]$; hence $3\overrightarrow{OP} = \overrightarrow{OP'}$ and the vectors $\overrightarrow{OP}$ and $\overrightarrow{OP'}$ are parallel.

Since it is often useful to refer to the angle between two vectors, we make the following definition.

Definition 13.6 The angle between any two nonzero vectors $\overrightarrow{AB}$ and $\overrightarrow{CD}$ is the angle θ $(0 \leq m\angle\theta \leq 180)$ whose sides are the rays $\overrightarrow{OP}$ and $\overrightarrow{OP'}$ where $\overrightarrow{OP} = \overrightarrow{AB}$ and $\overrightarrow{OP'} = \overrightarrow{CD}$.

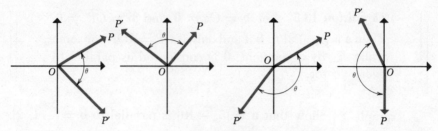

Figure 13-35

Note that, as indicated in Figure 13-35, θ is *not* a directed angle.

If the two directed segments $\overrightarrow{AB}$ and $\overrightarrow{CD}$ are contained in two lines $\overleftrightarrow{AB}$ and $\overleftrightarrow{CD}$ that are perpendicular as in Figure 13-36, we call $\overrightarrow{AB}$ and $\overrightarrow{CD}$ **perpendicular** vectors.

If $\overrightarrow{AB}$ is perpendicular to $\overrightarrow{CD}$, then the position vectors $\overrightarrow{OP}$ and $\overrightarrow{OR}$ such that $\overrightarrow{OP} = \overrightarrow{AB}$ and $\overrightarrow{OR} = \overrightarrow{CD}$ lie on rays that form a right angle as in Figure 13-37, and $m \angle \theta = 90$. The vector $\overrightarrow{RP}$ completes a right triangle. Accordingly, by the Pythagorean Theorem, we have

(1) $$|\overrightarrow{RP}|^2 = |\overrightarrow{OP}|^2 + |\overrightarrow{OR}|^2.$$

If the coordinates of P are (x_1, y_1) and of R are (x_2, y_2), we have

(2) $$(x_1 - x_2)^2 + (y_1 - y_2)^2 = (x_1{}^2 + y_1{}^2) + (x_2{}^2 + y_2{}^2).$$

The left-hand member of (2) may be rewritten as

$$(x_1{}^2 + y_1{}^2) + (x_2{}^2 + y_2{}^2) - 2(x_1 x_2 + y_1 y_2).$$

Hence

(3) $(x_1{}^2 + y_1{}^2) + (x_2{}^2 + y_2{}^2) - 2(x_1 x_2 + y_1 y_2)$
$$= (x_1{}^2 + y_1{}^2) + (x_2{}^2 + y_2{}^2)$$

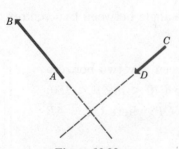

Figure 13-36

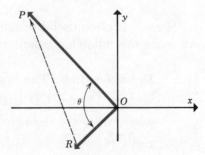

Figure 13-37

and thus

(4) $$x_1x_2 + y_1y_2 = 0.$$

Equation (1) implies Equation (4). Conversely, from (4) we can obtain Equation (2) and hence Equation (1). Thus we conclude that $\overrightarrow{OR}$ and $\overrightarrow{OP}$ are perpendicular if and only if $x_1x_2 + y_1y_2 = 0$. This number $x_1x_2 + y_1y_2$ is of special significance.

> **Definition 13.7**   If $\mathbf{a} = [x_1,\, y_1]$ and $\mathbf{b} = [x_2,\, y_2]$, then the **inner product** of the two vectors, denoted $\mathbf{a} \cdot \mathbf{b}$, is the number
>
> $$\mathbf{a} \cdot \mathbf{b} = x_1x_2 + y_1y_2.$$

Note that the inner product, $x_1x_2 + y_1y_2$, which is also called the **dot product**, is a number and not a vector, and is defined here only for vectors expressed as ordered pairs.

Using Definition 13.7, we can make the following assertion concerning perpendicular vectors.

Two vectors $\mathbf{a} = [x_1,\, y_1]$ and $\mathbf{b} = [x_2,\, y_2]$ are perpendicular if and only if $\mathbf{a} \cdot \mathbf{b} = 0$. We write $\mathbf{a} \perp \mathbf{b}$.

For convenience the zero vector is considered to be perpendicular to every other vector. Perpendicular vectors are sometimes referred to as **orthogonal** vectors.

Example 3   Show that the vectors $\overrightarrow{OP}$ and $\overrightarrow{OR}$ are perpendicular if P is the point $(4, 6)$ and R is the point $(-3, 2)$.

Solution: We have $\overrightarrow{OP} = [4, 6]$ and $\overrightarrow{OR} = [-3, 2]$. Then

$$x_1x_2 + y_1y_2 = 4(-3) + 6(2) = 0.$$

Hence the vectors are perpendicular.

Example 4   Show that the vectors $\mathbf{a} = [-6, 4]$ and $\mathbf{b} = [-3.5, -5]$ are not orthogonal.

Solution: We have

$$\mathbf{a} \cdot \mathbf{b} = -6(-3.5) + 4(-5) = 21 - 20 = 1 \neq 0.$$

Hence $\mathbf{a}$ and $\mathbf{b}$ are not orthogonal.

The properties of the inner product are similar in many ways to the properties of multiplication of real numbers. For example, let $\mathbf{a} = [x_1, y_1]$ and $\mathbf{b} = [x_2, y_2]$. Then

$$\mathbf{a} \cdot \mathbf{b} = x_1 x_2 + y_1 y_2.$$

But also

$$\mathbf{b} \cdot \mathbf{a} = x_2 x_1 + y_2 y_1 = x_1 x_2 + y_1 y_2.$$

Hence

$$\mathbf{a} \cdot \mathbf{b} = \mathbf{b} \cdot \mathbf{a}.$$

Again, suppose $\mathbf{a} = [x_1, y_1]$, $\mathbf{b} = [x_2, y_2]$, and $\mathbf{c} = [x_3, y_3]$. Then

$$\mathbf{a} \cdot (\mathbf{b} + \mathbf{c}) = x_1(x_2 + x_3) + y_1(y_2 + y_3)$$
$$= x_1 x_2 + x_1 x_3 + y_1 y_2 + y_1 y_3$$

since

$$\mathbf{b} + \mathbf{c} = [x_2 + x_3, y_2 + y_3].$$

But also

$$(\mathbf{a} \cdot \mathbf{b}) + (\mathbf{a} \cdot \mathbf{c}) = (x_1 x_2 + y_1 y_2) + (x_1 x_3 + y_1 y_3)$$
$$= x_1 x_2 + x_1 x_3 + y_1 y_2 + y_1 y_3.$$

Hence

$$\mathbf{a} \cdot (\mathbf{b} + \mathbf{c}) = (\mathbf{a} \cdot \mathbf{b}) + (\mathbf{a} \cdot \mathbf{c}).$$

Other properties of the inner product are investigated in the Exercises.

Example 5 The coordinates of points A, B, and C are $(-1, -1)$, $(2, 2)$, and $(-2, 6)$, respectively, as shown in Figure 13-38. Let $\overrightarrow{AB} = \mathbf{a}$, $\overrightarrow{BC} = \mathbf{b}$, and $\overrightarrow{CA} = \mathbf{c}$. Find (1) $\mathbf{a} \cdot \mathbf{b}$, (2) $\mathbf{b} \cdot \mathbf{a}$, (3) $\mathbf{a} \cdot (\mathbf{b} + \mathbf{c})$, (4) $\mathbf{a} \cdot \mathbf{c}$, (5) $2\mathbf{a} \cdot \mathbf{c}$, and (6) $(2\mathbf{a}) \cdot \mathbf{c}$.

Solution: $\mathbf{a} = [2 + 1,\ 2 + 1] = [3,\ 3]$, $\mathbf{b} = [-2 - 2,\ 6 - 2] = [-4, 4]$, and $\mathbf{c} = [-1 + 2, -1 - 6] = [1, -7]$. Thus

1. $\mathbf{a} \cdot \mathbf{b} = (3)(-4) + (3)(4) = -12 + 12 = 0$. (Note that $\mathbf{a} \perp \mathbf{b}$.)
2. $\mathbf{b} \cdot \mathbf{a} = (-4)(3) + (4)(3) = -12 + 12 = 0$. (Note that $\mathbf{a} \cdot \mathbf{b} = \mathbf{b} \cdot \mathbf{a}$.)
3. $\mathbf{b} + \mathbf{c} = [-4 + 1, 4 + (-7)] = [-3, -3]$.

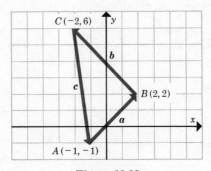

Figure 13-38

Hence $\mathbf{a} \cdot (\mathbf{b} + \mathbf{c}) = 3(-3) + 3(-3) = -18$.

4. $\mathbf{a} \cdot \mathbf{c} = 3(1) + 3(-7) = -18$. Thus $(\mathbf{a} \cdot \mathbf{b}) + (\mathbf{b} \cdot \mathbf{c}) = 0 + (-18) = -18$. (Note that $(\mathbf{a} \cdot \mathbf{b}) + (\mathbf{a} \cdot \mathbf{c}) = \mathbf{a} \cdot (\mathbf{b} + \mathbf{c})$.)

5. $2(\mathbf{a} \cdot \mathbf{c}) = 2(-18) = -36$.

6. $2\mathbf{a} = 2[3, 3] = [6, 6]$. Thus $(2\mathbf{a}) \cdot \mathbf{c} = 6(1) + 6(-7) = -36$. (Note that $2(\mathbf{a} \cdot \mathbf{c}) = (2\mathbf{a}) \cdot \mathbf{c}$.)

For convenient reference we list the basic properties of the inner product.

Properties III

If $\mathbf{u}$, $\mathbf{v}$, and $\mathbf{w}$ are vectors and k is a scalar, then

1. $\mathbf{u} \cdot \mathbf{v} = \mathbf{v} \cdot \mathbf{u}$.
2. $\mathbf{w} \cdot (\mathbf{u} + \mathbf{v}) = (\mathbf{w} \cdot \mathbf{u}) + (\mathbf{w} \cdot \mathbf{v})$.
3. $(\mathbf{u} + \mathbf{v}) \cdot \mathbf{w} = (\mathbf{u} \cdot \mathbf{w}) + (\mathbf{v} \cdot \mathbf{w})$.
4. $(k \cdot \mathbf{u}) \cdot \mathbf{v} = k(\mathbf{u} \cdot \mathbf{v})$.
5. $\mathbf{u} \cdot (k\mathbf{v}) = k(\mathbf{u} \cdot \mathbf{v})$.

EXERCISES 13.5

◀Exercises 1–5 refer to Figure 13-39.

1. Name a pair of perpendicular vectors. Show that they are perpendicular.
2. Name a pair of parallel vectors. Show that they are parallel.

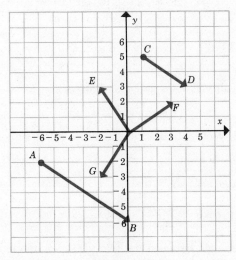

Figure 13-39

3. Name a pair of vectors that are neither parallel nor perpendicular. Justify your answer.

4. Name the position vector perpendicular to $\overrightarrow{CD}$.

5. Find the coordinates of P so that $|\overrightarrow{EP}| = |\overrightarrow{AB}|$ and $\overrightarrow{EP}$ is perpendicular to $\overrightarrow{CD}$.

◀ In Exercises 6–15, determine if the two given vectors are parallel, perpendicular, or neither.

6. [2, 4], [4, 8]   11. [5, 8], [−5, 8]

7. [1, 0], [0, 1]   12. [t, 2t], [−2t, t]

8. $\left[\dfrac{\sqrt{2}}{2}, \dfrac{\sqrt{2}}{2}\right], [-\sqrt{2}, -\sqrt{2}]$ 13. [a + b, c], [a, c − b]

9. [a, b], [2a, $\sqrt{2b}$]   14. [0, 0], [x, y]

10. [0, π], [π, 0]   15. [5, −8], [0.5, −0.8]

◀ In Exercises 16–21, find the value of k such that the two given vectors are parallel.

16. [3, 4], [10, k]   19. [$\sqrt{2}$, 5], [4, k]

17. [π, 2π], [k, 2]   20. [$-\sqrt{3}$, 1], [0, k]

18. $\left[\dfrac{1}{2}, \dfrac{2}{3}\right], \left[\dfrac{5}{8}, k\right]$ 21. [$-\sqrt{5}$, 5], [5, k]

◀ In Exercises 22–27, find the value of k such that the two given vectors are perpendicular.

22. [3, 4], [10, k]   25. [$\sqrt{2}$, 5], [4, k]

23. [π, 2π], [k, 2]   26. [$-\sqrt{3}$, 1], [0, k]

24. $\left[\dfrac{1}{2}, \dfrac{2}{3}\right], \left[\dfrac{5}{8}, k\right]$ 27. [$-\sqrt{5}$, 5], [5, k]

28. Show that for all vectors **u** and **v** and any scalar k

$$(k \cdot \mathbf{u}) \cdot \mathbf{v} = k(\mathbf{u} \cdot \mathbf{v}).$$

29. Show that two vectors **u** and **v** are parallel if and only if

$$\mathbf{u} \cdot \mathbf{v} = |\mathbf{u}||\mathbf{v}| \quad \text{or} \quad \mathbf{u} \cdot \mathbf{v} = -|\mathbf{u}||\mathbf{v}|.$$

30. **CHALLENGE PROBLEM.** Show that

$$\cos \theta = \frac{\mathbf{a} \cdot \mathbf{b}}{|\mathbf{a}||\mathbf{b}|} \quad \text{for } m\angle\theta \neq 90°,$$

where θ is the angle between the two vectors **a** and **b**.

13.6 BASIS VECTORS

The "parallelogram law" for the addition of vectors certainly applies to rectangles. Hence if $\overrightarrow{OR}$ lies along the nonnegative ray of the x-axis and $\overrightarrow{OS}$ lies along the nonnegative ray of the y-axis, the sum $\overrightarrow{OR} + \overrightarrow{OS}$ will be the diagonal of a rectangle as shown in Figure 13-40. We have

$$\overrightarrow{OR} + \overrightarrow{OS} = \overrightarrow{OP}.$$

If $\overrightarrow{OR'}$ also lies along the nonnegative ray of the x-axis and $|\overrightarrow{OR'}| = 1$, then, for some real number r, $r\overrightarrow{OR'} = \overrightarrow{OR}$. Likewise if $\overrightarrow{OS'}$ lies along the nonnegative ray of the y-axis and $|\overrightarrow{OS'}| = 1$, then for some real number s, $s\overrightarrow{OS'} = \overrightarrow{OS}$. Thus we have

$$r\overrightarrow{OR'} + s\overrightarrow{OS'} = \overrightarrow{OP}.$$

The vector $\overrightarrow{OP}$ is called a **linear combination** of the vectors $\overrightarrow{OR'}$ and $\overrightarrow{OS'}$. (Do you recall that we mentioned linear combinations when we were considering systems of equations in Chapter 8?) The vectors $\overrightarrow{OR'}$ and $\overrightarrow{OS'}$ are called **unit vectors** since their magnitude is 1. (See Figure 13-41.)

Conversely, any vector $\overrightarrow{OP}$ in the plane can be expressed as a linear combination of any two given nonzero vectors; one, $\overrightarrow{OR}$, along the nonnegative ray of the x-axis and the other, $\overrightarrow{OS}$, along the nonnegative ray of the y-axis. We demonstrate this in the following manner. A line parallel to the y-axis and through P will intersect the x-axis at

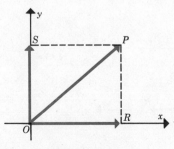

Figure 13-40

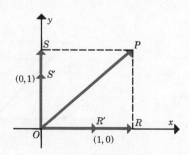

Figure 13-41

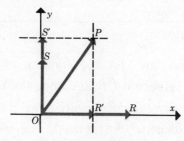

Figure 13-42

some point R' as in Figure 13-42. A line parallel to the x-axis and through P will intersect the y-axis at some point S'. For some real number r such that $|r| = \dfrac{|\overrightarrow{OR'}|}{|\overrightarrow{OR}|}$ we have

$$|r|\overrightarrow{OR'} = \overrightarrow{OR}$$

if $\overrightarrow{OR'}$ is in the same direction as $\overrightarrow{OR}$ and

$$-|r|\overrightarrow{OR} = \overrightarrow{OR'}$$

if $\overrightarrow{OR'}$ is in the opposite direction from $\overrightarrow{OR}$. Hence $r\overrightarrow{OR} = \overrightarrow{OR'}$. Likewise for some real number s, $s\overrightarrow{OS} = \overrightarrow{OS'}$. Thus

$$\overrightarrow{OP} = r\overrightarrow{OR} + s\overrightarrow{OS}$$

and hence the vector $\overrightarrow{OP}$ is expressed as a linear combination of $\overrightarrow{OR}$ and $\overrightarrow{OS}$.

The fixed vectors $\overrightarrow{OR}$ and $\overrightarrow{OS}$ are said to form a **basis** for the vector $\overrightarrow{OP}$. The same two vectors, $\overrightarrow{OR}$ and $\overrightarrow{OS}$, are a basis for all vectors in the plane as suggested by Figure 13-43. Here we see that

$$\overrightarrow{OP'} = r'\overrightarrow{OR} + s'\overrightarrow{OS} \qquad (r' < 0).$$

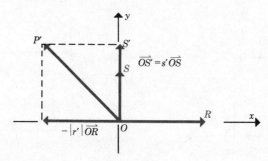

Figure 13-43

In general, we say that the vectors **u** and **v** form a basis if, for any vector **w**, there exist real numbers r and s such that $\mathbf{w} = r\mathbf{u} + s\mathbf{v}$.

Although, as suggested by Figure 13-44, any two nonzero vectors that are not parallel may be chosen as a basis, the usual and most convenient pair consists of the unit vectors, $\mathbf{i} = [1, 0]$ and $\mathbf{j} = [0, 1]$, as in Figure 13-45. Any vector **v**, whose magnitude $|\mathbf{v}| = 1$, is called a unit vector although only **i** and **j** have special symbols. Any vector $\mathbf{v} = [x, y]$ can be readily expressed as a linear combination of **i** and **j**. We have, in fact,

$$\mathbf{v} = x\mathbf{i} + y\mathbf{j}.$$

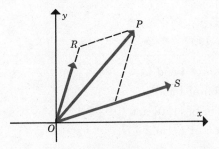

Figure 13-44

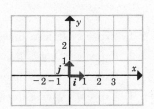

Figure 13-45

Example 1   Express $\mathbf{v} = [-2, 4]$ as a linear combination of **i** and **j**.

Solution:

$$\mathbf{v} = [-2, 4] = -2[1, 0] + 4[0, 1] = -2\mathbf{i} + 4\mathbf{j}.$$

In Figure 13-46, $\overrightarrow{OR} = -2\mathbf{i}$, $\overrightarrow{OS} = 4\mathbf{j}$, and $\overrightarrow{OR} + \overrightarrow{OS} = \overrightarrow{OP} = \mathbf{v}$.

The **components** of a vector **v** with respect to the basis vectors **i** and **j** are the numbers x and y such that

$$\mathbf{v} = x\mathbf{i} + y\mathbf{j}.$$

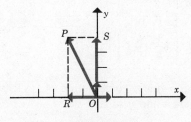

Figure 13-46

EXERCISES 13.6

◀Exercises 1–5 refer to Figure 13-47.

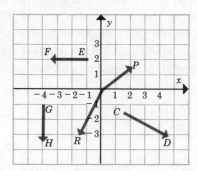

Figure 13-47

1. Express $\overrightarrow{OP}$ as a linear combination of **i** and **j**.
2. Express $\overrightarrow{CD}$ as a linear combination of **i** and **j**.
3. Express $\overrightarrow{OR}$ as a linear combination of [1, 0] and [0, −2].
4. Express $\overrightarrow{OP}$ as a linear combination of $\overrightarrow{EF}$ and $\overrightarrow{GH}$.
5. Express $-5\overrightarrow{CD}$ as a linear combination of **i** and **j**.

◀In Exercises 6–11, sketch each vector and state its magnitude.

6. $4\mathbf{i}$ 9. $-3\mathbf{i} - \dfrac{1}{2}\mathbf{j}$

7. $\mathbf{i} + \mathbf{j}$ 10. $\sqrt{2}\mathbf{i} - \mathbf{j}$

8. $3\mathbf{i} - 4\mathbf{j}$ 11. $\sqrt{2}\mathbf{i} + \sqrt{3}\mathbf{j}$

12. Find the real numbers x and y such that
$$2\mathbf{i} - \mathbf{j} = x(\mathbf{i} + \mathbf{j}) - y(4\mathbf{i} + 6\mathbf{j}).$$

◀In Exercises 13–18, express each vector as a linear combination of **i** and **j**.

13. $[-7, 3]$ 16. $[0, \pi]$
14. $[8, -5]$ 17. $[-6, 14]$
15. $[\sqrt{2}, 0]$ 18. $\left[4\dfrac{1}{2}, -1\dfrac{3}{4}\right]$

19. Express $[5, -4]$ as a linear combination of $[-1, 2]$ and $[-2, 2.5]$, that is, find the numbers x and y so that
$$[5, -4] = x[-1, 2] + y[-2, 2.5].$$

20. Express $\left[-\dfrac{7}{6}, -1\right]$ as a linear combination of $[2, 6]$ and $[4, 8]$.

21. Show that

$$|\mathbf{a} + \mathbf{b}|^2 = |\mathbf{a}|^2 + |\mathbf{b}|^2 + 2\mathbf{a} \cdot \mathbf{b}.$$

22. CHALLENGE PROBLEM. In three-space the coordinates of A and B are $(6, -4, 3)$ and $(4, 8, 1)$. Express $\overrightarrow{AB}$ as a linear combination of $\mathbf{i} = (1, 0, 0]$, $\mathbf{j} = [0, 1, 0]$, and $\mathbf{k} = [0, 0, 1]$.

13.7 VECTOR SPACES

The vectors that we have been discussing, together with the operations of addition and scalar multiplication, form a system called a vector space.

> *Definition* **13.8** A nonempty set V is called a **vector space** over the set of real numbers if there is a binary operation of addition of elements of V satisfying Properties I of Section 13.3. and an operation of scalar multiplication of elements of V by a real number satisfying Properties II of Section 13.4.

We have studied two sets which, with the operations of addition and scalar multiplication by real numbers, form vector spaces. The set D of directed line segments with initial point at the origin of the plane, with addition as defined by Definition 13.1 and scalar multiplication as defined by Definition 13.3, forms a vector space. Also the set G of ordered pairs $[x, y]$ together with addition defined as

$$[x_1, y_1] + [x_2, y_2] = [x_1 + x_2, y_1 + y_2]$$

and scalar multiplication defined as

$$k[x_1, y_1] = [kx_1, ky_1]$$

forms a vector space.

Since the elements of set D, together with the operations, form a vector space, we refer to the elements of D, namely the directed line segments, as vectors. In a similar manner, we refer to the ordered pairs $[a, b]$ as vectors. The elements of any set that serves as a model of a vector space are properly called vectors.

A completely trivial example of a vector space is the set $O = \{0\}$. Do you see that all the properties of vector addition and scalar multiplication are satisfied? Note that O is a subset of D.

A not so trivial but still very simple example of a vector space is the set P of all vectors parallel to a given vector $\overrightarrow{AB}$. Each vector in the set P has the form $k\overrightarrow{AB}$ for some $k \in R$. Note that P is also a subset of D.

The vector spaces O and P are called *subspaces* of D.

> **Definition 13.9**   A nonempty subset S of V is a **subspace** of V provided that the sum of any two elements of S is an element of S and that the scalar product of any element of S and a real number is also an element of S.

Example 1   Show that the set F of all vectors $[0, y]$ whose first coordinate is zero is a subspace of the vector space G of vectors $[x, y]$.

Solution: We need to show that, for all $a, b \in R$, the sum $[0, a] + [0, b]$ and the scalar product $k[0, a]$ are elements of F. Since

$$[0, a] + [0, b] = [0, a + b]$$

and

$$k[0, a] = [0, ka]$$

where $a + b \in R$ and $ka \in R$, we see that F satisfies Definition 13.9 and therefore is a subspace of G.

We have seen that any vector $[x, y]$ in the vector space G can be expressed as a linear combination of the two unit vectors **i** and **j**. Accordingly, the basis vectors **i** and **j** are said to **span** the vector space G. The minimum number of vectors that span a vector space is called the **dimension** of the particular vector space. Thus the dimension of G is 2.

Example 2   Find the dimension of the vector space F of Example 1.

Solution: Any vector $[0, y]$ that is an element of F can be expressed as a linear combination of the vector $[0, 1]$ since

$$[0, y] = y[0, 1] \qquad \text{for all } y \in R.$$

Therefore the one vector [0, 1] spans F and hence the dimension of F is 1.

We have confined our attention largely to two-dimensional vector spaces. The concepts, however, can be extended readily to vector spaces of any arbitrary dimension. For example, we can speak of a vector $[x_1, x_2, x_3]$ defined by three numbers, a vector $[x_1, x_2, x_3, x_4]$ defined by four numbers, or even of a vector $[x_1, x_2, x_3, \ldots, x_n]$ defined by n numbers. If $n > 3$, we cannot visualize these objects or represent them geometrically, but we can still consider them algebraically.

We have been exploring the beginnings of a branch of mathematics called vector analysis which is an important tool of modern mathematics. About 1900 the development of tensor analysis, an extension of vector analysis, provided the tools needed by Einstein in his work toward the theory of relativity.

EXERCISES 13.7

1. Express the vector [2, −8] as a linear combination of [4, 2] and [2, 4]. Show the addition geometrically.
2. Express the vector [9, −14] as a linear combination of [4, 2] and [−2, 4]. Show the addition geometrically.
3. Show that the vector [2, 4] cannot be expressed as a linear combination of [4, 2] and [−2, −1].
4. Show that the vector [5, 6] cannot be expressed as a linear combination of [6, 2] and [9, 3].
5. The two vectors [4, 2] and [−2, −1] are called **linearly dependent** vectors since $2[−2, −1] + 1[4, 2] = [0, 0]$. Give another pair of linearly dependent vectors.

◀In Exercises 6–11, show that each set, together with the operations of addition and multiplication by a real number as usually defined for the set, is a vector space over the real numbers.

6. The set A of all vectors [x, 0] for $x \in R$.
7. The set of all vectors [2k, 3k] for $k \in R$.
8. The set of all polynomials of the form $ax + b$, $a, b \in R$.
9. The set of all polynomials of the form $ax^2 + bx + c$, $a, b, c \in R$.
10. The set of complex numbers $a + bi$, $a, b \in R$.

11. The set of 2×2 matrices $\begin{pmatrix} a & b \\ c & d \end{pmatrix}$, $a, b, c, d \in R$. (For operations, see Chapter 4, Section 9.)

12. Show that the set of 2×2 matrices $\begin{pmatrix} a & 0 \\ 0 & d \end{pmatrix}$ $(a, d \in R)$ is a subspace of the vector space of Exercise 11.

◀ In Exercises 13–16, determine which of the following sets of vectors form a vector space over the real numbers.

13. The set of $[x, y]$ with $x \in I$ and $y \in R$.
14. The set of $[x, y]$ with $x + y = 2$.
15. The set of $[x, y]$ with $xy = 0$.
16. The set of $[x, y]$ with x and $y \in Q$ (the rational numbers).
17. Express $[-5, 8, 8]$ as a linear combination of $[3, 2, 0]$, $[4, 0, 2]$, and $[0, 2, 4]$.
18. Express $[4, 10, -2]$ as a linear combination of $[1, 0, 1]$, $[2, 2, 0]$, and $[-1, 1, -1]$.
19. **CHALLENGE PROBLEM.** Show that $[1, 4, 1]$ cannot be expressed as a linear combination of $[1, 0, 1]$, $[2, 2, 0]$, and $[7, 4, 3]$.
20. **CHALLENGE PROBLEM.** Show that the set $\{(x_1, x_2, x_3) \,|\, x_1, x_2, x_2 \in R\}$ forms a vector space over the real numbers if addition is defined by

$$(x_1, x_2, x_3) + (x'_1, x'_2, x'_3) = (x_1 + x'_1, x_2 + x'_2, x_3 + x'_3)$$

and scalar multiplication is defined by

$$k(x_1, x_2, x_3) = (kx_1, kx_2, kx_3).$$

CHAPTER SUMMARY

A directed segment $\overrightarrow{AB}$ is a VECTOR and is represented by an arrow. The length AB of the directed segment is the MAGNITUDE, $|\overrightarrow{AB}|$, of the vector. If the line $\overleftrightarrow{AB}$ containing $\overrightarrow{AB}$ forms an angle θ with the positive ray of the x-axis, $\cos \theta$ and $\sin \theta$ are called the DIRECTION COSINES of the vector and they are a measure of the DIRECTION of the vector.

If two vectors $\overrightarrow{AB}$ and $\overrightarrow{CD}$ have the same magnitude and direction, they are EQUIVALENT.

The set of all vectors having the same direction and magnitude is called an EQUIVALENCE CLASS of vectors. A POSITION VECTOR is the one vector in an equivalence class whose INITIAL point is at the origin of the Cartesian plane. A position vector $\overrightarrow{OP}$ is defined by naming the coordinates (a, b) of the TERMINAL point P. The vector $\overrightarrow{AB}$ from point (x_2, y_2) to point (x_1, y_1) is equivalent to the vector $[x_1 - x_2, y_1 - y_2]$.

The set of all vectors equivalent to the position vector $\overrightarrow{OP}$ (the coordinates of P being (a, b)) is denoted by $[a, b]$. If the set of vectors equivalent to $\overrightarrow{AB}$ is the same set as the set of vectors equivalent to $\overrightarrow{CD}$, we write $\overrightarrow{AB} = \overrightarrow{CD}$. Similarly, we write $\overrightarrow{AB} = [a, b]$ to indicate that the set of vectors equivalent to $\overrightarrow{AB}$ is the same set as the set of vectors $[a, b]$.

The SUM of the two vectors $\overrightarrow{AB}$ and $\overrightarrow{BC}$ is the vector $\overrightarrow{AC}$. The sum of the two vectors $[a, b]$ and $[c, d]$ is the vector $[a + c, b + d]$. The properties of vector addition are very similar to the properties of real number addition.

The SCALAR PRODUCT of a vector $\overrightarrow{AB}$ with a SCALAR k is a vector $\overrightarrow{CD}$ which has the same direction as $\overrightarrow{AB}$ and a magnitude $k|\overrightarrow{AB}|$. If $\overrightarrow{AB} = [a, b]$, then $\overrightarrow{CD} = [ka, kb]$. The properties of scalar multiplication are similar to the properties of real number multiplication.

The magnitude of a vector, $|\mathbf{v}|$, is equal to $\sqrt{a^2 + b^2}$ if $\mathbf{v} = [a, b]$.

The INNER PRODUCT of two vectors $\mathbf{a} = [a, b]$ and $\mathbf{b} = [c, d]$ is the real number $ac + bd$. We write $\mathbf{a} \cdot \mathbf{b} = ac + bd$.

Two vectors $\mathbf{a}$ and $\mathbf{b}$ are PARALLEL if and only if $\mathbf{a} = k\mathbf{b}$ for some $k \in R$. Two vectors are PERPENDICULAR if and only if $\mathbf{a} \cdot \mathbf{b} = 0$.

A vector $\mathbf{v}$ is called a UNIT VECTOR if and only if $|\mathbf{v}| = 1$. The unit vectors $[1, 0]$ and $[0, 1]$ are designated by the symbols $\mathbf{i}$ and $\mathbf{j}$, respectively. Any vector $\mathbf{v}$ can be expressed as a linear combination of $\mathbf{i}$ and $\mathbf{j}$. That is,

$$\mathbf{v} = x\mathbf{i} + y\mathbf{j} \qquad \text{for some } x \text{ and } y \in R.$$

The numbers x and y are called the COMPONENTS of $\mathbf{v}$, with respect to the BASIS vectors $\mathbf{i}$ and $\mathbf{j}$.

A VECTOR SPACE over the real numbers is a set of elements together with an operation of addition satisfying Properties I and an operation of scalar multiplication satisfying Properties II as listed on page 654.

Properties I

For all vectors **u**, **v**, and **w**,

u + **v** is a vector.	Closure
u + **v** = **v** + **u**.	Commutativity
(**u** + **v**) + **w** = **u** + (**v** + **w**).	Associativity

There exists a vector **0** such that for all vectors **u**,

$$\mathbf{u} + \mathbf{0} = \mathbf{u}. \qquad \text{Additive Identity}$$

For any vector **u** there exists a vector −**u** such that

$$\mathbf{u} + (-\mathbf{u}) = \mathbf{0}. \qquad \text{Additive Inverse}$$

For all vectors **u**, **v**, **w**, **a**, and **b**,

if **u** = **v** and **a** = **b**, then

u + **a** = **v** + **b**,   Substitution Property

if **u** + **v** = **u** + **w**, then

v = **w**.   Cancellation Property

Properties II

For all vectors **u** and **v** and real numbers *r* and *s*,

*r***u** is a vector.	Closure
$\left.\begin{array}{l} 0\mathbf{u} = \mathbf{0}. \\ r\mathbf{0} = \mathbf{0}. \end{array}\right\}$	Zero Properties
1**u** = **u**.	Multiplicative identity
r(*s***u**) = (*rs*)**u**.	Associativity
$\left.\begin{array}{l} r(\mathbf{u} + \mathbf{v}) = r\mathbf{u} + r\mathbf{v}. \\ (r + s)\mathbf{u} = r\mathbf{u} + s\mathbf{u}. \end{array}\right\}$	Distributivity

REVIEW EXERCISES

1. Find the magnitude and direction cosines of the directed segment $\overrightarrow{AB}$ if the coordinates of A are $(3, -5)$ and of B are $(6, 4)$.
2. Show geometrically the sum of the two vectors $\overrightarrow{OA}$ and $\overrightarrow{OB}$.
3. Show geometrically the difference of the two vectors $\overrightarrow{OA}$ and $\overrightarrow{OB}$.
4. Show geometrically that $(\mathbf{a} + \mathbf{b}) + \mathbf{c} = \mathbf{a} + (\mathbf{b} + \mathbf{c})$.

5. If $\mathbf{a} = [3, 5]$ and $\mathbf{b} = [-2, 6]$, find the position vectors equivalent to $\mathbf{a} + \mathbf{b}$ and $\mathbf{a} - \mathbf{b}$.

6. Show algebraically and geometrically that subtraction of vectors is not commutative.

7. Find the sum $8\mathbf{i} + 7\mathbf{j}$ and the difference $8\mathbf{i} - 7\mathbf{j}$.

8. Let $\mathbf{v} = [-3, 2]$. Find (a) $-\mathbf{v}$, (b) $4\mathbf{v}$, (c) $-\frac{1}{2}\mathbf{v}$, and (d) $0\mathbf{v}$.

9. Represent each of the vectors in Exercise 8 on the same Cartesian plane.

10. Show that $r[s(\mathbf{u} + \mathbf{v})] = r(s\mathbf{u}) + r(s\mathbf{v})$.

11. Express $[\sqrt{2}, 5]$ as a linear combination of $\mathbf{i}$ and $\mathbf{j}$.

12. Express $[9, 3]$ as a linear combination of $[-6, 2]$ and $[3, -1]$.

13. Graph the position vector equivalent to
   (a) $5\mathbf{i} - 3\mathbf{j}$ (c) $-2\mathbf{j}$
   (b) $8\mathbf{i}$ (d) $-3\mathbf{i} + 5\mathbf{j}$

14. Determine which of the following pairs of vectors are parallel.
   (a) $[2, 3]$, $[-8, -12]$ (c) $[4t, t]$, $[-t, 4t]$
   (b) $\left[\frac{5}{2}, -\frac{1}{3}\right]$, $\left[\frac{2}{3}, 5\right]$ (d) $[a + b, c]$, $[ca + cb, cab]$

15. Determine which of the pairs of vectors in Exercise 14 are perpendicular.

16. The coordinates of the four points A, B, C, and D are $(5, 4)$, $(-3, 2)$, $(2, -4)$, and $(-3, -5)$, respectively. Determine if $\overrightarrow{AB}$ is either parallel or perpendicular to $\overrightarrow{CD}$.

17. Determine if the set of vectors $[x, y]$ where $x, y \in R$ and $2x = 3y$ is a vector space.

18. Show that the dimension of the vector space consisting of the set of vectors $[x, y]$ $(x, y \in R)$ with $x = y$ is 1.

GOING FURTHER: READING AND RESEARCH

As it has been suggested several times in the Challenge Problems of Chapter 13, the concept of a vector is expandable from two-space, in which an ordered pair $[x_1, x_2]$ names an element, to three-space, in which an ordered triple $[x_1, x_2, x_3]$ names an element, to four-space, to five-space, even to n-space, in which an ordered n-tuple $[x_1, x_2, x_3, \ldots, x_n]$ names an element.

There may be some difficulty in picturing the elements of four-space, five-space, and so on. But this becomes less and less of a difficulty as we become more and more dependent on the algebraic structure. For example, the concept of magnitude does not need pictures. In two-space we define the magnitude $|\overrightarrow{AB}|$ of vector $\overrightarrow{AB}$ between two points (x_1, x_2) and (y_1, y_2) as

$$|AB| = \sqrt{(x_1 - y_1)^2 + (x_2 - y_2)^2}.$$

Note that we are using x and y in a slightly different sense than previously. For example, if we have two vectors in five-space, we denote them by

$$[x_1, x_2, x_3, x_4, x_5] \qquad \text{and} \qquad [y_1, y_2, y_3, y_4, y_5].$$

In three-space, we have

$$|\overrightarrow{AB}| = \sqrt{(x_1 - y_1)^2 + (x_2 - y_2)^2 + (x_3 - y_3)^2}.$$

The pattern holds for other spaces. In four-space we have

$$|\overrightarrow{AB}| = \sqrt{(x_1 - y_1)^2 + (x_2 - y_2)^2 + (x_3 - y_3)^2 + (x_4 - y_4)^2};$$

in n-space, we have

$$|\overrightarrow{AB}| = \sqrt{(x_1 - y_1)^2 + (x_2 - y_2)^2 + \cdots + (x_n - y_n)^2}.$$

Using this information, can you find the magnitude of the vector AB if

$$A = (2, 1, -1, 3) \text{ and } B = (-1, 0, 3, 1)?$$

The definition of orthogonal (perpendicular) vectors can be extended in the same manner. In two-space, two vectors $[x_1, x_2]$ and $[y_1, y_2]$ are orthogonal if and only if

$$x_1 y_1 + x_2 y_2 = 0.$$

In three-space, two vectors $[x_1, x_2, x_3]$ and $[y_1, y_2, y_3]$ are orthogonal if and only if

$$x_1 y_1 + x_2 y_2 + x_3 y_3 = 0.$$

Can you test whether the two vectors $[1, -1, 3, 2]$ and $[3, 0, 2, -4]$ are orthogonal?

Can you suggest a set of basis vectors for four-space?

All of the concepts we have studied in Chapter 13, such as addition, scalar multiplication, inner product, and linear combination, are applicable to three-space, four-space, and indeed n-space for any natural number $n \supset 1$. If you are interested in learning more than sketched here, the following books will provide help.

MURDOCH, D. C., *Linear Algebra For Undergraduates*. New York: John Wiley and Sons, 1957.

BARNETT, R. and J. FUJII, *Vectors*. New York: John Wiley and Sons, 1963.

NORTON, M. S., *Basic Concepts of Vectors*. St. Louis, Mo.: Webster Publishing Co., 1963.

CARMAN, R. A., *A Programmed Introduction to Vectors*. New York: John Wiley and Sons, 1963.

NASA

CHAPTER FOURTEEN

POWERS AND ROOTS OF COMPLEX NUMBERS

14.1 POLAR FORM OF COMPLEX NUMBERS

In Chapter 4 a graphical representation of complex numbers was produced by the use of Argand diagrams. The complex numbers, $3 + 4i, 5 - 2i$, and $-3i$, for example, can be made to correspond to points in the complex plane as shown in Figure 14-1. The point with coordinates $(3, 4)$ corresponds to the complex number $3 + 4i$, the point with coordinates $(5, -2)$ corresponds to $5 - 2i$, and the point with coordinates $(0, -3)$ corresponds to $-3i$. Such a correspondence of complex numbers with points in a plane is, as you know, one-to-one.

In Chapter 4 we used graphical representation to give a geometrical interpretation to addition and subtraction of complex numbers. We shall go several steps further in this chapter and use graphical representation to provide insights into the multiplication and the determination of powers and roots of complex numbers. The principal tool in this investigation will be trigonometry.

In order to apply the apparatus of trigonometry to complex numbers we first need to consider a different way of representing these numbers.

Let us consider the complex number $z = 4 + 3i$. Its representation in an Argand diagram is shown in Figure 14-2. The point $P(4, 3)$ corresponds to the complex number $4 + 3i$. If we construct the right triangle OPR as shown, with $\theta = \angle POR$, then $\cos \theta = \frac{4}{5}$ and $\sin \theta = \frac{3}{5}$.

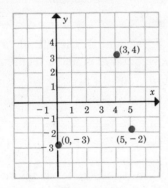

Figure 14-1

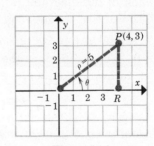

Figure 14-2

Here angle θ is in *standard position* as defined in Chapter 12. Unless specified otherwise, we shall assume that the unit of measure for angles is the radian and we shall, as before, use an abbreviated expression such as $\theta = \frac{\pi}{2}$ to mean "The measure of angle θ in radians is $\frac{\pi}{2}$." In this discussion, as in Chapter 12, we shall assign positive measure to an angle generated in a counterclockwise direction and negative measure to an angle generated in a clockwise direction.

Let us now consider Argand diagrams for various complex numbers $x + yi$ where x and y are real numbers. The diagrams in Figure 14-3 exhibit points in each of the four quadrants as well as angles of positive and negative measures.

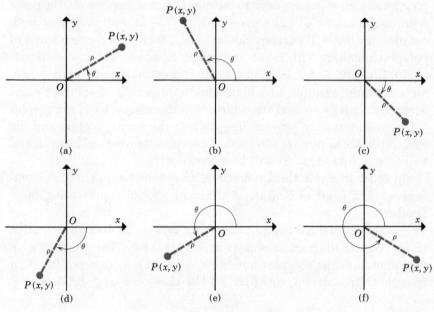

(a)   (b)   (c)

(d)   (e)   (f)

Figure 14-3

In each of the diagrams (a), (b), (c), (d), (e), and (f) the initial ray for angle θ is the nonnegative part of the horizontal axis, and a segment of the terminal ray is $\overline{OP}$. In (c) and (d), $\theta < 0$. In (a), (b), (e), and (f), $\theta > 0$. Let ρ represent the positive distance from O to P. Then we have, in every instance,

$$(1) \qquad \rho = \sqrt{x^2 + y^2}.$$

Furthermore, since in each case $\cos \theta = \dfrac{x}{\rho}$ and $\sin \theta = \dfrac{y}{\rho}$, it follows that

$$(2) \qquad x = \rho \cos \theta \qquad \text{and} \qquad y = \rho \sin \theta.$$

Thus we may write the complex number $x + yi$ as $\rho \cos \theta + (\rho \sin \theta)i$ or

$$(3) \qquad \rho(\cos \theta + i \sin \theta).$$

Since the distance ρ is measured from the origin (often called the **pole**), a complex number written in the form (3) is said to be written in **polar form**. The polar form $\rho(\cos \theta + i \sin \theta)$ is often abbreviated as ρ cis θ.

It is relatively easy to convert from a polar form to the more familiar form $a + bi$. For example, suppose that we are given the complex number z in polar form as

$$z = 4\left(\cos \frac{\pi}{6} + i \sin \frac{\pi}{6}\right).$$

From the discussions in Chapter 10, you know that $\cos \dfrac{\pi}{6} = \dfrac{\sqrt{3}}{2}$ and that $\sin \dfrac{\pi}{6} = \dfrac{1}{2}$. Hence

$$z = 4\left(\cos \frac{\pi}{6} + i \sin \frac{\pi}{6}\right) = 4\left(\frac{\sqrt{3}}{2} + \frac{1}{2}i\right) = 2\sqrt{3} + 2i.$$

When we write

$$4\left(\cos \frac{\pi}{6} + i \sin \frac{\pi}{6}\right) = 2\sqrt{3} + 2i,$$

we say that we are changing from polar form to algebraic form.

To change 5 cis $\frac{2}{3}\pi$ to algebraic form, we write

$$5 \text{ cis } \frac{2}{3}\pi = 5\left(\cos \frac{2}{3}\pi + i \sin \frac{2}{3}\pi\right)$$

$$= 5\left(-\frac{1}{2} + \frac{\sqrt{3}}{2}i\right)$$

$$= -\frac{5}{2} + \frac{5\sqrt{3}}{2}i.$$

As a final example, note that

$$3 \text{ cis } \frac{\pi}{2} = 3\left(\cos \frac{\pi}{2} + i \sin \frac{\pi}{2}\right)$$

$$= 3(0 + i)$$

$$= 3i.$$

EXERCISES 14.1A

◀ In Exercises 1–20, complex numbers are given in polar form. Write each of these in algebraic form.

1. $\cos \dfrac{\pi}{3} + i \sin \dfrac{\pi}{3}$

2. $2\left(\cos \dfrac{\pi}{4} + i \sin \dfrac{\pi}{4}\right)$

3. $4\left(\cos \dfrac{\pi}{6} + i \sin \dfrac{\pi}{6}\right)$

4. $5\left(\cos \dfrac{4}{3}\pi + i \sin \dfrac{4}{3}\pi\right)$

5. $2\left(\cos \dfrac{3}{4}\pi + i \sin \dfrac{3}{4}\pi\right)$

6. $\cos \dfrac{\pi}{2} + i \sin \dfrac{\pi}{2}$

7. $\cos \pi + i \sin \pi$

8. $6\left(\cos \dfrac{3}{2}\pi + i \sin \dfrac{3}{2}\pi\right)$

9. $2\left(\cos \dfrac{5}{6}\pi + i \sin \dfrac{5}{6}\pi\right)$

10. $4(\cos 0 + i \sin 0)$

11. $7 \operatorname{cis} \dfrac{5}{4}\pi$

12. $10 \operatorname{cis} \dfrac{5}{2}\pi$

13. $12 \operatorname{cis} \dfrac{7}{6}\pi$

14. $9 \operatorname{cis}\left(-\dfrac{\pi}{3}\right)$

15. $11 \operatorname{cis}\left(-\dfrac{3}{4}\pi\right)$

16. $15 \operatorname{cis} \dfrac{5}{3}\pi$

17. $6 \operatorname{cis}\left(-\dfrac{\pi}{4}\right)$

18. $10 \operatorname{cis} \dfrac{7}{2}\pi$

19. $18 \operatorname{cis}(-\pi)$

20. $\operatorname{cis} 3\dfrac{1}{3}\pi$

21. **CHALLENGE PROBLEM.** Write $1 + \sqrt{3}i$ in polar form.

The task of converting a complex number in algebraic form to polar form is somewhat more complicated. Because polar forms are very useful in the study of complex numbers, it is important to develop proficiency in this type of conversion.

First, we have noted that for the complex number $z = x + yi$, we have

$$\rho = \sqrt{x^2 + y^2}.$$

Thus we see that ρ is the absolute value of $x + yi$ as defined in Chapter 4, that is, if $z = x + yi$, then $\rho = |z|$. If $z = \rho(\cos \theta + i \sin \theta)$, θ is called an **amplitude** (or **argument**) of z.

Since

$$\sin \theta = \sin (\theta + 2n\pi)$$

and

$$\cos \theta = \cos (\theta + 2n\pi)$$

for any integer n, we must recognize at the outset that there are many different polar forms for the same complex number. For example, we have

$$\sqrt{3} + i = 2\left(\cos\frac{\pi}{6} + i \sin\frac{\pi}{6}\right)$$

$$= 2\left[\cos\left(2 + \frac{1}{6}\right)\pi + i \sin\left(2 + \frac{1}{6}\right)\pi\right]$$

$$= 2\left[\cos\left(4 + \frac{1}{6}\right)\pi + i \sin\left(4 + \frac{1}{6}\right)\pi\right]$$

$$= 2\left[\cos\left(-\frac{11}{6}\pi\right) + i \sin\left(-\frac{11}{6}\pi\right)\right],$$

and so on, since

$$\cos\frac{\pi}{6} = \cos\left(2 + \frac{1}{6}\right)\pi = \cos\left(4 + \frac{1}{6}\right)\pi = \cos\left(-\frac{11}{6}\pi\right) = \frac{\sqrt{3}}{2}$$

and

$$\sin\frac{\pi}{6} = \sin\left(2 + \frac{1}{6}\right)\pi = \sin\left(4 + \frac{1}{6}\right)\pi = \sin\left(-\frac{11}{6}\pi\right) = \frac{1}{2}.$$

Let us now make a few conversions from algebraic form to polar form. Suppose that we are given $z = 1 + \sqrt{3}i$ whose Argand diagram is shown in Figure 14-4. To find ρ we simply note that

$$\rho = \sqrt{1^2 + (\sqrt{3})^2} = \sqrt{1 + 3} = 2.$$

The problem of determining a value of θ is not quite so simple. Often,

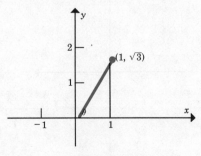

Figure 14-4

as in this instance, we can use the Argand diagram and some trigo-nometry to give us the clue.

Since $x = 1$ and $y = \sqrt{3}$, we have from (2) (page 660) that

$$1 = 2 \cos \theta \quad \text{and} \quad \sqrt{3} = 2 \sin \theta.$$

Hence

$$\cos \theta = \frac{1}{2} \quad \text{and} \quad \sin \theta = \frac{\sqrt{3}}{2}$$

and so

$$\theta = \frac{\pi}{3} + 2n\pi \quad \text{for } n \in I.$$

Hence

$$1 + \sqrt{3}i = 2\left(\cos \frac{\pi}{3} + i \sin \frac{\pi}{3}\right).$$

Check: $2\left(\cos \dfrac{\pi}{3} + i \sin \dfrac{\pi}{3}\right) = 2\left(\dfrac{1}{2} + i\dfrac{\sqrt{3}}{2}\right) = 1 + \sqrt{3}i.$

Also, of course,

$$1 + \sqrt{3}i = 2 \text{ cis}\left(\frac{\pi}{3} + 2\pi\right) = 2 \text{ cis}\left(\frac{\pi}{3} + 4\pi\right) = 2 \text{ cis}\left(\frac{\pi}{3} - 2\pi\right),$$

and so on. In practice, however, we usually choose the form in which $0 \leq \theta < 2\pi$, although there are times when it is more convenient to choose a value for θ outside this range.

Next consider the complex number $1 - \sqrt{3}i$. The Argand diagram for $1 - \sqrt{3}i$ is shown in Figure 14-5. Here again we see that $\rho = 2$. Two of the choices for a polar form are

$$2\left(\cos \frac{5}{3}\pi + i \sin \frac{5}{3}\pi\right)$$

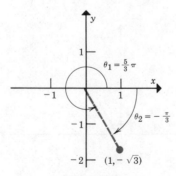

Figure 14-5

and

$$2\left[\cos\left(-\frac{\pi}{3}\right) + i\,\sin\left(-\frac{\pi}{3}\right)\right]$$

with, as we have said, 2 cis $\frac{5}{3}\pi$ being generally considered preferable.

Diagrams and inspection provide, as we have seen, useful clues for determining an amplitude. It is convenient, however, to have also a formula to fall back on—a formula which defines an amplitude θ explicitly for any complex number $x + yi$.

In stating this formula we shall use the arctangent function. Recall from Chapter 12 the following definition:

$$\theta = \tan^{-1} r \qquad (\text{or } \theta = \arctan r)$$

for a real number r if

$$(a) \qquad \tan \theta = r$$

and

$$(b) \qquad -\frac{\pi}{2} < \theta < \frac{\pi}{2}.$$

Our formula, in three parts, is as follows:

(4)   If $x > 0$, $\theta = \tan^{-1} \dfrac{y}{x}$.

(5)   If $x < 0$, $\theta = \tan^{-1} \dfrac{y}{x} + \pi$.

(6)   If $x = 0$, $\theta = \dfrac{\pi}{2}$ if $y > 0$, $\theta = -\dfrac{\pi}{2}$ if $y < 0$, and $\theta =$ any real number if $y = 0$.

A few illustrations will show how the formula can be applied. Let us first consider (4), that is, the case for $x > 0$. Figure 14-6 shows Argand diagrams for the complex numbers $\sqrt{3} + i$ and $\sqrt{3} - i$. From the diagrams and a knowledge of circular functions, we see that $\theta_1 = \frac{\pi}{6}$ and $\theta_2 = -\frac{\pi}{6}$.

Applying (4) to $\sqrt{3} + i$, we get

$$\theta_1 = \tan^{-1} \frac{y}{x} = \tan^{-1} \frac{1}{\sqrt{3}}.$$

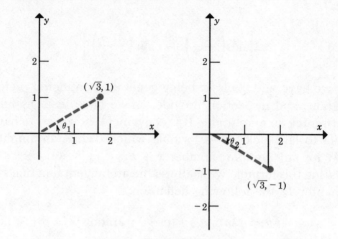

Figure 14-6

Although there are many angles with tangent $\dfrac{1}{\sqrt{3}}$, condition (b) specifies the unique value $\tan^{-1}\dfrac{1}{\sqrt{3}} = \dfrac{\pi}{6}$.

Applying (4) to $\sqrt{3} - i$, we get

$$\theta_2 = \tan^{-1}\frac{y}{x} = \tan^{-1}\frac{-1}{\sqrt{3}} = \tan^{-1}\left(-\frac{1}{\sqrt{3}}\right).$$

Again condition (b) specifies the unique value

$$\tan^{-1}\left(-\frac{1}{\sqrt{3}}\right) = -\frac{\pi}{6}.$$

Here we see that our formula has produced results that tally with information supplied by the Argand diagrams.

As an illustration of formula (5), let us look at two examples when $x < 0$. In Figure 14-7, point P corresponds to the complex number $-4 - 3i$. Since $\dfrac{y}{x} = \dfrac{-3}{-4} = \dfrac{3}{4}$, it appears from the graph that θ, the angle associated with $(-4, -3)$, has a measure of π radians more than $\theta' = \tan^{-1}\dfrac{3}{4}$. Thus

$$\theta = \theta' + \pi = \tan^{-1}\frac{3}{4} + \pi,$$

which is formula (5) with $x = -4$ and $y = -3$.

Let us now look at a point corresponding to $-4 + 3i$. In Figure

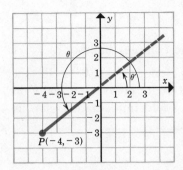

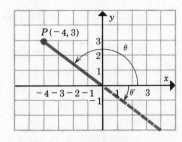

Figure 14-8

Figure 14-7

14-8, we see that $\theta' = \tan^{-1}\left(-\frac{3}{4}\right)$ and is negative. But the angle θ which is associated with point P has a measure which lies between $\frac{\pi}{2}$ and π. We have

$$\theta = \theta' + \pi = \tan^{-1}\left(-\frac{3}{4}\right) + \pi,$$

which is formula (5) with $x = -4$ and $y = 3$.

These two examples show the reason for adding π to $\tan^{-1}\dfrac{y}{x}$ in cases where $x < 0$.

We can see the reasonableness of (6) by examining Figure 14-9 which exhibits Argand diagrams for three complex numbers for which x is 0, that is,

$$2i, \qquad -3i, \qquad \text{and} \qquad 0.$$

We can justify the statement that for $z = 0 + 0i$, θ may be any angle whatsoever by simply noting that if $z = 0$, it follows that $x = 0$ and $y = 0$; hence $\rho = 0$. Thus

$$\rho(\cos\theta + i\sin\theta) = 0 + 0i$$

for any value of θ.

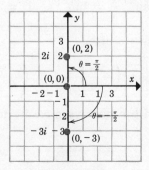

Figure 14-9

Let's try a few more test cases.

Example 1 Find a polar form for the complex number i.

Solution: In this case $x = 0$ and $y = 1$. Hence $\rho = \sqrt{0^2 + 1^2} = 1$ and $\theta = \frac{\pi}{2}$ from formula (6). In this situation, of course, we really do not need the formula since the picture would immediately give us the story.

In any event, a polar form of i is

$$\cos\frac{\pi}{2} + i \sin\frac{\pi}{2}.$$

Check: $\cos\frac{\pi}{2} + i \sin\frac{\pi}{2} = 0 + i \cdot 1 = i.$

Example 2 Give a polar form for $z = 4 - 4i$. (Figure 14-10 shows the Argand diagram.)

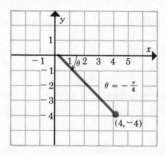

Figure 14-10

Solution: Here $x = 4$ and $y = -4$. Thus

$$\rho = \sqrt{16 + 16} = \sqrt{32} = 4\sqrt{2}.$$

Since $x > 0$, we can use

$$\theta = \tan^{-1}\frac{y}{x} = \tan^{-1}(-1) = -\frac{\pi}{4}.$$

Hence a polar form for $4 - 4i$ is

$$4\sqrt{2}\left[\cos\left(-\frac{\pi}{4}\right) + i \sin\left(-\frac{\pi}{4}\right)\right].$$

An alternative form with a positive measure for θ is

$$4\sqrt{2}\operatorname{cis}\frac{7}{4}\pi.$$

Example 3 Find a polar form for $-1 + i$. (Figure 14-11 shows the Argand diagram.)

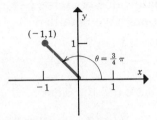

Figure 14-11

Solution: $\rho = \sqrt{(-1)^2 + 1^2} = \sqrt{2}$. Since $x < 0$, we use

$$\theta = \tan^{-1}\frac{y}{x} + \pi = \tan^{-1}(-1) + \pi = -\frac{\pi}{4} + \pi = \frac{3}{4}\pi.$$

Thus we have a polar form

$$\sqrt{2}\left(\cos\frac{3}{4}\pi + i\,\sin\frac{3}{4}\pi\right).$$

In Examples 1, 2, and 3 an amplitude, θ, has been fairly easy to determine. There are situations, however, in which the problem of finding θ may present a little more difficulty as in Example 4.

Example 4 Find a polar form for $-2 + 3i$. (See Figure 14-12.)

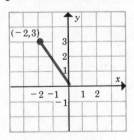

Figure 14-12

Solution: We first note that

$$\rho = \sqrt{(-2)^2 + 3^2} = \sqrt{4 + 9} = \sqrt{13}.$$

From the formulas, we see that, since $x < 0$,

$$\theta = \arctan\left(-\frac{3}{2}\right) + \pi.$$

The problem now is that there is no way to express arctan $\left(-\frac{3}{2}\right) + \pi$ in the form of a simple numeral such as $\frac{1}{2}$ or $\frac{\pi}{3}$. The best we can do is to find an approximate value of arctan $\left(-\frac{3}{2}\right)$ from a table of values of trigonometric functions. We note that

$$\arctan\left(-\frac{3}{2}\right) = -\arctan(1.5) \approx -0.98.$$

Hence $\theta \approx -0.98 + 3.14 = 2.16$. Thus

$$-2 + 3i \approx \sqrt{13}\,[\cos(2.16) + i\sin(2.16)].$$

An alternative is the somewhat awkward looking expression

$$\sqrt{13}\left\{\cos\left[\tan^{-1}\left(-\frac{3}{2}\right) + \pi\right] + i\sin\left[\tan^{-1}\left(-\frac{3}{2}\right) + \pi\right]\right\}.$$

At this point one might be tempted to call the whole thing off!

Under certain circumstances we may know an amplitude and the absolute value of a certain complex number z and wish to write z in algebraic form. The following example illustrates the operating principle.

Example 5   Write in algebraic form the complex number with absolute value 10 and an amplitude $\theta = \tan^{-1}\left(-\frac{3}{4}\right)$.

Solution: From the definition of arctan we know that $-\frac{\pi}{2} < \theta < \frac{\pi}{2}$. Hence for a negative number r, $\tan^{-1} r = \theta$ lies between $-\frac{\pi}{2}$ and 0, that is, the point in the Argand diagram lies in the fourth quadrant. Thus $\cos\theta > 0$ and $\sin\theta < 0$. This condition plus the fact that $\tan\theta = -\frac{3}{4}$ tells us that $\cos\theta = \frac{4}{5}$ and $\sin\theta = -\frac{3}{5}$. For the algebraic form, then, we have

$$10\left[\frac{4}{5} + \left(-\frac{3}{5}\right)i\right] = 8 + (-6)i$$

which we can also write as $8 - 6i$.

The following Exercises will provide practice in converting from algebraic form to polar form and in writing complex numbers in algebraic form given the absolute value and an amplitude of a complex number as in Example 5. Before you start these Exercises, however, a word of warning may be in order.

The general polar form for a complex number is

$$\rho(\cos\theta + i\sin\theta).$$

In Example 2, we found that

$$4 - 4i = 4\sqrt{2}\left[\cos\left(-\frac{\pi}{4}\right) + i\sin\left(-\frac{\pi}{4}\right)\right].$$

Since we know that for any real number r,

$$\sin(-r) = -\sin r$$

and

$$\cos(-r) = \cos r,$$

we might be tempted to write our "polar form" as

$$4\sqrt{2}\left(\cos\frac{\pi}{4} - i\sin\frac{\pi}{4}\right).$$

This is, however, *not* a polar form. The polar form specifies the "$+$" symbol between the two circular functional values. Also, since ρ is defined as the absolute value of z, it should be clear that an expression such as $-2\operatorname{cis}\frac{\pi}{6}$ is not a polar form.

EXERCISES 14.1B

◀In Exercises 1–20, write each of the given complex numbers in a polar form $\rho(\cos\theta + \sin\theta)$ with $0 \le \theta < 2\pi$.

1. $3 + 3i$ 11. $-8i$

2. $1 - \sqrt{3}i$ 12. $1 + i$

3. $-5 + 5i$ 13. -1

4. $\sqrt{3} - i$ 14. $2\sqrt{3} + 2i$

5. $-\sqrt{3} + i$ 15. $2\sqrt{3} - 2i$

6. $-i$ 16. $-1 + i$

7. $4i$ 17. $-2 - 2i$

8. 6 18. $-4 - 4\sqrt{3}i$

9. -4 19. $-\dfrac{1}{2} + \dfrac{\sqrt{3}}{2}i$

10. $3 - 3\sqrt{3}i$ 20. $\dfrac{1}{2} - \dfrac{\sqrt{3}}{2}i$

◀ In Exercises 21–25, values of ρ and θ are given as absolute values and amplitudes, respectively. Write the corresponding complex numbers in algebraic form.

21. $\rho = 5, \theta = \tan^{-1}\dfrac{4}{3}$ **24.** $\rho = 20, \theta = \tan^{-1}(-1.25)$

22. $\rho = 8, \theta = \tan^{-1}\sqrt{3}$ **25.** $\rho = 14, \theta = \tan^{-1}\dfrac{\sqrt{3}}{3}$

23. $\rho = 12, \theta = \tan^{-1}\dfrac{3}{4}$

◀ In Exercises 26–30, write each of the given complex numbers in a polar form with an approximate value of θ in the range $0 \leq \theta < 2\pi$ obtained from a table of trigonometric functions.

26. $2 + i$ **29.** $2 - 5i$
27. $3 - 2i$ **30.** $5 - 2i$
28. $4 + i$

14.2 PRODUCTS AND POWERS OF COMPLEX NUMBERS

If
$$z_1 = 2\sqrt{3} + 2i \qquad \text{and} \qquad z_2 = -3 + 3\sqrt{3}\,i,$$
then

$$
\begin{aligned}
z_1 z_2 &= (2\sqrt{3} + 2i)(-3 + 3\sqrt{3}\,i) \\
&= [2\sqrt{3}(-3) + (2i)(3\sqrt{3}\,i)] + [(2\sqrt{3})(3\sqrt{3})i + (2i)(-3)] \\
&= (-6\sqrt{3} + 6\sqrt{3}\,i^2) + (18i - 6i) \\
&= -12\sqrt{3} + 12i.
\end{aligned}
$$

Now let us multiply z_1 by z_2 again, this time using polar forms. For $2\sqrt{3} + 2i$ we have
$$\rho = \sqrt{(2\sqrt{3})^2 + 2^2} = \sqrt{12 + 4} = 4$$
and
$$\theta = \tan^{-1}\frac{2}{2\sqrt{3}} = \tan^{-1}\frac{1}{\sqrt{3}} = \tan^{-1}\frac{\sqrt{3}}{3} = \frac{\pi}{6}.$$

Thus, in polar form, $z_1 = 4(\cos\frac{\pi}{6} + i\sin\frac{\pi}{6})$.
For $-3 + 3\sqrt{3}\,i$ we have
$$\rho = \sqrt{3^2 + (3\sqrt{3})^2} = \sqrt{9 + 27} = 6.$$

Since $x = -3 < 0$, we use formula (5), that is,

$$\theta = \tan^{-1}\left(-\frac{3\sqrt{3}}{3}\right) + \pi.$$

Thus

$$\theta = \tan^{-1}\left(-\sqrt{3}\right) + \pi = -\frac{\pi}{3} + \pi = \frac{2}{3}\pi.$$

In polar form, then,

$$z_2 = 6\left(\cos\frac{2}{3}\pi + i\,\sin\frac{2}{3}\pi\right).$$

To form the product $z_1 z_2$ we have

$$z_1 z_2 = 4\left(\cos\frac{\pi}{6} + i\,\sin\frac{\pi}{6}\right) \cdot 6\left(\cos\frac{2}{3}\pi + i\,\sin\frac{2}{3}\pi\right)$$

$$= 4 \cdot 6\left[\left(\cos\frac{\pi}{6} \cdot \cos\frac{2}{3}\pi - \sin\frac{\pi}{6} \cdot \sin\frac{2}{3}\pi\right)\right.$$

$$\left. + i\left(\cos\frac{\pi}{6} \sin\frac{2}{3}\pi + \sin\frac{\pi}{6} \cos\frac{2}{3}\pi\right)\right].$$

Recall, now, that by Identity (X) of Chapter 10

$$\cos(\theta + \phi) = \cos\theta \, \cos\phi - \sin\theta \, \sin\phi$$

and by Identity (XVI)

$$\sin(\theta + \phi) = \cos\theta \, \sin\phi + \sin\theta \, \cos\phi.$$

If we apply these identities to the final form of the product by taking $\theta = \frac{\pi}{6}$ and $\phi = \frac{2}{3}\pi$, we can see that

$$\cos\frac{\pi}{6} \cos\frac{2}{3}\pi - \sin\frac{\pi}{6} \sin\frac{2}{3}\pi = \cos\left(\frac{\pi}{6} + \frac{2}{3}\pi\right)$$

and

$$\cos\frac{\pi}{6} \sin\frac{2}{3}\pi + \sin\frac{\pi}{6} \cos\frac{2}{3}\pi = \sin\left(\frac{\pi}{6} + \frac{2}{3}\pi\right).$$

Thus the product $z_1 z_2$ becomes

$$4 \cdot 6\left[\cos\left(\frac{\pi}{6} + \frac{2}{3}\pi\right) + i\,\sin\left(\frac{\pi}{6} + \frac{2}{3}\pi\right)\right]$$

or

$$24\left[\cos\frac{5}{6}\pi + i\,\sin\frac{5}{6}\pi\right].$$

Since, as you should check,

$$\cos\frac{5}{6}\pi = -\frac{\sqrt{3}}{2} \quad \text{and} \quad \sin\frac{5}{6}\pi = \frac{1}{2},$$

it follows that

$$z_1 \cdot z_2 = 24\left(-\frac{\sqrt{3}}{2} + \frac{1}{2}i\right)$$

$$= -12\sqrt{3} + 12i$$

which is the same result that we obtained by algebraic procedures.

From the process of multiplication using polar forms as in this example, can we derive a generalization? Let us review the operation in general terms. Let

$$z_1 = \rho_1(\cos\theta_1 + i\sin\theta_1)$$

and

$$z_2 = \rho_2(\cos\theta_2 + i\sin\theta_2).$$

Then

$$z_1 z_2 =$$
$$\rho_1\rho_2[(\cos\theta_1\cos\theta_2 - \sin\theta_1\sin\theta_2) + i(\cos\theta_1\sin\theta_2 + \sin\theta_1\cos\theta_2)],$$

which then becomes

$$\rho_1\rho_2[\cos(\theta_1 + \theta_2) + i\sin(\theta_1 + \theta_2)]$$

by the use of Identities (X) and (XVI) of Chapter 10.

In forming the product, then, we have multiplied ρ_1 by ρ_2 but have *added* the two amplitudes θ_1 and θ_2. We now state this result in somewhat more formal terms as a theorem.

THEOREM 14.1

If two complex numbers z_1 and z_2 have absolute values ρ_1 and ρ_2 and amplitudes θ_1 and θ_2, respectively, then

(1) the absolute value of the product $z_1 z_2$ is $\rho_1\rho_2$ and

(2) an amplitude of the product $z_1 z_2$ is $\theta_1 + \theta_2$;

that is,

$$(\rho_1 \text{ cis } \theta_1)(\rho_2 \text{ cis } \theta_2) = \rho_1\rho_2 \text{ cis }(\theta_1 + \theta_2).$$

Let us apply Theorem 14.1 to the product $i \cdot i$. In polar form, i can be written as

$$1\left(\cos\frac{\pi}{2} + i\sin\frac{\pi}{2}\right).$$

Hence

$$i \cdot i = (1 \cdot 1)\left[\cos\left(\frac{\pi}{2} + \frac{\pi}{2}\right) + i \sin\left(\frac{\pi}{2} + \frac{\pi}{2}\right)\right] = 1(\cos \pi + i \sin \pi).$$

Since $\cos \pi = -1$ and $\sin \pi = 0$, we have $i \cdot i = -1$. (Surprise!)

This result gives us some comforting reassurance concerning the validity of the theorem.

As we have seen, the polar form, together with Theorem 14.1, enables us to compute the product of two complex numbers by multiplying absolute values and adding amplitudes. This principle can, of course, be applied to the multiplication of a complex number by itself. For example, to find $(1 + i)^2$ we note that a polar form of $1 + i$ is $\sqrt{2}(\cos \frac{\pi}{4} + i \sin \frac{\pi}{4})$. Hence

$$(1 + i)^2 = (1 + i)(1 + i)$$
$$= \sqrt{2}\left(\cos\frac{\pi}{4} + i \sin\frac{\pi}{4}\right) \cdot \sqrt{2}\left(\cos\frac{\pi}{4} + i \sin\frac{\pi}{4}\right)$$
$$= 2\left[\cos\left(\frac{\pi}{4} + \frac{\pi}{4}\right) + i \sin\left(\frac{\pi}{4} + \frac{\pi}{4}\right)\right].$$

From this last result we see that

$$(1 + i)^2 = 2\left[\cos\left(\frac{\pi}{2}\right) + i \sin\left(\frac{\pi}{2}\right)\right] = 2(0 + i) = 2i.$$

This product, however, could obviously have been obtained very simply by multiplying in the ordinary way. On the other hand, the calculation of something like $(1 + i)^{10}$ might be quite burdensome by reiterated multiplication, or even by the Binomial Theorem.

Does this example suggest a method of finding $(1 + i)^3$? Does it seem reasonable to conjecture that

$$(1 + i)^3 = \left(\sqrt{2} \operatorname{cis}\frac{\pi}{4}\right) \cdot \left(\sqrt{2} \operatorname{cis}\frac{\pi}{4}\right) \cdot \left(\sqrt{2} \operatorname{cis}\frac{\pi}{4}\right)$$
$$= \sqrt{2} \cdot \sqrt{2} \cdot \sqrt{2}\left[\operatorname{cis}\left(\frac{\pi}{4} + \frac{\pi}{4} + \frac{\pi}{4}\right)\right]$$
$$= 2\sqrt{2} \operatorname{cis}\frac{3}{4}\pi?$$

Do you think that

$$(1 + i)^4 = \sqrt{2} \cdot \sqrt{2} \cdot \sqrt{2} \cdot \sqrt{2}\left[\operatorname{cis}\left(\frac{\pi}{4} + \frac{\pi}{4} + \frac{\pi}{4} + \frac{\pi}{4}\right)\right]$$
$$= 4 \operatorname{cis} \pi = 4(-1 + 0i) = -4?$$

The hoped for answer to these questions is provided by the following theorem which we shall prove by induction.

THEOREM 14.2

If z is a complex number with a polar form $\rho(\cos \theta + i \sin \theta)$, then z^n has a polar form

$$\rho^n(\cos n\theta + i \sin n\theta).$$

Proof: The proof by induction follows the usual pattern. We let

$$S = \{n : n \in N \text{ and } [\rho(\cos \theta + i \sin \theta)]^n = \rho^n(\cos n\theta + i \sin n\theta)\}.$$

It is indeed fairly obvious that $1 \in S$. All we need then is to show that S is an inductive set, that is, that $k \in S$ implies $k + 1 \in S$.

So we now assume that $k \in S$, that is, that

(1) $\qquad [\rho(\cos \theta + i \sin \theta)]^k = \rho^k(\cos k\theta + i \sin k\theta)$

and need to show that (1) implies that $k + 1 \in S$. That is, we must show that (1) implies

(2) $[\rho(\cos \theta + i \sin \theta)]^{k+1} = \rho^{k+1}[\cos (k + 1)\theta + i \sin (k + 1)\theta]$.

To do this we multiply both sides of (1) by $\rho(\cos \theta + i \sin \theta)$. This gives us

$$[\rho(\cos \theta + i \sin \theta)]^{k+1} = \rho(\cos \theta + i \sin \theta) \cdot \rho^k(\cos k\theta + i \sin k\theta).$$

We next apply Theorem 14.1 to the product on the right to get

$$\rho(\cos \theta + i \sin \theta) \cdot \rho^k(\cos k\theta + i \sin k\theta)$$
$$= \rho^{k+1}[\cos (k + 1)\theta + i \sin (k + 1)\theta].$$

This tells us that if (1) is a true statement, then

(2) $[\rho(\cos \theta + i \sin \theta)]^{k+1} = \rho^{k+1}[\cos (k + 1)\theta + i \sin (k + 1)\theta]$.

Thus we have shown that $k \in S$ implies $k + 1 \in S$ and our theorem is proved.

If we define, for example, $(x + yi)^{-5}$ to mean $\dfrac{1}{(x + yi)^5}$, and in general

$$(x + yi)^{-k} = \frac{1}{(x + yi)^k}$$

for any $k \in N$ and any nonzero complex number $x + yi$, it is also possible to prove that

$$[\rho(\cos\theta + i\,\sin\theta)]^n = \rho^n(\cos n\theta + i\,\sin n\theta)$$

when n is any integer. This proof will be reserved for the Exercises.

As an application of Theorem 14.2, let us find $(\sqrt{3} + i)^6$. In polar form

$$\sqrt{3} + i = 2\left(\cos\frac{\pi}{6} + i\,\sin\frac{\pi}{6}\right).$$

(Check this.) By Theorem 14.2,

$$(\sqrt{3} + i)^6 = 2^6\left[\cos 6\left(\frac{\pi}{6}\right) + i\,\sin 6\left(\frac{\pi}{6}\right)\right]$$

$$= 2^6(\cos\pi + i\,\sin\pi)$$

$$= 2^6(-1 + i\cdot 0)$$

$$= -64.$$

As a second illustration let us find $(-2 + 2i)^8$. The absolute value of $-2 + 2i$ is $\sqrt{(-2)^2 + 2^2}\,\sqrt{8} = 8^{\frac{1}{2}}$. For θ we have, by formula (5),

$$\theta = \tan^{-1}(-1) + \pi = -\frac{\pi}{4} + \pi = \frac{3}{4}\pi.$$

Hence

$$(-2 + 2i)^8 = \left[8^{\frac{1}{2}}\left(\cos\frac{3}{4}\pi + i\,\sin\frac{3}{4}\pi\right)\right]^8$$

$$= (8^{\frac{1}{2}})^8\left[\cos 8\left(\frac{3}{4}\pi\right) + i\,\sin 8\left(\frac{3}{4}\pi\right)\right]$$

$$= 8^4(\cos 6\pi + i\,\sin 6\pi)$$

$$= 8^4(\cos 0 + i\,\sin 0) = 8^4(1 + i\cdot 0)$$

$$= 8^4 = 4096.$$

To illustrate the case for $n < 0$, under the assumption that Theorem 14.2 holds for $n \in I$, let us evaluate $(1 - i)^{-4}$. We have

$$1 - i = \sqrt{2}\left[\cos\left(-\frac{\pi}{4}\right) + i\,\sin\left(-\frac{\pi}{4}\right)\right].$$

Hence

$$(1 - i)^{-4} = (\sqrt{2})^{-4}\left[\cos\left(-4 \cdot -\frac{\pi}{4}\right) + i \sin\left(-4 \cdot -\frac{\pi}{4}\right)\right]$$

$$= (\sqrt{2})^{-4}(\cos \pi + i \sin \pi)$$

$$= (2^{\frac{1}{2}})^{-4}(-1 + i \cdot 0)$$

$$= -\frac{1}{4}.$$

If we calculate $(1 - i)^4$, we get

$$1 - i = \sqrt{2} \operatorname{cis}\left(-\frac{\pi}{4}\right)$$

$$(1 - i)^4 = (\sqrt{2})^4 \operatorname{cis} 4\left(-\frac{\pi}{4}\right)$$

$$= 4 \operatorname{cis}(-\pi)$$

$$= -4.$$

Since the previous calculation gave us

$$(1 - i)^{-4} = -\frac{1}{4} = \frac{1}{-4},$$

we see that the answers are consistent.

To change a polar form to the algebraic form of a complex number it may be necessary to apply certain trigonometric conversion formulas. For example, suppose that a complex number has the polar form $2(\cos \frac{\pi}{12} + i \sin \frac{\pi}{12})$. Since the values of $\cos \frac{\pi}{6}$ and $\cos \frac{\pi}{4}$ are known, we can determine $\cos \frac{\pi}{12}$ by using the fact that

$$\frac{\pi}{12} = \frac{\pi}{4} - \frac{\pi}{6}$$

and

$$\cos\left(\frac{\pi}{4} - \frac{\pi}{6}\right) = \cos\frac{\pi}{4}\cos\frac{\pi}{6} + \sin\frac{\pi}{4}\sin\frac{\pi}{6}$$

$$= \frac{\sqrt{2}}{2} \cdot \frac{\sqrt{3}}{2} + \frac{\sqrt{2}}{2} \cdot \frac{1}{2}$$

$$= \frac{\sqrt{6} + \sqrt{2}}{4}.$$

Similarly, you can show that

$$\sin\frac{\pi}{12} = \frac{\sqrt{6} - \sqrt{2}}{4}.$$

Thus

$$2\left(\cos\frac{\pi}{12} + i\sin\frac{\pi}{12}\right) = \left(\frac{\sqrt{6} + \sqrt{2}}{2}\right) + \left(\frac{\sqrt{6} - \sqrt{2}}{2}\right)i.$$

Theorem 14.2 can be applied successfully whenever a simple polar form exists. However, as you have seen in Example 4 of Section 14.1, this is not always the case. Consider again the complex number $-2 + 3i$. As you recall, we can write an approximation to a polar form for $-2 + 3i$ as

$$-2 + 3i \approx \sqrt{13}(\cos 2.16 + i\sin 2.16).$$

If we then apply Theorem 14.2, we get

$$(-2 + 3i)^3 \approx 13\sqrt{13}(\cos 6.48 + i\sin 6.48).$$

To find an approximate value for $\cos 6.48$ from the Table, we recall that $\cos\theta = \cos(\theta - 2\pi)$. Using $\pi \approx 3.14$, we see that

$$\cos 6.48 = \cos(6.48 - 6.28) = \cos 0.20 \approx 0.9801.$$

Likewise

$$\sin 6.48 = \sin 0.20 \approx 0.1987.$$

Hence

$$(-2 + 3i)^3 \approx 13\sqrt{13}\,(0.9801 + 0.1987i)$$
$$\approx 45.94 + 9.31i$$

If we use the second form as shown in Example 4 of Section 14.1,

$$-2 + 3i = \sqrt{13}\,\cos\left[\tan^{-1}\left(-\frac{3}{2}\right) + \pi\right] + i\sin\left[\tan^{-1}\left(-\frac{3}{2}\right) + \pi\right],$$

we get

$$(-2 + 3i)^3 = 13\sqrt{13}\left\{\cos\left[3\tan^{-1}\left(-\frac{3}{2}\right) + 3\pi\right]\right.$$
$$\left. + i\sin\left[3\tan^{-1}\left(-\frac{3}{2}\right) + 3\pi\right]\right\}.$$

It is possible with the application of trigonometric identities and considerable maneuvering to convert this to

$$3\sqrt{13}\left(\frac{46\sqrt{13}}{169} + \frac{9\sqrt{13}}{169}i\right)$$

which becomes $46 + 9i$.

By ordinary multiplication we get

$$(-2 + 3i)(-2 + 3i) = -5 - 12i$$

and

$$(-5 - 12i)(-2 + 3i) = 46 + 9i.$$

Direct multiplication would appear to be the better choice!

The foregoing example illustrates a fundamental truth, namely that every mathematical method has its advantages and also its limitations.

EXERCISES 14.2

◀ In Exercises 1–13, form the products of the two given complex numbers (a) by using polar forms and Theorem 14.1 and (b) by ordinary complex number multiplication.

1. $3 + 3\sqrt{3}i, \; -\sqrt{3} + i$

2. $i + 1, \; i - 1$

3. $4i - 4, \; -3i + 3$

4. $5\sqrt{3} - 5i, \; 2\sqrt{3} + 2i$

5. $\sqrt{3} + i, \; 2 - 2i$

6. $3 + 3i, \; 2 - 2\sqrt{3}i$

7. $\sqrt{3} - i, \; 2 + 2i$

8. $3i - 3, \; 4i + 4$

9. $6i + 6, \; 3\sqrt{3} + 3i$

10. $4\sqrt{3} - 4i, \; 2i - 2$

11. $1 - \sqrt{3}i, \; 5i - 5$

12. $2\sqrt{3} - 2i, \; 3i - 3$

13. $1 - \sqrt{3}i, \; \sqrt{2} + \sqrt{2}i$

14. Check that $(\sqrt{3} + i)^6 = -64$ by using the Binomial Theorem.

15. Check that $(-2 + 2i)^8 = 4096$ by using the Binomial Theorem.

16. Find the value of $(1 + i)^{10}$ by using the Binomial Theorem.

17. Check the results of Exercise 16 by using Theorem 14.2.

18. Find the value of $(1 - i)^9$ by using the Binomial Theorem.

19. Check the result of Exercise 18 by using Theorem 14.2.

20. Find the value of $(\sqrt{3} - i)^8$ by using the Binomial Theorem.

21. Check the result of Exercise 20 by using Theorem 14.2.

◀ In Exercises 22–39, use Theorem 14.2 to determine the indicated power of the given complex number. Write final answers in algebraic form.

22. $(2 - 2i)^6$

23. $(\sqrt{3} + i)^9$

24. $(1 + \sqrt{3}i)^7$

25. $(3 - 3i)^5$

26. $(5 - 5\sqrt{3}i)^4$

27. $(-\sqrt{3} - i)^8$

28. $(-1 - i)^{10}$

29. i^{13}

30. $(2 - 2\sqrt{3}i)^{11}$

31. $\left(\dfrac{1}{2} - \dfrac{1}{2}\sqrt{3}i \right)^{12}$

32. $\left(\dfrac{1}{3} + \dfrac{\sqrt{3}}{3}i\right)^6$ **36.** $(1 + i)^{-10}$

33. $\left(\dfrac{2}{3} - \dfrac{2}{3}i\right)^8$ **37.** $(2 - 2i)^{-8}$

34. $\left(\dfrac{\sqrt{2}}{3} + \dfrac{\sqrt{2}}{3}i\right)^{10}$ **38.** $(2 - 2\sqrt{3}\,i)^{-4}$

35. $(\sqrt{3} + i)^{-6}$ **39.** $(1 - i)^{-10}$

40. Show that

$$[\rho(\cos\theta + i\sin\theta)]^n = \rho^n(\cos n\theta + i\sin n\theta)$$

if n is any integer by completing the following argument.

The equation is true for $n \in N$ by Theorem 14.2. If $n = 0$, we have

$$[\rho(\cos\theta + i\sin\theta)]^0 = 1 \qquad \text{by definition}$$

and

$$\rho^0(\cos 0 + i\sin 0) = \boxed{?}.$$

Furthermore, we know that

$$(1) \qquad (\cos\theta + i\sin\theta)(\cos\theta - i\sin\theta) = \cos^2\theta + \sin^2\theta = \boxed{?}.$$

We also know that

$$\cos(-\theta) = \cos\theta \qquad \text{and} \qquad \sin(-\theta) = -\sin\theta.$$

Assume now that n is negative, that is, that $-n \in N$. We see then that

$$[\rho(\cos\theta + i\sin\theta)]^{-n} = \left[\frac{1}{\rho(\cos\theta + i\sin\theta)}\right]^n$$

$$= \left[\rho^{-1}\left(\frac{1}{\cos\theta + i\sin\theta}\right)\right]^n.$$

But from (1) we know that $\dfrac{1}{\cos\theta + i\sin\theta} = \cos\theta - i\sin\theta$. Hence

$$[\rho(\cos\theta + i\sin\theta)]^n = [\rho^{-1}(\cos\theta - i\sin\theta)]^{-n}$$
$$= (\rho^{-1}[\cos(-\theta) + i\sin(-\theta)])^{-n}$$
$$= (\rho^{-1})^{-n}[\cos(-n)(-\theta) + i\sin(-n)(-\theta)],$$

by Theorem 14.2, since $(-n) \in N$. Therefore $[\rho(\cos\theta + i\sin\theta)]^n = \rho^n(\boxed{?})$, and the proof is complete.

14.3 ROOTS OF ORDER n

Theorem 14.2 is generally known as de Moivre's Theorem in honor of the English mathematician (with a French-sounding name), Abraham de Moivre (1667–1754). Since the computations in the previous set of Exercises could have all been done by ordinary complex number multiplication, even though some would have involved considerable labor, de Moivre's Theorem might appear to have little significance. The theorem, however, does come into its own because of the way it can be applied in the opposite direction.

According to Theorem 14.2,

$$[\rho(\cos\theta + i\sin\theta)]^n = \rho^n(\cos n\theta + i\sin n\theta)$$

for any positive integer n. Now suppose we are asked, for example, to find the cube root of a given complex number z.

In this "reverse" problem we seek a complex number c such that

$$c^3 = z = \rho(\cos\theta + i\sin\theta).$$

If c is written in polar form as

$$c = r(\cos\phi + i\sin\phi),$$

then we want

$$[r(\cos\phi + i\sin\phi)]^3 = \rho(\cos\theta + i\sin\theta).$$

Hence, by Theorem 14.2, we must have

$$r^3(\cos 3\phi + i\sin 3\phi) = \rho(\cos\theta + i\sin\theta).$$

This suggests strongly that we shall have a correct polar form for

$$c = r(\cos\phi + i\sin\phi)$$

if we let $r = \sqrt[3]{\rho}$ and $\phi = \frac{1}{3}\theta$.

Example 1   Find a cube root of $8i$.

Solution: (You probably can guess at an answer, but our interest here lies in testing a conjecture.) To find c such that $c^3 = 8i$ we first write $8i$ in polar form as

$$8\left(\cos\frac{\pi}{2} + i\sin\frac{\pi}{2}\right).$$

From previous considerations we conclude that we should be able to write c as

$$r(\cos\phi + i\sin\phi),$$

where

$$r = \sqrt[3]{\rho} = \sqrt[3]{8} = 2$$

and

$$\phi = \frac{1}{3}\theta = \frac{1}{3}\cdot\frac{\pi}{2} = \frac{\pi}{6}.$$

Thus we have good reason to conjecture that

$$2\left(\cos\frac{\pi}{6} + i\sin\frac{\pi}{6}\right)$$

is a cube root of $8i$.

We know that $\cos\dfrac{\pi}{6} = \dfrac{\sqrt{3}}{2}$ and $\sin\dfrac{\pi}{6} = \dfrac{1}{2}$. Thus in algebraic form,

$$c = 2\left(\frac{\sqrt{3}}{2} + i\,\frac{1}{2}\right)$$
$$= \sqrt{3} + i.$$

Now comes the moment of truth! Let us find $(\sqrt{3} + i)^3$ using the Binomial Theorem. We have

$$(\sqrt{3} + i)^3 = \binom{3}{0}\sqrt{3}^3 + \binom{3}{1}(\sqrt{3})^2 i + \binom{3}{2}\sqrt{3}\,i^2 + \binom{3}{3}i^3$$
$$= (\sqrt{3})^3 + 3(3)i + 3\sqrt{3}(-1) + (-i)$$
$$= 3\sqrt{3} + 9i - 3\sqrt{3} - i$$
$$= 8i.$$

There is something surprising here! Had we been asked to guess at the answer, we would probably have taken a stab at $-2i$, and would have been right, since

$$(-2i)(-2i)(-2i) = (-2)^3(i^3)$$
$$= (-8)(-i)$$
$$= 8i.$$

Using de Moivre's Theorem in reverse, however, we came up with an answer $\sqrt{3} + i$, which is also correct.

Let us look at the situation from another standpoint. We are essentially trying to solve an equation

$$x^3 = 8i.$$

Because of the Fundamental Theorem of Algebra (Chapter 5), we know that such an equation can have at most three roots. We have used de Moivre's Theorem in reverse to determine one of these roots. The next question is: Can we, by a similar procedure, determine any other roots?

Let us have another look at $8i$ in polar form, that is,

$$8i = 8\left(\cos\frac{\pi}{2} + i\,\sin\frac{\pi}{2}\right).$$

We have already acknowledged the existence of many polar forms for the same complex number. Among the alternatives for $8i$ are

$$8\left[\cos\left(\frac{\pi}{2} + 2n\pi\right) + i\,\sin\left(\frac{\pi}{2} + 2n\pi\right)\right]$$

for any $n \in N$.

Letting $n = 1$ and 2, we get

$$8i = 8\left[\cos\left(\frac{\pi}{2} + 2\pi\right) + i\,\sin\left(\frac{\pi}{2} + 2\pi\right)\right] = 8\left(\cos\frac{5}{2}\pi + i\,\sin\frac{5}{2}\pi\right)$$

and

$$8i = 8\left[\cos\left(\frac{\pi}{2} + 4\pi\right) + i\,\sin\left(\frac{\pi}{2} + 4\pi\right)\right] = 8\left(\cos\frac{9}{2}\pi + i\,\sin\frac{9}{2}\pi\right),$$

respectively.

Recall now that we obtained a polar form for c, a cube root of $8i$, by letting $r = \sqrt[3]{\rho}$ and $\phi = \frac{1}{3}\theta$.

Repeating the process on the two new forms of $8i$, and noting that $\frac{1}{3}(\frac{5}{2}\pi) = \frac{5}{6}\pi$ and $\frac{1}{3}(\frac{9}{2}\pi) = \frac{3}{2}\pi$, we get

$$c' = 2\left(\cos\frac{5}{6}\pi + i\,\sin\frac{5}{6}\pi\right)$$

and

$$c'' = 2\left(\cos\frac{3}{2}\pi + i\,\sin\frac{3}{2}\pi\right).$$

Since

$$\cos\frac{5}{6}\pi = -\frac{\sqrt{3}}{2}, \qquad \sin\frac{5}{6}\pi = \frac{1}{2},$$

$$\cos\frac{3}{2}\pi = 0, \qquad \text{and} \qquad \sin\frac{3}{2}\pi = -1,$$

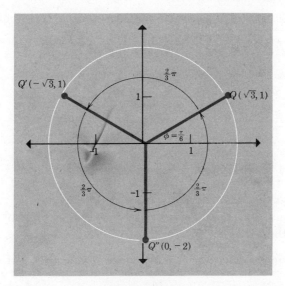

Figure 14-13

we see that

$$c' = 2\left(-\frac{\sqrt{3}}{2} + i \cdot \frac{1}{2}\right) = -\sqrt{3} + i$$

and

$$c'' = 2(0 - i) = -2i.$$

We have already verified the fact that $-2i$ is a cube root of $8i$. By the Binomial Theorem you can also check that $(-\sqrt{3} + i)^3 = 8i$.

Finally, we ask two more questions: (1) Are there any more solutions? (2) What happens to $8[\cos\left(\frac{\pi}{2} + 2n\pi\right) + i \sin \frac{\pi}{2} + 2n\pi)]$ when $n = 3, 4, \ldots,$ and so on? By the Fundamental Theorem of algebra we know that the answer to the first question is "No" since a polynomial equation of degree 3 with complex coefficients can have at *most* three roots.

To answer the second question, let us first construct a graphical representation of the process we have just completed. Figure 14-13 gives the picture. It shows points Q, Q', and Q'' corresponding, respectively, to $c = \sqrt{3} + i$, $c' = -\sqrt{3} + i$, and $c'' = -2i$, the three cube roots of $8i$. From the graph we see that the amplitude of c' is obtained by adding $\frac{2}{3}\pi$ to the amplitude $\frac{\pi}{6}$ of c. Likewise, the amplitude of c'' is obtained by adding $\frac{2}{3}\pi$ to the amplitude of c'.

The following table illustrates in outline form the process of obtaining amplitudes for c, c', and c''.

Amplitudes of z	Amplitudes of Cube Roots of z
$\dfrac{\pi}{2}$	$\dfrac{1}{3}\left(\dfrac{\pi}{2}\right) = \dfrac{\pi}{6} = \phi_1$
$\dfrac{\pi}{2} + 2\pi$	$\dfrac{1}{3}\left(\dfrac{\pi}{2} + 2\pi\right) = \dfrac{\pi}{6} + \dfrac{2}{3}\pi = \dfrac{5\pi}{6} = \phi_2$
$\dfrac{\pi}{2} + 4\pi$	$\dfrac{1}{3}\left(\dfrac{\pi}{2} + 4\pi\right) = \dfrac{\pi}{6} + \dfrac{4}{3}\pi = \dfrac{3}{2}\pi = \phi_3$

It should be clear from the graph that if we continue the process of adding $\frac{2}{3}\pi$ to the amplitudes (that is, the measure of the angles), we return to the points already plotted.

Analogously, if we introduce new amplitudes for z by adding 2π to each of the previous ones, we get the following table.

Amplitudes of z	Amplitudes of Cube Roots of z
$\dfrac{\pi}{2} + 6\pi$	$\dfrac{1}{3}\left(\dfrac{\pi}{2} + 6\pi\right) = \dfrac{\pi}{6} + 2\pi = \phi_1'$
$\dfrac{\pi}{2} + 8\pi$	$\dfrac{1}{3}\left(\dfrac{\pi}{2} + 8\pi\right) = \dfrac{5\pi}{6} + 2\pi = \phi_2'$
$\dfrac{\pi}{2} + 10\pi$	$\dfrac{1}{3}\left(\dfrac{\pi}{2} + 10\pi\right) = \dfrac{3}{2}\pi + 2\pi = \phi_3'$

Clearly, the terminal rays of angles ϕ_1', ϕ_2', and ϕ_3' coincide with the terminal rays of angles ϕ_1, ϕ_2, and ϕ_3, respectively. Hence $\cos \phi_1 = \cos \phi_1'$ and $\sin \phi_1 = \sin \phi_1'$; $\cos \phi_2 = \cos \phi_2'$ and $\sin \phi_2 = \sin \phi_2'$; $\cos \phi_3 = \cos \phi_3'$ and $\sin \phi_3 = \sin \phi_3'$.

From all of this we see that additional polar forms of $8i$, that is,

$$8\left[\cos\left(\frac{\pi}{2} + 2n\pi\right) + i \sin\left(\frac{\pi}{2} + 2n\pi\right)\right]$$

for $n = 3, 4, \ldots$ will yield no additional cube roots; similarly, for $n = -1, -2, \ldots$. (The latter can be seen by noting, for example, that

$$\frac{1}{3}\left(\frac{\pi}{2} - 2\pi\right) = \frac{\pi}{6} - \frac{2}{3}\pi = -\frac{\pi}{2},$$

which is coterminal with $\frac{3}{2}\pi$. On the graph this would be represented as a *clockwise* move from c to c'', that is, adding $-\frac{2}{3}\pi$ to $\frac{\pi}{6}$.)

From the detailed discussion of this example we can make these two significant conjectures:

(7) $[\rho(\cos\theta + i\sin\theta)]^{\frac{1}{n}} = \sqrt[n]{\rho}\left(\cos\dfrac{1}{n}\theta + i\sin\dfrac{1}{n}\theta\right)$ for $n \in N$.

(8) Every equation of the form

$$x^n = z$$

with $n \in N$ and z a nonzero complex number has exactly n solutions.

In essence, Statement (8) says that every nonzero complex number has n distinct nth roots. Statement (7) provides the clue as to how these roots can be exhibited.

For example, the five fifth roots of a nonzero complex number z can be written as follows. Let

$$z = \rho(\cos\theta + i\sin\theta).$$

For the five fifth roots we have

$$\sqrt[5]{\rho}\left(\cos\dfrac{1}{5}\theta + i\sin\dfrac{1}{5}\theta\right),$$

$$\sqrt[5]{\rho}\left[\cos\dfrac{1}{5}(\theta + 2\pi) + i\sin\dfrac{1}{5}(\theta + 2\pi)\right],$$

$$\sqrt[5]{\rho}\left[\cos\dfrac{1}{5}(\theta + 4\pi) + i\sin\dfrac{1}{5}(\theta + 4\pi)\right],$$

$$\sqrt[5]{\rho}\left[\cos\dfrac{1}{5}(\theta + 6\pi) + i\sin\dfrac{1}{5}(\theta + 6\pi)\right],$$

and

$$\sqrt[5]{\rho}\left[\cos\dfrac{1}{5}(\theta + 8\pi) + i\sin\dfrac{1}{5}(\theta + 8\pi)\right].$$

Here $\sqrt[5]{\rho}$ is the positive (real) fifth root of ρ.

As indicated earlier, a continuation of the process of adding 2π to θ fails to produce any new values. For instance, we can see that

$$\sqrt[5]{\rho}\left[\cos\dfrac{1}{5}(\theta + 10\pi) + i\sin\dfrac{1}{5}(\theta + 10\pi)\right]$$

$$= \sqrt[5]{\rho}\left[\cos\left(\dfrac{1}{5}\theta + 2\pi\right) + i\sin\left(\dfrac{1}{5}\theta + 2\pi\right)\right]$$

$$= \sqrt[5]{\rho}\left[\cos\dfrac{1}{5}\theta + i\sin\dfrac{1}{5}\theta\right].$$

For the general case we can state the following theorem.

THEOREM 14.3

The n distinct nth roots of any complex number $z = \rho \operatorname{cis} \theta$ are given by the formula

$$\sqrt[n]{\rho}\left[\operatorname{cis}\left(\frac{\theta + 2k\pi}{n}\right)\right] \quad \text{for } k = 0, 1, 2, \ldots, n-1$$

where $\sqrt[n]{\rho}$ is the positive (real) nth root of ρ.

Proof: By Theorem 14.2 we know that if $z = \rho(\cos\theta + i \sin\theta)$, then

$$\left\{\sqrt[n]{\rho}\left[\cos\left(\frac{\theta + 2k\pi}{n}\right) + i \sin\left(\frac{\theta + 2k\pi}{n}\right)\right]\right\}^n = z$$

for $k = 0, 1, 2, \ldots, n-1$, since $\cos(\theta + 2k\pi) = \cos\theta$ and $\sin(\theta + 2k\pi) = \sin\theta$ for $k = 0, 1, 2, \ldots, n-1$. Thus for any value of k in the set $\{0, 1, 2, \ldots, n-1\}$, the formula in Theorem 14.3 yields an nth root of z.

It remains to show that these n roots are distinct. Let us consider the two complex numbers

$$A = \operatorname{cis}\left(\frac{\theta}{n} + k \cdot \frac{2\pi}{n}\right)$$

and

$$B = \operatorname{cis}\left(\frac{\theta}{n} + j \cdot \frac{2\pi}{n}\right)$$

where k and j are integers such that $0 \le k \le n-1$ and $0 \le j \le n-1$. We must prove that $A \ne B$ except when $k = j$.

Let us use an indirect proof, that is, suppose that $A = B$ but $k \ne j$. Without loss of generality we can let $k > j$, that is, we can assume that $k - j > 0$. We know that $\operatorname{cis}\alpha \ne \operatorname{cis}\beta$ unless $\alpha = \beta + 2l\pi$ for some integer l. For, if not, then two distinct points on a circle of radius 1 in the Argand plane would be the coordinates of the same complex number. (See Figure 14-14.) But this contradicts the one-to-one correspondence between complex numbers and points in the Argand plane.

Our assumption that $A = B$, then, is equivalent to saying that

(1) $$\frac{\theta}{n} + k\frac{2\pi}{n} = \frac{\theta}{n} + j\frac{2\pi}{n} + 2l\pi$$

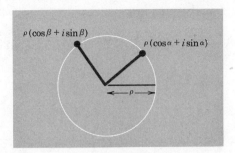

Figure 14-14

for some integer l. But if (1) is true, then

$$\left(\frac{\theta}{n} + k\frac{2\pi}{n}\right) - \left(\frac{\theta}{n} + j\frac{2\pi}{n}\right) = 2l\pi,$$

$$k \cdot \frac{2\pi}{n} - j\frac{2\pi}{n} = 2l\pi,$$

$$\left(\frac{k-j}{n}\right)2\pi = 2l\pi,$$

and

$$k - j = ln.$$

Now since $k - j > 0$, l must be a positive integer. Hence

$$k - j \geq n$$

so that

$$k \geq n + j.$$

But since j and k are nonnegative integers, this would mean that $k > n - 1$, contrary to the condition that $0 \leq k \leq n - 1$. Hence we have arrived at a contradiction of the hypothesis that $k > j$. Thus if $k \neq j$, then $A \neq B$; that is, $A \neq B$ except when $k = j$.

Thus we have shown that for the n values of k, 0, 1, 2, . . . , $n - 1$, the roots

$$\sqrt[n]{\rho}\, \text{cis}\left(\frac{\theta + 2k\pi}{n}\right)$$

are distinct.

Example 2   Find the four fourth roots of -2.

Solution: $-2 = 2(\cos \pi + i \sin \pi)$. Hence the four fourth roots are

$$\sqrt[4]{2}\left[\cos\left(\frac{\pi}{4} + k \cdot \frac{2\pi}{4}\right) + i \sin\left(\frac{\pi}{4} + k \cdot \frac{2\pi}{4}\right)\right] \quad \text{for } k = 0, 1, 2, 3.$$

That is, the four fourth roots are

$$\sqrt[4]{2}\left(\cos\frac{\pi}{4} + i\,\sin\frac{\pi}{4}\right),\quad \sqrt[4]{2}\left(\cos\frac{3\pi}{4} + i\,\sin\frac{3\pi}{4}\right),$$

$$\sqrt[4]{2}\left(\cos\frac{5}{4}\pi + i\,\sin\frac{5}{4}\pi\right),\quad\text{and}\quad \sqrt[4]{2}\left(\cos\frac{7}{4}\pi + i\,\sin\frac{7}{4}\pi\right).$$

We know that

$$\cos\frac{\pi}{4} = \sin\frac{\pi}{4} = \frac{\sqrt{2}}{2},\ \cos\frac{3\pi}{4} = -\frac{\sqrt{2}}{2},\ \sin\frac{3\pi}{4} = \frac{\sqrt{2}}{2},$$

$$\cos\frac{5\pi}{4} = \sin\frac{5\pi}{4} = -\frac{\sqrt{2}}{2},\ \cos\frac{7\pi}{4} = \frac{\sqrt{2}}{2},$$

and

$$\sin\frac{7\pi}{4} = -\frac{\sqrt{2}}{2}.$$

Furthermore,

$$\sqrt[4]{2}\cdot\frac{\sqrt{2}}{2} = \frac{2^{\frac{1}{4}}\cdot 2^{\frac{1}{2}}}{2} = 2^{\frac{3}{4}-1} = 2^{-\frac{1}{4}} = \frac{1}{\sqrt[4]{2}} = \sqrt[4]{\frac{1}{2}}.$$

Thus the four fourth roots of -2 in algebraic form are

$$\sqrt[4]{\frac{1}{2}} + i\sqrt[4]{\frac{1}{2}},\ -\sqrt[4]{\frac{1}{2}} + i\sqrt[4]{\frac{1}{2}},\ -\sqrt[4]{\frac{1}{2}} - i\sqrt[4]{\frac{1}{2}},$$

and

$$\sqrt[4]{\frac{1}{2}} - i\sqrt[4]{\frac{1}{2}}.$$

Example 3   Find the three cube roots of 7.

Solution: $7 = 7(\cos 0 + i\,\sin 0)$. The three cube roots are

$$\sqrt[3]{7}(\cos 0 + i\,\sin 0),$$

$$\sqrt[3]{7}\left(\cos\frac{2}{3}\pi + i\,\sin\frac{2\pi}{3}\right),$$

$$\sqrt[3]{7}\left(\cos\frac{4}{3}\pi + i\,\sin\frac{4\pi}{3}\right),$$

or

$$\sqrt[3]{7},\ -\frac{1}{2}\sqrt[3]{7} + i\frac{1}{2}\sqrt[3]{7}\sqrt{3},\quad\text{and}\quad -\frac{1}{2}\sqrt[3]{7} - i\frac{1}{2}\sqrt[3]{7}\sqrt{3}.$$

Once again we should point out that the use of Theorem 14.3 for finding nth roots of complex numbers can become very involved if a simple polar form does not exist. In this connection recall the example in Section 14.2 involving the determination of $(-2 + 3i)^3$.

For the cube roots of $-2 + 3i$ we would have to use an approximation such as

$$-2 + 3i \approx \sqrt{13}(\cos 2.16 + i \sin 2.16)$$

to obtain the following approximate roots:

$$\sqrt[6]{13}(\cos 0.72 + i \sin 0.72),$$
$$\sqrt[6]{13}(\cos 2.81 + i \sin 2.81),$$

and

$$\sqrt[6]{13}(\cos 4.90 + i \sin 4.90)$$

because $\frac{2\pi}{3} \approx 2.09$, $0.72 + 2.09 = 2.81$, and $2.81 + 2.09 = 4.90$.

EXERCISES 14.3

◀ In Exercises 1–15, find the n nth roots of the given complex number as indicated. Express results in algebraic form if a simple algebraic form exists, otherwise leave in polar form.

1. Three cube roots of i
2. Three cube roots of $-i$
3. Three cube roots of 8
4. Three cube roots of $-8i$
5. Four 4th roots of 4
6. Four 4th roots of $-4i$
7. Two square roots of $2 + \sqrt{3}i$
8. Three cube roots of $1 - \sqrt{3}i$
9. Four 4th roots of $2 - 2i$
10. Four 4th roots of 9
11. The six 6th roots of $-i$
12. The six 6th roots of 2
13. The six 6th roots of $\sqrt{3} - i$
14. The eight 8th roots of 8
15. The eight 8th roots of 1

◀ In Exercises 16–20, express the indicated roots in polar form.

16. The five 5th roots of -1
17. The five 5th roots of $1 + \sqrt{3}i$
18. The five 5th roots of $5 - 5i$
19. The seven 7th roots of 7
20. The five 5th roots of $\sqrt{3} - i$

14.4 THE nth ROOTS OF 1

In Chapter 5 we obtained three cube roots of 1 by solving the equation $x^3 - 1 = 0$ through synthetic division and the quadratic formula.

The process went as follows:

$$\underline{1\rfloor} \quad \begin{array}{cccc} 1 & 0 & 0 & -1 \\ & 1 & 1 & 1 \\ \hline 1 & 1 & 1 & 0 \end{array}$$

Reduced equation:

$$x^2 + x - 1 = 0.$$

By the quadratic formula,

$$x = -\frac{1}{2} + \frac{\sqrt{3}}{2}i \text{ or } x = -\frac{1}{2} - \frac{\sqrt{3}}{2}i.$$

The latter two roots were denoted by the special symbols ω and ω^2, respectively.

We can, of course, obtain these results using the method of Section 14.3. We have $1 = \cos 0 + i \sin 0$ and hence the three cube roots of 1 are

$$\cos\left(\frac{1}{3} \cdot 0\right) + i \sin\left(\frac{1}{3} \cdot 0\right) = \cos 0 + i \sin 0 = 1,$$

$$\cos\frac{2\pi}{3} + i \sin\frac{2\pi}{3} = -\frac{1}{2} + \frac{\sqrt{3}}{2}i,$$

and

$$\cos\frac{4\pi}{3} + i \sin\frac{4\pi}{3} = -\frac{1}{2} - \frac{\sqrt{3}}{2}i.$$

Let us now consider the more general question of finding the n nth roots of 1. From the polar form

$$\cos 0 + i \sin 0$$

we obtain the nth roots which are

$$\cos\left(\frac{0}{n} + k \cdot \frac{2\pi}{n}\right) + i \sin\left(\frac{0}{n} + k \cdot \frac{2\pi}{n}\right) = \cos\frac{2k\pi}{n} + i \sin\frac{2k\pi}{n}$$

for $k = 0, 1, 2, \ldots, n - 1$.

It should be clear that when $k = 0$, the resulting nth root is 1. For $k = 1$, we get

$$\cos\frac{2\pi}{n} + i \sin\frac{2\pi}{n}.$$

Let us call this R_n. For $k = 2$, we get

$$\cos \frac{4\pi}{n} + i \sin \frac{4\pi}{n}$$

and for $k = 3$, we get

$$\cos \frac{6\pi}{n} + i \sin \frac{6\pi}{n}.$$

Notice now that

$$\left(\cos \frac{2\pi}{n} + i \sin \frac{2\pi}{n} \right)^2 = \cos \frac{4\pi}{n} + i \sin \frac{4\pi}{n},$$

$$\left(\cos \frac{2\pi}{n} + i \sin \frac{2\pi}{n} \right)^3 = \cos \frac{6\pi}{n} + i \sin \frac{6\pi}{n},$$

and, in general,

$$\left(\cos \frac{2\pi}{n} + i \sin \frac{2\pi}{n} \right)^k = \cos \frac{2k\pi}{n} + i \sin \frac{2k\pi}{n}.$$

Thus we may designate the n nth roots of 1 as 1, $R_n = \cos \frac{2\pi}{n} + i \sin \frac{2\pi}{n}$, R_n^2, R_n^3, . . . , R_n^{n-1}. The symbol ω, appearing earlier, is customarily used for R_3.

With this in mind we can say that the n nth roots of 1 can be *generated* by the root R_n, the root with the smallest positive amplitude, $\frac{2\pi}{n}$.

For the six 6th roots of 1 we have

$$R_6 = \cos \frac{\pi}{3} + i \sin \frac{\pi}{3} = \frac{1}{2} + \frac{\sqrt{3}}{2} i.$$

The other roots are R_6^2, R_6^3, R_6^4, R_6^5, and 1 ($= R_6^6$).

It is interesting to examine graphically these nth roots of 1. Figure 14-15 is an Argand diagram showing each of the six 6th roots of 1. From this figure we see that the points in the plane corresponding to the roots of 1 are equally spaced around the unit circle. In this instance they constitute the vertices of a regular inscribed hexagon.

It should be clear that the points corresponding to each of the n nth roots of 1, often called the nth **roots of unity,** will similarly constitute vertices of a regular polygon of n sides inscribed in a unit circle.

We can use the nth roots of unity to find the nth roots of any given complex number. To see how this is done consider the complex

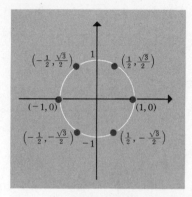

Figure 14-15

number z with a polar form

$$\rho(\cos \theta + i \sin \theta).$$

We have already seen that an nth root of z is

$$\sqrt[n]{\rho}\left(\cos \frac{\theta}{n} + i \sin \frac{\theta}{n}\right).$$

To form another root we would write

$$\sqrt[n]{\rho}\left[\cos \left(\frac{\theta}{n} + \frac{2\pi}{n}\right) + i \sin \left(\frac{\theta}{n} + \frac{2\pi}{n}\right)\right].$$

But this can also be written as the product of two factors:

$$\sqrt[n]{\rho}\left(\cos \frac{\theta}{n} + i \sin \frac{\theta}{n}\right) \cdot \left(\cos \frac{2\pi}{n} + i \sin \frac{2\pi}{n}\right).$$

What does that factor on the right look like? Does it have a striking resemblance to R_n?

It should be easy to develop the idea that if

$$r = \sqrt[n]{\rho}\left(\cos \frac{\theta}{n} + i \sin \frac{\theta}{n}\right)$$

is an nth root of z, then the n nth roots of z are

$$r, \ rR_n, \ rR_n^2, \ rR_n^3, \ \ldots, \ rR_n^{n-1}.$$

In one of our earlier examples we sought the cube roots of $8i$. The first to turn up was $\sqrt{3} + i$. The other two were $-\sqrt{3} + i$ and $-2i$. Our discussion now reveals that these last two could be obtained as $(\sqrt{3} + i)\omega$ and $(\sqrt{3} + i)\omega^2$, where, as before, $\omega = -\dfrac{1}{2} + \dfrac{\sqrt{3}}{2}i$ is one of the cube roots of 1.

In the use of R_n to determine nth roots of a complex number it is sometimes more convenient to use degree rather than radian measure.

Let us now consider the matter of finding the cube roots of $1 + i$. The polar form of $1 + i$ is

$$\sqrt{2}(\cos 45 + i \sin 45)$$

where "45" represents the amplitude measured in degrees. A cube root of 1, which we have called ω, is

$$\cos 120 + i \sin 120.$$

A cube root, r, of $1 + i$ is

$$\sqrt[6]{2}(\cos 15 + i \sin 15).$$

For the three cube roots r, $r\omega$, $r\omega^2$, we have, respectively,

$$\sqrt[6]{2}(\cos 15 + i \sin 15),$$
$$\sqrt[6]{2}(\cos 135 + i \sin 135),$$

and

$$\sqrt[6]{2}(\cos 255 + i \sin 255).$$

These can be converted into algebraic form by the use of identities (for example, $\cos 15 = \cos(60 - 45)$) or by the use of tables, if an approximation is satisfactory.

An nth root R_n^k of unity is called a **primitive** nth root of unity if $(R_n^k)^m \neq 1$ for $0 < m < n$. Consider, for example, the 10th roots of unity, that is,

$$1, R_{10}(= R_{10}^1), R_{10}^2, R_{10}^3, \ldots, R_{10}^9.$$

As we have shown, $(R_{10}^1)^m \neq 1$ for $m = 1, 2, 3, \ldots, 9$ since all 10 roots are distinct. Hence $R_{10}^1 = R_{10}$ is a primitive 10th root of unity.

What about the root R_{10}^2? We know that $(R_{10}^2)^5 = R_{10}^{10} = 1$. Since, obviously, $0 < 5 < 10$, it follows that R_{10}^2 is not a primitive 10th root of unity.

Is R_{10}^3 a primitive 10th root of unity? Is there a natural number m in the interval $0 < m < 10$ for which $(R_{10}^3)^m = 1$? If $m = 2$, we have

$$(R_{10}^3)^2 = R_{10}^6 \neq 1,$$

and if $m = 3$, then

$$(R_{10}^3)^3 = R_{10}^9 \neq 1.$$

For $m = 4$ we have

$$(R_{10}^3)^4 = R_{10}^{12} = (R_{10}^{10})(R_{10}^2) = 1 \cdot R_{10}^2 \neq 1.$$

By showing that, also, $(R_{10}^3)^k \neq 1$ for $k = 5, 6, 7, 8$, and 9, we may prove that R_{10}^3 is a primitive 10th root of unity.

Finally, consider R_{10}^4. Since

$$(R_{10}^4)^5 = R_{10}^{20} = (R_{10}^{10})(R_{10}^{10}) = 1 \cdot 1 = 1,$$

we conclude that R_{10}^4 is not a primitive 10th root of unity. Can you suggest a means, other than trial and error, for determining whether or not R_n^k is a primitive nth root of unity for a given n and $k \in N$ with $k < n$? You will have an opportunity to deal with this question in the Exercises.

EXERCISES 14.4

1. Show by multiplication that $(\sqrt{3} + i)\omega$ is a cube root of $8i$.
2. Show by multiplication that $(\sqrt{3} + i)\omega^2$ is a cube root of $8i$.
3. Find the eight 8th roots of 1 and express these in algebraic form.
4. Plot the points corresponding to the results of Exercise 3 using a unit circle.
5. Find the five 5th roots of unity, expressing answers in polar form with amplitudes given in degree measure.
6. Plot the points corresponding to the results of Exercise 5 using a unit circle.
7. Convert the results of Exercise 5 to algebraic form using approximations from a table of trigonometric functions.
8. Show that the five 5th roots of 1 are all powers of R_5^2. (*Hint:* Use the fact that $R_5^5 = 1$.)
9. Are the five 5th roots of 1 all powers of any other roots beside R_5 and R_5^2? Explain.
10. Give the primitive 6th roots of unity expressed as powers of R_6.
11. Give the primitive 10th roots of unity expressed as powers of R_{10}.
12. Give the primitive 12th roots of unity expressed as powers of R_{12}.
13. How many primitive 13th roots of unity are there?
14. Give the number of primitive 20th roots of unity.

15. Copy and complete the statement of the following theorem.

THEOREM. An nth root R_n^k of 1 is a primitive nth root of unity if and only if the greatest common divisor of n and k is $\boxed{?}$.

16. CHALLENGE PROBLEM. Prove the theorem stated in Exercise 15.

CHAPTER SUMMARY

Argand diagrams have been used to help determine a POLAR FORM, $\rho(\cos\theta + i\sin\theta)$, abbreviated cis θ, for a complex number $x + yi$. In the process it has been shown that the following relations hold.

$$\rho = \sqrt{x^2 + y^2}.$$
$$x = \rho\cos\theta \text{ and } y = \rho\sin\theta.$$
$$\text{If } x > 0, \theta = \tan^{-1}\frac{y}{x}.$$
$$\text{If } x < 0, \theta = \tan^{-1}\frac{y}{x} + \pi.$$

If $x = 0$, $\theta = \frac{\pi}{2}$ if $y > 0$, $\theta = -\frac{\pi}{2}$ if $y < 0$, and, finally, the measure of θ is any real number if $x = 0$ and $y = 0$.

The real number ρ is the ABSOLUTE VALUE of the complex number z, and the measure of θ is called an AMPLITUDE of z.

With respect to polar forms the following theorems hold.

THEOREM 14.1
If two complex numbers z_1 and z_2 have absolute values ρ_1 and ρ_2 and amplitudes θ_1 and θ_2, respectively, then

(1) The absolute value of the product z_1z_2 is $\rho_1\rho_2$ and
(2) an amplitude of the product z_1z_2 is $\theta_1 + \theta_2$;
that is, $(\rho_1 \text{ cis } \theta_1)(\rho_2 \text{ cis } \theta_2) = \rho_1\rho_2 \text{ cis }(\theta_1 + \theta_2)$.

THEOREM 14.2
(*De Moivre's Theorem*) was proved by induction. It states that if z is a complex number with a polar form $\rho(\cos\theta + i\sin\theta)$, then z^n for $n \in N$ has a polar form

$$\rho^n(\cos n\theta + i\sin n\theta).$$

Theorem 14.2 was used to deduce another useful theorem.

THEOREM 14.3

The n distinct nth roots of any complex number $z = \rho \operatorname{cis} \theta$ are given by the formula

$$\sqrt[n]{\rho} \operatorname{cis}\left(\frac{\theta + 2k\pi}{n}\right)$$

for $k = 0, 1, 2, \ldots, n - 1$, where $\sqrt[n]{\rho}$ is the positive (real) nth root of ρ.

The n nth roots of 1, also called the nth roots of unity, may be written as $1, R_n, R_n^2, \ldots, R_n^{n-1}$, where

$$R_n = \cos\frac{2\pi}{n} + \sin\frac{2\pi}{n}.$$

The n nth roots of any complex number z may be written as r, $rR_n, rR_n^2, \ldots, rR_n^{n-1}$, where

$$r = \sqrt[n]{\rho}\left(\cos\frac{\theta}{n} + i\sin\frac{\theta}{n}\right),$$

ρ is the absolute value of z, and θ is an amplitude of z.

It has been observed that the n nth roots of unity can be represented graphically as the vertices of a regular polygon of n sides inscribed in a unit circle.

An nth root R_n^k of 1 is called a PRIMITIVE nth root of unity if $(R_n^k)^m \neq 1$ for $0 < m < n$.

REVIEW EXERCISES

◀In Exercises 1–5, write each of the given complex numbers in a polar form such that $0 \leq \theta < 2\pi$.

1. $4 + 4i$
2. $-2i$
3. 10
4. $3\sqrt{3} - 3i$
5. $\dfrac{1}{4} - \dfrac{\sqrt{3}}{4}i$

6. Write $12\left(\cos\dfrac{5}{6}\pi + i\sin\dfrac{5}{6}\pi\right)$ in algebraic form.

7. Write $\cos\dfrac{5\pi}{12} + i\sin\dfrac{5\pi}{12}$ in algebraic form.

8. Write in algebraic form the complex number z with absolute value 6 and an amplitude $\theta = \tan^{-1}\left(-\dfrac{3}{4}\right)$.

◀ In Exercises 9 and 10, give a polar form with an approximate amplitude for the given complex number by using a table of trigonometric functions.

   9. $2 - 3i$ **10.** $5 + 2i$

◀ In Exercises 11 and 12, find the product of the two given complex numbers by using (a) polar forms and (b) direct multiplication.

   11. $3i - 3, 4 - 4i$ **12.** $6\sqrt{3} - 6, 2 + 2\sqrt{3}$

◀ In Exercises 13–18, use de Moivre's Theorem to determine the indicated powers of the given complex numbers. Write answers in algebraic form.

   13. $(1 - \sqrt{3}i)^7$ **16.** $(1 - i)^8$

   14. $(5 + 5\sqrt{3}i)^6$ **17.** $\left(\dfrac{1}{2} + \dfrac{1}{2}\sqrt{3}i\right)^6$

   15. $(2 + 2\sqrt{3}i)^{10}$ **18.** $(7 - 7i)^{11}$

   19. Find the three cube roots of $-27i$.
   20. Find the four 4th roots of 12.
   21. Find the two square roots of $2 - 2\sqrt{3}i$.
   22. Express the five 5th roots of $1 - \sqrt{3}i$ in polar form.
   23. Express the seven 7th roots of i in polar form.
   24. Write the eight 8th roots of 1 in algebraic form.
   25. Draw the Argand diagram of the eight 8th roots of 1 using a unit circle.

◀ In Exercises 26–30, express as powers of R_n the primitive nth roots of unity for each of the given values of n.

   26. $n = 15$ **29.** $n = 30$
   27. $n = 21$ **30.** $n = 41$
   28. $n = 23$

GOING FURTHER: READING AND RESEARCH

   You may have wondered how tables for the trigonometric functions have been constructed. At the conclusion of Chapter 1 there was some discussion about the convergence of an infinite series; in Chapter 6 the Binomial Theorem was discussed; and in this chapter you have been introduced to de Moivre's Theorem. All three ideas can be put together to form the basis for approximating values of $\cos \theta$ and $\sin \theta$.

   Here is a small sampling of how it is done. We start with $\cos n\theta + i \sin n\theta$, which, by de Moivre's Theorem, is equal to

$$(\cos \theta + i \sin \theta)^n.$$

By the Binomial Theorem we get

$$(\cos\theta + i\sin\theta)^n = \cos^n\theta + n\cos^{n-1}\theta\, i\sin\theta$$
$$+ \frac{n(n-1)}{2!}\cos^{n-2}\theta i^2\sin^2\theta + \frac{n(n-1)(n-2)}{3!}\cos^{n-3}\theta i^3\sin^3\theta + \cdots.$$

Since $i^2 = -1$, $i^3 = -i$, $i^4 = 1$, and so on, this becomes

$$(1) \qquad \cos^n\theta + in\cos^{n-1}\theta\sin\theta - \frac{n(n-1)}{2!}\cos^{n-2}\theta\sin^2\theta$$
$$- \frac{in(n-1)(n-2)}{3!}\cos^{n-3}\theta\sin^3\theta$$
$$+ \frac{n(n-1)(n-2)(n-3)}{4!}\cos^{n-4}\theta\sin^4\theta + \cdots.$$

The larger the value of n, the more terms in the series.

In our original number,

$$\cos n\theta + i\sin n\theta,$$

the real part is $\cos n\theta$. Since equal complex numbers must have equal real parts, then $\cos n\theta$ must equal the real parts in the binomial expansion (1), that is,

$$(2) \qquad \cos n\theta = \cos^n\theta - \frac{n(n-1)}{2!}\cos^{n-2}\theta\sin^2\theta$$
$$+ \frac{n(n-1)(n-2)(n-3)}{4!}\cos^{n-4}\theta\sin^4\theta + \cdots.$$

We now let $n\theta = \alpha$. Then $\theta = \dfrac{\alpha}{n}$ and $n = \dfrac{\alpha}{\theta}$. (Here, of course, we are considering α as a real number, that is, the radian measure of an angle.) Substituting $\dfrac{\alpha}{\theta}$ for n, we can write (2) as

$$(3) \qquad \cos\alpha = \cos^n\theta - \frac{\frac{\alpha}{\theta}\left(\frac{\alpha}{\theta}-1\right)}{2!}\cos^{n-2}\theta\sin^2\theta$$
$$+ \frac{\frac{\alpha}{\theta}\left(\frac{\alpha}{\theta}-1\right)\left(\frac{\alpha}{\theta}-2\right)\left(\frac{\alpha}{\theta}-3\right)}{4!}\cos^{n-4}\theta\sin^4\theta + \cdots.$$

Since $\dfrac{\alpha}{\theta}\left(\dfrac{\alpha}{\theta}-1\right) = \dfrac{\alpha}{\theta}\left(\dfrac{\alpha-\theta}{\theta}\right) = \dfrac{\alpha(\alpha-\theta)}{\theta^2}$, we can write

$$-\frac{\frac{\alpha}{\theta}\left(\frac{\alpha}{\theta} - 1\right)}{2!} \cos^{n-2}\theta \, \sin^2\theta$$

as

$$-\frac{\alpha(\alpha - \theta)}{2!} \cos^{n-2}\theta \frac{(\sin\theta)^2}{\theta^2}.$$

With other similar manipulations the series (3) becomes

$$(4) \quad \cos\alpha = \cos^n\theta - \frac{\alpha(\alpha - \theta)}{2!} \cos^{n-2}\theta \left(\frac{\sin\theta}{\theta}\right)^2$$

$$+ \frac{\alpha(\alpha - \theta)(\alpha - 2\theta)(\alpha - 3\theta)}{4!} \cos^{n-4}\theta \left(\frac{\sin\theta}{\theta}\right)^4 \cdots.$$

Now we want α to remain constant as n increases. Then, since $\theta = \dfrac{\alpha}{n}$,

we see that as n increases, θ decreases.

We will now take a deep mathematical plunge. (You will be taking many of these as you explore higher mathematics!) We let n become larger and larger without bound, which means that θ will approach 0. Now as θ approaches 0, $\cos\theta$ approaches 1, that is, the closer θ is to 0, the closer $\cos\theta$ is to 1. (Indeed, $\cos 0 = 1$.) Moreover, it can be shown that $\dfrac{\sin\theta}{\theta}$ also approaches 1, as θ approaches 0, a result which becomes intuitively plausible if one studies the table of $\sin\theta$ for θ in radians. From this table we see that $\sin\theta$ and θ differ by less and less as θ approaches 0. (In fact, the table shows that to four significant figures, $\sin\theta = \theta$ for $0 \leq \theta \leq 0.06$.) Hence the quotient $\dfrac{\sin\theta}{\theta}$ gets closer and closer to 1.

Since $\alpha - \theta$ approaches α as θ approaches 0, our series finally becomes

$$(5) \qquad \cos\alpha = 1 - \frac{\alpha^2}{2!} + \frac{\alpha^4}{4!} - \frac{\alpha^6}{6!} + \cdots.$$

Let us try the series using $\alpha = 0.2$ radians. Then (1) gives us

$$\cos 0.2 \approx 1 - \frac{(0.2)^2}{2!} + \frac{(0.2)^4}{4!} = 1 - 0.02 + 0.00007 \approx 0.98007.$$

Our table gives us $\cos 0.2 = 0.9801$ to four significant figures. By equating the imaginary parts in the binomial expansion (and you should try doing this) we get

$$(6) \qquad \sin\alpha = \alpha - \frac{\alpha^3}{3!} + \frac{\alpha^5}{5!} + \cdots.$$

Using (6) we get

$$\sin 0.2 \approx 0.2 - \frac{(0.2)^3}{3!} + \frac{(0.2)^5}{5!} \approx 0.1987.$$

Compare this result with the table entry.

Many functions can have their functional values expressed in terms of an infinite series such as the two given above for the sine and cosine functions. Further details can be found in the chapters on infinite series in almost any calculus text.

TABLE I. FOUR-PLACE LOGARITHMS

	0	1	2	3	4	5	6	7	8	9
10	0000	0043	0086	0128	0170	0212	0253	0294	0334	0374
11	0414	0453	0492	0531	0569	0607	0645	0682	0719	0755
12	0792	0828	0864	0899	0934	0969	1004	1038	1072	1106
13	1139	1173	1206	1239	1271	1303	1335	1367	1399	1430
14	1461	1492	1523	1553	1584	1614	1644	1673	1703	1732
15	1761	1790	1818	1847	1875	1903	1931	1959	1987	2014
16	2041	2068	2095	2122	2148	2175	2201	2227	2253	2279
17	2304	2330	2355	2380	2405	2430	2455	2480	2504	2529
18	2553	2577	2601	2625	2648	2672	2695	2718	2742	2765
19	2788	2810	2833	2856	2878	2900	2923	2945	2967	2989
20	3010	3032	3054	3075	3096	3118	3139	3160	3181	3201
21	3222	3243	3263	3284	3304	3324	3345	3365	3385	3404
22	3424	3444	3464	3483	3502	3522	3541	3560	3579	3598
23	3617	3636	3655	3674	3692	3711	3729	3747	3766	3784
24	3802	3820	3838	3856	3874	3892	3909	3927	3945	3962
25	3979	3997	4014	4031	4048	4065	4082	4099	4116	4133
26	4150	4166	4183	4200	4216	4232	4249	4265	4281	4298
27	4314	4330	4346	4362	4378	4393	4409	4425	4440	4456
28	4472	4487	4502	4518	4533	4548	4564	4579	4594	4609
29	4624	4639	4654	4669	4683	4698	4713	4728	4742	4757
30	4771	4786	4800	4814	4829	4843	4857	4871	4886	4900
31	4914	4928	4942	4955	4969	4983	4997	5011	5024	5038
32	5051	5065	5079	5092	5105	5119	5132	5145	5159	5172
33	5185	5198	5211	5224	5237	5250	5263	5276	5289	5302
34	5315	5328	5340	5353	5366	5378	5391	5403	5416	5428
35	5441	5453	5465	5478	5490	5502	5514	5527	5539	5551
36	5563	5575	5587	5599	5611	5623	5635	5647	5658	5670
37	5682	5694	5705	5717	5729	5740	5752	5763	5775	5786
38	5798	5809	5821	5832	5843	5855	5866	5877	5888	5899
39	5911	5922	5933	5944	5955	5966	5977	5988	5999	6010
40	6021	6031	6042	6053	6064	6075	6085	6096	6107	6117
41	6128	6138	6149	6160	6170	6180	6191	6201	6212	6222
42	6232	6243	6253	6263	6274	6284	6294	6304	6314	6325
43	6335	6345	6355	6365	6375	6385	6395	6405	6415	6425
44	6435	6444	6454	6464	6474	6484	6493	6503	6513	6522
45	6532	6542	6551	6561	6571	6580	6590	6599	6609	6618
46	6628	6637	6646	6656	6665	6675	6684	6693	6702	6712
47	6721	6730	6739	6749	6758	6767	6776	6785	6794	6803
48	6812	6821	6830	6839	6848	6857	6866	6875	6884	6893
49	6902	6911	6920	6928	6937	6946	6955	6964	6972	6981
50	6990	6998	7007	7016	7024	7033	7042	7050	7059	7067
51	7076	7084	7093	7101	7110	7118	7126	7135	7143	7152
52	7160	7168	7177	7185	7193	7202	7210	7218	7226	7235
53	7243	7251	7259	7267	7275	7284	7292	7300	7308	7316
54	7324	7332	7340	7348	7356	7364	7372	7380	7388	7396

TABLE I. FOUR-PLACE LOGARITHMS—*continued*

	0	1	2	3	4	5	6	7	8	9
55	7404	7412	7419	7427	7435	7443	7451	7459	7466	7474
56	7482	7490	7497	7505	7513	7520	7528	7536	7543	7551
57	7559	7566	7574	7582	7589	7597	7604	7612	7619	7627
58	7634	7642	7649	7657	7664	7672	7679	7686	7694	7701
59	7709	7716	7723	7731	7738	7745	7752	7760	7767	7774
60	7782	7789	7796	7803	7810	7818	7825	7832	7839	7846
61	7853	7860	7868	7875	7882	7889	7896	7903	7910	7917
62	7924	7931	7938	7945	7952	7959	7966	7973	7980	7987
63	7993	8000	8007	8014	8021	8028	8035	8041	8048	8055
64	8062	8069	8075	8082	8089	8096	8102	8109	8116	8122
65	8129	8136	8142	8149	8156	8162	8169	8176	8182	8189
66	8195	8202	8209	8215	8222	8228	8235	8241	8248	8254
67	8261	8267	8274	8280	8287	8293	8299	8306	8312	8319
68	8325	8331	8338	8344	8351	8357	8363	8370	8376	8382
69	8388	8395	8401	8407	8414	8420	8426	8432	8439	8445
70	8451	8457	8463	8470	8476	8482	8488	8494	8500	8506
71	8513	8519	8525	8531	8537	8543	8549	8555	8561	8567
72	8573	8579	8585	8591	8597	8603	8609	8615	8621	8627
73	8633	8639	8645	8651	8657	8663	8669	8675	8681	8686
74	8692	8698	8704	8710	8716	8722	8727	8733	8739	8745
75	8751	8756	8762	8768	8774	8779	8785	8791	8797	8802
76	8808	8814	8820	8825	8831	8837	8842	8848	8854	8859
77	8865	8871	8876	8882	8887	8893	8899	8904	8910	8915
78	8921	8927	8932	8938	8943	8949	8954	8960	8965	8971
79	8976	8982	8987	8993	8998	9004	9009	9015	9020	9025
80	9031	9036	9042	9047	9053	9058	9063	9069	9074	9079
81	9085	9090	9096	9101	9106	9112	9117	9122	9128	9133
82	9138	9143	9149	9154	9159	9165	9170	9175	9180	9186
83	9191	9196	9201	9206	9212	9217	9222	9227	9232	9238
84	9243	9248	9253	9258	9263	9269	9274	9279	9284	9289
85	9294	9299	9304	9309	9315	9320	9325	9330	9335	9340
86	9345	9350	9355	9360	9365	9370	9375	9380	9385	9390
87	9395	9400	9405	9410	9415	9420	9425	9430	9435	9440
88	9445	9450	9455	9460	9465	9469	9474	9479	9484	9489
89	9494	9499	9504	9509	9513	9518	9523	9528	9533	9538
90	9542	9547	9552	9557	9562	9566	9571	9576	9581	9586
91	9590	9595	9600	9605	9609	9614	9619	9624	9628	9633
92	9638	9643	9647	9652	9657	9661	9666	9671	9675	9680
93	9685	9689	9694	9699	9703	9708	9713	9717	9722	9727
94	9731	9736	9741	9745	9750	9754	9759	9763	9768	9773
95	9777	9782	9786	9791	9795	9800	9805	9809	9814	9818
96	9823	9827	9832	9836	9841	9845	9850	9854	9859	9863
97	9868	9872	9877	9881	9886	9890	9894	9899	9903	9908
98	9912	9917	9921	9926	9930	9934	9939	9943	9948	9952
99	9956	9961	9965	9969	9974	9978	9983	9987	9991	9996

TABLE II

Four-Place Values of Trigonometric Functions
Angle θ in Degrees and Radians

Angle θ		sin θ	csc θ	tan θ	cot θ	sec θ	cos θ		
Degrees	Radians								
0° 00′	.0000	.0000	No value	.0000	No value	1.000	1.0000	1.5708	90° 00′
10	029	029	343.8	029	343.8	000	000	679	50
20	058	058	171.9	058	171.9	000	000	650	40
30	.0087	.0087	114.6	.0087	114.6	1.000	1.0000	1.5621	30
40	116	116	85.95	116	85.94	000	.9999	592	20
50	145	145	68.76	145	68.75	000	999	563	10
1° 00′	.0175	.0175	57.30	.0175	57.29	1.000	.9998	1.5533	89° 00′
10	204	204	49.11	204	49.10	000	998	504	50
20	233	233	42.98	233	42.96	000	997	475	40
30	.0262	.0262	38.20	.0262	38.19	1.000	.9997	1.5446	30
40	291	291	34.38	291	34.37	000	996	417	20
50	320	320	31.26	320	31.24	001	995	388	10
2° 00′	.0349	.0349	28.65	.0349	28.64	1.001	.9994	1.5359	88° 00′
10	378	378	26.45	378	26.43	001	993	330	50
20	407	407	24.56	407	24.54	001	992	301	40
30	.0436	.0436	22.93	.0437	22.90	1.001	.9990	1.5272	30
40	465	465	21.49	466	21.47	001	989	243	20
50	495	494	20.23	495	20.21	001	988	213	10
3° 00′	.0524	.0523	19.11	.0524	19.08	1.001	.9986	1.5184	87° 00′
10	553	552	18.10	553	18.07	002	985	155	50
20	582	581	17.20	582	17.17	002	983	126	40
30	.0611	.0610	16.38	.0612	16.35	1.002	.9981	1.5097	30
40	640	640	15.64	641	15.60	002	980	068	20
50	669	669	14.96	670	14.92	002	978	039	10
4° 00′	.0698	.0698	14.34	.0699	14.30	1.002	.9976	1.5010	86° 00′
10	727	727	13.76	729	13.73	003	974	981	50
20	756	756	13.23	758	13.20	003	971	952	40
30	.0785	.0785	12.75	.0787	12.71	1.003	.9969	1.4923	30
40	814	814	12.29	816	12.25	003	967	893	20
50	844	843	11.87	846	11.83	004	964	864	10
5° 00′	.0873	.0872	11.47	.0875	11.43	1.004	.9962	1.4835	85° 00′
10	902	901	11.10	904	11.06	004	959	806	50
20	931	929	10.76	934	10.71	004	957	777	40
30	.0960	.0958	10.43	.0963	10.39	1.005	.9954	1.4748	30
40	989	987	10.13	992	10.08	005	951	719	20
50	.1018	.1016	9.839	.1022	9.788	005	948	690	10
6° 00′	.1047	.1045	9.567	.1051	9.514	1.006	.9945	1.4661	84° 00′
		cos θ	sec θ	cot θ	tan θ	csc θ	sin θ	Radians	Degrees
								Angle θ	

TABLE II—*continued*

Angle θ Degrees	Radians	sin θ	csc θ	tan θ	cot θ	sec θ	cos θ		
6° 00′	.1047	.1045	9.567	.1051	9.514	1.006	.9945	1.4661	84° 00′
10	076	074	9.309	080	9.255	006	942	632	50
20	105	103	9.065	110	9.010	006	939	603	40
30	.1134	.1132	8.834	.1139	8.777	1.006	.9936	1.4573	30
40	164	161	8.614	169	8.556	007	932	544	20
50	193	190	8.405	198	8.345	007	929	515	10
7° 00′	.1222	.1219	8.206	.1228	8.144	1.008	.9925	1.4486	83° 00′
10	251	248	8.016	257	7.953	008	922	457	50
20	280	276	7.834	287	7.770	008	918	428	40
30	.1309	.1305	7.661	.1317	7.596	1.009	.9914	1.4399	30
40	338	334	7.496	346	7.429	009	911	370	20
50	367	363	7.337	376	7.269	009	907	341	10
8° 00′	.1396	.1392	7.185	.1405	7.115	1.010	.9903	1.4312	82° 00′
10	425	421	7.040	435	6.968	010	899	283	50
20	454	449	6.900	465	6.827	011	894	254	40
30	.1484	.1478	6.765	.1495	6.691	1.011	.9890	1.4224	30
40	513	507	6.636	524	6.561	012	886	195	20
50	542	536	6.512	554	6.435	012	881	166	10
9° 00′	.1571	.1564	6.392	.1584	6.314	1.012	.9877	1.4137	81° 00′
10	600	593	277	614	197	013	872	108	50
20	629	622	166	644	084	013	868	079	40
30	.1658	.1650	6.059	.1673	5.976	1.014	.9863	1.4050	30
40	687	679	5.955	703	871	014	858	1.4021	20
50	716	708	855	733	769	015	853	992	10
10° 00′	.1745	.1736	5.759	.1763	5.671	1.015	.9848	1.3963	80° 00′
10	774	765	665	793	576	016	843	934	50
20	804	794	575	823	485	016	838	904	40
30	.1833	.1822	5.487	.1853	5.396	1.017	.9833	1.3875	30
40	862	851	403	883	309	018	827	846	20
50	891	880	320	914	226	018	822	817	10
11° 00′	.1920	.1908	5.241	.1944	5.145	1.019	.9816	1.3788	79° 00′
10	949	937	164	974	066	019	811	759	50
20	978	965	089	.2004	4.989	020	805	730	40
30	.2007	.1994	5.016	.2035	4.915	1.020	.9799	1.3701	30
40	036	.2022	4.945	065	843	021	793	672	20
50	065	051	876	095	773	022	787	643	10
12° 00′	.2094	.2079	4.810	.2126	4.705	1.022	.9781	1.3614	78° 00′
10	123	108	745	156	638	023	775	584	50
20	153	136	682	186	574	024	769	555	40
30	.2182	.2164	4.620	.2217	4.511	1.024	.9763	1.3526	30
40	211	193	560	247	449	025	757	497	20
50	240	221	502	278	390	026	750	468	10
13° 00′	.2269	.2250	4.445	.2309	4.331	1.026	.9744	1.3439	77° 00′
		cos θ	sec θ	cot θ	tan θ	csc θ	sin θ	Radians	Degrees
								Angle θ	

TABLE II—*continued*

Angle θ Degrees	Radians	sin θ	csc θ	tan θ	cot θ	sec θ	cos θ		
13° 00′	.2269	.2250	4.445	.2309	4.331	1.026	.9744	1.3439	77° 00′
10	298	278	390	339	275	027	737	410	50
20	327	306	336	370	219	028	730	381	40
30	.2356	.2334	4.284	.2401	4.165	1.028	.9724	1.3352	30
40	385	363	232	432	113	029	717	323	20
50	414	391	182	462	061	030	710	294	10
14° 00′	.2443	.2419	4.134	.2493	4.011	1.031	.9703	1.3265	76° 00′
10	473	447	086	524	3.962	031	696	235	50
20	502	476	039	555	914	032	689	206	40
30	.2531	.2504	3.994	.2586	3.867	1.033	.9681	1.3177	30
40	560	532	950	617	821	034	674	148	20
50	589	560	906	648	776	034	667	119	10
15° 00′	.2618	.2588	3.864	.2679	3.732	1.035	.9659	1.3090	75° 00′
10	647	616	822	711	689	036	652	061	50
20	676	644	782	742	647	037	644	032	40
30	.2705	.2672	3.742	.2773	3.606	1.038	.9636	1.3003	30
40	734	700	703	805	566	039	628	974	20
50	763	728	665	836	526	039	621	945	10
16° 00′	.2793	.2756	3.628	.2867	3.487	1.040	.9613	1.2915	74° 00′
10	822	784	592	899	450	041	605	886	50
20	851	812	556	931	412	042	596	857	40
30	.2880	.2840	3.521	.2962	3.376	1.043	.9588	1.2828	30
40	909	868	487	994	340	044	580	799	20
50	938	896	453	.3026	305	045	572	770	10
17° 00′	.2967	.2924	3.420	.3057	3.271	1.046	.9563	1.2741	73° 00′
10	996	952	388	089	237	047	555	712	50
20	.3025	979	357	121	204	048	546	683	40
30	.3054	.3007	3.326	.3153	3.172	1.048	.9537	1.2654	30
40	083	035	295	185	140	049	528	625	20
50	113	062	265	217	108	050	520	595	10
18° 00′	.3142	.3090	3.236	.3249	3.078	1.051	.9511	1.2566	72° 00′
10	171	118	207	281	047	052	502	537	50
20	200	145	179	314	018	053	492	508	40
30	.3229	.3173	3.152	.3346	2.989	1.054	.9483	1.2479	30
40	258	201	124	378	960	056	474	450	20
50	287	228	098	411	932	057	465	421	10
19° 00′	.3316	.3256	3.072	.3443	2.904	1.058	.9455	1.2392	71° 00′
10	345	283	046	476	877	059	446	363	50
20	374	311	021	508	850	060	436	334	40
30	.3403	.3338	2.996	.3541	2.824	1.061	.9426	1.2305	30
40	432	365	971	574	798	062	417	275	20
50	462	393	947	607	773	063	407	246	10
20° 00′	.3491	.3420	2.924	.3640	2.747	1.064	.9397	1.2217	70° 00′
		cos θ	sec θ	cot θ	tan θ	csc θ	sin θ	Radians	Degrees
									Angle θ

TABLE II—*continued*

Angle θ Degrees	Radians	sin θ	csc θ	tan θ	cot θ	sec θ	cos θ		
20° 00′	.3491	.3420	2.924	.3640	2.747	1.064	.9397	1.2217	70° 00′
10	520	448	901	673	723	065	387	188	50
20	549	475	878	706	699	066	377	159	40
30	.3578	.3502	2.855	.3739	2.675	1.068	.9367	1.2130	30
40	607	529	833	772	651	069	356	101	20
50	636	557	812	805	628	070	346	072	10
21° 00′	.3665	.3584	2.790	.3839	2.605	1.071	.9336	1.2043	69° 00′
10	694	611	769	872	583	072	325	1.2014	50
20	723	638	749	906	560	074	315	985	40
30	.3752	.3665	2.729	.3939	2.539	1.075	.9304	1.1956	30
40	782	692	709	973	517	076	293	926	20
50	811	719	689	.4006	496	077	283	897	10
22° 00′	.3840	.3746	2.669	.4040	2.475	1.079	.9272	1.1868	68° 00′
10	869	773	650	074	455	080	261	839	50
20	898	800	632	108	434	081	250	810	40
30	.3927	.3827	2.613	.4142	2.414	1.082	.9239	1.1781	30
40	956	854	595	176	394	084	228	752	20
50	985	881	577	210	375	085	216	723	10
23° 00′	.4014	.3907	2.559	.4245	2.356	1.086	.9205	1.1694	67° 00′
10	043	934	542	279	337	088	194	665	50
20	072	961	525	314	318	089	182	636	40
30	.4102	.3987	2.508	.4348	2.300	1.090	.9171	1.1606	30
40	131	.4014	491	383	282	092	159	577	20
50	160	041	475	417	264	093	147	548	10
24° 00′	.4189	.4067	2.459	.4452	2.246	1.095	.9135	1.1519	66° 00′
10	218	094	443	487	229	096	124	490	50
20	247	120	427	522	211	097	112	461	40
30	.4276	.4147	2.411	.4557	2.194	1.099	.9100	1.1432	30
40	305	173	396	592	177	100	088	403	20
50	334	200	381	628	161	102	075	374	10
25° 00′	.4363	.4226	2.366	.4663	2.145	1.103	.9063	1.1345	65° 00′
10	392	253	352	699	128	105	051	316	50
20	422	279	337	734	112	106	038	286	40
30	.4451	.4305	2.323	.4770	2.097	1.108	.9026	1.1257	30
40	480	331	309	806	081	109	013	228	20
50	509	358	295	841	066	111	001	199	10
26° 00′	.4538	.4384	2.281	.4877	2.050	1.113	.8988	1.1170	64° 00′
10	567	410	268	913	035	114	975	141	50
20	596	436	254	950	020	116	962	112	40
30	.4625	.4462	2.241	.4986	2.006	1.117	.8949	1.1083	30
40	654	488	228	.5022	1.991	119	936	054	20
50	683	514	215	059	977	121	923	1.1025	10
27° 00′	.4712	.4540	2.203	.5095	1.963	1.122	.8910	1.0996	63° 00′
		cos θ	sec θ	cot θ	tan θ	csc θ	sin θ	Radians	Degrees
								Angle θ	

TABLE II—*continued*

Angle θ									
Degrees	**Radians**	**sin θ**	**csc θ**	**tan θ**	**cot θ**	**sec θ**	**cos θ**		
27° 00′	.4712	.4540	2.203	.5095	1.963	1.122	.8910	1.0996	63° 00′
10	741	566	190	132	949	124	897	966	50
20	771	592	178	169	935	126	884	937	40
30	.4800	.4617	2.166	.5206	1.921	1.127	.8870	1.0908	30
40	829	643	154	243	907	129	857	879	20
50	858	669	142	280	894	131	843	850	10
28° 00′	.4887	.4695	2.130	.5317	1.881	1.133	.8829	1.0821	62° 00′
10	916	720	118	354	868	134	816	792	50
20	945	746	107	392	855	136	802	763	40
30	.4974	.4772	2.096	.5430	1.842	1.138	.8788	1.0734	30
40	.5003	797	085	467	829	140	774	705	20
50	032	823	074	˙505	816	142	760	676	10
29° 00′	.5061	.4848	2.063	.5543	1.804	1.143	.8746	1.0647	61° 00′
10	091	874	052	581	792	145	732	617	50
20	120	899	041	619	780	147	718	588	40
30	.5149	.4924	2.031	.5658	1.767	1.149	.8704	1.0559	30
40	178	950	020	696	756	151	689	530	20
50	207	975	010	735	744	153	675	501	10
30° 00′	.5236	.5000	2.000	.5774	1.732	1.155	.8660	1.0472	60° 00′
10	265	025	1.990	812	720	157	646	443	50
20	294	050	980	851	709	159	631	414	40
30	.5323	.5075	1.970	.5890	1.698	1.161	.8616	1.0385	30
40	352	100	961	930	686	163	601	356	20
50	381	125	951	969	675	165	587	327	10
31° 00′	.5411	.5150	1.942	.6009	1.664	1.167	.8572	1.0297	59° 00′
10	440	175	932	048	653	169	557	268	50
20	469	200	923	088	643	171	542	239	40
30	.5498	.5225	1.914	.6128	1.632	1.173	.8526	1.0210	30
40	527	250	905	168	621	175	511	181	20
50	556	275	896	208	611	177	496	152	10
32° 00′	.5585	.5299	1.887	.6249	1.600	1.179	.8480	1.0123	58° 00′
10	614	324	878	289	590	181	465	094	50
20	643	348	870	330	580	184	450	065	40
30	.5672	.5373	1.861	.6371	1.570	1.186	.8434	1.0036	30
40	701	398	853	412	560	188	418	1.0007	20
50	730	422	844	453	550	190	403	977	10
33° 00′	.5760	.5446	1.836	.6494	1.540	1.192	.8387	.9948	57° 00′
10	789	471	828	536	530	195	371	919	50
20	818	495	820	577	520	197	355	890	40
30	.5847	.5519	1.812	.6619	1.511	1.199	.8339	.9861	30
40	876	544	804	661	501	202	323	832	20
50	905	568	796	703	1.492	204	307	803	10
34° 00′	.5934	.5592	1.788	.6745	1.483	1.206	.8290	.9774	56° 00′
		cos θ	**sec θ**	**cot θ**	**tan θ**	**csc θ**	**sin θ**	**Radians**	**Degrees**
								Angle θ	

TABLE II—*continued*

Angle θ									
Degrees	Radians	sin θ	csc θ	tan θ	cot θ	sec θ	cos θ		
34° 00′	.5934	.5592	1.788	.6745	1.483	1.206	.8290	.9774	56° 00′
10	963	616	781	787	473	209	274	745	50
20	992	640	773	830	464	211	258	716	40
30	.6021	.5664	1.766	.6873	1.455	1.213	.8241	.9687	30
40	050	688	758	916	446	216	225	657	20
50	080	712	751	959	437	218	208	628	10
35° 00′	.6109	.5736	1.743	.7002	1.428	1.221	.8192	.9599	55° 00′
10	138	760	736	046	419	223	175	570	50
20	167	783	729	089	411	226	158	541	40
30	.6196	.5807	1.722	.7133	1.402	1.228	.8141	.9512	30
40	225	831	715	177	393	231	124	483	20
50	254	854	708	221	385	233	107	454	10
36° 00′	.6283	.5878	1.701	.7265	1.376	1.236	.8090	.9425	54° 00′
10	312	901	695	310	368	239	073	396	50
20	341	925	688	355	360	241	056	367	40
30	.6370	.5948	1.681	.7400	1.351	1.244	.8039	.9338	30
40	400	972	675	445	343	247	021	308	20
50	429	995	668	490	335	249	004	279	10
37° 00′	.6458	.6018	1.662	.7536	1.327	1.252	.7986	.9250	53° 00′
10	487	041	655	581	319	255	969	221	50
20	516	065	649	627	311	258	951	192	40
30	.6545	.6088	1.643	.7673	1.303	1.260	.7934	.9163	30
40	574	111	636	720	295	263	916	134	20
50	603	134	630	766	288	266	898	105	10
38° 00′	.6632	.6157	1.624	.7813	1.280	1.269	.7880	.9076	52° 00′
10	661	180	618	860	272	272	862	047	50
20	690	202	612	907	265	275	844	.9018	40
30	.6720	.6225	1.606	.7954	1.257	1.278	.7826	.8988	30
40	749	248	601	.8002	250	281	808	959	20
50	778	271	595	050	242	284	790	930	10
39° 00′	.6807	.6293	1.589	.8098	1.235	1.287	.7771	.8901	51° 00′
10	836	316	583	146	228	290	753	872	50
20	865	338	578	195	220	293	735	843	40
30	.6894	.6361	1.572	.8243	1.213	1.296	.7716	.8814	30
40	923	383	567	292	206	299	698	785	20
50	952	406	561	342	199	302	679	756	10
40° 00′	.6981	.6428	1.556	.8391	1.192	1.305	.7660	.8727	50° 00′
10	.7010	450	550	441	185	309	642	698	50
20	039	472	545	491	178	312	623	668	40
30	.7069	.6494	1.540	.8541	1.171	1.315	.7604	.8639	30
40	098	517	535	591	164	318	585	610	20
50	127	539	529	642	157	322	566	581	10
41° 00′	.7156	.6561	1.524	.8693	1.150	1.325	.7547	.8552	49° 00′
		cos θ	sec θ	cot θ	tan θ	csc θ	sin θ	Radians	Degrees
								Angle θ	

TABLE II—*continued*

Angle θ		sin θ	csc θ	tan θ	cot θ	sec θ	cos θ		
Degrees	**Radians**								
41° 00′	.7156	.6561	1.524	.8693	1.150	1.325	.7547	.8552	49° 00′
10	185	583	519	744	144	328	528	523	50
20	214	604	514	796	137	332	509	494	40
30	.7243	.6626	1.509	.8847	1.130	1.335	.7490	.8465	30
40	272	648	504	899	124	339	470	436	20
50	301	670	499	952	117	342	451	407	10
42° 00′	.7330	.6691	1.494	.9004	1.111	1.346	.7431	.8378	48° 00′
10	359	713	490	057	104	349	412	348	50
20	389	734	485	110	098	353	392	319	40
30	.7418	.6756	1.480	.9163	1.091	1.356	.7373	.8290	30
40	447	777	476	217	085	360	353	261	20
50	476	799	471	271	079	364	333	232	10
43° 00′	.7505	.6820	1.466	.9325	1.072	1.367	.7314	.8203	47° 00′
10	534	841	462	380	066	371	294	174	50
20	563	862	457	435	060	375	274	145	40
30	.7592	.6884	1.453	.9490	1.054	1.379	.7254	.8116	30
40	621	905	448	545	048	382	234	087	20
50	650	926	444	601	042	386	214	058	10
44° 00′	.7679	.6947	1.440	.9657	1.036	1.390	.7193	.8029	46° 00′
10	709	967	435	713	030	394	173	.7999	50
20	738	988	431	770	024	398	153	970	40
30	.7767	.7009	1.427	.9827	1.018	1.402	.7133	.7941	30
40	796	030	423	884	012	406	112	912	20
50	825	050	418	942	006	410	092	883	10
45° 00′	.7854	.7071	1.414	1.000	1.000	1.414	.7071	.7854	45° 00′
		cos θ	sec θ	cot θ	tan θ	csc θ	sin θ	**Radians**	**Degrees**
								Angle θ	

INDEX

Addition of vectors, 626
  associative property, 627
  cancellation property, 628–629
  commutative property, 627
  properties, 629
  substitution property, 626
Analytic geometry, 203, 254–255
  absolute values, 225
  algebraic proof, 241
  approximate solutions (irrational numbers),
    232, 236, 238
Cartesian (rectangular coordinate) plane, 203,
    312–313
  distance, 216, 218
  families of lines, 211
  general form of equations, 209
  graphs of polynomials, 220–224
  inequalities, 227–230
  linear equations, 204, 206–207
  linear interpolation, 237
  maximum point, 222
  minimum point, 221
  parallel lines, 213–214
  parameter, 211
  proofs using coordinates, 244, 245
  quadrants, 436
  slope, 207, 208, 256–257
  three-dimensional, 247–250
  y-intercept, 209
Angles, 555, 567–568, 607–608
  coterminal, 556
  directed (signed), 555
  of elevation, 574
  of inclination, 580
  measurement (degrees vs. radians), 557–560,
    561
  perpendicular lines and, 585
  reference, 567
  standard position, 556
Apoleonius of Perga, 261
Argand, J. R., 115

Argand diagrams, 141–143, 659
Arithmetic functions, 13
  common difference and, 13
Arithmetic progression, 13, 30
  arithmetic mean and, 14–15
Arithmetic series, 26
Arrangements, 377, 384, 385, 392
  combinations, 389
  principle of choice, 378, 391, 415–416
  probability, 395
  subsets and, 379, 388, 392

Basis vectors, 646–647
Bell, E. T., 200
Binary numerals, 381
Binomial Theorem, 402, 403, 407–409, 416
  induction and, 411

Cartesian plane, 203, 312–313
Circle, 264
Circular functions, 433, 440, 477–478
  addition formulas, 459–463
  amplitude, 448, 478
  asymptotes, 450
  composite functions, 452, 456, 478
  equations vs. identities, 472–473
  graphs of, 447, 452
  identities, 434, 468–469, 471, 478–479
  quadrantal values, 437
  secant and cosecant, 440
  sine and cosine, 433
  sine wave, 448
  tangent and cotangent, 440
Complex numbers, 115, 158
  absolute value, 141
  amplitude (argument), 662, 697
  complex conjugate, 158
  conjugates, 129
  division, 127, 129
  geometrical interpretation, 141–143
  isomorphic fields and, 148, 159

matrix representation, 150–151
  operations on, 118, 120
  as ordered pairs, 145
  polar form, 661, 662–671, 697
  powers of, 676
  products of, 674
  properties, 131
  roots of, 681–687, 688, 691
  roots of unity (*n*th root of 1), 692, 693–696
  square root of, 123, 125
  standard form, 130
Composite functions, 452, 456, 457
  composition, 457
Conic sections, 259, 260, 318–321
  circle, 264
  eccentricity, 322
  ellipse, 268, 270, 279
  equations for, 259, 262, 279, 286, 294
  graph of, 274–277, 290
  hyperbola, 292
  inequalities, 312–313
  nappes, 259
  parabola, 284–285
  quadratic functions, 308–311
  special cases (limiting or degenerate), 315
  symmetric curves, 274
  translation, 301–302, 304
Convergent series, 32, 33
Coordinate systems, *see* Analytic geometry
Cosecant, 440
Cosine, 433
Cotangent, 440
Coterminal angles, 556

de Moivre, Abraham, 681–682
de Moivre's Theorem, 676, 682, 697
Descartes, Rene, 203, 238
Directed (signed) angles, 555
Divergent series, 33
Division Algorithm Theorem, 175, 198

Ellipse, 268, 270
  equation for, 279
  focal radii, 268
  graph of, 274–277
Equivalent equations, 105
Exponential functions, 499–501, 503, 504,
    547–548
  composite, 519
  inverse, 515, 517, 520, 521–524, 527
  linear interpolation, 507–509
  logarithmic, 531, 535, 536, 538
  monotonic, 524
  properties, 503

Exponents, 547–548
  exponential functions, 499–501, 503, 504
  of integers, 487, 489
  of natural numbers, 485
  properties, 485
  of rational numbers, 492, 493, 494–495, 497
  recursive definition, 485

Factorial, 383
Factor Theorem (polynomials), 177, 198
Fibonacci sequence, 12
Fields, 61–62, 63, 92, 93, 96
  associative property, 62
  closure property, 62
  commutative property, 62, 63
  completeness property, 77, 79, 83, 89
  complex numbers, 131
  density property, 72
  distributive property, 63
  identity properties, 62
  inverse properties, 62
  isomorphic, 148, 159
  least upper bound, 78
  ordered, 67
  order properties, 67–68, 69, 71–72
  subfields, 83, 88, 96
  transitive property, 68, 71
  trichotomy principle, 68, 71
Fourier, Joseph, 481
Fourier series, 481–482
Fractional equations, 105
Function, 1, 30
  arithmetic, 13
  circular, 433, 440, 477–478
  constant, 3
  domain of, 1
  image and, 1
  inverse and, 8
  linear, 8
  as mapping, 4–5
  periodic, 421
  polynomial, 172, 198
  quadratic, 308–311
  range of, 1
  sequence and, 9–13
  sequence functions, 8–9, 30
  translation, 301–302
  wrapping, 422
Fundamental Theorem of Algebra, 195, 198

Galois, Evariste, 194, 200
Gauss, Karl Friedrich, 195, 200
Geometric progression, 17, 18, 30
  geometric mean, 19

Geometric series, 26
Group, 98

Hamilton, William R., 161
Heron's Formula, 588
  semiperimeter, 589
Hyperbola, 292, 294
  asymptote, 296
  equation for, 294
  equilateral (rectangular), 314
  graph of, 295–296

Identities (circular functions), 434, 468–469,
    471, 478–479
Inductive set, 47, 55
Inner product of vectors, 641
  properties, 642, 643
Integers, 92, 94, 96
Integral domain, 93, 94, 96
  cancellation property, 93, 169
Integral exponents, 487
  properties, 489
"Invented" numbers, 111
  complex numbers, 115, 118, 120, 123
  pure imaginary numbers, 116
Inverse functions, 515, 517, 520, 521–524,
    547–548
  graphs of, 527
  logarithmic, 531, 535, 536, 538, 543, 547
  monotonic, 524
  trigonometric, 596–597, 599–604
Irrational numbers, *see* "Invented" numbers
Isomorphic fields, 148, 159

Law of cosines, 587
Law of sines, 592
Linear equations, 325
  linear combination, 330
Linear functions, 8
  slope of, 8
Logarithms, 531, 535, 547–548
  common, 536
  natural, 547
  properties, 535
  tables, 538, 543

Mathematical induction property, 36, 38, 39,
    41, 47, 52, 53
  inductive set, 47, 55
  well-ordering, 51, 52, 55
Matrices, *m* X*n*, 360–366, 369
  2 X 2, 150–151, 372–374
Meserve, B. F., 200
m X*n* matrix, 360–366, 369

elementary row operations, 361
  row equivalence, 361

Natural exponents, properties, 485
  recursive definition, 485
Natural numbers, 37, 55, 56–57
  associative property, 38
  cancellation property, 38
  closure property, 37
  commutative property, 37
  distributive property, 38
  identity property, 38
  successors, 57
  trichotomy property, 38
  well-ordering property, 51, 52, 53, 55

Ordered pairs, 144–145

Parabola, 284–285
  axis of symmetry, 285
  equation for, 286
  graph of, 290
  vertex, 285
Peano, Giuseppe, 57
Periodic function, 421
  period, 421
Permutations (arrangements), 387
Polar form of complex numbers, 661, 662–671,
    697
Polynomial function, 172, 198
  zeros of, 172, 173, 188, 198
Polynomials, addition of, 163
  degree, 170
  division, 174–175, 176
  Division Algorithm Theorem, 175, 198
  factorization of, 165
  Factor Theorem, 177, 198
  function, 172
  Fundamental Theorem of Algebra, 195, 198
  graphs of, 220–224
  as integral domain, 166, 168
  multiplication of, 163–164
  reduced, 187
  Remainder Theorem, 176, 198
  synthetic division, 182, 186
  Theorem of Rational Roots, 188, 198
  zero polynomial, 168, 174
Position vector, 621
Principle of choice, 378, 391, 415–416
  binary numerals, 381
  Binomial Theorem, 402, 403, 407–409, 416
  with subsets, 379
Principle of Mathematical Induction, 36, 38, 41,
    47, 52, 53, 55

inductive set, 47, 55
well-ordering, 51, 52, 55
Probability, 395
event, 395
sample space, 395
Pure imaginary numbers, 116, 146

Quadratic equations, 101
completing the square, 101–102
discriminant, 136
factoring, 101
quadratic formula, 103
quadratic functions, 308–311
solutions, 133–134

Rational exponents, 492, 493, 497
properties, 494–495
Real numbers, 61, 96
completeness property, 77, 79, 83, 89
least upper bound, 78
order properties, 67–68, 69, 71–72
rational numbers and, 80, 88
scalar quantities, 615
upper bound, 77
Rational number system, 63, 83, 93, 96
completeness property and, 80
as subfield, 88
Reduced polynomial, 187
Remainder Theorem (division of polynomials),
176, 198
Roots of unity (nth root of 1), 692, 693–696,
698
primitive, 695

Scalar multiplication of vectors, 633
as ordered pairs, 634
properties, 635
Scalar quantities, 615
Secant, 440
Sequence, 9–11, 30
arithmetic, 13
Fibonacci, 12
finite, 9
geometric, 17
infinite, 9, 10
iteration formula, 11
recursion formula, 11
sum of, 22–23, 26, 31
terms of, 9
Sequence functions, 8–9, 30
sequence of partial sums, 26
sum of, 22–23, 26
Series, 26
arithmetic, 26

convergent, 32, 33
divergent, 33
Fourier, 481–482
geometric, 26
sequence of partial sums, 26
terms of, 26
Sine, 433
Subtraction of vectors, 628
Succesor (in natural numbers), 57
Sum of a sequence, 22–23, 26, 31
Synthetic division (of polynomials), 182, 186
reduced polynomial, 187
Systems of equations, 325, 330, 369
algebraic methods and, 327, 329, 334–337
dependent and consistent, 326
equivalent, 325
inconsistent, 325
independent and consistent, 327
inequalities, 349–352
linear combination, 330, 354–359, 369
linear and second degree, 333–338
$m \times n$ matrix, 360–366
two second-degree equations, 340–347
Systems of inequalities, 349
linear and quadratic, 350–351
more than two, 352–353
two linear, 349–350
two second-degree, 351–352
Systems of linear equations, 354
elementary row operations, 361
$m \times n$ matrix, 360–366
row equivalence, 361
three, 355–359

Tangent, 440
Theorem of Rational Roots, 188, 198
Translation, 301–302, 304, 454
invariant distance, 302
Triangles, ambiguous case, 593
Heron's Formula, 588
law of cosines, 587
law of sines, 592
measurement, 586, 591, 593
semiperimeter, 589
Trigonometric functions, 563, 607–609
angles, 555–561, 567–568, 574, 580
arccosine, 603
arcsine, 599
arctangent, 604
Heron's Formula, 588
inverse, 596–597, 599–604
law of cosines, 587
law of sines, 592
principal value, 597

reference angle, 567
right triangles, 569–574
solving triangles, 572
tables, 565
triangular measure, 586, 591, 593
triangulation, 574
trigonometric ratios, 570–571
2 X 2 matrix, 150–151
  in mathematical operations, 150
  in multiplication, 150, 372–374

Vectors, 615, 652–654
  addition of, 626–627
  angle between vectors, 639
  basis vectors, 646–647
  cancellation property for vector addition, 629
  directed line segment, 615
  direction cosines, 616
  equivalence class, 621
  equivalent, 620–621
  equivalent displacements, 618

inner product (dot product), 641–643
linear combination of, 645
magnitude, 616
parallel, 639
perpendicular (orthogonal), 640, 641
position vector, 621
properties, 629, 635, 643, 654
scalar multiplication, 633, 634
spaces, 649
subspaces, 650
subtraction of, 628
substitution property for addition, 626
unit vectors, 645
x and y components, 615, 647
zero vector, 621
Vector spaces, 649, 651, 655–656
  dimension of, 650
  spanning, 650
  subspaces, 650

Weiss, M. J., 200
Wrapping function, 422, 477
  standard mapping, 422